Lecture Notes on Coastal and Estuarine Studies

Managing Editors:
Malcolm J. Bowman Richard T. Barber
Christopher N. K. Mooers John A. Raven

30

P. B. Crean
T. S. Murty
J. A. Stronach

Mathematical Modelling of Tides and Estuarine Circulation

The Coastal Seas of Southern British Columbia and Washington State

Springer-Verlag
New York Berlin Heidelberg London Paris Tokyo

Managing Editors

Malcolm J. Bowman
Marine Sciences Research Center, State University of New York
Stony Brook, N.Y. 11794, USA

Richard T. Barber
Monterey Bay Aquarium Research Institute
Pacific Grove, CA 93950, USA

Christopher N.K. Mooers
Institute for Naval Oceanography
National Space Technology Laboratories
MS 39529, USA

John A. Raven
Dept. of Biological Sciences, Dundee University
Dundee, DD1 4HN, Scotland

Contributing Editors

Ain Aitsam (Tallinn, USSR) · Larry Atkinson (Savannah, USA)
Robert C. Beardsley (Woods Hole, USA) · Tseng Cheng-Ken (Qingdao, PRC)
Keith R. Dyer (Merseyside, UK) · Jon B. Hinwood (Melbourne, AUS)
Jorg Imberger (Western Australia, AUS) · Hideo Kawai (Kyoto, Japan)
Paul H. Le Blond (Vancouver, Canada) · L. Mysak (Montreal, Canada)
Akira Okuboi (Stony Book, USA) · William S. Reebourgh (Fairbanks, USA)
David A. Ross (Woods Hole, USA) · John H. Simpson (Gwynedd, UK)
Absornsuda Siripong (Bangkok, Thailand) · Robert L. Smith (Covallis, USA)
Mathias Tomczak (Sydney, AUS) · Paul Tyler (Swansea, UK)

Authors

P.B. Crean
T.S. Murty
Institute of Ocean Sciences
Department of Fisheries and Oceans
P.O. Box 6000, Sydney
British Columbia, V8L 4B2, Canada

J.A. Stronach
Pacific Ocean Sciences Limited
301A-3700 Gilmore Way, Burnaby
British Columbia, V5G 4M1, Canada

ISBN 3-540-96897-0 Springer-Verlag Berlin Heidelberg New York
ISBN 0-387-96897-0 Springer-Verlag New York Berlin Heidelberg

This work is subject to copyright. All rights are reserved, whether the whole or part of the material is concerned, specifically the rights of translation, reprinting, re-use of illustrations, recitation, broadcasting, reproduction on microfilms or in other ways, and storage in data banks. Duplication of this publication or parts thereof is only permitted under the provisions of the German Copyright Law of September 9, 1965, in its version of June 24, 1985, and a copyright fee must always be paid. Violations fall under the prosecution act of the German Copyright Law.

© by Springer-Verlag New York, Inc. 1988
Printed in Germany

Printing and binding: Druckhaus Beltz, Hemsbach/Bergstr.
2837/3140-543210 – Printed on acid-free paper

Norman Stuart Heaps

IN MEMORIAM

At the inception of this program in the late 1960's, the senior author visited a number of establishments engaged in tidal and coastal sea research. It was in the course of that tour that Norman Stuart Heaps, as advisor and subsequently as teacher and friend, became associated with this project.

As the phase of the work was drawing to a close, we planned to repeat a highly successful trip of some fifteen years earlier from the Canadian Rockies to the Pacific Coast when approaches to so much of the work described here was discussed – but as Norman said, he had to get his health back first.

It is with much respect and affection that this book is dedicated to his memory.

PREFACE

Of the few major ports on the western seaboard of North America, two are located on the extensive complex of waters contained between Vancouver Island and the mainland coasts of British Columbia and the State of Washington. Prolific in marine life and supporting major fisheries, the importance of these waters is presently being enhanced by extensive developments in aquaculture. Increases in the discharge of domestic and industrial effluents and in the density of marine traffic, both commercial and recreational, emphasize the need for a quantitative understanding of the basic circulation and predictive capability with respect to major contingencies likely to occur. This work attempts a broad overview ranging from tidal and estuarine circulation, including the dynamical simulation of a major river plume and influences propagating in from the open boundaries, to the effects of storm surges and tsunamis.

Strongly tidal, variously stratified, and topographically complex, the system presents an immense variety of fluid mechanical problems varying widely with respect to spatial and temporal scales, the entirety incapable of detailed resolution in the forseeable future. On the other hand, observations show behaviour which is certainly not merely random. The basic thesis underlying this work is that it is possible to simulate the large-scale motions without detailed treatment of the small-scale semi-turbulent motions, the interaction between the two scales being approximated by the use of appropriate parameters. Significant large-scale organized flows are generally detectable in vector and scalar observations. Such observations are costly and thus deployments are much subject to both spatial and temporal limitations. Physical models can provide graphic insight into a range of flow phenomena but when applied to large coastal systems suffer from two major disadvantages. One of these concerns the effects of the earth's rotation and consequent requirement that the model be constructed on a rotating and virtually vibration-free, rigid platform. The other concerns the requirement for a practicable depth of working fluid and the need to greatly exaggerate the vertical scale with respect to the horizontal scale of the model.

Numerical models employ some schematization of the system into an array of computational meshes with respect to each of which the mass and momentum conservation equations are applied. Here again, limitations occur with respect to phenomena resolvable by the mesh scale employed and with the possibility of numerical artifacts being added to the computed solutions or the occurrence of catastrophic non-linear instability. Such models are, however, particularly suited to the purpose in hand.

The work described below might be described as a numerical/observational exploration of the tidal and estuarine processes in these waters in an ongoing dialectic of theory and observation.

The work described in this monograph began in the mid 1960's and spans an era when the mechanical current meter lowered from a vessel was being replaced by the sophisticated moored instruments of today, the thermometer-festooned sampling bottle with the CSTD, and the electro-mechanical desk calculator with the computer. Scientific reports generally purport to present impeccability of sequence in the movement from description to explanation. This is not, of course, the way in which it actually happens. The problem generally exceeds, by an order of magnitude, the resources available, begged, borrowed, or out of order. The working ambience will almost always contain at least one person to make life impossible. The lucky fluke that made all the difference can always be reworked to reflect favourably on the intellectual gifts of the writer.

When the senior author was encouraged to investigate the currents in the waters between Vancouver Island and the mainland, the primary source of information was a technical report by Mr. W.S. Huggett on the results obtained from the first current meter moorings undertaken by the Canadian Hydrographic Service in 1963 and 1964 at various locations in the Georgia/Fuca system. Eloquent of the results is the following extract.

> "Thus it has been shown in the foregoing diagrams that the directions and speeds of the current may vary wildly from day to day and from place to place, both horizontally and vertically, and at times appear to behave completely independently of the tide."

If scarcely sanguine about the outcome, there was, however, the welcome prospect of indefinite employment.

Armed with some dexterity in the handling of Nansen bottles and garbled recollections from a hydraulics course undertaken, as a minor requirement in a Chemical Engineering Master's program a decade earlier. authorization was given to visit a number of establishments engaged in relevant work. There remains remarkable in terms of prescience, kindness, and patience, the names of Dr. J.J. Dronkers of the Netherlands Rijkswaterstaat, Dr. W. Hansen of the Institut für Meereskunde, University of Hamburg, and Drs. R.J. Rossiter and N.S. Heaps of the Liverpool Observatory and Tidal Institute. It was with the latter, however, that an enduring association with the project was formed.

A one-day course in Fortran brought about precarious access to the new IBM 1130 computer acquired by the Pacific Biological Station, Nanaimo, British Columbia. Later, the machine had to be used at night because of the prodigious amounts of time required by the one-dimensional model GF1, tenderly nurtured as each increase in core saw the growth of an additional side channel. Field studies generally resulted from a particularly happy, if somewhat unofficial collusion with our confreres in the Tides and Currents Branch of the Canadian Hydrographic Service in Victoria. Thus, the 1968 program of monthly oceanographic surveys was in full swing using the then new research vessel, Parizeau, before the paint was hardly dry. When the tide gauges duly arrived after our allocations of personnel and ship time had expired, the installations were undertaken using a borrowed 28 foot launch, its bows perennially pointed skyward at the insistence of a large mobile diesel air compressor (required to drill pilings) lashed down in the stern-well, ends of the 20-foot stilling wells trailing in a reluctant wake.

Further developments in the project required ready access to relevant skills and a large computer. Granted educational leave of absence, the senior author continued the work as part of a particularly happy and rewarding program of doctoral studies at the Liverpool Observatory and Tidal Institute, associated with the University of Liverpool.

Subsequent access to a large computer was obtained through the extended hospitality of the Department of Oceanography at the University of British Columbia, thus enabling the development of further models consonant with the magnitude of the problem. It was through this Department that Dr. J.A. Stronach carried out the basic work on the buoyant-spreading upper layer model GF4.

The relevant field observations were obtained with the use of a survey launch made available to us by the Canadian Hydrographic Service. The extensive drogue tracking operations in the Fraser River plume and use of the Vessel Traffic Management radar system were enabled through the enthusiastic co-operation of the Canadian Coast Guard.

It became apparent, however, that the continuing demands of field and numerical work had led to an unconscionably large accumulation of results requiring coherent presentation prior to the senior author's retirement. It was thus appropriate that the inheritor of the models and an author more experienced and prolific in the publishing world, Dr. T.S. Murty, should collaborate in the preparation of this volume.

Plans for the implementation of the final three-dimensional models GF8 (fixed level) and GF9 (hybrid layer–fixed level) are now in an advanced state of development. It is anticipated that work will soon commence in a collaborative project between the second and third authors.

Prolonged immurement in any task generally presupposes some underlying appropriation of meaning. For the senior author, encounter with elements of such potential meaning occurred in the late 1950's in a chance attendance at a seminar in Halifax, Nova Scotia and led to subsequent studies of the speaker's extensive and demanding writings, notably "Insight: A Study of Human Understanding", described in a 1965

Time magazine article as a "general field theory of mind". An imperfect but viable response to its writer's invitation to ground his statements in personal experience and to the personal appropriation of intellectual and affective processes given in consciousness was to take some two and a half decades. Thus, ideas can be derived from reflection of the mind's operations upon sensation as well as from sensation itself, for within each of us, there is an entity not reducible to its neural substrate which must be considered in its own terms. It has become clear that the objects of scientific enquiry need not be restricted to imaginable entities moving through imaginable processes in an imaginable space-time. Such an appreciation constitutes a significant movement towards apprehending the nature of thought and life in the universe towards that to which we are made privy by the gift of consciousness. Thus, the seeking mind tends towards the theological insofar as it becomes explicitly conscious of the norms of its own procedures. In this regard, the senior author expresses his deep indebtedness to the thought of B.J.F. Longergan, S.J. Such considerations lead to an intellectual appreciation of the transforming power of symbol. In this regard, a further deep indebtedness is expressed to the Benedictine monks of Mount Angel Abbey, Oregon.

The immense task of preparing and editing the text and diagrams, with many revisions was carried out by Ms. Rosalie M. Rutka with an enthusiasm, dedication, and patience much exceeding any conventional involvement in these tasks. It is with particular gratitude that the authors acknowledge this major contribution to the overall project. An additional major contribution by Mr. Do Kyu Lee was in the preparation of the computer drawn diagrams. Preparation of this monograph would not have been possible but for the constant encouragement and resource allocation by Dr. J.R. Garrett, Head of Ocean Physics, Institute of Ocean Sciences. When, in something of a state of desperation, a draft manuscript is assembled, the authors are fortunate indeed if the demanding process of review can be undertaken by a critic excelling both in competence and attentivity (and hopefully a viable but restrained sense of humour!). In this regard, the authors would particularly like to thank Dr. R.E. Thomson, both for the thoroughness of his endeavour and profusion of helpful comments. Further particular acknowledgment is due to the collaboration in this extended program of Mr. P.J. Richards and Mr. Do Kyu Lee for programming and operating the models, and for associated data analysis and display, and to Mr. S. Huggett and Mr. A. Ages for extensive participation in field operations. The initial phase of the three-dimensional model studies was carried under the bilateral scientific and technological agreement between Canada and the Federal Republic of Germany in collaboration with Dr. J. Backhaus of the Institut für Meereskunde, University of Hamburg. In this context, the senior author would like to thank Dr. J. Sünderman for enabling this co-operation, and in particular, Dr. and Mrs. Backhaus for kindness and hospitality during the course of his visits to Hamburg. For much advice, support, and encouragement we thank Drs. R.W. Burling,

P.W. Nasmyth, T.R. Parsons, S. Pond, R.W. Stewart, and M. Waldichuk. We wish to express our appreciation for assistance received from the Canadian Hydrographic Service and Canadian Coast Guard. Meteorological data and advice was provided by the Atmospheric Environment Service, Vancouver. In this regard, we would particularly like to acknowledge the help of Mr. D. Faulkner. We are also pleased to acknowledge the bilateral agreement between Canada and the Federal Republic of Germany in the development of the three-dimensional model GF6. We are grateful to our colleagues in the Pacific Marine Environmental Laboratory, Seattle for fruitful collaboration and discussion. We finally thank the various authors and publishing companies for granting us permission to use material from their publications.

The second author thanks his wife, Kamla, and his daughter, Hima, for their patience and understanding while he was away most of the time in Vancouver during a two-year period working with his co-authors.

CONTENTS

1 INTRODUCTION 1
  1.1 Physical Characteristics 1
      1.1.1 Winds 4
      1.1.2 Tides 4
      1.1.3 Estuarine Character 7
  1.2 Oceanographic Problems 12
  1.3 Regional Oceanographic Description 14
      1.3.1 Adjacent Continental Shelf 14
      1.3.2 Juan de Fuca Strait 14
      1.3.3 Puget Sound 20
      1.3.4 San Juan and Gulf Island Passages 22
      1.3.5 Surface Waters of the Strait of Georgia 26
            1.3.5.1 River Mouth 27
            1.3.5.2 River Plume 29
      1.3.6 Subsurface Waters 36
      1.3.7 Northern Passages 40
      1.3.8 Freshwater Volumes and Flushing Rates 43

2 FIELD OBSERVATIONS 45

BAROTROPIC MODELS 51

3 THE ONE-DIMENSIONAL: MODEL GF1 53
  3.1 Introduction 53
  3.2 Fundamental Equations 54
  3.3 Finite-Difference Equations 54
  3.4 The Computational Scheme 56
  3.5 Results 60
      3.5.1 M_2 Tidal Harmonic Constants in the Main Channels . . 61
      3.5.2 M_2 Tidal Harmonic Constants in the Side Channels . . 64
      3.5.3 M_2 Tidal Stream Harmonic Constants 65
      3.5.4 K_1 Tidal Harmonic Constants in the Main Channels . . 66

3.5.5 K_1 Tidal Harmonic Constants in the Side Channels 68

3.5.6 K_1 Tidal Stream Harmonic Constants 68

3.6 Discussion 69

4 THE COMBINED ONE- AND TWO-DIMENSIONAL MODEL: GF2 . 71

4.1 Introduction 71

4.2 Fundamental Hydrodynamic Equations 71

    4.2.1 The Two-Dimensional Scheme 71

    4.2.2 The First One-Dimensional Scheme 74

    4.2.3 The Second One-Dimensional Scheme 75

    4.2.4 Junctions Between One- and Two-Dimensional Schemes 76

    4.2.5 Approximate Representation of Narrow Passes
         in the Two-Dimensional Scheme 77

    4.2.6 Frictional Adjustments 79

4.3 Description of the Model 79

    4.3.1 The Strait of Georgia and Juan de Fuca Strait Scheme 79

    4.3.2 The Puget Sound Scheme 81

    4.3.3 The Burrard Inlet Scheme 81

    4.3.4 The Howe Sound Scheme 82

    4.3.5 The Jervis Inlet Scheme 82

    4.3.6 The Northern Passages Scheme 82

4.4 Summarized Sequence of Model Development 83

4.5 Non-Linear Tidal Interactions 90

4.6 The Simulation of Mixed Tides 93

4.7 Volume Transports 105

4.8 Energy Calculations 107

    4.8.1 Alternative Estimates for the Energy Balance 110

    4.8.2 Work Done by or Against the Equilibrium Tide 113

5 THE FINE GRID MODEL: GF3 115

5.1 Introduction 115

5.2 Description of the Model 115

5.3 Simulation of Mixed Tides 117

5.4 Applications of Results from GF2 and GF3 124

6 THE LIMITED AREA MODEL (LAM) AND
SMALL-SCALE LIMITED AREA MODEL (SSLAM) 129

6.1 Local Area Modelling 129

6.2 The Limited Area Model (LAM) 130

6.3 The Small-Scale Limited Area Model (SSLAM) 133

7 THE OVERALL FINE-GRID MODEL: GF7 137
   7.1 Introduction 137
   7.2 Numerical Grid Scheme 137
   7.3 Sensitivity Trials 139
   7.4 Simulation of Mixed Tides 140
   7.5 Horizontal Tidal Residual Circulation 146

8 NORMAL MODES 167
   8.1 Introduction 167
   8.2 Gravitational and Rotational Normal Modes . . . 167
   8.3 Method of Determination 168
   8.4 Oscillatory Response of the System 168
   8.5 Internal Modes 172

9 STORM SURGES 178
   9.1 Introduction 178
   9.2 Model Development Considerations 178
   9.3 Discussion of Results 181
   9.4 Conclusions 186

10 TSUNAMIS 187
   10.1 Introduction 187
   10.2 Simulation of the Tsunami of 23 June 1946 . . . 187
   10.3 Future Tsunami Estimates 193

BAROCLINIC MODELS 203

11 THE UPPER LAYER MODEL: GF4 205
   11.1 Introduction 205
   11.2 Governing Equations 206
   11.3 Boundary Conditions 210
   11.4 Finite-Difference Equations 210
       11.4.1 Free Surface and Interfacial Boundaries . . 214
       11.4.2 Lateral Boundaries 216
       11.4.3 Initial Conditions 221
   11.5 Results 222
       11.5.1 River and Tide Interactions in the Central Strait of Georgia 222
       11.5.2 River, Tide, and Wind Interactions in the Overall Strait of Georgia 223
           11.5.2.1 No Wind Forcing 228
           11.5.2.2 Wind Forcing Effects 233
       11.5.3 Residual Circulation 290

    11.5.4 Comparison with Observations 291
        11.5.4.1 Comparison of Observed and Computed Velocity Vectors . . 291
  11.6 Concluding Discussion 300

12 THE LATERALLY-INTEGRATED MODEL: GF5 302

  12.1 Introduction 302
  12.2 Governing Equations 303
  12.3 Model Description 304
  12.4 Finite-Difference Equations 304
  12.5 Boundary Conditions 308
  12.6 Review of Longitudinal Circulation 308
  12.7 Model Results 310
    12.7.1 Tidal Residual Circulation for a Homogeneous Density Field 310
    12.7.2 Baroclinic Residual Circulation 311
    12.7.3 Diagnostic Trials 313
        12.7.3.1 Winter Circulation 313
        12.7.3.2 Summer Circulation 317
        12.7.3.3 Volume Fluxes and Recirculation 319
        12.7.3.4 Prognostic Salinity Calculations 322
        12.7.3.5 Sea Levels 328
  12.8 Prescribed Lid Numerical Trials 329
    12.8.1 The Numerical Application of the Prescribed Lid 331
    12.8.2 Numerical Experiments 332
  12.9 Concluding Summary 334

13 THE THREE-DIMENSIONAL MODEL: GF6 336

  13.1 Introduction 336
  13.2 Fundamental Hydrodynamical Equations 337
    13.2.1 Explicit Finite-Difference Equations 338
  13.3 General Trials of the Model 344
    13.3.1 Tidal Simulation — Short Version 346
    13.3.2 Tidal Simulation — Long Version 346
    13.3.3 Energy Fluxes 351
    13.3.4 Puget Sound Model Trials 354
    13.3.5 Numerical Diffusion 357
    13.3.6 Tidally-Induced Residual Circulation in a Vertical Plane 358
    13.3.7 Vertical Eddy Viscosity 361
    13.3.8 Prescription of Pre-Computed Advective Accelerations 362
  13.4 Residual Circulation 363
    13.4.1 Barotropic Tidal Residual Circulation 364

    13.4.2 General Baroclinic and Tidal Residual Circulation 369
    13.4.3 Baroclinic Circulation in the Strait of Georgia 382
    13.4.4 Residual Surface Elevations 382
    13.4.5 Residual Density Distribution 385
    13.4.6 Residual Vertical Velocities 387
    13.4.7 Dynamics of Modelled Circulation 403
    13.4.8 Critical Assessment of the Model 408
    13.4.9 Effects of Winds 420
    13.4.10 Net Volume Transports 428
    13.4.11 Freshwater Volumes and Flushing Rates 429
13.5 Conclusions 430

14 CONCLUDING DISCUSSION 432
14.1 Concluding Summary 443

REFERENCES 446
APPENDIX I 455
AUTHOR INDEX 463
SUBJECT INDEX 465

1 INTRODUCTION

The waters between Vancouver Island and the mainland coasts of British Columbia and Washington State constitute a deep, topographically complex and strongly tidal estuarine system. Contiguous to the main population centres, Vancouver and Seattle, these waters are characterized by high traffic density, both commercial and pleasure, and in addition, support major fisheries. Pressing questions involve the effects of commercial and domestic effluent, provision of data for biological research programs, and shore and underwater coastal engineering installations. An excellent general account of the physical oceanography of the region may be found in Thomson (1981).

This study is primarily concerned with the numerical modelling studies undertaken to achieve an understanding of elements controlling the basic circulation in the system, starting out from features that stand out clearly against the background of less systematic variability. Such models can provide good approximations to both the actual topographic and non-linear effects which are particularly important in this system.

The work has spanned an era in which Nansen bottles and reversing thermometers have largely been replaced by the conductivity, salinity, temperature, and depth (CSTD) probe while the mechanical current meter, lowered from a vessel, has been replaced by the sophisticated moored recording instruments and acoustic Doppler current profiles of today. The first numerical model (GF1) could only be operated in a shortened version, and then only at night, to avoid monopoly of the small computer then available. Such a model can execute in seconds on even one of the more modest computers now available.

The basic general features — topography, winds, tides, and estuarine character, of the overall system are now discussed. These will be followed by more detailed regional oceanographic descriptions.

1.1 PHYSICAL CHARACTERISTICS

The basic geographical features of the region (Fig. 1.1) consist essentially of an outer strait (Juan de Fuca) 140 km long with a surface area of $3,700$ km^2 and an inner strait (Georgia) 200 km long with widths 20–40 km and surface area about $7,000$ km^2. The major axes of both straits are approximately parallel, joined by a 180° bend containing a complex of islands (the Gulf and San Juan Islands) which lie at the head of a major system of side channels (Puget Sound) that have an area about $2,600$ km^2. The inner strait is connected to the open sea at its northern end through a series of narrow channels (referred to later as the northern passages) containing relatively shallow sills of depth 30–100 m where tidal velocities can exceed 6 m/s. In the aggregate, the combined cross-sectional areas available for seaward egress of freshwater northward constitute less than 7% of the comparable cross-sectional area available to the south in Haro and Rosario Straits. A number of fjords up to 800 m deep and with sills at their mouths open out from the mainland coast.

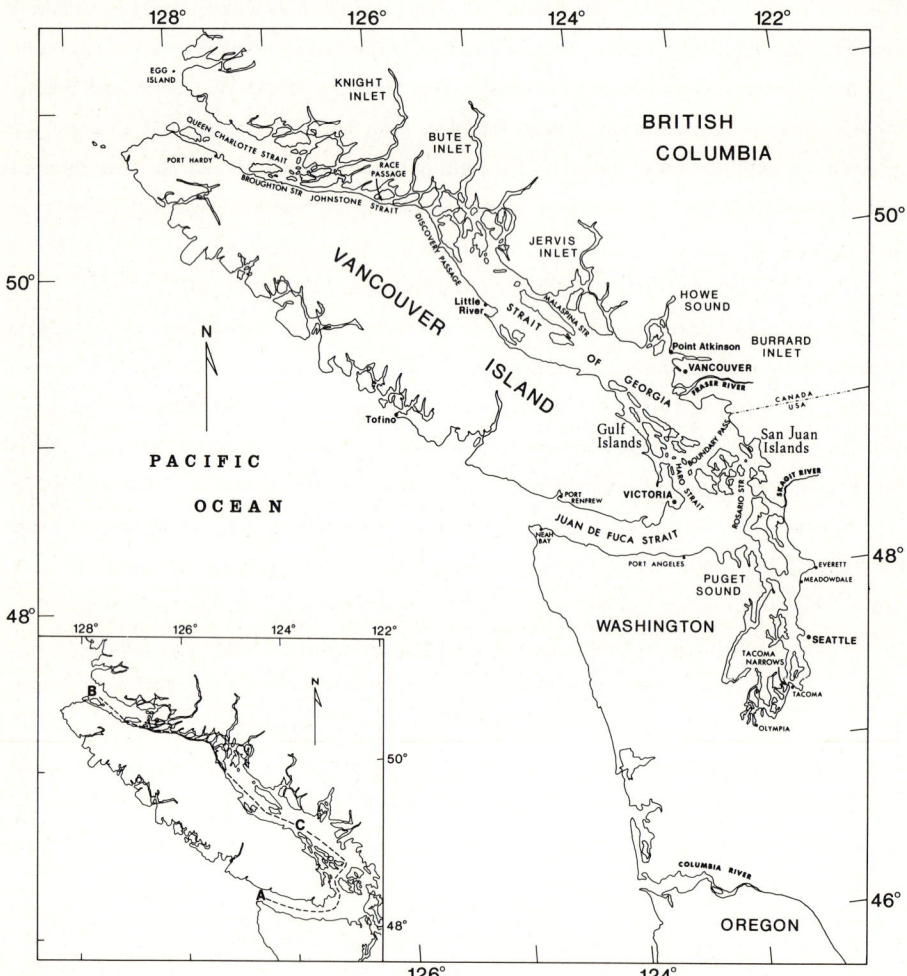

Fig. 1.1 Location map of the study area for British Columbia and Washington waters. Inset indicates the site locations referred to in the text. The dashed line from A to B denotes the locations of the hydrographic stations used in the contoured vertical sections of salinity and temperature (Figs. 1.4 and 1.5).

The Fraser River which drains the largest watershed in the province of British Columbia (230,000 km^2), discharges into the southern end of the inner strait. It constitutes by far, the single major source of freshwater entering the system and accounts for some 70 to 80% of total freshwater input.

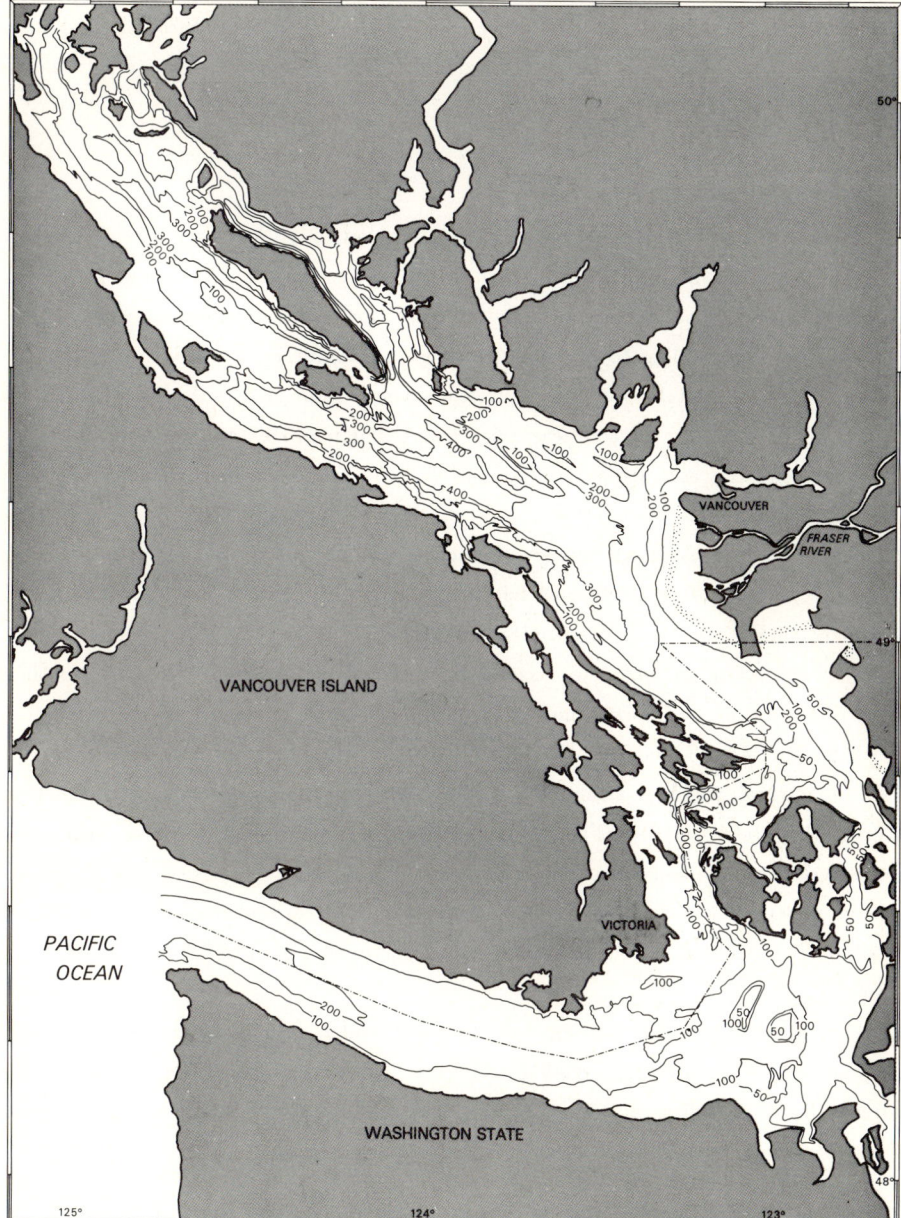

Fig. 1.2 Bathymetry of the Strait of Georgia, Gulf and San Juan Islands region, and Juan de Fuca Strait. Depth is in metres.

The bottom topography of the inner and outer straits (Fig. 1.2) also illustrates the complex nature of the two major conveying channels connecting them. Haro Strait is a deep trench with depths of 180–250 m and a shallow irregular sill about 60 m deep at its northern end (Boundary Pass) which contains a short (0.5 km), narrow (0.5 km), and deep (250 m) opening. This strait plays a major role in the fresh- and salt water exchanges of the inner basin. The other major conveying channel through the islands, Rosario Strait, is shallower (60–70 m) and hence is limited in its capacity to replenish deep saline water in the inner basin.

1.1.1 Winds

Winds in the northeastern Pacific Ocean are dominated by two major atmospheric pressure systems, the North Pacific high and the Aleutian low. Variations in location and intensity of these systems strongly influence the prevailing winds over the offshore waters, the high tending to dominate the region in summer, the low during the winter. Thus, from October to March cyclonic winds associated with the Aleutian low result in a dominance of southeasterlies whereas in summer weaker winds from the west or northwest, associated with the anti-cyclonic North Pacific high, tend to prevail (Kendrew and Kerr, 1955).

Winds over the waters between Vancouver Island and the mainland are strongly influenced by mountainous terrain. In Juan de Fuca Strait, the cyclonic winds over the ocean in winter tend to be associated with easterlies (Fig. 1.3a). In summer inward-blowing westerlies are likely to accompany anti-cyclonic systems. Over the southern part of the Strait of Georgia and the San Juan Islands, funnelling effects of Juan de Fuca Strait, Puget Sound, and the Fraser Valley (which opens out onto the Vancouver-lower mainland area) play an important role, favouring a counter-clockwise wind pattern during the winter months (Fig. 1.3b). In summer the winds are lighter with a less regular pattern but still tend to dominate from the southeast and southwest. Over the northern part of the Strait of Georgia there is a closer approximation to the offshore seasonal wind patterns, southeasterlies again dominating in winter, northwesterlies in summer.

Further complications of these basic patterns tend to result from differential heating of land and sea during fine weather and from outbreaks of polar continental air that stream down the valleys and inlets leading from the interior of British Columbia to the coast in winter.

1.1.2 Tides

The oceanic tide is of the mixed type and moves northward past the mouth of Juan de Fuca Strait and the western coast of Vancouver Island. Tidal co-oscillations within the system result in the semi-diurnal constituents having degenerate amphidromes in the eastern end of Juan de Fuca Strait while diurnal amplitudes and phases increase monotonically from the mouth of the Strait to the northern end of the Strait of Georgia.

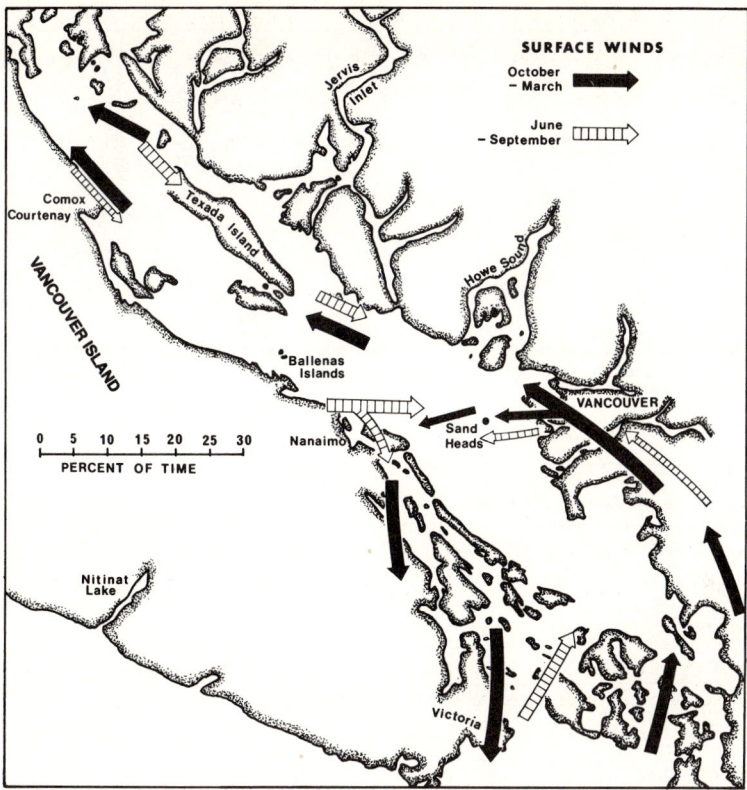

Fig. 1.3a Prevailing surface wind patterns over the Strait of Georgia in winter (solid arrows) and summer (hatched arrows). Thick arrows denote speeds 4.5–9 m/s and thin arrows less than 4.5 m/s. Comparison of arrow length with scale shows frequency of occurrence of a particular wind. (from Barker, 1974)

While the passage of the ocean's semi-diurnal tides along the outer coast of Vancouver Island takes approximately 20 min, the same tide takes about 5 h to traverse the shallower confines of the outer and inner straits, leading to significant differences in tidal elevations through the narrow passages that link the northern end of the Strait of Georgia to the open sea. As a consequence, the largest tidal streams in the world are to be found associated with constrictions in these passages.

The earliest level measurements on the B.C. coast were obtained between 1858 and 1870 by survey vessels of the Royal Navy. From 1898 to 1910 HMS Egeria was on permanent station. Analysis of these and drift-pole measurement data by the Liverpool Observatory and Tidal Institute led to a publication of the first tide tables in 1901 and current tables in 1908. The first permanent tide gauge in the region was installed in Seattle in 1899 and the first description of tides was published in 1904 (Harris, 1904). An analysis of tide gauge data then available was undertaken by Redfield (1950a) who provided gross estimates of frictional dissipation obtained by the fitting of damped waves

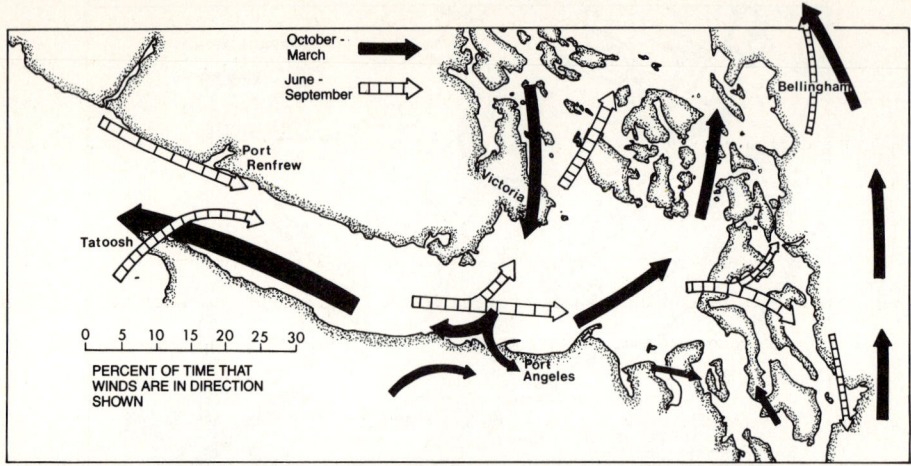

Fig. 1.3b Most frequent pattern of surface winds over Juan de Fuca Strait in winter (solid arrows) and summer (hatched arrows). Thick arrows correspond to wind speeds over 9 m/s, medium arrows 4.5–9 m/s, and thin arrows less than 4.5 m/s. Comparison of arrow length with scale gives frequency of wind occurrence from a given direction. (from Barker, 1974)

to the observed longitudinal distributions of M_2 and K_1 tidal harmonic constants. This indicated that the dissipation of tidal energy in the system was significantly greater than that obtained when the same method was applied to either the Bay of Fundy or Long Island Sound. Extensive analyses of tide and stream data from Juan de Fuca Strait and the Strait of Georgia have been reported by Parker (1977) and from the Puget Sound system by Möfjeld and Larsen (1984).

A comprehensive tidal current atlas covering the Strait of Georgia and Juan de Fuca Strait, based on numerical model studies described below, is now in common use (Canadian Hydrographic Service, 1983). A numerical model of tides along the open coast has been developed by Flather (1987).

The basic character of the tide within the system can be described in terms of the principal harmonic constituents M_2, S_2, N_2, K_1, and O_1 having periodicities deriving from the changing relative position of the earth, moon, and sun. Tides of the mixed type can be described by the ratio of the summed major diurnal constituent amplitudes to those of the major semi-diurnal constituents $(K_1 + O_1)/(M_2 + S_2)$. Values of this ratio based on observations at various locations throughout the world can vary fron the strongly semi-diurnal (*e.g.*, 0.1 at Immingham, England) to strongly diurnal (18.9 at Do San, Korea). At the entrance to Juan de Fuca Strait this ratio is 0.81 and at Point Atkinson (Fig. 1.1) 1.17. In the inner part of Juan de Fuca Strait, however, there exist semi-diurnal degenerate tidal amphidromes. Thus, minimal values of the semi-diurnal amplitudes increase the value of this ratio to 2.6 at Victoria (Fig. 1.1).

The first major variation occurring in the semi-diurnal tide is associated with the shift between those occurring when the earth, moon, and sun are approximately in line (spring tides) and those when the moon is located on line orthogonal to that joining the earth and sun (neap tides). Thus, during spring tides the M_2 and S_2 constituents are in phase, and at neaps they are out of phase. For this system the semi-diurnal spring tides actually occur about 1 day after the new or full moon.

The second major variation in the semi-diurnal tides concerns the variation in the distance of the moon from the earth. This is described by movement of the M_2 and N_2 constituents in and out of phase. Since the N_2 constituent is similar in magnitude to the S_2, this effect has an importance equal to that of the spring–neap cycle for this system. Maximal values actually occur about 3 days after the time when the moon is closest to the earth and the M_2 and N_2 constituents are in phase.

The major variation occurring in the diurnal tides is associated with the monthly movement of the moon in its orbit above and below the plane through the earth's equator. Thus, tropic tides should occur when the moon is at maximum declination above this plane and equatorial tides when it is located in this plane. The resulting variation in the magnitude of the diurnal tide is described by the K_1 and O_1 tidal constituents which move in and out of phase over an interval of about a fortnight (13.36 days). For this system, diurnal tides are minimal and tides in the Strait of Georgia almost semi-diurnal, about 1 day after the moon passes through the plane of the equator.

1.1.3 Estuarine Character

In these waters, density is primarily determined by salinity although temperature may have an appreciable dynamical effect under unusually cold winter conditions when saline stratification tends to be least. The persistent overall estuarine character of the system, its tidal modification, and major seasonal differentiation between winter and summer are represented by contoured vertical sections of temperature (Fig. 1.4a and b) and salinity (Fig. 1.5a and b) along the major conveying channels of the system denoted by the line AB (inset, Fig. 1.1).

The channels (Haro and Juan de Fuca Straits to the south; Discovery Passage, Johnstone and Broughton Straits to the north) linking the main inner basin (Georgia) to the open sea have sills characterized by marked vertical and horizontal salinity gradients, a consequence of strong tidal mixing in the vicinity of the sills. This mixing tends to favour recirculation of mixed water in the Strait of Georgia and limit access of higher salinity ocean water. It has been estimated that the volume of freshwater contained in the Strait of Georgia is equivalent to about 1.3 yr of discharge from the Fraser River, assuming a mixture with saline ocean water of $33.8^0/_{00}$ (Waldichuk, 1957). Contrasting with the weakened vertical stratification of the winter, there is a marked summer dilution of the surface water in the Strait of Georgia while in Juan de Fuca

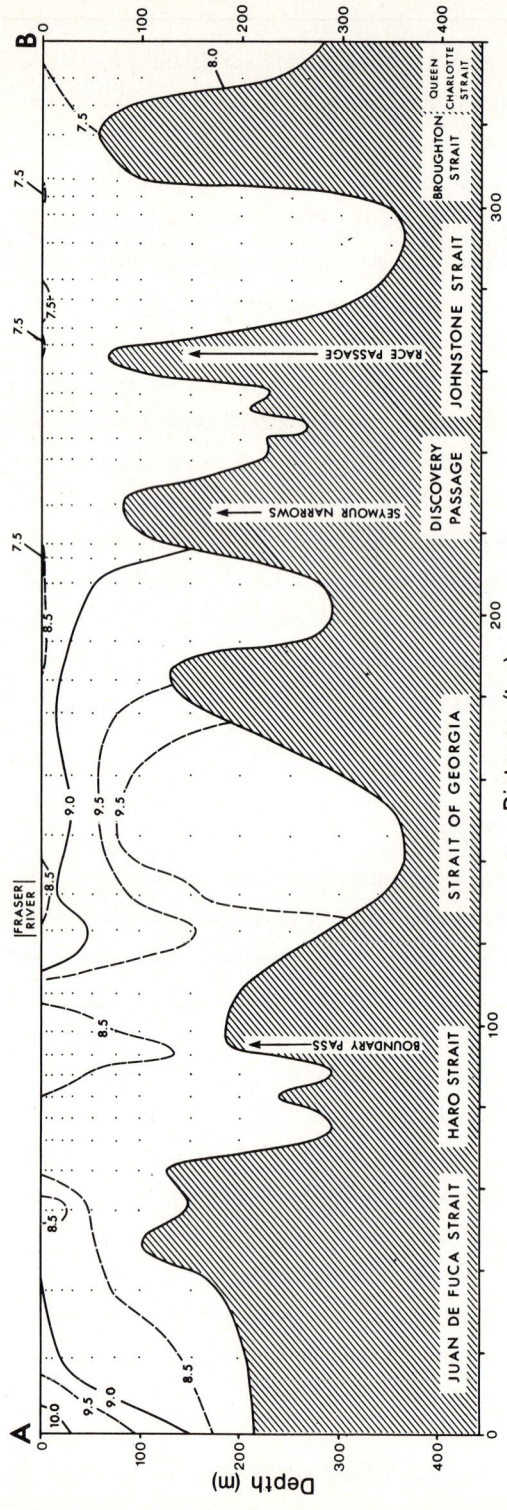

Fig. 1.4a Winter contoured temperature (°C) distributions over a vertical section through the major conveying channels of the system. Data used for plotting from Juan de Fuca Strait to the northen end of the Strait of Georgia was taken in December 1967 and from Discovery Passage to the northern end of Queen Charlotte Strait in January 1977.

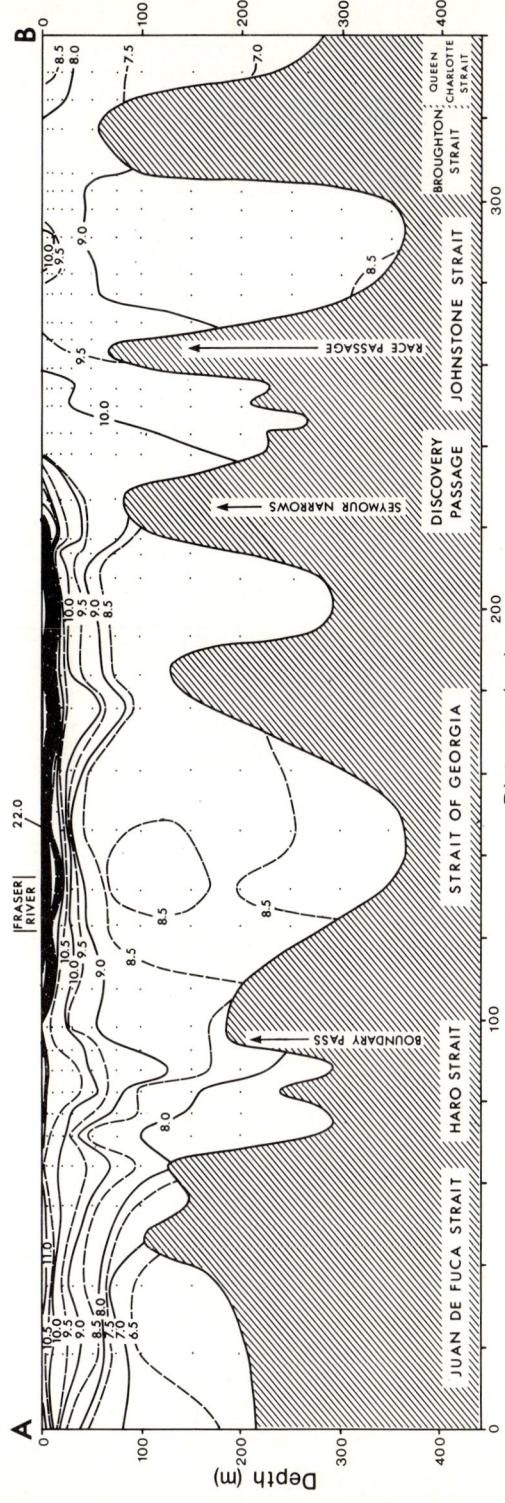

Fig. 1.4b Summer contoured temperature (°C) distributions over a vertical section through the major conveying channels of the system. Data used for Juan de Fuca Strait to the northern end of the Strait of Georgia was taken in July 1968 and from Discovery Passage to the northern end of Queen Charlotte Strait in July 1977.

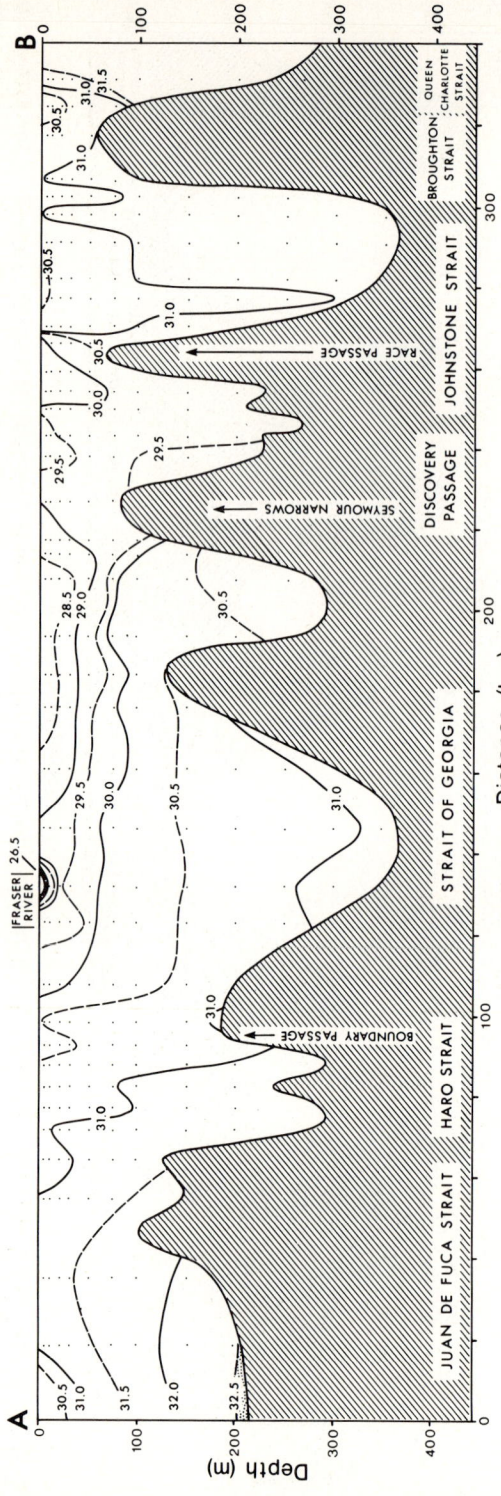

Fig. 1.5a Winter contoured salinity (°/₀₀) distributions over a vertical section through the major conveying channels of the system. Winter data used for plotting are for the same locations and times as those in Fig. 1.4a.

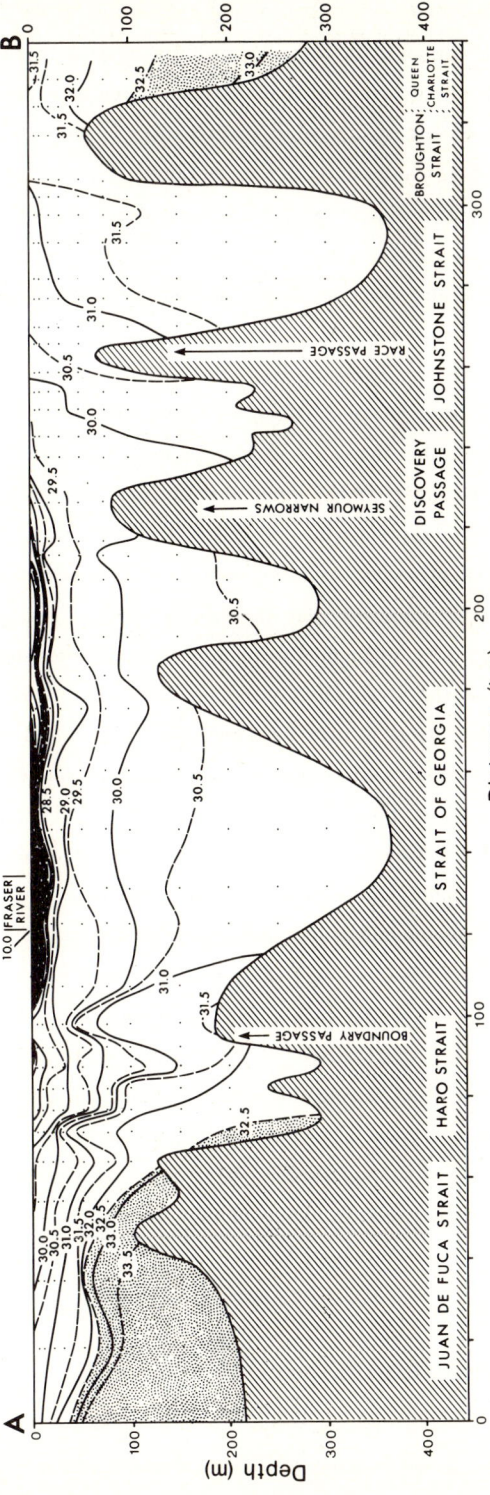

Fig. 1.5b Summer contoured salinity (‰) distributions over a vertical section through the major conveying channels of the system. Data used are for the same locations and times as those in Fig. 1.4b.

Strait increased salinity at depth indicates a renewal of the deep water which may be anticipated, in turn, to enhance a renewal of the deep water in the Strait of Georgia.

Although the data employed in these longitudinal sections serve to illustrate the distinction between winter and summer distributions of temperature and salinity, it is desirable to take note of any strong wind disturbances shortly before or during the surveys in the course of which these data were obtained.

Considering initially the winter data, the survey stations in the Strait of Georgia were occupied over the period 4–6 December 1967. At the three instrumental locations then best representing winds over the length of the Strait of Georgia (Sand Heads, Ballenas Islands, and Comox, Fig. 1.3a) the prevailing wind direction was from the southeast from 1–6 December, rising at times to a range 70–80 km/h. Observations in Juan de Fuca Strait were made 8 December. Over the offshore waters moderate southerly winds on 4 December increased to gale force by 6 December and slackened to westerlies on 7 and 8 December. In the inner part of Juan de Fuca Strait, observations on 6 December at Victoria showed winds from the southeast rising to 74 km/h in the evening, light winds then prevailing over the period 7 and 8 December.

It may reasonably be supposed that the winter salinity section in the Strait of Georgia, depicted in Fig. 1.5a, will reflect some movement of lower salinity surface water into the northern part of the Strait and consequent depression of isopycnal surfaces at depth. In Juan de Fuca Strait it may be anticipated that southeasterly winds have enhanced seaward movement of surface water while water of relatively low salinity from the Columbia River has been moved towards the vicinity of the mouth, as indeed is evident in the section. In Chapter 12, this salinity field is applied in a diagnostic context to a simplified single channel, baroclinic model of the inner and outer straits to illustrate dominant features in the longitudinal vertical circulation that results.

The summer data were observed over the period 2–6 July 1968. Winds were generally much lighter than in the winter case and only one episode of likely oceanographic significance occurred in the southern Strait of Georgia, as recorded at Sand Heads. Thus, immediately prior to and during the observations in the region of the San Juan Islands, winds from the northwest of order 35 km/h persisted for about 24 h. An appreciable displacement of low salinity surface water from the southern Strait of Georgia into the San Juan Island passages, in excess of that normally asociated with the estuarine circulation, may thus be anticipated.

1.2 OCEANOGRAPHIC PROBLEMS

Such a variously stratified system, subject to forcing at its free surface and open boundaries, can undergo a wide range of possible responses. These include the effects of the pulsed entrant jet of freshwater into the Strait of Georgia, the movements of estuarine surface flow towards the strong tidal mixing regions at its boundary openings, and the returning undercurrents of mixed water having some salinity intermediate between that of water at comparable depths in the adjacent ocean and the brackish outflow. These processes are subject to strong tidal and seasonal modulations, in the latter case due to changes both in river discharge and in the salinity of intruding ocean water. It may thus be anticipated that internal oscillations at tidal periods will occur. Since the internal Rossby radius is of order 10 km, the effects of rotation will strongly influence the character of such oscillations as indeed will the seasonal change in stratification. Superimposed upon these motions will be the internal responses to wind events. Studies of responses in stratified lakes (Mortimer, 1963, 1974; Simons, 1980) suggest that shore zones can be dominated by long internal Kelvin waves while away from the shore, near-inertial Poincaré modes can occur.

In the other major water body of the system, Juan de Fuca Strait, the vigorous tidal streams may be expected to interact with the complex topography to bring about strong eddy formation and associated residual circulation. In the outer part of Juan de Fuca Strait, the estuarine circulation is subject to a variety of influences deriving from processes in the waters seaward of its entrance. In the barotropic context, the primary feature is the ocean tide that is predominantly responsible for the tidal co-oscillations within the region. Surge elevations generated by strong offshore winds can propagate into the system although these are rarely of such magnitude as to cause loss of life. Tsunamis deriving from seismic activity elsewhere around the North Pacific can also propagate into the system although apparently being attenuated to insignificance in the highly dissipative channels of the San Juan Islands (Waldichuk, 1964).

In the case of baroclinic effects, the most notable is the movement of higher salinity water into Juan de Fuca Strait at depth in summer, a consequence of seasonal upwelling on the adjacent continental shelf. This will affect the generation of internal tides in the vicinity of sills or marked topographical changes the Strait. Southerly winds which tend to prevail in winter can move surface waters diluted by discharge from the Columbia River to the mouth of the Strait. This water contrasts with the higher salinity surface waters in the Strait itself, a consequence of the vigorous tidal mixing in the channels opening off its inner end. The surface intrusions that result (Holbrook et al., 1983) can be superimposed upon a more complex process in which internal Kelvin waves generated in the stratified shelf waters by wind events, can propagate into the system as long internal Kelvin waves (Proehl and Rattray, 1984). These effects which can also occur in fall and summer, can reverse the basic estuarine circulation in the Strait.

Simulation of the basic features controlling the circulation in such a system is clearly a major task. The effects of the earth's rotation and vertical exaggeration effectively prohibit the use of physical models. Numerical models, however, provide a graded approach in which the limitations of one model provide the design criteria for

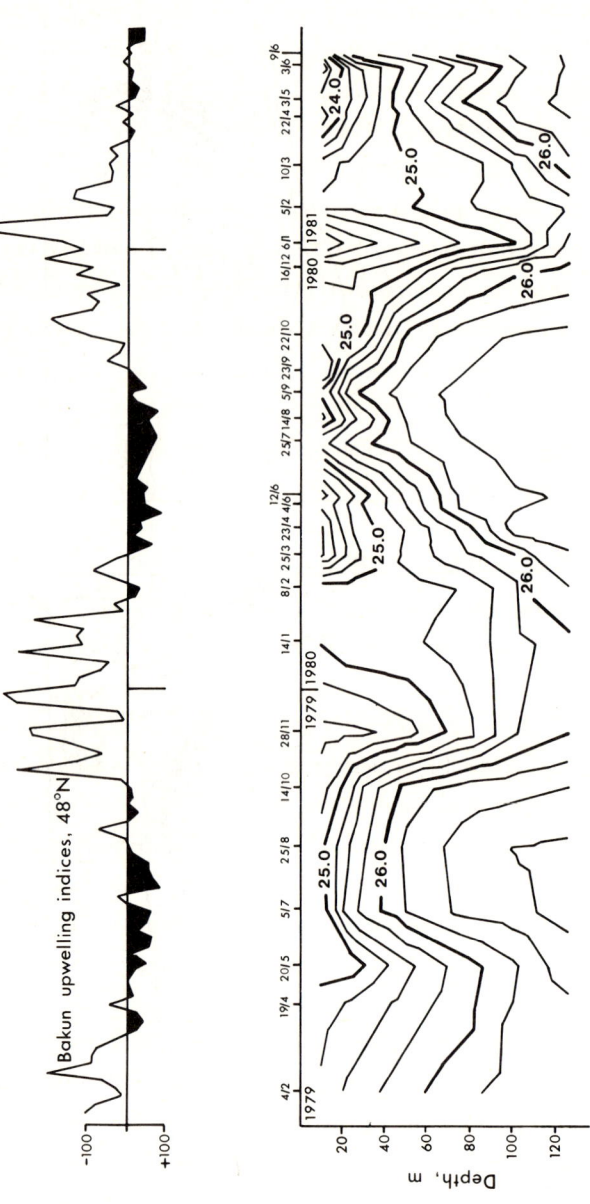

Fig. 1.7 Upwelling index computed for 48°N (top) and time series of σ_t at a midshelf station near the entrance to Juan de Fuca Strait. (from Freeland and Denman, 1982)

water from the Strait of Georgia lies around the southern coast of Vancouver Island. Extended records from moored current meters (8 m deep) in Race Passage (Fig. 1.8) and (18 m deep) at a location in Juan de Fuca Strait 6.4 km south of Race Rocks show a strong net seaward flow with evidence of both fortnightly and seasonal modulation with maximal flows occurring in early summer (Fig. 1.9). It should be noted that this is a region of strong tidal residual circulation. This will be discussed in Chapters 5 and 7 in connection with the net circulation occurring in the barotropic tidal models, GF3 and GF7, respectively. Extended records of daily surface salinity measurements at Race Rocks (Fig. 1.10) attest to the high degree of mixing with deeper sea water undergone by the Fraser River water in its seaward egress. There is also a roughly fortnightly modulation of surface salinity with some appreciable dilution occurring in summer following the Fraser's spring freshet, as well as during the winter months.

Comparison of the winter and summer longitudinal salinity distributions in Juan de Fuca Strait (Fig. 1.5a and b) shows an intrusion of higher salinity water at depth. This is consistent with upwelling on the continental shelf (Fig. 1.7).

In March and April 1973, in association with the numerical studies described in this book, an array of current meters was located over a cross-section of Juan de Fuca Strait (line 11, Fig. 2.1, p. 46). While the meters were in position, a slow net intrusion of saline water occurred (Crean et al., 1979). This is shown by the vertical longitudinal salinity sections through Juan de Fuca Strait (Fig. 1.11a, b, and c) where, over a 5-wk period, water originally at depth at the ocean entrance has moved up onto the sill in the Strait's inner end. The distribution of residual velocities over the cross-section (Fig. 1.12) shows net inflow at depth favouring the southern shore of Juan de Fuca Strait and net outflow near the free surface, with the contour of no net motion sloping downward from the Washington to the Vancouver Island shore. Superimposed upon this advance of the saline tongue are cross-channel internal oscillations of tidal periods (Fig. 1.13). These are illustrated by a contoured time series of salinity profiles from the Washington (Station B) and Vancouver Island (Station D) sides of Juan de Fuca Strait (Fig. 2.3, p. 48) over a 36-h interval. The approximate matching of peaks and troughs in the isohalines across the channel are consistent with the excitation of an internal cross-channel tidal oscillation. A rough calculation of the natural period of the first cross-channel internal mode is close to that of the semi-diurnal tide. In at least one instance the vertical displacement is in excess of 50 m along the Washington shore. Analyses of these current meter data (Godin et al., 1981) effectively indicate compensatory surface outflows and deep inflows having residual flow rates that vary from 7.78–13 8 km^3/day over the period of the observations. Fluctuations in the two flow rates occur in phase.

As noted in connection with Fig. 1.6b, relatively low salinity surface water moves northward along the Washington coast to the mouth of Juan de Fuca Strait in winter.

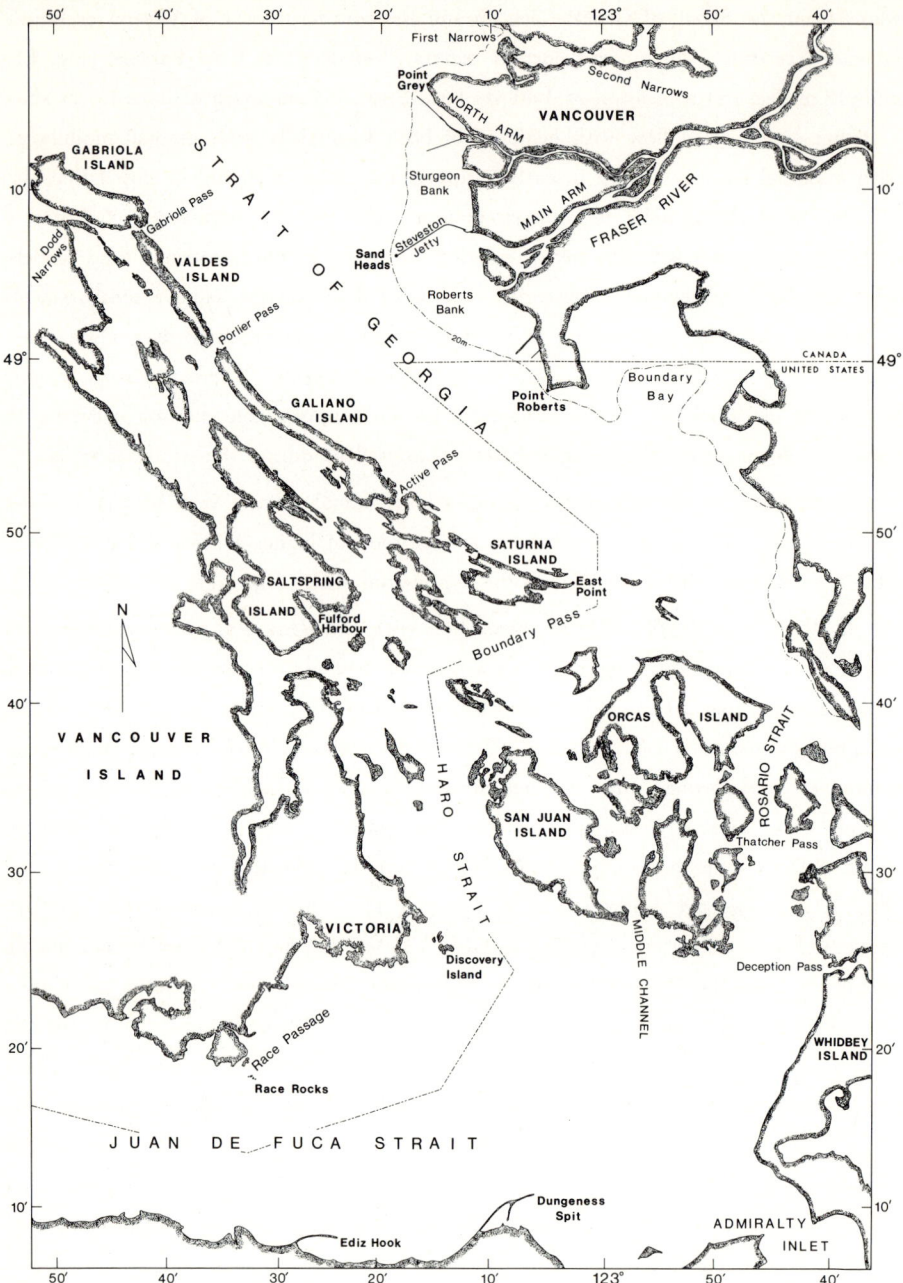

Fig. 1.8 Study region from the southern end of the Strait of Georgia to Juan de Fuca Strait.

The presence of such surface water having salinities much lower than those observed at Race Rocks (Fig. 1.10) is apparent in the first longitudinal salinity section shown in Fig. 1.11a. Such an intrusion was a persistent feature in contoured cross-sections

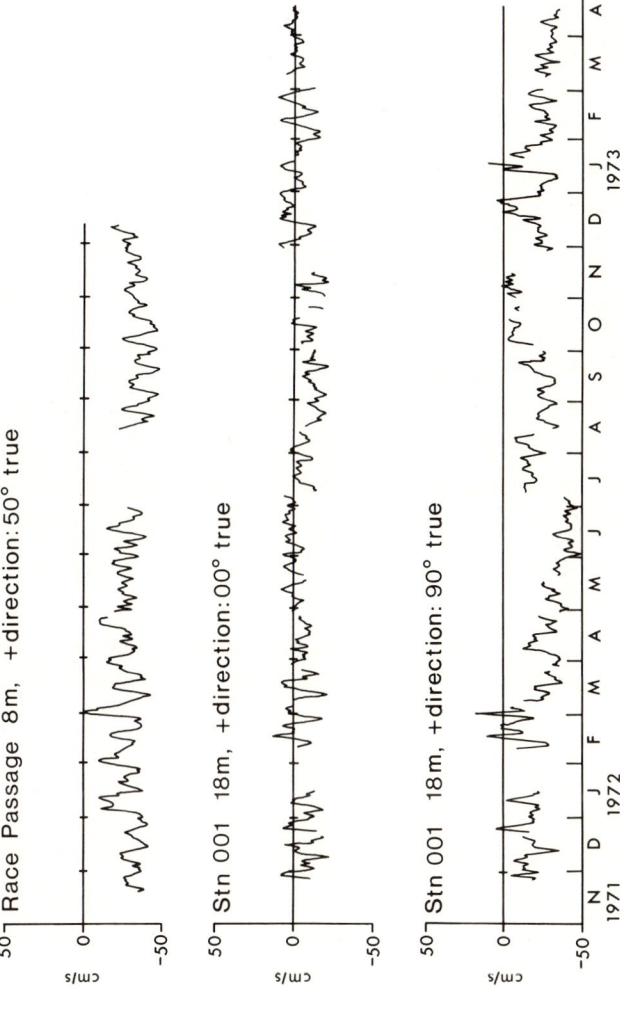

Fig. 1.9 Time series of observed velocities at a central location and depth of 8 m in Race Passage, the direction being parallel to the median line of the Passage and at a location (Stn 001) 6.4 km south of Race Rocks and depth of 18 m. The positive values of the upper trace for Station 001 denote the northerly component while the positive values of the lower trace denote the easterly component.

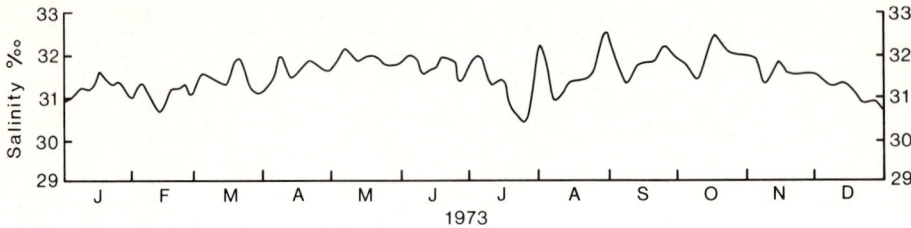

Fig. 1.10 Time series of daily salinity (‰) observations at Race Rocks for 1 yr.

of salinities (Fig. 1.14), near the position of line 11 (Fig. 2.3) over a 24-h period in December 1973. The intrusion of Columbia River water, Coriolis-favouring the Washington shore, contrasts with the seaward-moving, more saline estuarine outflow from the Strait of Georgia, which Coriolis favours the Vancouver Island shore. Analyses of line 11 current meter data (Godin *et al.*, 1981) indicate that two major reversals of the normal pattern of surface outflow and deep inflow occurred, each lasting about 5 days and following within 2 days of the passage of the intense low pressure systems over the exposed coastal waters.

Continuous observations of currents, temperatures, and salinities over a 2-yr period further confirmed that the normally vigorous estuarine circulation can frequently be reversed by surface water intrusions due to Ekman drift driven by southerly winds over the exposed coastal waters (Holbrook *et al.*, 1983) (Fig. 1.15). These intruding surface waters, warmer in summer due to thermal heating and less saline in winter due to the presence of Columbia River water (Barnes *et al.*, 1972), are invariably less dense than the ambient water in Juan de Fuca Strait. This results from the fact that the intruding waters are not subjected to the strong tidal mixing processes which are present in the channels conveying waters from the Strait of Georgia and Puget Sound into the eastern end of Juan de Fuca Strait.

1.3.3 Puget Sound

Of several fjords located along the mainland coast, the largest is the Puget Sound system which covers an area one third the size of the Strait of Georgia. The annual freshwater inflow is some 10–20% of that entering the Strait of Georgia. Of this inflow, 60% is associated with the Skagit River in the northeast corner of the system (Fig. 1.1). Only about half of this freshwater is thought to enter the main basin of Puget Sound with the remainder moving through Deception Pass into Rosario Strait (Fig. 1.8).

Puget Sound is a fjord type estuary with depths typically of order 200 m, its fresher and less stratified waters separated from the more saline and strongly stratified waters of Juan de Fuca Strait by a sill 30 km in length (Admiralty Inlet) where the tidal streams are strong. Another sill (Tacoma Narrows, Fig. 1.1), also characterized by

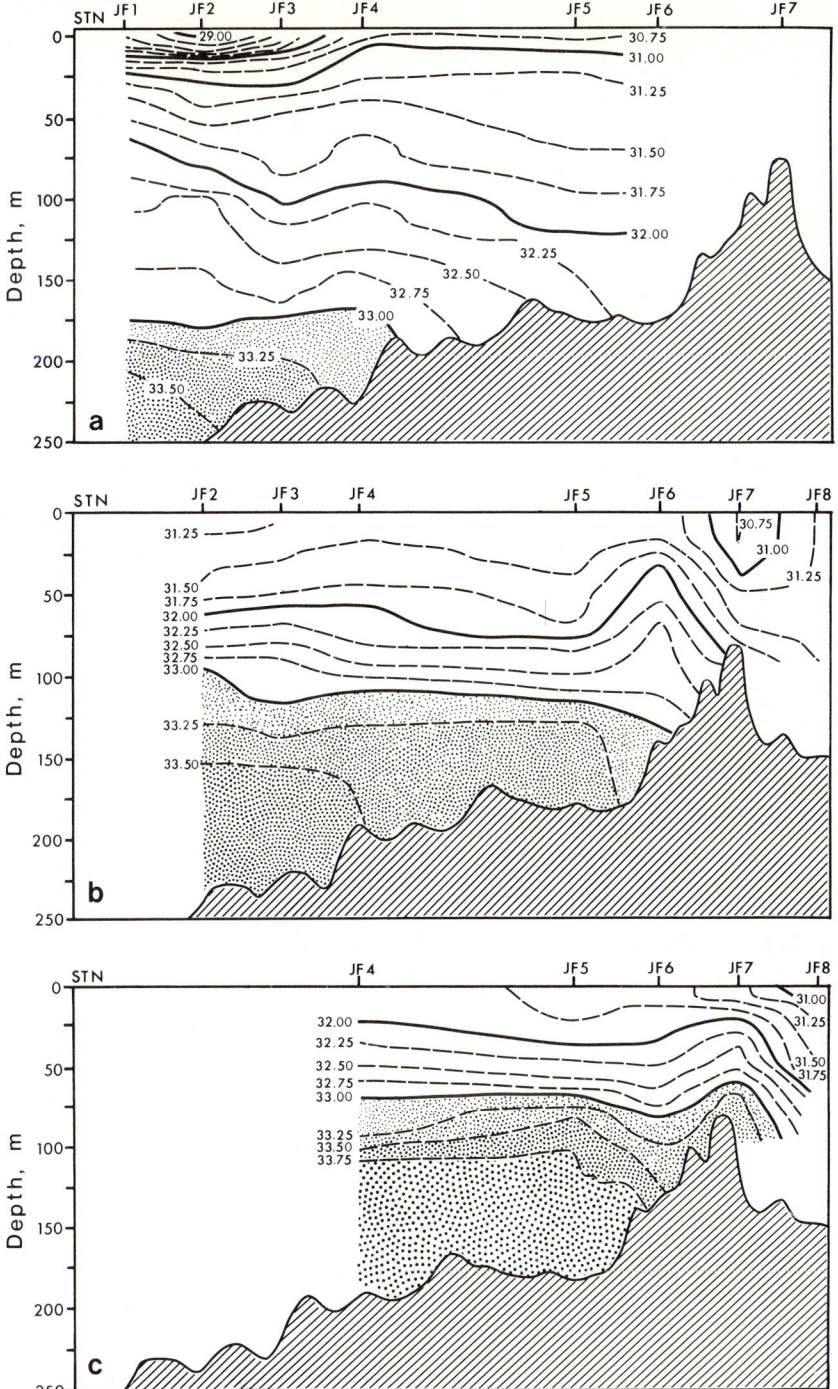

Fig.1.11 Contoured vertical sections of salinity through the median line of Juan de Fuca Strait for (a) March 5, (b) March 16, and (c) April 13, 1973. For station locations in these profiles, see Fig. 2.4, Chapter 2.

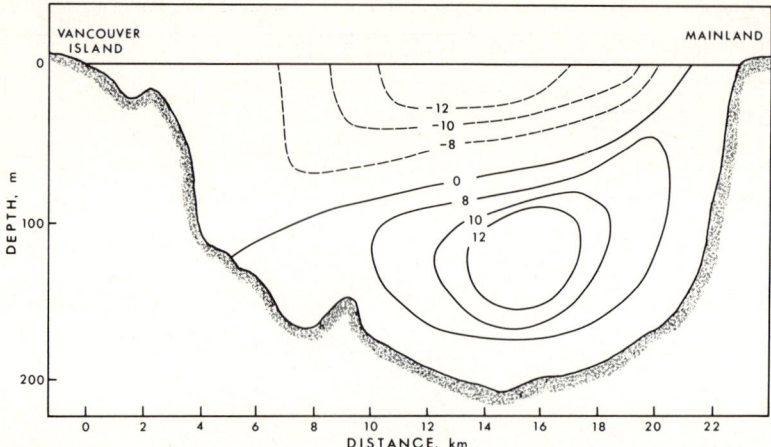

Fig. 1.12 Contoured residual velocities (cm/s) normal to a cross-section of Juan de Fuca Strait (line 11, Fig. 2.1, Chapter 2) observed in March–April 1973. Negative values denote seaward flow. (from Huggett *et al.*, 1976b)

strong tidal streams, connects its southern end to an extensive irregular storage area where the tidal amplitudes are large. Tidal action at the sills maintains a vigorous two-layer vertical circulation with a depth of no net motion roughly equal to that of the sills (Ebbesmeyer and Barnes, 1980; Möfjeld and Larsen, 1984). Episodes of deep water renewal by denser water intruding over the outer sill can occur throughout the year, the densest intrusions occurring in early fall (Cannon and Laird, 1978). A marked fortnightly modulation in these intrusions is attributed to changes in the degree of turbulent mixing by the tidal streams with large intrusions occurring when the tidal range is great (Geyer and Cannon, 1982). Residual velocities associated with these bottom currents entering Puget Sound are about 20 cm/s.

1.3.4 San Juan and Gulf Island Passages

The Strait of Georgia is connected to Juan de Fuca Strait through several passages (San Juan and Gulf Island passages). Most of these passages are small or narrow except the two most important, Haro and Rosario Straits. The effective sill depth in Rosario Strait is about 50 m. In addition to describing the physical oceanography of the southern part of the Strait of Georgia, including the San Juan Island passages, Schumacher *et al.* (1978) also discussed the results of a moored current meter program conducted from February through April 1975. Observations showed a strong southeasterly flow on the western side of the southern Strait of Georgia and a weaker northwesterly flow on the eastern side. At depth, inflowing saline water from Haro Strait augmented the flooding tide on the eastern, and the ebb on the western sides of the Strait.

The tide wave entering the Strait of Georgia from Juan de Fuca Strait is split into two major branches by the San Juan Islands. The larger portion passes through Haro

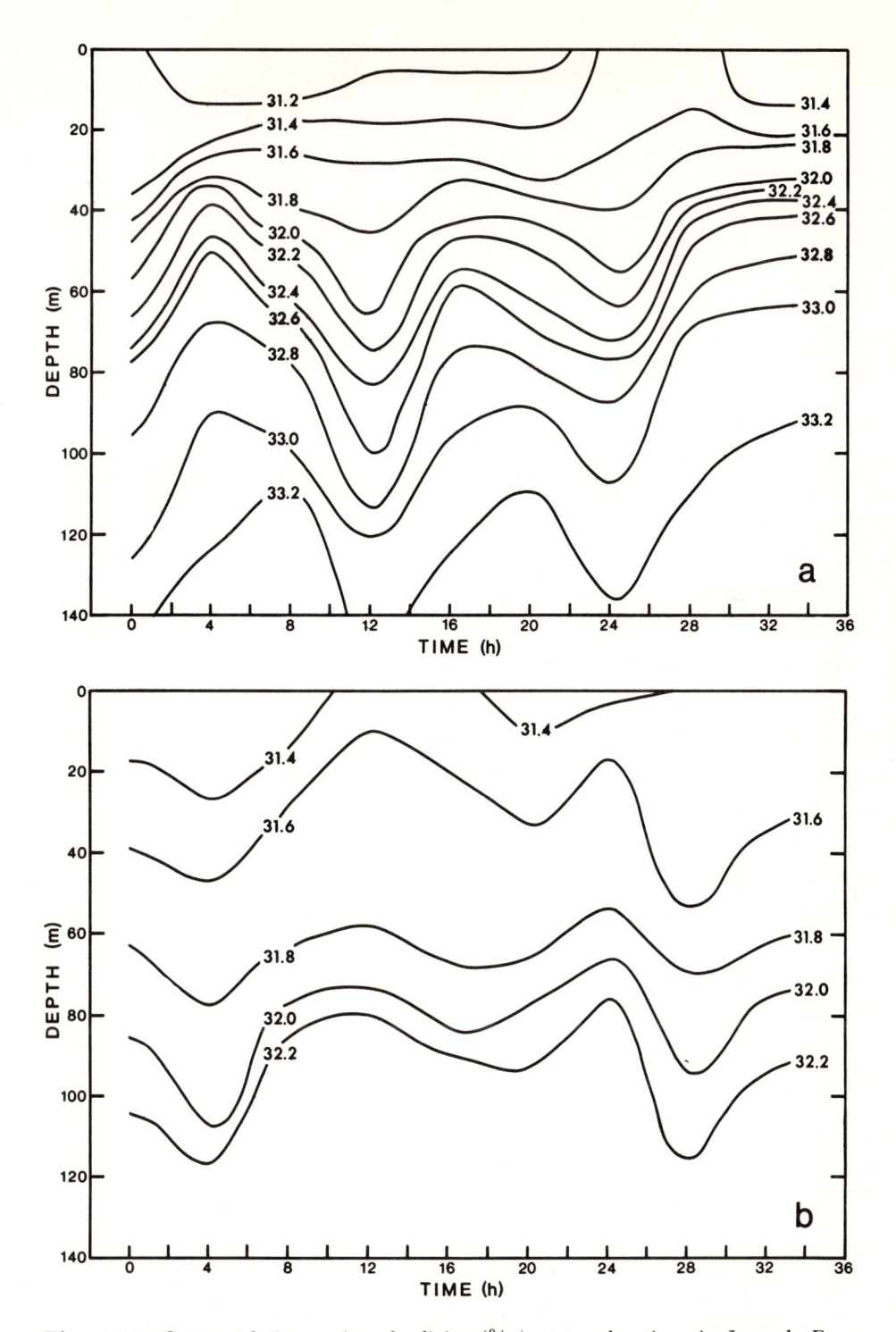

Fig. 1.13 Contoured time series of salinity (⁰/₀₀) at two locations in Juan de Fuca Strait near line 11 at (a) Station B near the Washington shore and (b) Station D near the Vancouver Island shore (Fig. 2.3) over a 36-h interval.

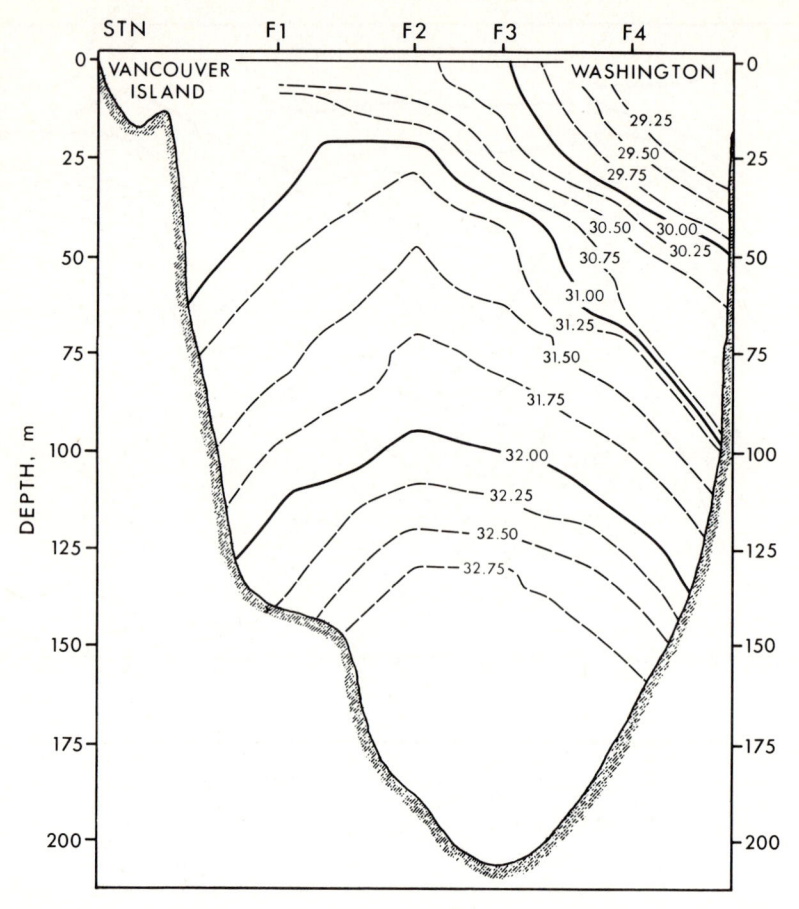

Fig. 1.14 Contoured cross-section of salinities (⁰/₀₀) near line 11 on 4 December 1973. Station positions are shown in Fig. 2.4 (Chapter 2).

Strait with the smaller portion going through Rosario Strait. Parker's (1977) estimate of flood volume transport during half of an M_2 cycle showed that of the total M_2 flood volume entering the eastern part of Juan de Fuca Strait, 24% enters Admiralty Inlet, 20% flows into Rosario Strait, 51% goes to Haro Strait, and 5% enters Middle Channel. The tide wave in Rosario Strait is delayed by about an hour relative to that in Haro Strait. The Rosario stream dominates the eastern side of the Strait of Georgia while the Haro stream, with its larger volume and higher velocities, moves into the main body of the Strait. The two waves meet around the north side of Orcas Island (Fig. 1.8). Generally, M_2 and K_1 tidal current phases lead the local tidal elevation phases by 50–70°, indicating a predominantly standing wave character. The general tidal hydrodynamics for the Juan de Fuca–Strait of Georgia region are also discussed by Parker (1977).

Haro Strait is an irregular body of water approximately 50 km long and varying

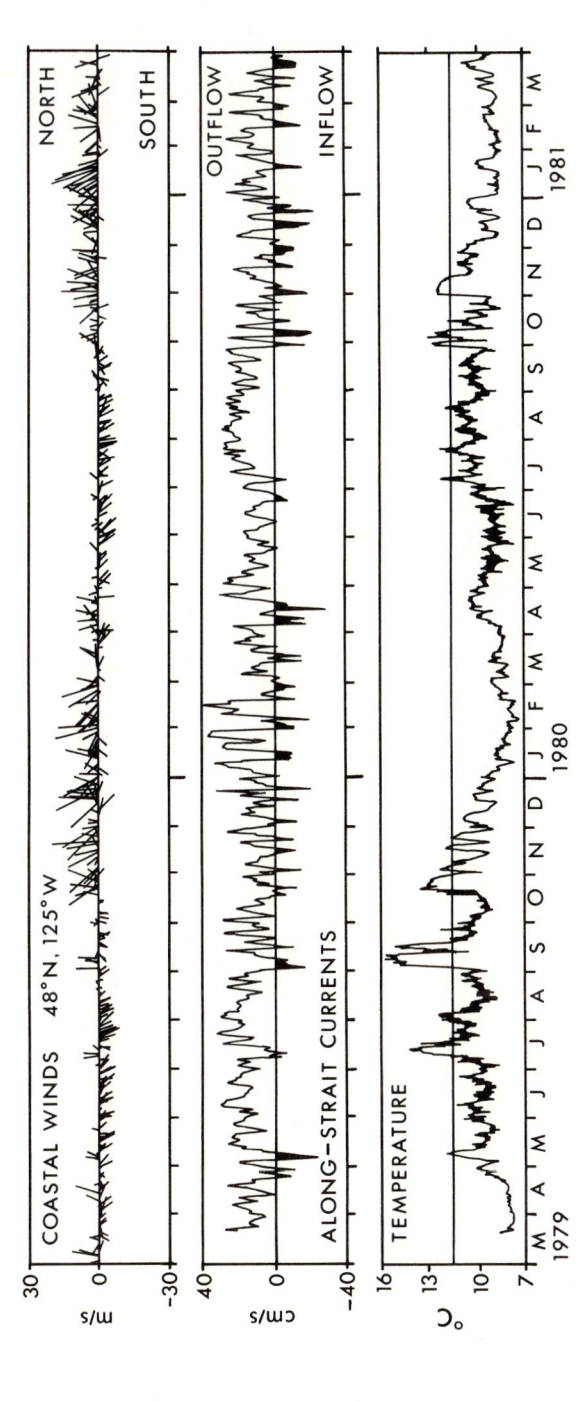

Fig. 1.15 Time series of coastal winds, low-pass filtered, along-strait currents, and unfiltered temperature at Station 15 (4 m deep). See Fig. 2.3 for station location. Wind vectors indicate the direction towards which the wind is blowing. (from Holbrook et al., 1983)

from 4–11 km in width that connects Juan de Fuca Strait with the Strait of Georgia. It is important for the major role it plays in the fresh- and salt water exchange between the Strait of Georgia and the open ocean.

During the period 5 July to 10 September 1976, a physical oceanographic study was carried out in Haro Strait (Webster, 1977) to determine major features of the water characteristics and flow in that region for late summer, generally the most active time of the year in terms of estuarine exchange and deep water renewal. The results were consistent with earlier work showing the estuarine type circulation that exists in Haro Strait, *i.e.*, a relatively fresh surface layer flows southward out of Haro Strait over saline water flowing in. During the summer of 1976 the depth of the dividing line between the two flow regimes was observed to vary between 40 and 100 m on the eastern side of Haro Strait while retaining a value of about 100 m on the western side. At each location average salinities, temperatures, and longitudinal and transverse velocities were found to vary over a 2-wk cycle associated with the fortnightly cycle of tides. Minimal vertical gradients of velocity, temperature, and salinity were found to be associated with large tides and maximal gradients with small tides. This is due to greater vertical mixing throughout the water column during strong tidal flows. This accords with the findings of Geyer and Cannon (1982) for Admiralty Inlet. Water movements in Haro Strait were dominated by tidal currents with strengths well over 1 m/s. The magnitudes of the two largest components, M_2 and K_1, were greatest in the upper 100 m on the eastern side of the Strait and smallest on the western side. The tidal phases showed pronounced changes over a cross-section of the Strait. The smallest phases of both the M_2 and K_1 components occurred in the surface layers of the eastern side but the phases increased with depth towards the western side. In fact, for the M_2 component a phase difference of almost 100° was observed between two current meters in diametrically opposed positions (shallow eastern versus deep western) of the cross-section. The existence of a large tidal residual eddy in this vicinity is shown in numerical model GF3 results which will be discussed in Chapter 5. Pronounced changes occurred in the water characteristics in the whole water column of Haro Strait over the period of a day.

1.3.5 Surface Waters of the Strait of Georgia

In describing the oceanographic features of the Strait of Georgia, it is convenient to deal initially with the surface circulation which is dominated by the discharge from the Fraser River. The deeper circulation, influenced largely by mixed water intruding from the strong tidal mixing regions to the north and south of the Strait will then be discussed.

The Fraser River drains the largest watershed in British Columbia and is the primary source of freshwater, accounting for some 70% of the total **freshwater input** to the system. Near the river mouth the channel divides into the Main and North Arms (Fig. 1.8), the former carrying 87% of the total discharge. The seasonal cycle in discharge is dominated by a single large peak corresponding to the freshet following spring thaw (Fig. 1.16). Near the river mouth saline water from the Strait of Georgia intrudes to form a long sloping wedge below the outgoing freshwater. Figure 1.16 also shows the penetration of the salt wedge as a function of discharge rate, large penetrations occurring in winter and small ones during the freshet (A. Ages, pers. comm.). Salinities in the wedge are higher (about $29^0/_{00}$) in winter although with considerable dispersion of the isolines over the extended length of the wedge. In summer, salinities average about $25^0/_{00}$, the isolines being crowded together close to the toe of the wedge.

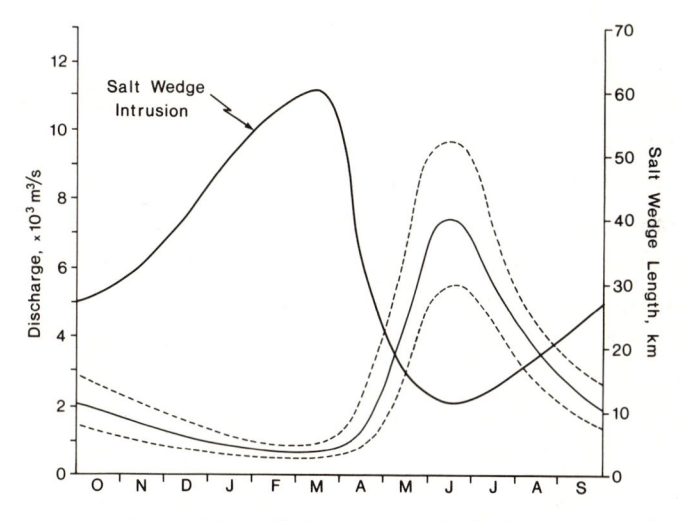

Fig. 1.16 Average Fraser River discharge for a hydrological year (from October to September) and the corresponding observed penetration maxima of the salt wedge upriver. Eighty percent of the river discharge rates fall between the dashed lines.

1.3.5.1 River mouth

The rate of discharge from the Fraser River is strongly modulated by the tidal rise and fall of the water surface in the Strait of Georgia near the river mouth. The changes in speed and direction of the surface flow at the mouth can be illustrated by data (Fig 1.17) from a current meter moored just south of the Main Arm near Sand Heads (location 1815, Fig. 1.18). About an hour after high water the surface flow starts to increase rapidly. The direction is about 250°, that is, predominantly seaward but with a small component towards the main river channel due to the ebb flow off the banks. About 2.5 h before low water there is a small secondary minimum. This is due to the destruction of the salt wedge in the Fraser River when some critical value of the vertical shear is attained, the outgoing momentum in the upper part of the water column being then redistributed over the whole depth down to the bed. This was confirmed by observations (Crean and Stronach, unpubl.; A. Ages, pers. comm.). The velocity maximum is attained about 0.75 h before low water, thereafter decreasing rapidly as the tide turns and commences to flood. Although freshwater flow in the Main Arm is directed primarily towards the Strait of Georgia, significant cross-channel flows onto the shallow banks near the river mouth can occur at certain phases of the tide. About 4 h after low water the direction shifts to indicate a strong southerly flow onto the banks at about high water.

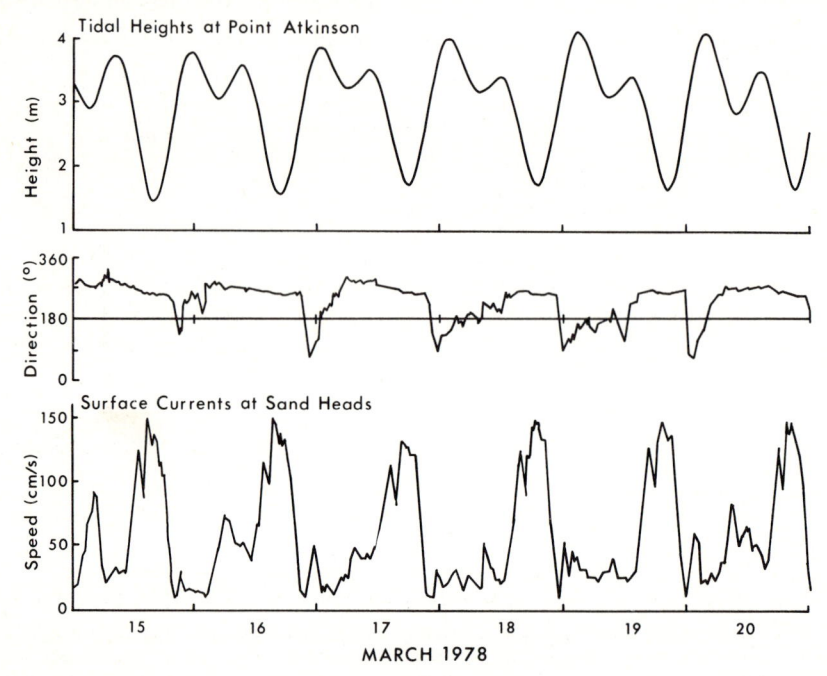

Fig. 1.17 Tidal elevations at Point Atkinson (Fig. 1.1), a representative sequence of directions (true) and speeds from a moored current meter (1 m below the water surface) at location 1815 (Fig. 1.18).

CTD and current measurements made at a later date show the nature of the change in surface flow at the river mouth on the latter part of the flooding tide (Fig. 1.18). An important feature is the formation of a front between the river water (the vertical structure being represented by the profile denoted by the number 1800) and the water flooding onto the banks from the Strait of Georgia (profile 1720). Solid arrows indicating surface flow show the northward flood (1720) along the edge of the banks while at the river mouth the flow is deflected from its seaward path, curving around and onto the banks (1740, 1800, 1815). This deflection is due to the opposing directions of river flow and the advancing flood tide from the Strait. The resulting convergence and presence of a retaining wall (Steveston jetty) give rise to a cross-channel pressure gradient which deflects the river flow onto the banks. Near the bottom, the

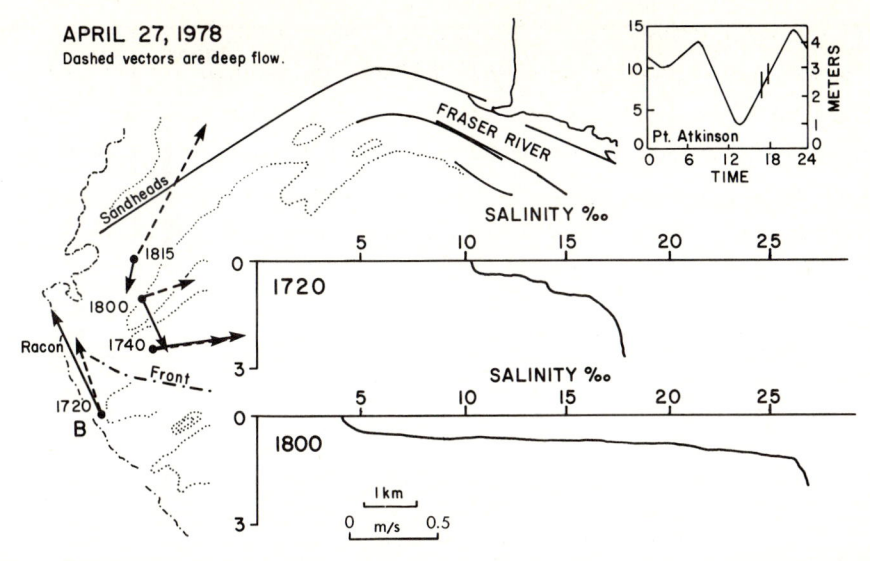

Fig. 1.18 Surface (solid arrows) and near-bottom (dashed arrows) velocity vectors near the mouth of the Fraser River at maximum flood (see tidal elevation inset) under conditions of moderate river flow (3,000 m³/s). Also represented are salinity profiles and location of the freshwater front (dash-dot lines). Numbers 1720 to 1815 denote time of observation in hours and minutes.

dashed lines show the advancing flow of salt water into the Fraser River and onto the banks.

1.3.5.2 River plume

As the tide falls in the Strait of Georgia and the surface flow from the Fraser River increases, the front described in the previous section intensifies and expands out into the ebbing streams of the open Strait. In the absence of winds the tendency to a southerly displacement of the plume by the ebbing stream in the Strait is, to some extent, offset by a northward Coriolis deflection (Giovando and Tabata, 1970). With the change to flood, a northerly set becomes predominant. The effect of this tidal pulsing of the river discharge on surface currents may be illustrated by a 3-day time series at an anchor station (A in Fig. 1.19) located 6.3 km from Sand Heads on a line determined by extending the median along the Main Arm to the edge of the banks at Sand Heads (Tabata *et al.*, 1970).

If it is assumed that there is no river discharge, a representative distribution of the ebb velocity vectors in the general vicinity of the river mouth, including the anchor station location, may be obtained from the vertically-integrated model GF3 (described in Chapter 5). These vectors are shown in the figure and illustrate the curvature of the local tidal flow field. The streams are not subject to significant rotation of the stream vector. It is conceptually convenient to resolve the observed velocities at this anchor station into a first component having its positive direction toward the river mouth and

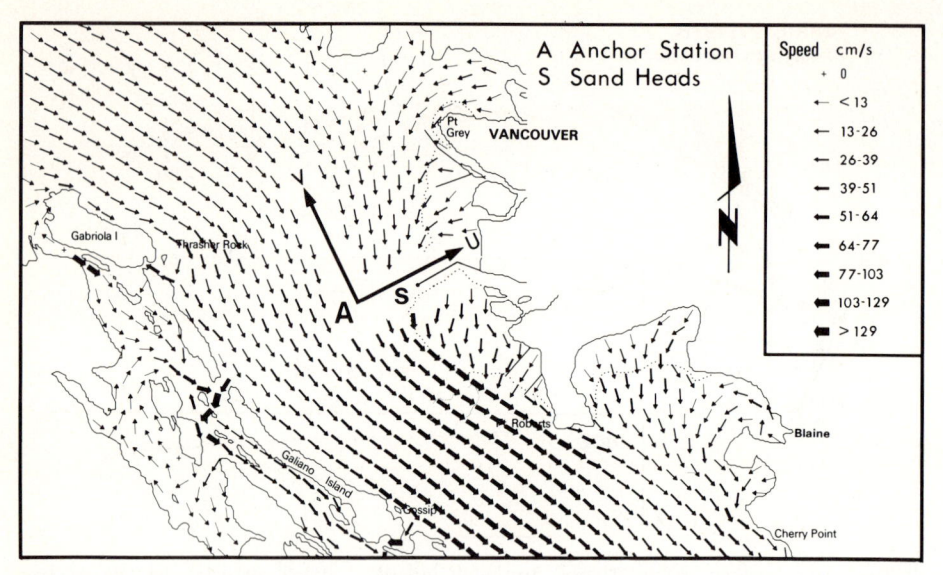

Fig. 1.19 Representative ebb velocity vectors in the general vicinity of the Fraser River mouth and the location (A) of an anchor station. The axes denote directions into which the observed velocity components were resolved.

a second orthogonal component, positive northward which is, in essence, tangential to the predominant tidal stream direction below the region affected by the river discharge.

For the top 10 m of the water column these currents (Fig. 1.20) were observed at approximately hourly intervals. The resulting time series were smoothed by a Butterworth filter (suppressing periodicities in the record of less than 5 h). The river discharge at the time was $7,500 \text{ m}^3/\text{s}$, typical of the lower limit of observed freshet maxima.

Also shown in the figure are the contoured time series of observed temperatures and salinities. For reference purposes the predicted tidal elevations at Point Atkinson (tide gauge #7795, Fig. 2.1, p. 46), which may be taken as representative of the tidal elevations near the Fraser River mouth, are included. Considering initially the time series of temperature and salinity, marked depressions in the isolines correlate strongly with times shortly before the occurrence of lower low water. Weaker depressions correspond to the times of higher low water. In studies of the Connecticut River plume, Garvine and Monk (1974) observed marked downward vertical velocities associated with the strong horizontal convergences that accompany the passage of freshwater fronts. This is reflected in the depressions of isolines referred to above. Of particular interest is the fact that the supply of freshwater from the river mouth completely dominates any vertical diffusive mechanism, resulting in the appearance of an advancing front of an intact body of freshwater.

The velocity time series at 1, 2, and 3 m display a distinctively different character from that at 5, 7, and 10 m. In the upper three traces, both components show a marked tidal modulation and coherence between the corresponding component at each depth.

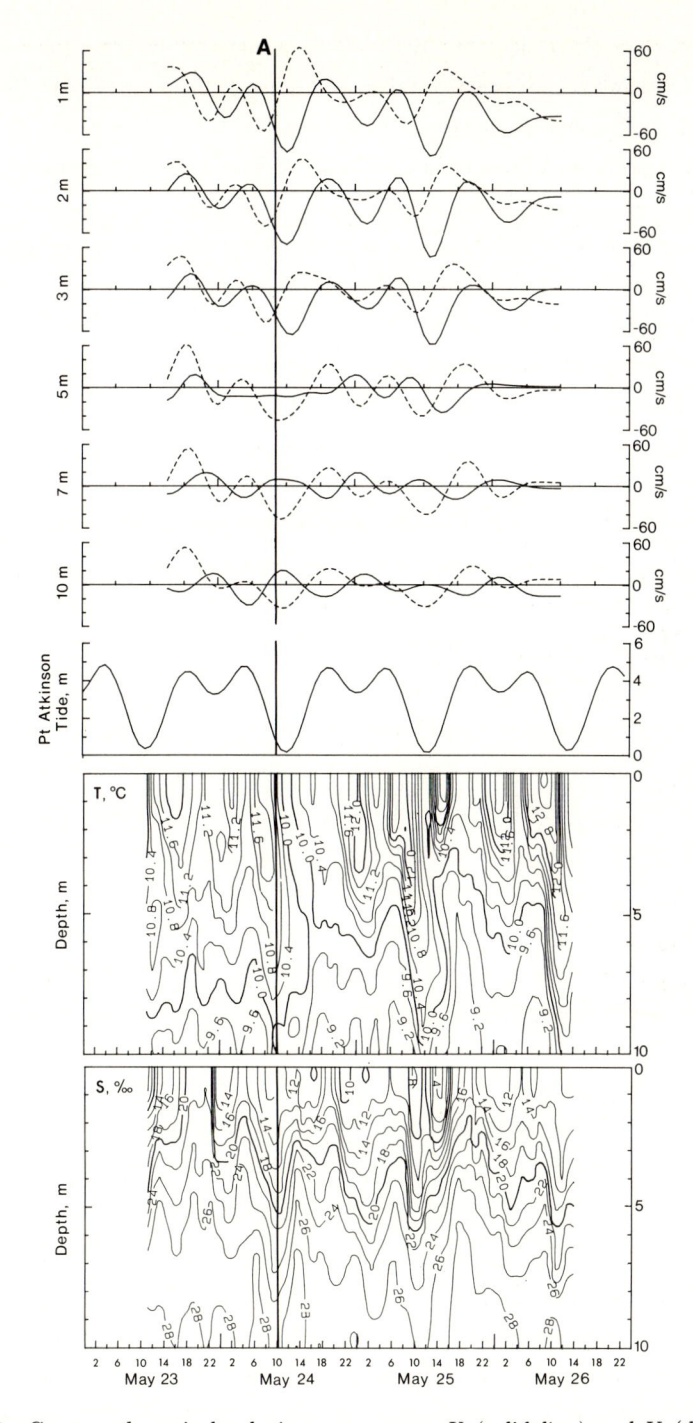

Fig. 1.20 Contoured vertical velocity components U (solid line) and V (dashed line) for the upper 10 m observed at Anchor Station A. Also shown are the tidal elevation at Point Atkinson and contoured time series of temperature and salinity for the period 23–26 May 1967. The vertical line at 1005 h, 24 May denotes the passage of the front referred to in Fig. 1.22.

The component resolved along the river axis is dominantly negative (flow away from the mouth) with maximum negative values coinciding with low waters in the Strait of Georgia. The orthogonal components directed parallel to the median line of the Strait conform reasonably well to the general sequence of flood and ebb streams in the Strait itself. At 5, 7, and 10 m, the components directed along the river axis are weaker and more irregular than those at 1, 2, and 3 m. It is evident from the concomitant temperature and salinity records that strong vertical displacements, associated with outgoing pulses of river discharge and passage of fronts, extend down to a depth of at least 10 m. The components directed along the river axis are weaker and more irregular than those above. The orthogonal components display a reasonably coherent modulation, basically of tidal character with phase shifted from the corresponding component traces near the surface. Thus, maximum values (flood direction in the Strait) are in reasonable accord with local high waters, similarly the minimum values with local low waters. Such a phase relationship between flow and elevation maxima accords more with progressive wave behaviour. However, the modelled depth-averaged currents are approximately consistent with standing waves. This discrepancy is presumably a local influence of the Fraser River discharge.

The filtering applied to the time series presented above emphasizes the tidal nature of the surface flows but eliminates events such as the passage of the front. As the tide falls in the Strait of Georgia, the front moves out from the Fraser River mouth. On 24 May 1967 such a front passed the anchored vessel 1.5 h before lower low water (line A, Fig. 1.20). The change in speed and direction associated with the passage of this front is illustrated by part of the trace from the surface current meter record (Fig. 1.21). Prior to the arrival of the front, the velocities were about 60–70 cm/s along the line of a seaward extension of the river axis. When the front passed, the speed increased to about 125 cm/s and the flow direction shifted 20–30° south of the line of the river axis. At this stage of the ebb, the part of the water surface slope deriving from the overall tidal co-oscillation within the Strait of Georgia was near its minimal value. Likely contributing factors to this southward deflection were the static head and the entrainment of water having the southerly-directed momentum of the ebbing tide below the river plume. As noted by Cordes *et al.* (1980) strong entrainment near the mouth of the Fraser River results in significant deceleration of the plume.

For 2 h after the passage of the front, the surface velocity (Fig. 1.22) increased to a maximum value of 150 cm/s, thereafter decreasing gradually for the next 4 h to 25 cm/s. Between 3 and 4 h after lower low water, the surface flow direction in the Strait of Georgia changed sharply to that of the flooding tide.

Drogues released at the river mouth about 2 h before lower low water on 26 May followed tracks (Fig. 1.23) reasonably consistent with the direction of the surface flow at the anchor station after the passage of the front on 24 May. The speeds of the various

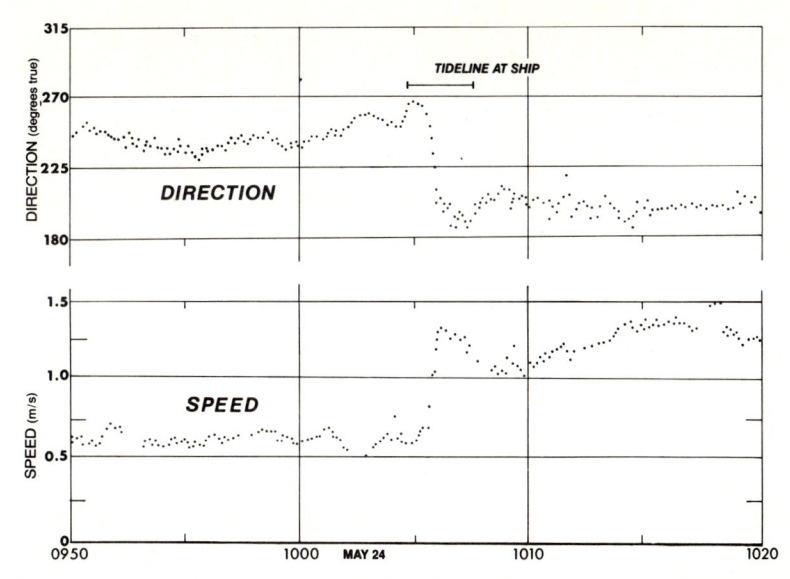

Fig. 1.21 Observed direction and speed of the surface current at about the time of passage of the river water front at Anchor Station A, 24 May 1967. (Thomson, 1981)

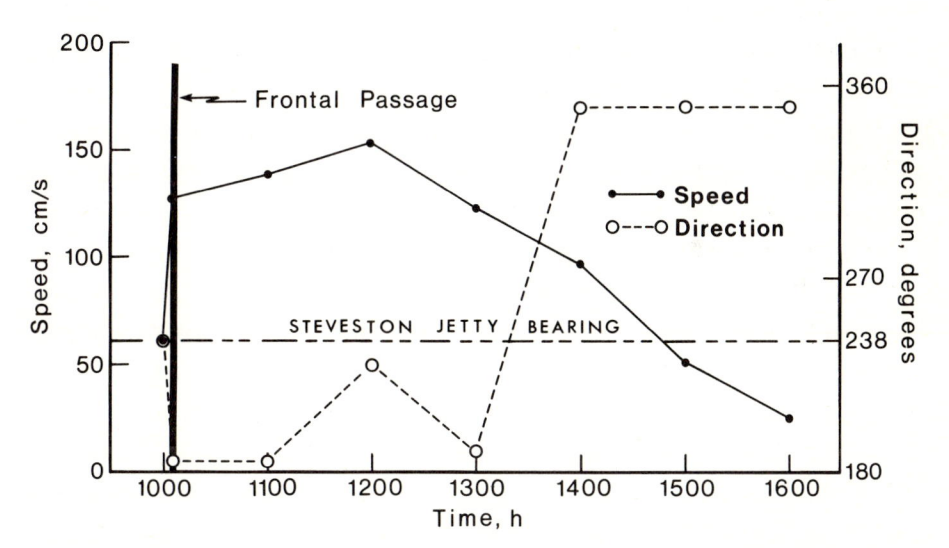

Fig. 1.22 Observed direction and speed of the surface current immediately before and following passage of the river water front at 1005 h, 24 May 1967 at Anchor Station A. Also shown is the bearing of the Steveston jetty indicating the major axis of the plume at its source.

drogued floats varied within 118–175 cm/s, the speed increasing the farther south the tracks were located.

Moving farther afield from the river mouth, the subsequent movements of the plume may be illustrated by representative tracks of free-floating surface current followers (1 m-deep drag elements) described by Giovando and Tabata (1970). The coherent movement

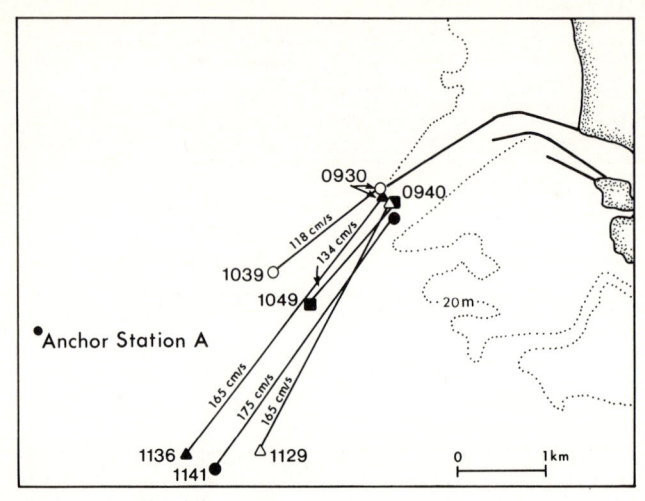

Fig. 1.23 Tracks of surface current drogues released at the mouth of the Fraser River (Sand Heads) at the time the freshwater front is moving towards Anchor Station A on the morning of 24 May 1967.

of the drifters suggests a flow field which can be described in terms of a relatively small group of well defined features. These, however, can give rise to a large number of possible realizations depending on tide, wind, and river discharge.

Tracks A, B, and C (Fig. 1.24) were obtained over a 30-h interval which included a small ebb and flood followed by a large ebb and flood sequence. The Fraser River discharge was moderately high (7,800 m^3/s) and over the first 8 h winds were about 2.5–8 m/s from the southeast. The relatively close pattern of the drifters was maintained and illustrates the net northerly deflection of the plume, with some assistance from southeasterly winds, into the broad central part of the Strait of Georgia where the tidal streams are relatively weak. Tracks D, E, F, G, and H were obtained over a 22-h period which included a large flood, small ebb small flood, and partial large ebb sequence. The river flow was moderately low (2,700 m^3/s) and southeasterly winds of 5–9 m/s prevailed. Again, much the same type of behaviour was evident though rather more closely confined to the eastern shores. Tracks I and J were observed spanning an 8-h interval over essentially a moderate flood tide. The river flow was fairly low (2,700 m^3/s) and the winds were negligible. These tracks are suggestive of a large back eddy having a radius of curvature r consistent with Coriolis deflection, i.e., $r = u/f$ where u is an average tangential velocity and f is the Coriolis constant. This feature is confirmed by several other sequences of drift observations (Giovando and Tabata, 1970). Tracks K, L, and M were observed over a 16-h interval. Conditions were a moderately high flow (7,800 m^3/s) and a large flood and small ebb. Of particular interest however, is the effect of northwesterly winds with speeds between 8–13 m/s during the tracking period. The consequent southerly drift illustrates how winds can offset the northward

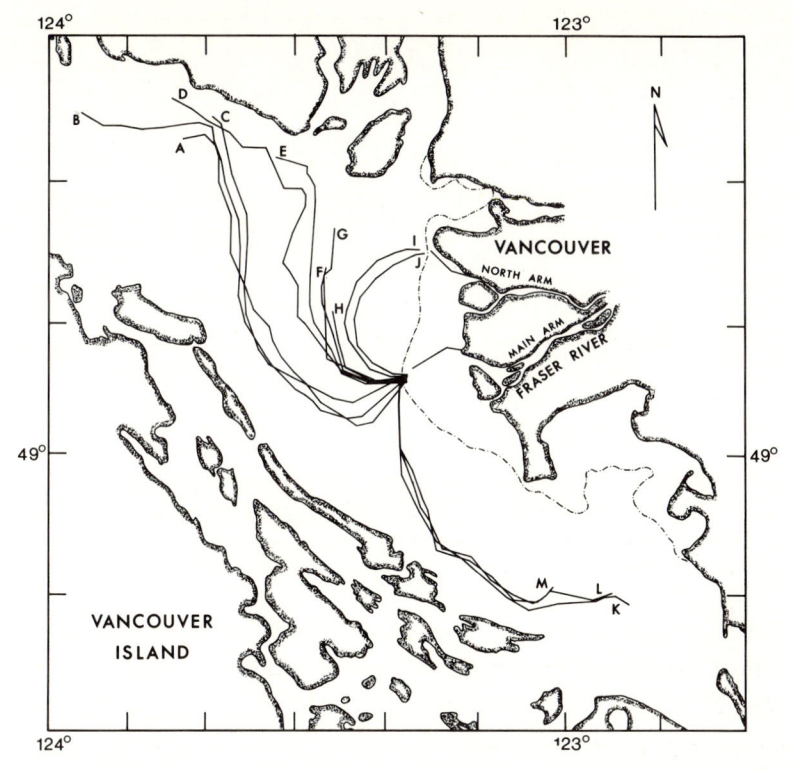

Fig. 1.24 Representative tracks of surface drogues (1 m deep) released near the mouth of the Fraser River under various conditions of river discharge and wind. The significance of the lettering is explained in the text.

Coriolis deflection of the river flow and flooding tide.

The most extensive overall coverage of the plume in terms of free-drifting surface current followers (Stronach *et al.*, 1984) is obtained by combining tracks observed on two consecutive days that have essentially identical tides (lower low water at about midday), high river flow (11,200 m^3/s), and negligible wind effects (Fig. 1.25). In this case the drift elements were 2 m deep. The reversal of the surface flows with the change from ebb to flood in the southern part of the Strait of Georgia where the tidal streams are strongest, is clearly noted. The changing orientation of the emergent jet can be seen in comparing the two successive deployments, one on the ebb, and one on the flood, off the mouth of the Fraser River. The main northward sweep of the plume back towards, and its division along, the mainland coast is well depicted.

Under conditions of weaker river discharge (2,400 m^3/s) following 2 days of northwesterly winds acting over the exposed fetch of open strait to the north, quite a different pattern of surface flow results. Thus the net southerly surface drift and formation of a large eddy off Valdes Island by the arrested river plume dominate the flow field (Fig. 1.26).

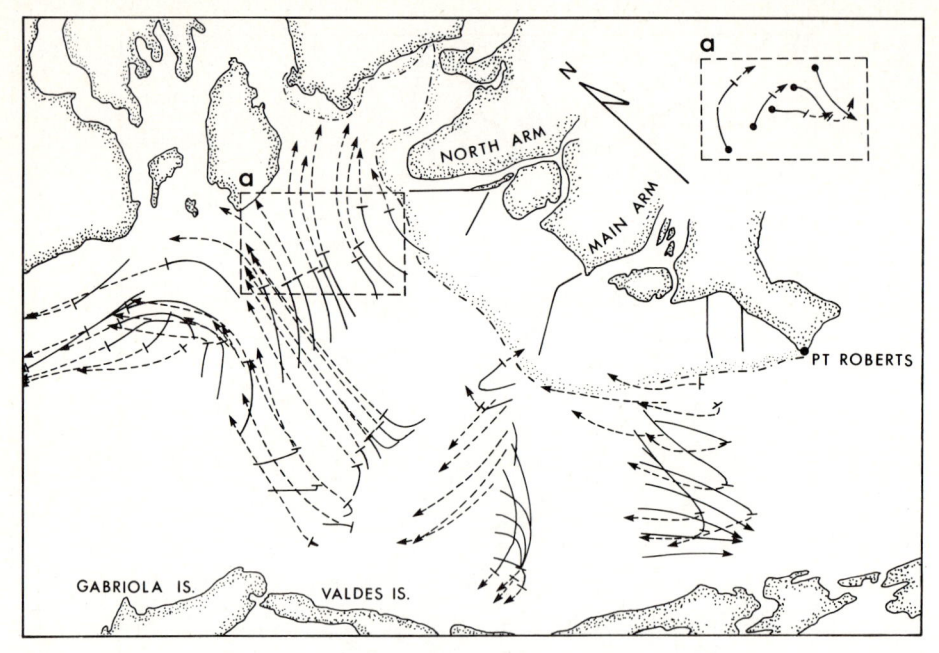

Fig. 1.25 Drogue track composite (2 m-deep drag elements) for 22 and 23 June 1982 under conditions of negligible winds, lower low water at noon, and strong daily mean river flow (11,200 m^3/s). Solid lines denote tracks obtained before lower low water; dashed lines show tracks after lower low water. In inset a, tracks depict an eddy that forms at the time of higher high water. Large dots denote start positions.

1.3.6 Subsurface Waters

Turning now to the deeper waters in the Strait of Georgia, it is of interest to consider initially the net flow immediately below the Fraser River plume. The evidence indicates that, in absence of northwesterly winds, the predominant direction followed by the plume is northward towards and along the mainland coast. The question arises as to whether or not the bulk of the Fraser River discharge leaves the system through the northern passages. Ultimately, however, it would appear that the southerly route is the favoured means of egress.

Estuarine outflow from the system must pass through either of two passages, Juan de Fuca Strait to the south or Johnstone Strait to the north, both of which are located between Vancouver Island and the mainland coast. Freshwater from the Fraser River has to pass through regions of strong tidal mixing to reach these passages. Surface salinities in both passages are similar as indeed are the available observed distributions of residual velocity. It is reasonable to suppose that the vertical distribution of velocity and density adjust to establish stable patterns of residual flow yielding a value of an overall internal Froude number (the ratio of the residual current to the square root of reduced gravity and water depth where reduced gravity is the product of gravity and the density difference between top and bottom layers) close to unity.

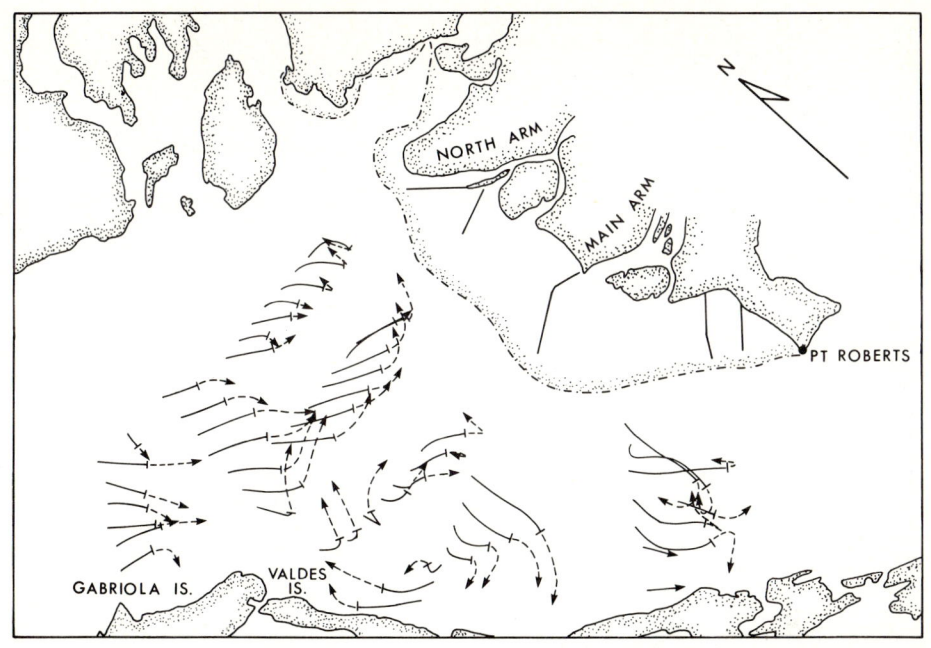

Fig. 1.26 A composite of observed drogue tracks (2 m-deep drag elements) obtained 14 and 15 October 1981 under similar conditions: negligible winds, lower low water, about noon, and low daily mean river discharge (2,400 m^3/s) following 2 days of northwesterly winds acting over the exposed fetch of open strait to the north.

Observations from current meters moored in the Strait of Georgia (lines 3, 4, 5, and 6, Fig. 2.1, p. 46) at a depth of 15 m below the water surface and hence below the immediate dynamical influence of forces acting on the upper layer, indicate a predominantly southerly residual flow (Fig. 1.27) except at three locations near the mainland coast (line 4) and also at a location southeast of Point Roberts where the circulation is dominated by a large residual tidal eddy. Residual current vectors at 20 m obtained from moored current meter and cyclesonde current observations (Stacey et al., 1987) are included in Fig. 1.27. These vectors have been supplemented by further residuals measured at 5 and 21 m (Schumacher et al., 1978).

From an analysis of temperature and salinity data collected in 1931 and 1932, Redfield (1950b) considers the Strait of Georgia to be a deep estuary in which five major water masses within or immediately seaward of the system can be identified. Three of these were employed in the analyses of data accruing from subsequent observations in the years 1949 to 1953 (Waldichuk, 1957). This analysis conceives the system as consisting essentially of an inner basin (Georgia) and an outer strait (Juan de Fuca) separated by a system of "mixing baffles" (San Juan Island passages) with the major freshwater source (Fraser River) being located near the "mixing baffles" in the inner basin. Brackish surface water of the inner basin entering the mixing region encounters higher salinity water incoming at depth through the outer strait to form water of

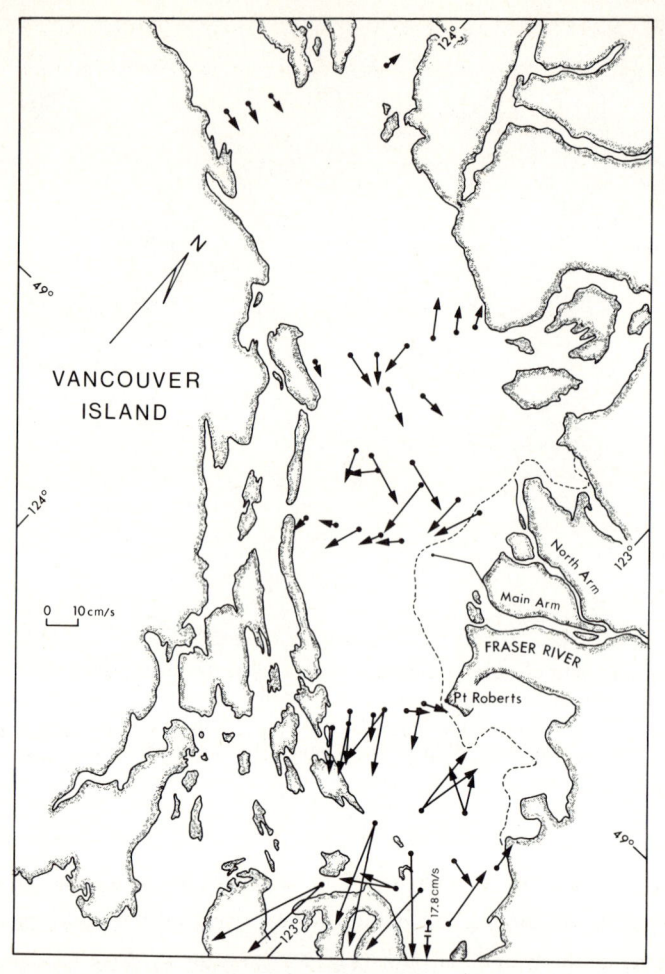

Fig. 1.27 Residual velocities obtained from current meters moored at depths of some 5–20 m below the region of Fraser River discharge and thus below the region where the dominant dynamical effect is the net Coriolis deflection northward of the river discharge. (from the Canadian Hydrographic Service, 1972a,b; 1973a,b; Schumacher et al., 1978)

intermediate salinity. The latter in part, descends to replenish bottom water in the inner strait, the remainder moving seaward through the outer strait. At intermediate depths in the inner basin, renewal of water can occur throughout the year (Herlinveaux and Giovando, 1969).

The major seasonal change in the deep water is associated with the intrusion of upwelled water from the adjacent continental shelf into Juan de Fuca Strait (Section 1.3.2). Thus, over a period of 3–4 months in the summer and fall a persistent net increase in the salinity of deep water in the Strait of Georgia occurs. This is illustrated for a central location (C, Fig. 1.1) in the Strait by the contoured vertical time series of salinity (Fig. 1.28). In the vicinity of this location, further data, although somewhat intermittent, have been obtained over the last 20 yr. Seasonal and fortnightly cycles dominate the variability (Thomson and Louttit, 1986). The data further suggest an interannual variability that correlates with Pacific atmospheric and oceanic circulations at El Niño periods (from years to decades).

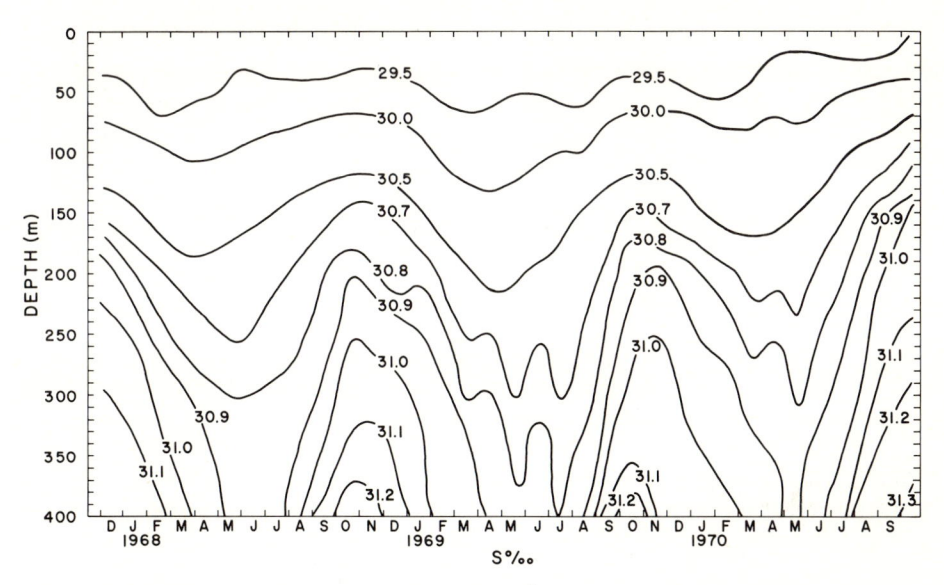

Fig. 1.28 Contoured time series of salinities ($^0/_{00}$) in a vertical located in the deep basin of the Strait of Georgia (location C, Fig. 1.1, inset) over the period December 1967 to September 1970.

Velocity and temperature records were obtained from a line of current meters moored for 1 yr over a cross-section in the Strait of Georgia (Tabata and Strickland, 1972a,b,c). Analyses of these data show significant low frequency baroclinic fluctuations of about a monthly period (Chang et al., 1976). In general, these fluctuations are characterized by a short horizontal correlation scale of less than 10 km. Variations in current ellipses with time suggest that internal tides may be a contributary factor. Cyclesonde velocity data indicate the presence of low frequency (10- to 30-day period) current fluctuations in the Strait of Georgia (Stacey et al., 1987). An empirical orthogonal function analysis of the vertical structure of the current fluctuations gives evidence for the existence of wind-forced Ekman spirals.

In considering the origin of baroclinic oscillations in the Strait of Georgia, the large eddy driven by the flood tidal streams from Boundary Pass is a feature of considerable importance. This eddy is discussed in detail in Chapters 5 and 7 in connection with the models GF3 and GF7, respectively. A contoured time series of vertical salinity profiles at a central location in the eddy (Station BP2, Fig. 2.5a, p. 50) shows displacements in the depth of isolines of 75 m over tidal periods (Fig. 1.29). In one instance, in a shorter data set a vertical displacement of 150 m was observed to occur over a 6-h interval. Also evident are the tidally modulated pulses of low salinity surface water

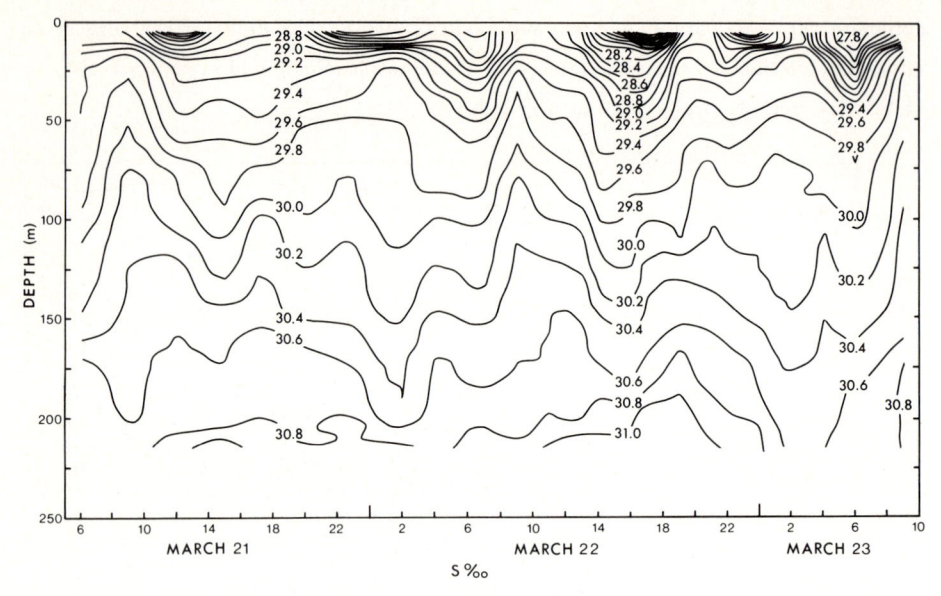

Fig. 1.29 Contoured time series of salinities ($^0/_{00}$) of vertical velocity profiles at a central location where a large tidal eddy forms in the southern Strait of Georgia (Station BP2, Fig. 2.5a, Chapter 2).

moving seaward.

The deep circulation in the Strait of Georgia is thus a topic of considerable complexity. In the interests of clarity, further considerations and presentation of field data are relegated to discussion of results obtained from the three-dimensional model GF6 (Chapter 13).

1.3.7 Northern Passages

The primary conveying channels (Discovery Passage, Johnstone, Broughton, and Queen Charlotte Straits) to the north of the Strait of Georgia are generally deep and narrow. Four major constrictions, Seymour Narrows, Okisollo Channel, Yuculta and Arran Rapids (Fig. 1.30), characterized by very strong tidal mixing, inhibit fresh- and salt water exchanges with the open sea. They have a combined cross-sectional area ($0.98 \times 10^5 \, \text{m}^2$) which is only 7% of the limiting cross-sectional area ($14.45 \times 10^5 \, \text{m}^2$) available for such exchange in the San Juan Island passages to the south (Waldichuk, 1957). The fact that the northern passages are long and have a relatively large volume should theoretically reduce their efficiency in flushing the Strait of Georgia. On every tidal cycle much of the Strait of Georgia water that enters the passages on the ebb tide probably returns to the Strait with the flood. Ebbing water leaving the passages to the south, on the other hand, should be rapidly swept away and new water should enter on the flood. Figure 1.31 includes the contoured distribution of salinity over a vertical section extending from the northern Strait of Georgia into the southern end of Discov-

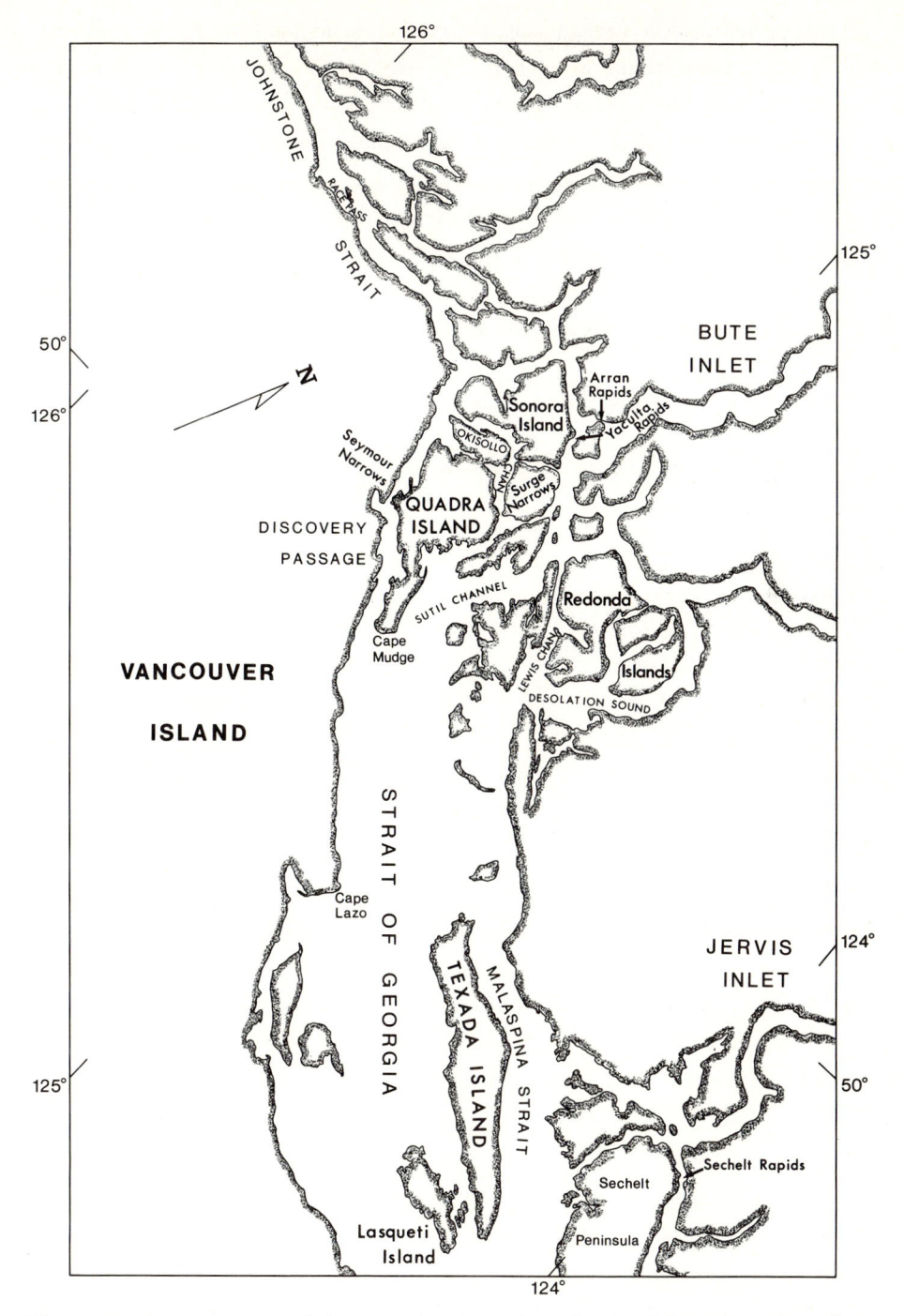

Fig. 1.30 Location map of the central and northern Strait of Georgia and northern passages.

ery Passage and also the strong mixing region of Seymour Narrows. The main features of the plot, though varying in intensity, generally persist throughout the year. The marked dilution of surface water from the Strait of Georgia entering the mixing system is evident while below, denser and more saline Strait of Georgia water is present. Vertical homogeneity exists over the sill while seaward, a weaker vertical density gradient exists. The densities seaward and below the level of the sill are considerably less dense (*i.e.*, fresher) than on the Strait of Georgia side which suggests that the primary net surface flow direction is predominantly northward. This would appear to hold for the other channel constrictions (Waldichuk, 1957).

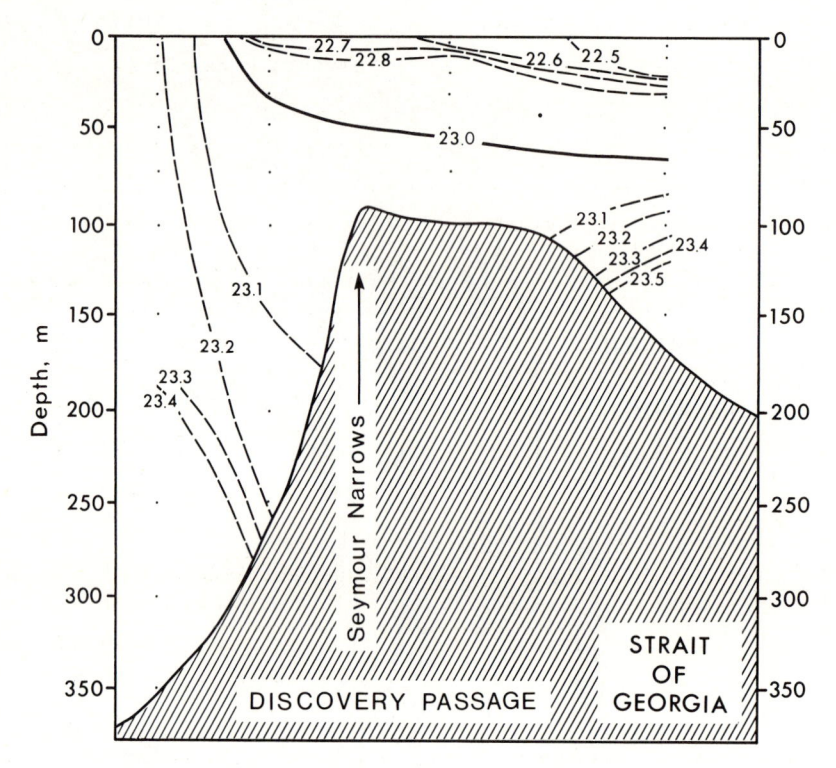

Fig. 1.31 Contoured salinity distributions ($^0/_{00}$) over longitudinal section of Seymour Narrows.

It is worth anticipating that residual current vectors obtained from the model GF7, discussed in Chapter 7, indicate that the flooding streams over the sill expelling mixed water into the Strait of Georgia form a large eddy. In contrast, the ebbing stream will favour advection of further dilute surface water onto the sill. The formation of a net southerly flow of mixed water from Discovery Passage at middepth is consistent with recent moored current meter measurements (M. Woodward, pers. comm.) and results obtained from the three-dimensional numerical model GF6 (Chapter 13).

Thomson and Huggett (1980) studied the M_2 constituent of the baroclinic tide and showed that the M_2 tide and currents possess great variability over intervals of order a few weeks and that their seasonal character is related quantitatively to the changes in the general density structure. Thomson (1976, 1977) studied the tidal currents and estuarine circulation in Johnstone Strait. The distribution of residual velocities normal to a cross-section of the Strait illustrates the net flow occurring in a single passage, between Vancouver Island and the mainland, that connects the northern part of the Strait of Georgia to the open sea (Fig. 1.32). These data were obtained over the period February to June 1973 and might thus be regarded as reasonably representative of a mean annual distribution. The mean rate of residual surface outflow is 0.17 km^3/day. The net transport through the section is, however, small (of order ±0.09 km^3/day) and thus indicative of a weak estuarine circulation. Thomson suggested that for the western part of the basin strong currents in the lower layer are associated with baroclinic motions generated by the barotropic tide propagating over the rapidly shoaling bathymetry to the east. The baroclinic amplitudes of the first three modes were about 40% of the barotropic which is indicative of a significant baroclinic tide.

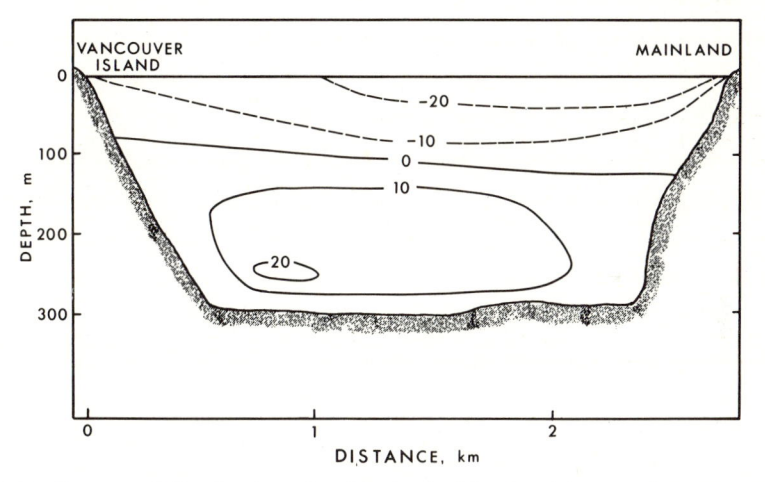

Fig. 1.32 Contoured distribution of residual velocities (cm/s) normal to a cross-section of Johnstone Strait (line 1, Fig. 2.1, Chapter 2) observed over the period February to June 1973. Negative values denote seaward flow. (from Huggett *et al.*, 1976a)

1.3.8 Freshwater Volumes and Flushing Rates

An important practical feature of any estuary concerns the rate at which freshwater, and possible pollutants, are removed from the system. In general, increased river discharge results in increased seaward efflux, the rate of freshwater volume increase in the estuary being less than the increased rate of river discharge. The flushing time is taken as the time required to replace the existing freshwater volume in the estuary at a rate equal to that of the river discharge. This presupposes invariance of river discharge and source salinity. In the Georgia/Fuca system there is a marked seasonal cycle both in river discharge and in the salinity at depth at the entrance to Juan de Fuca Strait. Further considerations include the existence of two openings leading seaward, a high degree of recirculation in the Strait of Georgia, and occasional surface intrusions of Columbia River water entering Juan de Fuca Strait.

Some useful estimates of freshwater volumes in the Strait of Georgia have been made (Waldichuk, 1957). If it is assumed that the salinity of any water particle in the system is determined by its relative proportions of freshwater and sea-water of salinity $33.3^0/_{00}$, it is possible to estimate the volume of freshwater. Using salinity data from eight cruises spaced over the year (December 1949–December 1950), the freshwater volume in the Strait of Georgia varied from 110 km^3 in December 1949 to 132 km^3 the following August. Between April and July, a severalfold increase in river flow resulted in only a 22% increase in the net freshwater volume within the Strait. The total freshwater inflow over the year from the Fraser River was 86 km^3. Thus, a rough estimate of the flushing time is 1.3 yr. Estimates of the changes in flushing rate over a year indicated that there was a slow accumulation of freshwater during much of the early part of the year while over the remainder, the freshwater outflow from the system exceeded the inflow. Thus, from February to June the average ratio of freshwater input to outflow is 0.7 while from July to January, the average ratio is 1.3.

The residual flows normal to cross-sections of Juan de Fuca and Johnstone Straits that have been presented in Sections 1.3.2 and 1.3.7, respectively, may be combined with observed salinity data to yield an estimate of the proportioning of freshwater outflow between the northern and southern boundary openings of the system, again assuming some fixed salinity (33.8%) of incoming deep ocean water. This indicates that some 92% of the freshwater leaves through Juan de Fuca Strait. Using the variations in surface outflow referred to above (7.78–13.8 km^3/day) the percentage varies from 88–93%.

2 FIELD OBSERVATIONS

A major feature of the numerical model studies concerned the aquisition of a data base adequate to the operation and verification of the models. The main aspects of the program, which in large part were made possible through the collaboration of the Canadian Hydrographic Service, are reviewed in the basic sequence of occurrence. It is perhaps worth noting that the timing when ships, instruments, and personnel became available infrequently coincides with the demands of an evolving modelling program. In light of available information, it appeared important to plan the acquisition of a basic data base as follows.

The first specific objective of these field operations was to obtain simultaneous water level data at locations throughout the main conveying channels over a 1-yr period. A second objective was to acquire time series records of velocities over cross-sections, both in the Strait of Georgia and in the straits to the north (Johnstone) and south (Juan de Fuca) which join the system to the open sea. A third objective was the provision of data on deep intruding sea-water in Juan de Fuca Strait and associated baroclinic tidal oscillations. A fourth objective concerned observations characterizing the highly labile plume formed in the Strait of Georgia by the Fraser River. A fifth objective was to verify the existence of large tidal eddies predicted by the model and associated residual circulations.

Figure 2.1 shows the locations of tide gauges and lines of moored current meters located over cross-sections of the main conveying channels. Also shown are the locations of other moorings referred to in the text.

Measurements of sea surface elevations are important not only in determining the nature of the tidal oscillations but also as an indicator of subsurface oceanographic processes (Barber, 1958; Waldichuk, 1957, 1964). Therefore in the late 1960's, the permanent tide gauges within the system were augmented by an extensive array of additional gauges. These latter were retained in operation for at least a year to provide a measure of seasonal changes in monthly mean sea levels.

To simulate baroclinic processes, it was also necessary to define the seasonal change in the distribution of scalar properties through the main conveying channels of the system. Thus, synoptic oceanographic surveys, each of a 3- to 4-day duration, were carried out at monthly intervals for a year. The cruise plan (Fig. 2.2) included some 76 oceanographic stations (Crean and Ages, 1971) where temperatures and salinities were

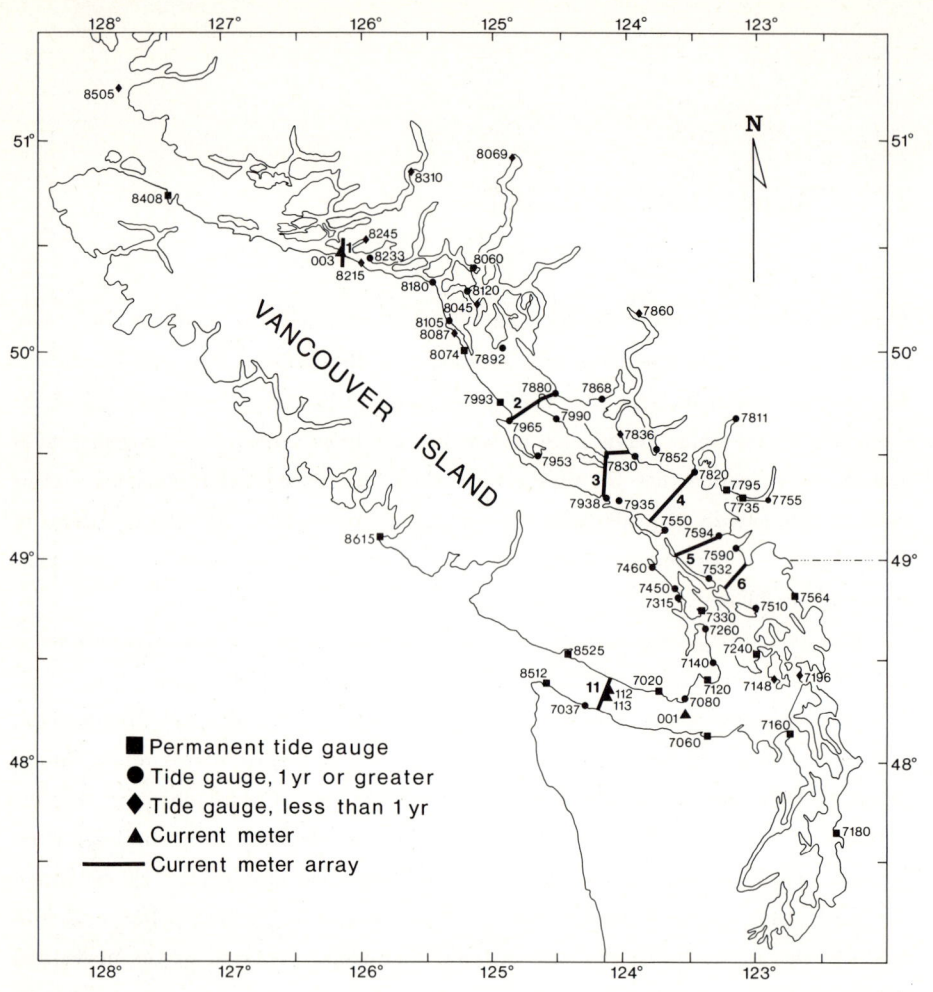

Fig. 2.1 Location of tide gauges and current meter arrays previous to and including 1976.

observed. In addition, oxygen was monitored at further selected stations. An atlas of monthly contoured temperatures, salinities, densities, and oxygen over a longitudinal vertical section through the main conveying channels was prepared.

A further program of temperature and salinity measurements was carried out in Juan de Fuca Strait in 1973 when the array of current meters was in place over a cross-section (line 11, Fig. 2.1) (Crean and Miyake, 1976). These cruises, concomitant with the current meter measurements, were intended to provide a data base for the three-dimensional simulation of the seasonal deep intrusion of saline water. A further intention concerned the acquisition of data to be used in studies of baroclinic oscillations at the main tidal frequencies. (Such studies will require the additional spatial resolution afforded by the planned three-dimensional 2-km mesh model GF8.) These data obtained in March and April 1973 included hydrocasts over longitudinal sections of the Strait as

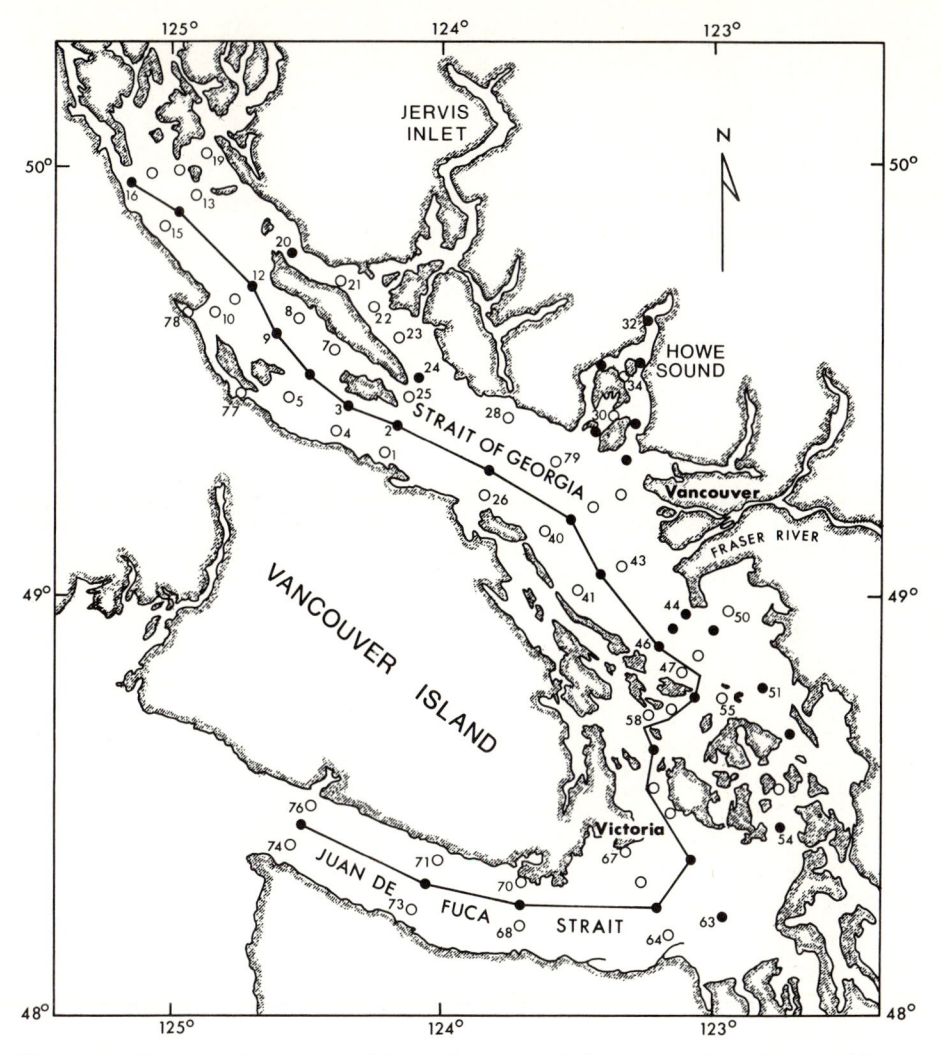

Fig. 2.2 Pattern of oceanographic stations occupied at approximately monthly intervals where temperature, salinity (open dots), and, at selected stations, also oxygen (black dots), were measured over the period December 1967 to December 1968.

well as cross-sections in the vicinity of the current meters (line 11) (Figs. 2.3 and 2.4) (Crean and Miyake, 1976). Of particular interest is a 72-h time series in which, with the collaboration of the University of Washington, two vessels were employed. Each vessel repeatedly occupied one of the triangular patterns of stations shown in Fig. 2.3. This gave a 4-h sampling frequency at each station, minimally sufficient to resolve tidal features. These data illustrated the tidal modulations of a salt tongue advancing from the ocean at depth into Juan de Fuca Strait. At monthly intervals throughout the remainder of 1973, 24-h time series of hydrocasts over the cross-section were continued (Crean and Lewis, 1976).

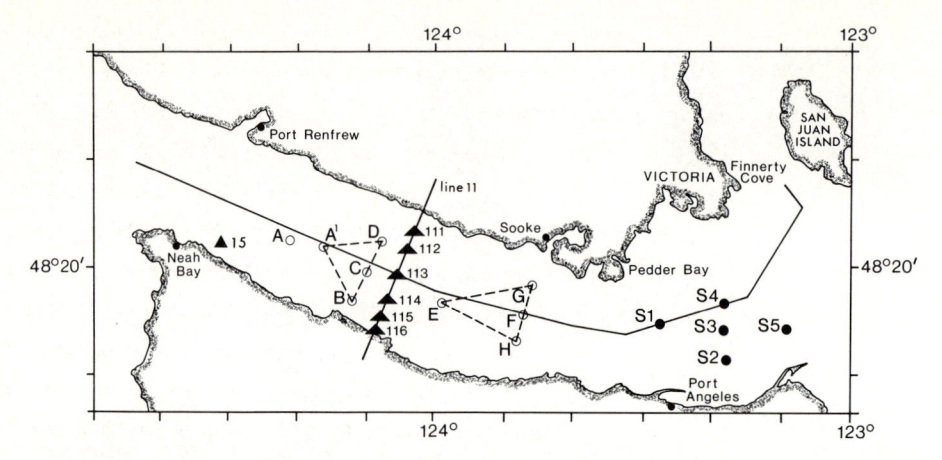

Fig. 2.3 Location of current meter moorings (line 11) and triangular patterns of oceanographic stations each occupied by a research vessel for a 72-h period. Single letters denote station positions in the eastern part of Juan de Fuca Strait. Station positions denoted by S1 to S5, inclusive, were located in the vicinity of the effective sill and were occupied sequentially over a 24-h period.

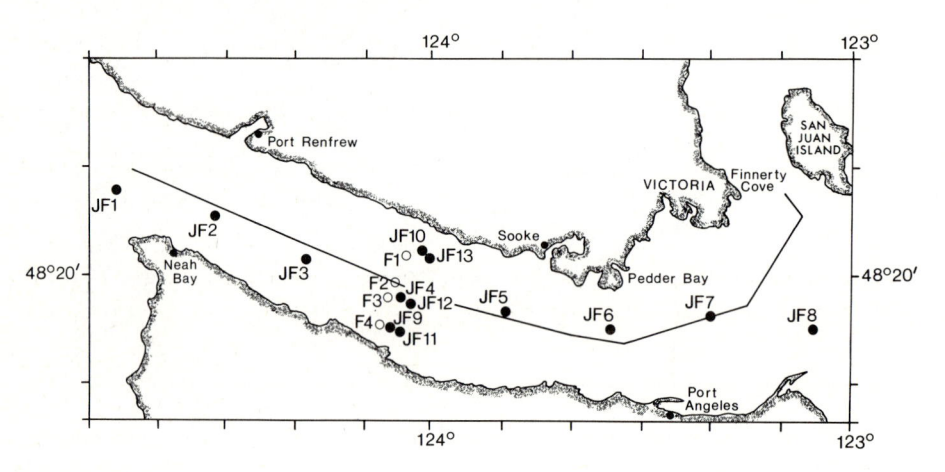

Fig. 2.4 Oceanographic stations located on longitudinal and transverse sections of Juan de Fuca Strait.

A particularly demanding problem concerned the acquisition of field observations pertinent to the adjustment and verification of the upper layer buoyant-spreading model GF4 of the Fraser River plume. Initial programs of observations were carried out for various states of river flow and tide using a hydrographic survey launch (Stronach and Crean, 1978). Data included both CSTD and velocity profiles. Attention was also directed to the determination of processes near the mouth of the Fraser River and the adjoining shallow tidal banks. Further measurements, including dye (Rhodamine B) experiments, were carried out in the vicinity of freshwater fronts moving out from the river mouth into the salt water of the Strait of Georgia.

On the basis of this work and earlier drogue tracking studies just referred to, it appeared feasible to determine the changing distributions of salinities and velocities in the plume over a large ebb-flood tidal sequence at various river discharge rates. These operations were carried out with the co-operation of the Canadian Hydrographic Service (survey launch), Canadian Coast Guard (helicopter, hovercraft, and rescue launch), and the Pacific Biological Station (fisheries launch) and in collaboration with Beak Consultants Limited (Dr. J. Stronach). In view of the large area to be covered, each operation was carried out during daylight hours on days having lower low water at midday. The general distribution of temperature and salinity was determined by lowering a CSTD from a helicopter. The remaining vehicles each deployed a 16-km line of free-drifting surface floats at 1.6-km intervals. Each float supported a large weighted sail 2 m deep and 3 m long and a radar reflector on a short mast. The drogues were deployed over the southern part of the plume region on one day and over the northern part on the other day. Drogue positions were determined using Vessel Traffic Management radar made available by the Canadian Coast Guard. In the absence of significant wind interference, river flow and tides could be assumed the same on the two consecutive days and, using lower low water as a common reference time, the aggregate of tracks could be assembled to provide coverage of the whole plume region over a single large ebb-large flood sequence. The 2-day operations were repeated approximately once a month through a sequence from April to November. Representative composite diagrams illustrative of the drogue tracks obtained have already been presented in Chapter 1.

Major tidal eddies and associated residual circulation occurring in the vertically-integrated tidal models led to other field observations including drogue tracking, moored current meter observations, and hydrocasts. Figure 2.5a shows the pattern of oceanographic stations where time series of hydrocasts were carried out and, along with Fig. 2.5b, the location of current meter moorings in the southern Strait of Georgia. These data confirmed dramatically the residual tidal eddy and fluctuations in the depths of isopycnal surfaces of 75–150 m at tidal frequencies. The existence of other major eddies was corroborated by drogue tracks on the appropriate tidal stream at the southern end of Haro Strait and in the vicinity of Race Rocks.

In Chapter 1 the observational data has been examined including data developed as part of the model development program presented in this chapter. In the remainder of the book, a sequence of numerical models increasing in complexity, is described. Although having some mathematical interest, such models essentially furnish the other pole in an ongoing dialectic between theory and data crucial to the movement from description to explanation.

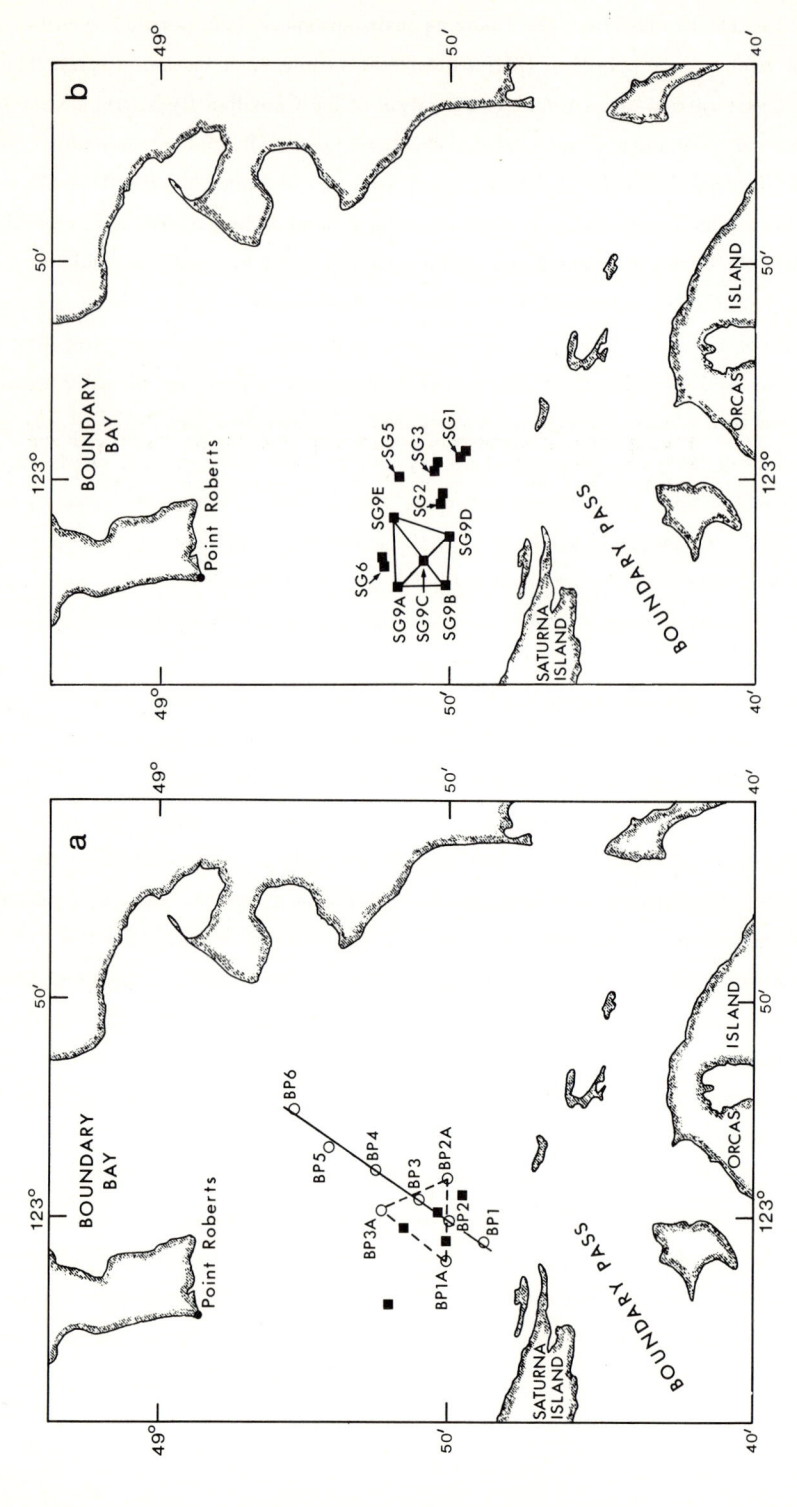

Fig. 2.5 Locations of (a) current meter moorings (black squares) and hydrographic stations (circles) and (b) current meter moorings, only, used in studies of the residual tidal circulation in the Strait of Georgia.

BAROTROPIC MODELS

3 THE ONE-DIMENSIONAL MODEL: GF1

3.1 INTRODUCTION

In recent years considerable success has been achieved in the use of hydrodynamical numerical methods to simulate barotropic tides in coastal seas and estuaries. Although the basic procedures are relatively simple, difficulties are encountered in the application of such techniques to problems concerning strong tidal flows in topographically complex systems. It is usually presumed that variations in phase speed and partial reflections of a long wave propagating through the system can be adequately represented by a sufficiently detailed finite-difference approximation to the actual geometry. Energy losses from the wave are then empirically accounted for by adjusting the frictional coefficients in the momentum equations to optimize the agreement between the observed and computed tidal elevations around the interior of the sea. Before conclusions are drawn concerning the dynamics of wave propagation, it is important to ensure, as far as possible, that such adjustments do not compensate for inadequacies in the theoretical assumptions, schematic representation of topography, or in prescribing boundary conditions. A prerequisite is thus a sufficiency of observations, both of elevations and velocities, to establish standards of performance which could in principle be achieved by the model. If necessary, the sensitivity of the computed fields to the various approximations is then systematically investigated and improvements made until these standards are attained or the need for a radically different type of model has been demonstrated.

The work commenced with some preliminary calculations of the tidal currents likely to be encountered in the Strait of Georgia. Numerical evaluation of the M_2 tidal current ellipses due to Kelvin wave reflections (Taylor, 1921) which occur in a rotating rectangular gulf of depth and width roughly that of the Strait of Georgia, indicated that currents of the rectilinear type should dominate except in a highly limited area near the (assumed) closed end. Using known surface areas and cross-sections and assuming a single M_2 tidal amplitude at a closed northern end, the average rectilinear tidal currents over each cross-section could readily be determined (Proudman, 1953) along the Strait of Georgia. Any attempt to extend these calculations beyond the San Juan Islands led to large discrepancies between the observed and calculated amplitudes. These results suggested that further exploration should be undertaken with the employment of a multi-channel one-dimensional model.

To preserve continuity in the historical development of these models, the original description of this work has largely been retained in this chapter.

3.2 FUNDAMENTAL EQUATIONS

The equations of continuity and motion (Proudman, 1953) in a channel where the velocity is sensibly in one direction and the displacements of the water surface are small with respect to the undisturbed depth are

$$\frac{\partial(Au)}{\partial x} + b\frac{\partial \zeta}{\partial t} = 0, \tag{3.1}$$

$$\frac{\partial u}{\partial t} + u\frac{\partial u}{\partial x} + \frac{ku|u|}{h} + g\frac{\partial \zeta}{\partial x} = 0 \tag{3.2}$$

where

- x = distance measured along the medial line of the channel measured positively from the mouth,
- t = time,
- $u(x, t)$ = mean velocity over a cross-section,
- $\zeta(x, t)$ = elevation of the water surface above the equilibrium depth,
- $b(x)$ = width of the channel,
- $A(x)$ = cross-sectional area,
- g = acceleration due to gravity,
- k = coefficient of friction,
- $h(x)$ = mean depth of water below the undisturbed surface.

A detailed discussion of the assumptions implied by these equations is available in Dronkers (1964). It will be noted in the above equations that no account is taken of the Coriolis effect. The model should thus give representative tidal amplitudes in Juan de Fuca Strait intermediate in magnitude between the observed values for either side of the Strait. These latter values display differences in range of the type normally associated with the superposition of Kelvin waves travelling in the opposite directions.

3.3 FINITE-DIFFERENCE EQUATIONS

The equations of continuity and motion are written in finite-difference form following essentially the same scheme as that employed by Rossiter and Lennon (1965). The specification of the variables on the grid is shown in Fig. 3.1a. Central differences in space and time are used. For computational convenience, the same time subscript is used to denote the elevation $Z_{m,n}$ at $t = m\Delta t$ and the velocity $U_{m,n}$ at time $t = (m + \frac{1}{2})\Delta t$. The finite-difference form of Eq. (3.1) is

$$Z_{m,n} = Z_{m-1,n} - \Delta t \left(\frac{U_{m-1,n} A_n - U_{m-1,n-1} A_{n-1}}{b_n \Delta s} \right) \tag{3.3}$$

where

Δt = increment of time,

Δs = increment of distance,

b_n = mean width of a section of the channel contained between two velocity sections, $U_{m-1,n}$ and $U_{m-1,n-1}$,

A_n = the cross-sectional area of the velocity section $U_{m-1,n}$.

The denominator in the second term of Eq. (3.3) can be replaced by the appropriate surface area.

It is reasonable to suppose that over the greater part of the model the convective term in Eq. (3.2) is small. If this term is neglected, the value of the velocity $U_{m,n}$ can be found from the finite-difference form of the equation of motion (3.2),

$$U_{m,n} = -g\Delta t\left(\frac{Z_{m,n+1} - Z_{m,n}}{\Delta s}\right) + U_{m-1,n} - k\Delta t\left(\frac{U_{m-1,n}|U_{m-1,n}|}{h_n}\right). \tag{3.4}$$

If, for a specific location it is desirable to include the convective acceleration, a suitable finite-difference form of Eq. (3.2) is

$$\begin{aligned} U_{m,n} = &-g\Delta t\left(\frac{Z_{m,n+1} - Z_{m,n}}{\Delta s}\right) \\ &- U_{m-1,n}\left[\Delta t\left(\frac{U_{m-1,n+1} - U_{m-1,n-1}}{4\Delta s}\right) - 1 + \frac{k\Delta t|U_{m-1,n}|}{h_n}\right] \\ &\times \left[1 + \Delta t\left(\frac{U_{m-1,n+1} - U_{m-1,n-1}}{4\Delta s}\right)\right]^{-1}. \end{aligned} \tag{3.5}$$

A section length Δs of 22.2 km was found to be a convenient choice to locate velocity sections near channel junctions.

Numerical stability (O'Brien et al., 1950) requires that

$$\frac{\Delta t}{\Delta s} \leq \frac{1}{\sqrt{gh_{\max}}}$$

when applied to the greatest mean depth giving the resulting time step of 6 min. Velocity sections are located at a junction where the condition of continuity is applied, assuming a particular flow direction as positive (Fig. 3.1b). Thus,

$$Q_a + Q_b + Q_c = 0.$$

The elevation of the water surface in a section of channel terminated by a junction can then be found using Eq. (3.3).

To determine the velocity at the entrance to one arm of a junction, Eq. (3.4) is employed. Thus, for example (Fig. 3.1b), the velocity through the section denoted by Q_b is

$$U_{m,b} = -g\Delta t\left(\frac{Z_{m,b} - Z_{m,a}}{\Delta s}\right) + U_{m-1,b} - k\Delta t\left(\frac{U_{m-1,b}|U_{m-1,b}|}{h_b}\right).$$

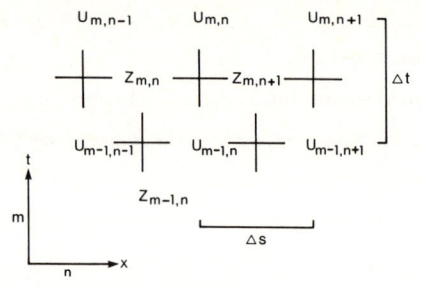

Fig. 3.1a Notation defining the grid scheme in the x, t plane.

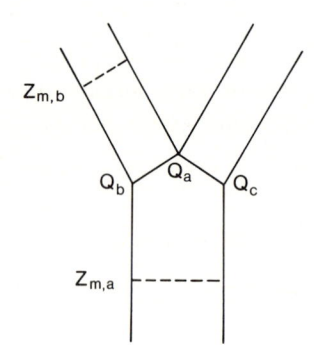

Fig. 3.1b Notation defining conditions to be applied at a channel junction.

The assumptions involved in applying these conditions at a junction are discussed by Dronkers (1964).

3.4 THE COMPUTATIONAL SCHEME

The schematization of such a system necessarily involves a good deal of approximation. The major tidal conveying channels were identified using sources including the British Columbia Pilot, Pacific Coast Tide and Current Tables, and tidal data supplied by the Canadian Hydrographic Service. Further information was obtained from the Tidal Current Tables for the Pacific Coast of North America and Asia (U.S. Coast and Geodetic Service).

Median lines were drawn and a section length was selected so that velocity sections were close to the junctions of major conveying channels. In instances where difficulties arose regarding the location of velocity sections at junctions, preference was given to the assignment of correct spacing in the major channel. A further difficulty concerned the location of sections where channels divide and rejoin, the respective branches being of somewhat different lengths as, for example, in Haro and Rosario Straits. In the latter instance the channels were considered to be of equal length. In certain sections there are appreciable areas of drying banks, for example, in the vicinity of the Fraser River delta. Such areas are small compared to the overall sea area and a representative perimeter midway between the edge of the tidal flats and the coast was taken for the fluid boundary. The surface area of a channel section located between two velocity cross-sections $S_n = b_n \Delta s$ was determined from charts by planimeter. The cross-sectional area was obtained in a similar manner from plotted bottom profiles. It was possible to check cross-sectional areas obtained in this manner for three sections in the Strait of Georgia against those obtained from actual echo-sounder profiles. The agreement in all cases was within 4%. However, many of the deep side channels have few soundings listed on the charts and the cross-sectional areas are therefore only approximate. In critical cases such as certain narrow sections in Discovery Passage and adjoining channels, data were obtained from the original plot sheets of the Canadian Hydrographic Service.

The appropriate frictional dissipation of energy in a model of this type is difficult to determine. The law generally employed assumes that the frictional force is proportional to the square of some representative mean velocity. In a simple channel the constant of proportionality can be deduced from direct field observations (Dronkers, 1964). In the present instance, the topographical complexity of the system led to the adoption of the present course. A mean depth was determined for each velocity cross-section and from these a representative mean depth appropriate to a section of channel between any two adjacent elevation sections was determined graphically. This depth was then included in the friction term of the equation of motion. The appropriate coefficient of friction was determined so that the differences between the observed and calculated tidal elevations and phase lags within the sea were minimal.

The schematization employed is shown in Fig. 3.2. Dashed lines refer to sections where the elevation is calculated; solid lines refer to velocity sections. Heavy lines through groups of islands have been drawn where it has been assumed that the flow between the islands is negligible. Variable names used in the computer program are prefixed by Z or U and refer to elevation or velocity sections, respectively. The remaining letters of the name designate a particular group of sections representing one of the tide-conveying channels. Capital letters AA' to JJ' designate the velocity sections where the main channels divide.

The overall computational scheme in the x, t plane for values $m = 1, 2$ is shown in Fig. 3.3. Each square bracket encloses the variables appropriate to a particular segment of channel and the positive flow directions through the model are denoted by arrows from one channel segment to another. The letters located to the bottom right of the brackets refer to the remainder of the variable name by which a channel segment is designated. Dotted lines drawn through a sequence of velocity points indicate that the velocities at these points enter into the continuity condition applied at a junction. The capital letters and dashed lines refer to a particular junction. At elevation or velocity points where a boundary condition is prescribed, the appropriate Z or U in the diagram is enclosed in a square box. The mean depths are specified in centimetres and the areas

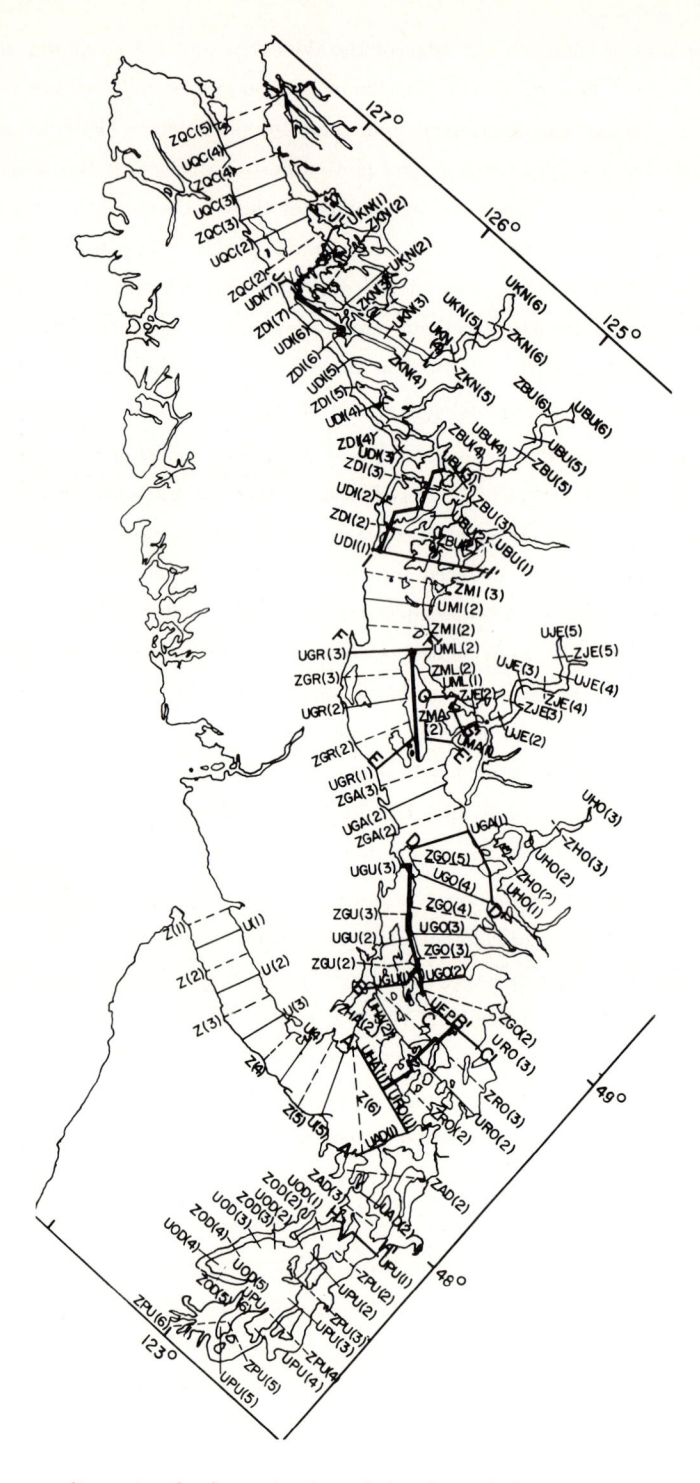

Fig. 3.2 The one-dimensional schematization of the channels between Vancouver Island and the mainland.

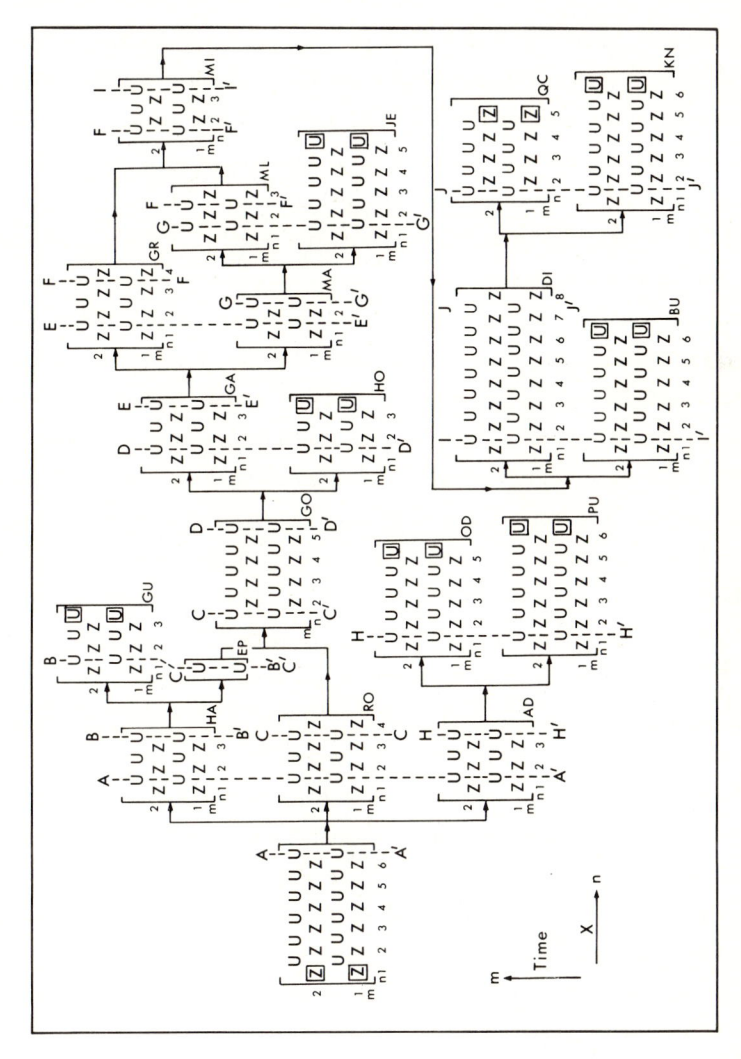

Fig. 3.3 Multi-channel grid scheme in the x, t plane for $m = 1, 2$. Variables appropriate to a particular channel segment are enclosed in square brackets. Arrows denote the positive flow direction through the model. Where a boundary condition is prescribed, the variable is enclosed in a square.

of velocity cross-sections and the surface areas of channel sections contained between adjacent velocity sections in km² are prescribed. All values of $U_{m,n}$ at $t = 0$ and of $Z_{m,n}$ at time $t = -\Delta t/2$ are set equal to zero.

The operation of the model is based on the repeated applications of Eqs. (3.3) and (3.4), except for the Discovery Passage and Johnstone Strait section where Eq. (3.5), which includes the convective acceleration, is employed.

At the entrance to Juan de Fuca Strait the prescribed amplitude and phase lag were the averages of those determined from observations at Neah Bay (#8512, see Fig. 2.1) and Port Renfrew (#8525). At the northern entrance to Queen Charlotte Strait the selection of an appropriate amplitude and phase lag to be prescribed was more difficult due to the scarcity of suitable gauge data. The final compromise was to average the amplitudes and phase lags obtained from observations at Port Hardy (#8408) and Egg Island (#8805) (see Fig. 1.1). At the closed ends of the side channels the velocity was set equal to zero.

The frictional coefficient chosen should permit good agreement between observed and calculated amplitudes and phase lags for both the diurnal and semi-diurnal constituents. The coefficient of friction employed did not entirely satisfy these criteria. However, one was finally selected to optimize the agreement between the observed and calculated M_2 tidal amplitudes in the Strait of Georgia since this particular constituent dominates the motion. The value of the coefficient thus determined was 0.028 and was applied throughout the greater part of the system except in the southern part of Discovery Passage where a value of 0.0028 was required to give a satisfactory agreement between observed and calculated amplitudes. It is of interest to note that this latter value is much closer to that normally employed in calculations of this type for coastal seas.

In general, amplitudes and phases were determined from direct inspection of output. Plotted values of water elevations and velocities at selected locations in the main conveying channels showed negligible asymmetry and in earlier work the above method was employed except for the vicinity of Discovery Passage where harmonics were in evidence. In the case of the M_2 tide, subsequent harmonic analyses showed the general prevalence throughout the model of a small M_6 harmonic, 1–2 cm in amplitude, which could lead to errors of up to 10° in the calculated phase lags. The results presented were, however, determined by harmonic analyses.

3.5 RESULTS

Summarizing briefly the sequence of presentation, attention is first directed to the distribution of the M_2 tidal constants along the main conveying channels, which includes Juan de Fuca and Haro Straits, the Strait of Georgia, Discovery Passage, Johnstone and Queen Charlotte Straits. Observed and calculated tidal amplitudes are presented together with the velocity amplitudes. Observed and calculated tidal phase lags and the calculated velocity phase lags are then shown. The observed and calculated harmonic constants of the tides and entrance streams in the important side channels are next considered. Stream harmonic constants are presented in tables for individual consideration. Corresponding aspects of the K_1 tide are then considered in a similar sequence.

3.5.1 M_2 Tidal Harmonic Constants in the Main Channels

Figure 3.4 shows the observed and calculated M_2 tidal and velocity amplitudes along the main conveying channels. The calculated amplitudes decrease to a degenerate nodal point located in the inner part of Juan de Fuca Strait, increasing thereafter through Haro Strait and the Strait of Georgia. The agreement between observed and calculated values is reasonably good. The considerable scatter in the observed values in Juan de Fuca Strait is due to the increase in tidal range across the Strait associated with the Coriolis effect. This will be considered further in Chapter 4 in connection with the two-dimensional model. It would appear that the degenerate nodal point deriving from the calculated values is located too far inward from the mouth of Juan de Fuca Strait. In the case of a channel closed at one end, an important factor in determining the location of this point is the effective length of channel between the mouth and the plane of zero velocity where the reflection of the incoming wave can be assumed to occur. The location of such a hypothetical plane is contingent on the volume of water entering the Strait of Georgia from the northern passages. If it is assumed that the northern end is closed and that Bute Inlet (see Fig. 1.1) receives its tidal complement through Juan de Fuca Strait and the Strait of Georgia, the model results show a further distinct displacement of the degenerate nodal point inwards so that the point is located in the San Juan Islands. It would thus appear that there is a modest deficiency in the volume inflow of water from the northern end in the course of a tidal cycle.

Turning now to the northern passages, *i.e.*, those of Discovery Passage and beyond, the tidal amplitudes increase in magnitude from the mouth to the southern end of Queen Charlotte Strait and thereafter decrease to a minimum value in the southern part of Discovery Passage. The agreement between observed and calculated values is quite satisfactory, particularly in view of the simplicity of the finite-difference representation. Harmonic analyses showed that M_4 and M_6 harmonics, each having an amplitude of about 6 cm, were superimposed on the M_2 tidal elevations at the southern end of Discovery Passage. The M_4 and M_6 harmonics arise from the non-linear terms in the equation of motion. There is also a net depression of mean sea level of 5 cm at the elevation section corresponding to Seymour Narrows. Such a change in mean sea level results from the inclusion of the convective acceleration in the equation of motion. Since the depths are assumed independent of time, the friction term will not lead to a comparable change in mean sea level. These effects have been discussed by

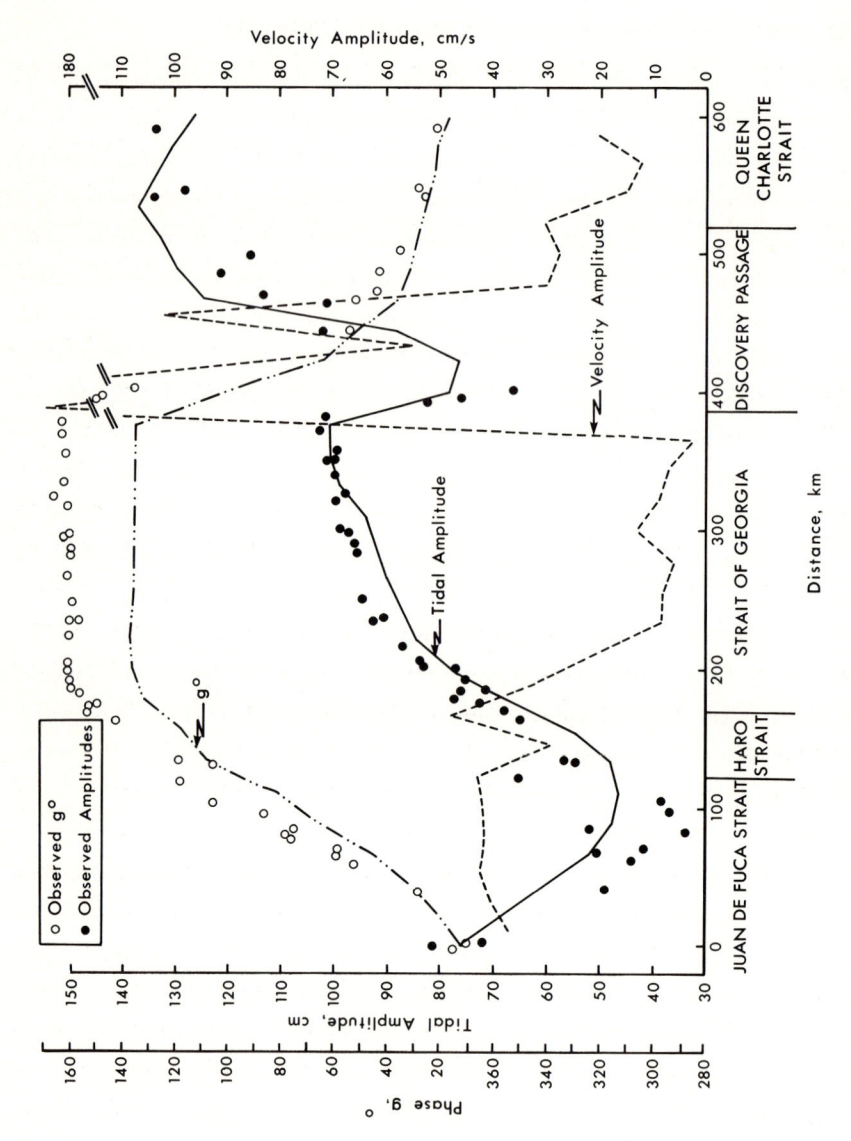

Fig. 3.4 Computed and observed M_2 tidal amplitudes and Greenwich phases along the main conveying channels between the northern and southern boundary openings of the one-dimensional model GF1. The solid line refers to the computed tidal amplitude; dash-dot line indicates the computed Greenwich phase and dashed line shows the computed M_2 velocity amplitude.

Proudman (1953). The velocity amplitudes remain fairly high through Juan de Fuca Strait and decrease rapidly to minimal values in the southern part of the Strait of Georgia. Very high values occur in the southern end of Discovery Passage and generally decrease thereafter towards the mouth of Queen Charlotte Strait.

Observed and calculated phase lags of both elevations and velocities for the M_2 tide are considered next. In the case of the velocities, the phase lag shown is such that velocities in the flood direction are counted positive. Here the flood is assumed to be directed inwards through Juan de Fuca Strait to the closed ends of the inlets opening off Juan de Fuca Strait and the Strait of Georgia. This coincides with the direction of x increasing in the model. For the northern passages, the flood is considered to be directed southward towards the Strait of Georgia and thus corresponds to negative velocities in the model. The flood stream entering the Strait of Georgia through Discovery Passage is then assumed to join the north-going flood stream entering Bute Inlet.

There is a difference in the observed tidal phase lags of approximately 150° between the mouth of Juan de Fuca Strait and the northern end of the Strait of Georgia. On the other hand, there is a difference in the observed phase lags at the mouth of Juan de Fuca Strait and the central part of Discovery Passage of about 50°. Observations show a phase lag of about 100° at the time of high water between the southern and central parts of Discovery Passage. It is evident that the high tidal velocities shown for this vicinity are due to the surface gradients which result from these phase differences.

The calculated phase lags in the Strait of Georgia are smaller than those observed by about 20-30°. This error becomes particularly noticeable in the inner part of Juan de Fuca Strait, and remains roughly constant throughout the Strait of Georgia. A consequence of this will be a diminished flow in the model through Discovery Passage into the Strait of Georgia. This may account for the apparent inward displacement of the degenerate M_2 nodal point mentioned earlier.

The calculated phase lags of the M_2 velocities increase slightly from the ocean entrance to the vicinity of Haro Strait and thereafter decrease slightly towards the northern Strait of Georgia. A marked phase change occurs between velocity sections in this area. The velocity amplitudes are very small, however. It would therefore appear that the tide in Bute Inlet derives primarily from the flow through the northern passages. It is of interest to note that the meeting of the incoming tides from the northern and southern entrances occurs in this vicinity (British Columbia Pilot, 1965).

In addition to the main channels described above there are two other important conveying channels, Rosario Strait which is connected 'in parallel' with Haro Strait, and Malaspina Strait which is connected 'in parallel' with part of the Strait of Georgia (see Fig. 1.1). The calculated M_2 amplitudes in Rosario Strait differ negligibly from those of Haro Strait. The calculated M_2 phase lags show that both high water and the flood maximum in the model occur about 15° later in Haro Strait than in Rosario Strait. A comparison of the observed M_2 amplitudes at Finnerty Cove (#7140) (Fig. 2.1) and Reservation Bay (#7196) suggest that the amplitudes should be higher in Rosario Strait than in Haro Strait while the differences in phase lags should be smaller. The fact that the calculated amplitudes are almost identical is probably due to the omission of Coriolis effects from the model. The calculated and observed tidal amplitudes and phase lags for Malaspina Strait differ negligibly from the calculated values for the northern part of the Strait of Georgia.

3.5.2 M_2 Tidal Harmonic Constants in the Side Channels

The principle side channel openings off the main conveying channels are the Puget Sound system, Howe Sound, Jervis, Bute, and Knight Inlets (Fig. 1.1). Depths in the first two are of the order of 90–200 m while in the last three, depths range from about 180–550 m.

From a general description of tides in these waters (British Columbia Pilot, 1965), it is noted that the streams decrease in magnitude from the entrance to the head and that the turn of the tide occurs about the time of high or low water. With one exception, tide gauge observations (#7811, #7860, #8069, and Olympia) have been made near the head of each inlet. In the case of Knight Inlet (#8310), the gauge was located roughly half way along the inlet.

The observed and calculated M_2 tidal harmonic constants and the calculated harmonic constants for the velocities at the respective entrances to these inlets are shown in Table 3.1. With one important exception, the agreement between the observed and calculated tidal amplitudes is reasonably good. The calculated amplitude at the southern end of the Puget Sound system is only about half the observed value. This is probably due to the fact that the model cannot reproduce the increase in tidal range between Victoria (#7120) and Port Townsend (#7160) due to the omission of the Coriolis effect from the inner part of Juan de Fuca Strait.

The calculated phase lags at the southern end of the Puget Sound system (Olympia) and in Knight Inlet (#8310) show good agreement. The calculated phase lags for the heads of Howe Sound (#7811), Jervis (#7860) and Bute Inlets (#8069) show discrepancies of 20–30° which are similar to those encountered in the Strait of Georgia. With the exception of the Puget Sound system the difference in the calculated phase lags of the velocities at the mouth of the inlets and high waters at their respective heads, indicates that the current leads high water by 90°. This is consonant with dynamical expectation. Deep water and small velocities lead to minimal frictional dissipation. Simple standing wave theory should thus provide an adequate approximation to the tides and streams in these inlets. Note that the large discrepancy between the observed and calculated tidal amplitudes at the head of the Puget Sound system may reasonably

Table 3.1 Observed and calculated harmonic constants, amplitudes H (cm) for M_2 tides near the heads (except Knight Inlet) of inlets and calculated harmonic constants, velocity amplitudes V (cm/s) for the streams at their respective entrances. Greenwich phases g are in degrees.

Location	Tide gauge number	Observed Tides		Calculated Tides		Calculated Streams	
		H	g	H	g	V	g
Puget Sound		145	158	68	156	43	52
Howe Sound	7811	94	159	89	137	4	45
Jervis Inlet	7860	101	160	110	136	7	44
Bute Inlet	8069	101	166	106	135	8	44
Knight Inlet	8310	147	27	155	26	21	-62

be attributed to the inability of the one-dimensional model to include the Coriolis effect, thereby eliminating cross-channel slopes associated with the tidal stream in Juan de Fuca Strait where the Puget Sound system entrance is located.

3.5.3 M_2 Tidal Stream Harmonic Constants

Observed and calculated M_2 tidal current harmonic constants are presented in Table 3.2. The agreement between the observed and calculated velocity amplitudes in Juan de Fuca Strait (line 11, Fig. 2.1) and the Strait of Georgia (line 4) is excellent.

Table 3.2 Observed and calculated M_2 tidal stream harmonic constants, velocity amplitudes V (cm/s) and Greenwich phases g (°). Lines referred to are taken from Fig. 2.1 (Chapter 2).

Location	Observed		Calculated	
	V	g	V	g
Juan de Fuca Strait, line 11	46	63	41	32
Admiralty Inlet	100	66	33	52
Strait of Georgia, line 4	6	85	8	36
Seymour Narrows	466	61	184	70

As previously noted, direct comparison of the observed and calculated velocity amplitudes at Seymour Narrows in the northern passages is dubious. This is attributed to a number of smaller channels leading to the northern Strait of Georgia, characterized in places by 'rapids' where the observed current phase lags are about 30° less than in Seymour Narrows. It is also unlikely that the observed velocity is representative of the mean velocity over a cross-section of the model. However, the calculated velocities are much higher than at any other location in the system and this is qualitatively consistent with the observations. The calculated velocity amplitude in Admiralty Inlet is much smaller than that observed. This is consistent with the fact that the calculated tidal amplitudes in the Puget Sound system are much too small.

Considering now the observed and calculated velocity phase lags, the calculated value for Juan de Fuca is low by about 30°. This implies that the flood stream is too early in the model, as indeed so is the time of high water in the Strait of Georgia as noted in Section 3.5.1. In the case of the Strait of Georgia section, the calculated phase lag is low by 49°. At Seymour Narrows the observed and calculated M_2 phase lags are in good agreement while in Admiralty Inlet the calculated phase lag is low by 14°. The discrepancies between the observed and calculated phases of tides and tidal streams serve to illustrate the inherent limitation of a one-dimensional model in reproducing the tidal regime in this system.

3.5.4 K_1 Tidal Harmonic Constants in the Main Channels

Figure 3.5 shows observed and calculated values of K_1 tidal and velocity amplitudes along the main conveying channel. The tidal amplitudes increase in magnitude from the entrance to Juan de Fuca Strait to the northern end of the Strait of Georgia, decreasing thereafter through Discovery Passage, Johnstone and Queen Charlotte Straits. The agreement between the observed and calculated values is good in the northern passages but in the remainder of the system the calculated amplitudes are generally low. In the northern Strait of Georgia this deficiency amounts to about 10 cm. There is a marked decrease in observed K_1 tidal amplitudes between the southern and central parts of Discovery Passage. In the model this decrease is less than that observed. The calculated amplitudes of the tidal velocities are smaller than those of the M_2 tide but display a similar distribution along the main conveying channel.

The observed phase lags show a considerably smaller difference between the values in the central part of Discovery Passage and those at its southern end than in the case of the M_2 tide. The calculated K_1 phase lags are generally low in the Strait of Georgia. Since both the amplitude and phase differences along the southern part of Discovery Passage are less than those observed, it would appear that, as in the case of the M_2 constituent, the volume flow into the Strait of Georgia through this channel must be somewhat less than is shown by the model.

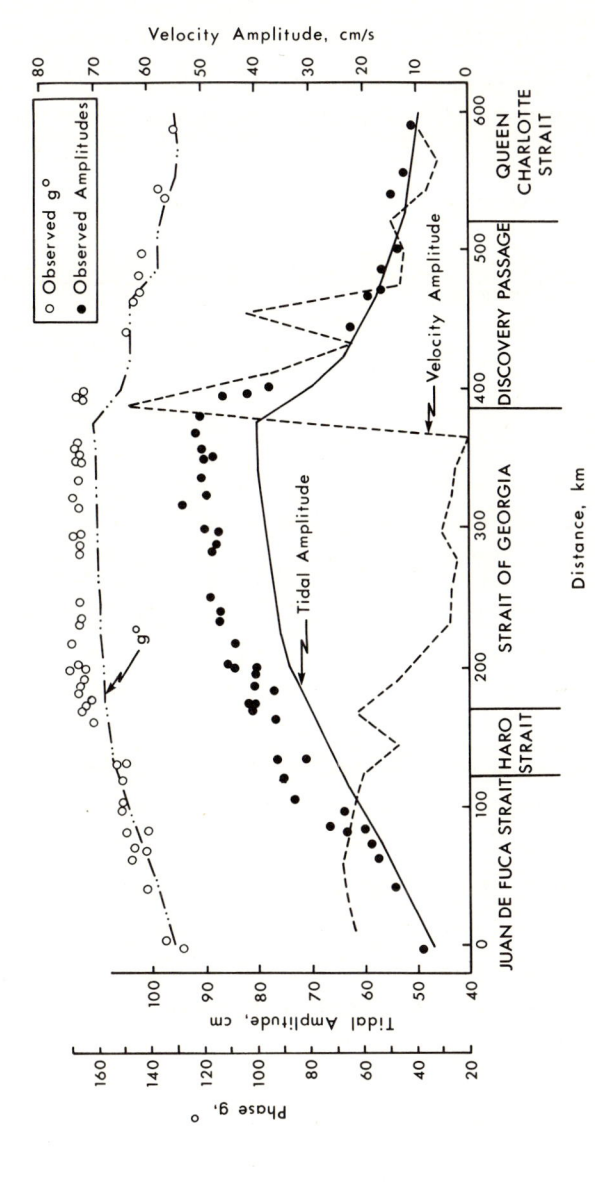

Fig. 3.5 Computed and observed K_1 tidal amplitudes and Greenwich phases along the main conveying channels between the northern and southern boundary openings of the one-dimensional model GF1. The solid line refers to the computed tidal amplitude; dash-dot line indicates the computed Greenwich phase and dashed line shows the computed K_1 velocity amplitude.

3.5.5 K_1 Tidal Harmonic Constants in the Side Channels

Considering now the calculated K_1 tidal constants in the inlets (Table 3.3) the agreement between the observed and calculated tidal amplitudes in Knight Inlet is good but in the remaining inlets the calculated amplitudes are persistently lower. This is consistent with the fact that the calculated amplitudes through Juan de Fuca Strait and the Strait of Georgia are also low.

Table 3.3 Observed and calculated harmonic constants, amplitudes H (cm) for the K_1 tides near the heads (except for Knight Inlet) of inlets and calculated harmonic constants, velocity amplitudes V (cm/s) for the streams at their respective entrances. Greenwich phases g are in degrees.

Location	Tide gauge number	Observed		Calculated			
		Tides		Tides		Streams	
		H	g	H	g	V	g
Puget Sound	—	88	167	80	167	28	79
Howe Sound	7811	87	167	78	156	2	64
Jervis Inlet	7860	88	165	82	156	3	71
Bute Inlet	8069	93	170	81	157	3	65
Knight Inlet	8310	52	131	53	134	5	46

The large velocity amplitude at the entrance to the Puget Sound system reflects a much larger surface area compared to those of the other mainland inlets included in the model.

The observed and calculated tidal phase lags are in good agreement for Knight Inlet and Puget Sound but for Howe Sound, Jervis and Bute Inlets, the phase lags, as in the Strait of Georgia, are low by 9–13°. Again, it will be noted that the tidal motions in these inlets correspond to those of standing waves.

3.5.6 K_1 Tidal Stream Harmonic Constants

Observed and calculated harmonic constants for the K_1 tidal streams are shown in Table 3.4. The agreement between the observed and calculated velocity amplitudes in Juan de Fuca Strait and the Strait of Georgia section is excellent. In Admiralty Inlet the calculated velocity amplitude is low, again consistent with omission of the effects of rotation in Juan de Fuca Strait.

Table 3.4 Observed and calculated K_1 tidal stream harmonic constants, velocity amplitudes V (cm/s) and Greenwich phases g (°). Lines referred to are taken from Fig. 2.1 (Chapter 2).

Location	Observed		Calculated	
	V	g	V	g
Juan de Fuca, line 11	28	67	25	57
Admiralty Inlet	44	73	28	79
Strait of Georgia, line 4	4	75	4	56
Seymour Narrows,	101	37	64	75

The agreement between observed and calculated phase lags at line 11 is fairly good. There is, however, a large discrepancy (19°) between observed and calculated phase lags in the Strait of Georgia. In this instance the model value is probably closer to reality since there is much scatter in the observed values. There is also a large discrepancy (38°), though in the opposite sense, between the observed and calculated phase lags at Seymour Narrows. This is due to the the calculated K_1 tidal amplitude and phase lag deficiencies in the Strait of Georgia. The agreement between the observed and calculated phase lags in Admiralty Inlet is satisfactory.

3.6 DISCUSSION

The above results demonstrate that the major features of the M_2 and K_1 tidal regimes in a system of great topographical complexity can be reasonably well reproduced by using a simple finite-difference scheme and elementary dynamical assumptions. In the Strait of Georgia and in the inlets, with the possible exception of a relatively small departure in the Puget Sound system, the basic character of the motion is that of a standing wave. This is also true of the diurnal tide in Juan de Fuca Strait. The semi-diurnal tide in this latter Strait, however, displays behaviour somewhat closer to that of a progressive wave as shown by the observed and calculated phase lags of the tide and tidal streams.

The major deficiencies of the model are as follows. The friction factor employed throughout the greater part of the model is higher, roughly greater by a factor of 10, than that normally employed in other work on coastal seas. Similar values have, however, been reported for certain sections of the English Channel which, when compared with values obtained in the same manner for the Bristol Channel, were found to be considerably larger (Grace, 1936). It is reasonable to suppose that in the case of the Georgia/Fuca system, a contributory factor may be the energy loss associated with its complex geometry. In applying solutions of the linearized equation of motion to the Georgia/Fuca system, assuming the northern end of the Strait of Georgia closed, Redfield (1950b) found that a reasonable correlation between observed and calculated wave properties could only be attained when employing a considerably higher damping factor compared to those required in applying the same technique to the Bay of Fundy and Long Island Sound. A numerical comparison of the value used by Redfield with that just cited is nebulous since some arbitrary assumption must be made concerning a typical velocity amplitude for the system.

A friction factor was selected to optimize the agreement between the observed and calculated M_2 tidal amplitudes in the Strait of Georgia. This led to M_2 tidal phase lags in that area of the model which are too small by about 20–30°. The calculated M_2 tidal current phase lags in Juan de Fuca Strait appear to be too small by 31°. Thus the M_2 tide in the southern part of the model is occurring earlier than in nature. The calculated K_1 tidal amplitudes in the Strait of Georgia are low by about 10 cm and the phase lags too small by about 10°. The M_2 tidal amplitudes in the Puget Sound system are much less than those observed. To a lesser degree this is also true of the K_1 tide. The difference between the observed tidal amplitudes in Haro Strait is not evident in the model.

Since rotation was the only major dynamical component left out of the equation of motion, it is reasonable to conclude that its omission is to a large degree responsible for much of the inaccuracies in the model, and that the Coriolis effect in the inner part of Juan de Fuca Strait has an important role in determining the general tidal regime within the system. This effect could readily be included in a two-dimensional vertically-integrated model of the region. This two-dimensional model would also provide a more detailed understanding of the tides and tidal streams. Subsequent chapters will be concerned with the design criteria, development, and performance of such a model.

4 THE COMBINED ONE- AND TWO-DIMENSIONAL MODEL: GF2

4.1 INTRODUCTION

With access to a larger computer in 1970, it became possible to transcend the inherent limitations of the one-dimensional numerical tidal model GF1. A system of joined one- and two-dimensional models which well simulated the complex mixed tidal propagation in the overall system was eventually established. In addition to describing the model and its performance, the various limitations encountered and subsequent modifications that led to the final version, GF2, are also summarized.

4.2 FUNDAMENTAL HYDRODYNAMIC EQUATIONS

Three types of numerical schemes are employed. The Strait of Georgia, Juan de Fuca Strait, and the intervening system of islands and passages are represented by a two-dimensional, vertically-integrated numerical scheme. To represent narrower passages where Coriolis effects are not important, a one-dimensional scheme, which is an adaptation of the two-dimensional one, is applied to the Puget Sound system. A second more versatile one-dimensional scheme is applied to the the northern passages and mainland inlets.

4.2.1 The Two-Dimensional Scheme

The vertically-integrated equations of continuity and momentum applicable to tidal motions in a flat rotating sea (Proudman, 1953) may be written as

$$\frac{\partial \zeta}{\partial t} + \frac{\partial U}{\partial x} + \frac{\partial V}{\partial y} = 0, \tag{4.1}$$

$$\frac{\partial U}{\partial t} + \frac{\partial}{\partial x}\left[\frac{U^2}{\zeta+h}\right] + \frac{\partial}{\partial y}\left[\frac{UV}{\zeta+h}\right] - fV + g(\zeta+h)\frac{\partial}{\partial x}(\zeta-\bar{\zeta}) + \frac{kU\sqrt{U^2+V^2}}{(\zeta+h)^2} = 0, \tag{4.2}$$

$$\frac{\partial V}{\partial t} + \frac{\partial}{\partial x}\left[\frac{UV}{\zeta+h}\right] + \frac{\partial}{\partial y}\left[\frac{V^2}{\zeta+h}\right] + fU + g(\zeta+h)\frac{\partial}{\partial y}(\zeta-\bar{\zeta}) + \frac{kV\sqrt{U^2+V^2}}{(\zeta+h)^2} = 0 \tag{4.3}$$

where

x, y = Cartesian co-ordinates in the plane of the undisturbed surface,

t = time,

$\zeta(x, y, t)$ = elevation of the sea surface, about $z = 0$,

$U(x, y, t), V(x, y, t)$ = components of vertically-integrated velocity in directions x and y,

$h(x, y)$ = depth of water below the undisturbed surface,

f = Coriolis parameter, assumed uniform over the region,

$\bar{\zeta}(t)$ = equilibrium form of the surface elevation corresponding to the tide-generating body forces, assumed spatially uniform over the region,

$k(x, y)$ = coefficient of friction,

g = acceleration due to gravity.

In the derivation of these equations it is assumed that the velocities are essentially independent of depth. Equations (4.1), (4.2), and (4.3) are solved numerically for the sea. The finite-difference scheme is spatially similar to that used by Hansen (1961) but a modified numerical procedure in time is employed (Flather and Heaps, 1975).

The area is represented on a system of orthogonal grid lines where the grid notation is shown in Fig. 4.1. Each mesh side is 2 km long and is characterized by a central elevation point while the normal components of vertically-integrated velocity are evaluated at the midpoints of the sides.

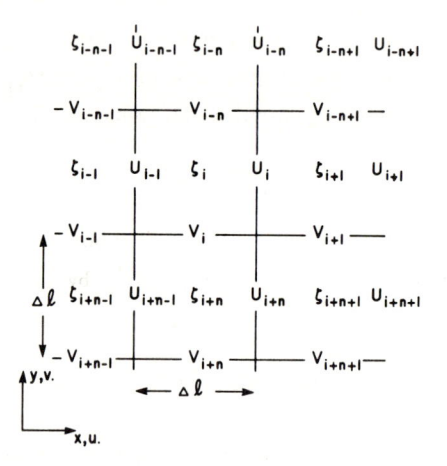

Fig. 4.1 The two-dimensional grid. The ζ points denote locations where elevations are calculated and depths specified. The velocity components are calculated at U and V points, respectively.

To calculate the elevation $\zeta_i(t + \Delta t)$ at the centre of a mesh, Eq. (4.1) is replaced by

$$\frac{\zeta_i(t + \Delta t) - \zeta_i(t)}{\Delta t} = -\frac{U_i(t) - U_{i-1}(t) + V_{i-n}(t) - V_i(t)}{\Delta \ell} \qquad (4.4)$$

where n is the number of columns in the numerical grid scheme.

The momentum equations (4.2) and (4.3) are replaced by

$$[U_i(t+\Delta t) - U_i(t)]\frac{1}{\Delta t} = f\overline{\overline{V}}_i(t) - g\frac{\overline{H}_i^x}{\Delta \ell}[\zeta_{i+1}(t+\Delta t) - \zeta_i(t+\Delta t)]$$

$$- kU_i(t)\frac{[U_i^2(t) + \overline{\overline{V}}_i^2(t)]^{1/2}}{[\overline{H}_i^x]^2}$$

$$- \left[\frac{[\overline{U}_i^x(t)]^2}{\frac{1}{2}[\overline{H}_i^x + \overline{H}_{i+1}^x]} - \frac{[\overline{U}_{i-1}^x(t)]^2}{\frac{1}{2}[\overline{H}_{i-1}^x + \overline{H}_i^x]}\right]\frac{1}{\Delta \ell}$$

$$- \left[\frac{\overline{U}_i^y(t)\overline{V}_{i-n}^x(t)}{\overline{\overline{H}}_i} - \frac{\overline{U}_{i+n}^y(t)\overline{V}_i^x(t)}{\overline{\overline{H}}_{i+n}}\right]\frac{1}{\Delta \ell}$$

$$+ \frac{\varepsilon_i}{\Delta \ell^2}[U_{i-n} + U_{i+n} + U_{i-1} + U_{i+1} - 4U_i], \quad (4.5)$$

$$[V_i(t+\Delta t) - V_i(t)]\frac{1}{\Delta t} = -f\overline{\overline{U}}_i(t) - g\frac{\overline{H}_{i-n}^y}{\Delta \ell}[\zeta_i(t+\Delta t) - \zeta_{i-n}(t+\Delta t)]$$

$$- kV_i(t)\frac{[\overline{\overline{U}}_i^2(t) + V_i^2(t)]^{1/2}}{[\overline{H}_{i+n}^y]^2}$$

$$- \left[\frac{\overline{U}_{i+n}^y(t)\overline{V}_i^x(t)}{\overline{\overline{H}}_{i+n}} - \frac{\overline{U}_{i+n-1}^y(t)\overline{V}_{i-1}^x(t)}{\overline{\overline{H}}_{i+n-1}}\right]\frac{1}{\Delta s}$$

$$- \left[\frac{[\overline{V}_i^y(t)]^2}{\frac{1}{2}[\overline{H}_i^y + \overline{H}_{i+n}^y]} - \frac{[\overline{V}_{i+n}^y(t)]^2}{\frac{1}{2}[\overline{H}_{i+n}^y + \overline{H}_{i+2n}^y]}\right]\frac{1}{\Delta s}$$

$$+ \frac{\varepsilon_i}{\Delta \ell^2}[V_{i-n} + V_{i+n} + V_{i-1} + V_{i+1} - 4V_i] \quad (4.6)$$

where the averaging of a grid variable with the neighbouring variable of that type in the direction x or y increasing is denoted, respectively, by expressions such as

$$\overline{U}_i^x = \tfrac{1}{2}[U_i + U_{i+1}],$$
$$\overline{U}_i^y = \tfrac{1}{2}[U_i + U_{i-n}].$$

To denote the value at a point of a variable obtained by averaging the four values of that variable surrounding the point, the notation is

$$\overline{\overline{H}}_i = \tfrac{1}{4}[H_i + H_{i-n} + H_{i-n+1} + H_{i+1}],$$
$$\overline{\overline{U}}_i = \tfrac{1}{4}[U_i + U_{i+n} + U_{i+n-1} + U_{i-1}],$$
$$\overline{\overline{V}}_i = \tfrac{1}{4}[V_i + V_{i-n} + V_{i-n+1} + V_{i+1}]$$

where $H_i = \zeta_i + h_i$.

In laying out the model grid, coastal boundaries are represented by mesh sides where the normal component of velocity is zero. Open boundaries are taken along a line of elevation points where the values are determined from observations. Equations (4.4), (4.5), and (4.6) are solved explicitly at their respective locations in the interior of the grid from initial conditions $\zeta = U = V = 0$. Under the influence of friction, the effect of the initial conditions becomes negligibly small after about two tidal cycles and the solution responds to the values prescribed at boundary openings. In earlier stages of the work, trials undertaken with this computational scheme proceeded from simpler forms of the Eqs. (4.5) and (4.6) from which the advective accelerations and gravitational tide-producing forces were omitted. The influence of the latter were subsequently determined more carefully when it appeared that they might significantly reduce the relatively small differences between computed and observed tides.

4.2.2 The First One-Dimensional Scheme

The equations of continuity and momentum (Dronkers, 1964) applied at a cross-section of a one-dimensional channel are, respectively,

$$\frac{\partial \zeta}{\partial t} + \frac{1}{b}\frac{\partial Q}{\partial s} = 0, \qquad (4.7)$$

$$\frac{1}{A}\frac{\partial Q}{\partial t} + 2\frac{Q}{A^2}\frac{\partial Q}{\partial s} - \frac{Q^2}{A^3}\frac{\partial A}{\partial s} + g\frac{\partial \zeta}{\partial s} + \frac{k\,Q|Q|}{A^2(\zeta+h)} = 0 \qquad (4.8)$$

where

s = distance along the median line of the channel,
$Q(s,t)$ = volume flow rate through a cross-section,
$A(s,t)$ = area of a cross-section,
$h(s)$ = average depth of a cross-section below the mean level,
$b(s)$ = average storage width of the cross-section,
k = coefficient of friction.

Relatively small storage areas off the main channels are included in the average storage width. Changes in momentum flux deriving from temporal variations in cross-sectional areas and spatial gradients of the cross-sectional areas are included in Eq. (4.8). In applying the first finite-difference scheme to the Puget Sound system, the assumptions that the advective accelerations were negligible and that the cross-sectional areas were independent of time led to satisfactory tidal simulations for present purposes. Thus, the second and third terms were omitted from Eq. (4.8).

The finite-difference formulation of these equations conform to that used in the two-dimensional model. The computational grid is shown in Fig. 4.2. The finite-difference forms of Eqs. (4.7) and (4.8) are

$$\frac{\zeta_i' - \zeta_i}{\Delta t} = -\frac{Q_i - Q_{i-1}}{b_i \Delta s}, \qquad (4.9)$$

$$\frac{Q'_i - Q_i}{A_i \Delta t} = -g \left[\frac{\zeta'_{i+1} - \zeta'_i}{\Delta s} \right] - \frac{k_i Q_i(t)|Q_i(t)|}{A_i^2 (\zeta_i + h_i)} \qquad (4.10)$$

where primes denote values taken at that time $t + \Delta t$ and

Δs = distance between velocity or elevation sections,
b_i = average channel width between velocity sections,
h_i = average channel depth between elevation points,
A_i = cross-sectional area evaluated at a section where Q_i is calculated.

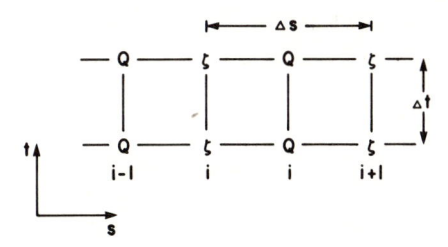

Fig. 4.2 The first one-dimensional scheme.

Where junctions of conveying channels occur in the one-dimensional model, the equation of continuity is applied at the junction and it is assumed that there is no significant difference in the elevations of the water surfaces at the entrances to the branching channels (Dronkers, 1964).

It is assumed that A_i is reasonably typical of the section of channel between the elevation points used to calculate Q_i. In principle, the empirical adjustment of the friction factor to optimize the agreement between computed and observed elevations tends to compensate for discrepancies, although in the present context, where the depths of the conyeying channels are in general much greater than those encountered in rivers, fine adjustment of the friction is not critical to the successful operation of the model.

4.2.3 The Second One-Dimensional Scheme

An alternative finite-difference formulation (Dronkers, 1969) of Eqs. (4.7) and (4.8) was employed in channels where constrictions and large spatial changes in the cross-sectionally averaged velocities occur (*e.g.*, Seymour Narrows, Sechelt Rapids, see Fig. 1.30). Advective acceleration is included in the momentum equation. The distance between sections may be selected arbitrarily to facilitate the schematic representation of a channel using sections of reasonably typical average depth and cross-sectional area or for the numerically-convenient spacing of sections near proposed junctions of one- and two-dimensional schemes.

Since the equations are of implicit form, the stability restrictions on the time step occassioned by large depths in certain mainland inlets, are mitigated. Surface storage and cross-sectional areas are again assumed independent of time.

Using the computational scheme shown in Fig. 4.3, the finite-difference forms of Eqs. (4.7) and (4.8) are

$$\frac{Q'_{i+1} - Q'_i}{\Delta s_i} = -b_i \left[\frac{(\zeta'_{i+1}) + \zeta'_i - (\zeta_{i+1} + \zeta_i)}{2\Delta t} \right], \qquad (4.11)$$

$$g\left[\frac{\zeta'_{i+1} - \zeta'_i}{\Delta s_i}\right] = -\frac{[(Q'_{i+1} + Q'_i) - (Q_{i+1} + Q_i)]}{2A_i \Delta t} - \frac{[(Q_{i+1} + Q_i)(Q'_{i+1} - Q'_i)]}{A_i^2 \Delta s_i}$$

$$+ \frac{[(Q_{i+1} + Q_i)(Q'_{i+1} + Q'_i)(A_{i+1} - A_i)]}{4 A_i^3 \Delta s_i}$$

$$- k_i \frac{[(Q'_{i+1} + Q'_i)|Q_{i+1} + Q_i|]}{4 A_i^2 (\zeta_i + h_i)}. \qquad (4.12)$$

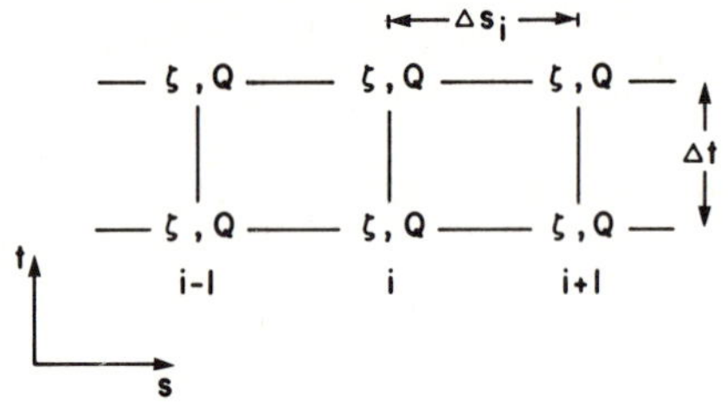

Fig. 4.3 The second one-dimensional scheme.

These equations are solved implicitly using the "double sweep" method (Dronkers, 1969). A detailed account of the method may be found in Mungall (1973).

4.2.4 Junctions between One- and Two-Dimensional Schemes

For reasons to be discussed later, it is desirable to join the various one- and two-dimensional schemes. At each junction it is assumed that the entrance width to the side channel is the same as the two-dimensional mesh size. The correct cross-sectional area can be obtained by adjusting the depth. If this is significantly different from the average depth of the entrance, some error is introduced into the frictional dissipation but is not significant in the process of overall adjustment of the model.

Considering initially the first one-dimensional scheme, the spatial arrangement of variables in the vicinity of the junction is shown in Fig. 4.4. The section length employed in the side channel is considerably longer (12 km) than the two-dimensional mesh size (4 km). When the one- or two-dimensional scheme is operated independently, the value $\zeta_1(t + \Delta t)$ or $\zeta_{60}(t + \Delta t)$ is prescribed from observations. When joined, $\zeta_1(t + \Delta t)$ is found from Eq. (4.9) where the volume flow entering the one-dimensional scheme $Q_1(t) = V_{60}(t)\Delta \ell$ and V_{60} is the vertically-integrated velocity through the corresponding side of the two-dimensional mesh. Then $\zeta_{60}(t + \Delta t)$ is found from the following six-point Lagrangian interpolation using three values of elevation along the normal to the section at the mouth of the channel from each computational scheme.

$$\zeta_{60} = 0.1000 \cdot \zeta_{171} - 0.5185 \cdot \zeta_{134} + 1.0500 \cdot \zeta_{97} + 0.3889 \cdot \zeta_1 - 0.222 \cdot \zeta_2 + 0.0019 \cdot \zeta_3$$

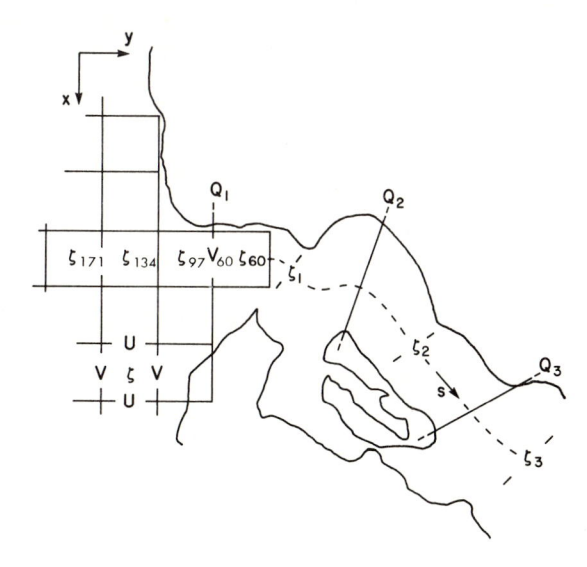

Fig. 4.4 Junction of the first one- and two-dimensional schemes at the entrance to the Puget Sound system.

In the two-dimensional scheme the velocity field is dominated by the component along this normal. When the values of these elevations have been determined, the volume transports in each scheme are computed using Eqs. (4.5), (4.6), and (4.10) in the usual manner. In joining the second one- and two-dimensional schemes, the need for a spatial interpolation is eliminated by selecting a section length equal to the mesh size at the channel entrance (Fig. 4.5). In this case $\zeta_i(t + \Delta t)$ is determined from Eq. (4.4) in which the substitution

$$V_{i-n}\Delta \ell = -\frac{Q_{b-1} + Q_b}{2}$$

has been made.

Conservation of volume of the various junctions was verified by numerical integrations of volume flows over several tidal cycles.

4.2.5 Approximate Representation of Narrow Passes in the Two-Dimensional Scheme

There occur within the region of the two-dimensional scheme a number of navigationally important narrow passes (*e.g.*, Active Pass, see Fig. 1.8) which cannot be adequately represented since the grid size is much larger than the width of the pass.

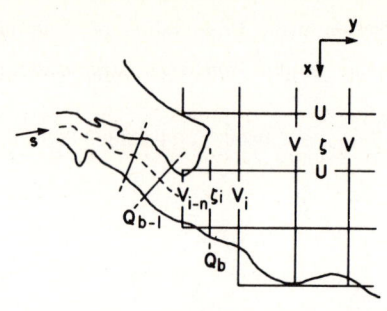

Fig. 4.5 Junction of the second one-dimensional scheme with the two-dimensional scheme at the junction of Juan de Fuca Strait and the Strait of Georgia.

Depending on local coastal configuration and bathymetry, improved simulations can be achieved in some cases as follows.

If a pass is similar in length to the mesh size (Fig. 4.6) and the local flows in the adjoining mesh squares are dominantly parallel to the longitudinal axis through the pass, the change in momentum flux due to the change in width may be approximated for the case of a vertically-integrated velocity at one end of the pass by replacing the fourth term on the right hand side of Eq. (4.5) with the following expression:

$$\frac{4}{(\Delta\ell + b)\Delta\ell} \left[\frac{\Delta\ell [\overline{U}_i^x(t)]^2}{\overline{H}_i^x(t) + \overline{H}_{i+1}^x(t)} - \frac{b[\overline{H}_{i-1}^x(t)]^2}{\overline{H}_{i-1}^x(t) + \overline{H}_i^x(t)} \right]. \tag{4.13}$$

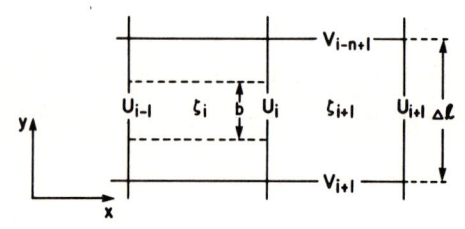

Fig. 4.6 Grid notation at a narrow pass.

To determine the elevation in a mesh adjacent to the pass, Eq. (4.4) is rearranged using volume transports through the mesh sides as follows,

$$\zeta_{i+1}(t + \Delta t) = \zeta_{i+1}(t) - \frac{\Delta t}{\Delta\ell} \left[U_{i+1}(t) - \frac{b}{\Delta\ell} U_i(t) + V_{i-n+1}(t) - V_{i+1}(t) \right]. \tag{4.14}$$

When the pass is much shorter than a mesh width, a small opening is left in the side of the appropriate mesh. The velocity through the pass is then determined using Eq. (4.5) with an empirically-adjusted friction coefficient to give optimal agreement between computed values and such local observations of tides and streams as may be available for that vicinity. Elevations in meshes adjoining the pass are then found from an equation similar to (4.14), depending on the orientation of the pass.

4.2.6 Frictional Adjustments

A major problem in the development of this model concerns the determination of the proper distribution of frictional dissipation. Ideally, the friction coefficient appropriate to a particular mesh or channel section is determined by direct measurement of surface slopes and velocities entering into a particular finite-difference momentum equation. This can be done, at least in principle, for sections in a one-dimensional river model (Dronkers, 1964) but becomes impractical for broad estuaries and coastal seas. In addition to the work of Redfield (1950b), various methods have been employed. Following Taylor's (1919) work, Heisanken (1921), Jeffreys (1921), and Miller (1966) have estimated the frictional dissipation in coastal seas using the net energy flux through boundary openings where elevations and velocities are known. Grace (1936, 1937) divided the Bristol and English Channels into one-dimensional sections in which the elevations were recorded. Using the equation of continuity to obtain the velocities between sections and observed velocity data for one cross-section of the channel, the friction coefficients appropriate to particular sections can be determined. A similar method has been employed by Filloux (1973).

In the present work, the approach taken involves the empirical fitting of the computed water levels to those obtained from extensive measurements of elevations around the interior of the modelled region. Through regional adjustments of frictional dissipation using the M_2 tidal constituent, it was subsequently found necessary to carry out such adjustments with the inclusion of at least the major tidal constituents because of non-linear interactions. Thus, no explicit account is taken of energy lost from the barotropic tides to increase the overall potential energy of this stratified system through vertical mixing, a topic that will be considered in future three-dimensional model studies.

4.3 DESCRIPTION OF THE MODEL

4.3.1 The Strait of Georgia and Juan de Fuca Strait Scheme

Since the system lies within a relatively small span of latitude a two-dimensional Cartesian co-ordinate system was employed. A mesh size of 4 km was selected to provide adequate representation of the cross-channel slopes of the water surface associated with Kelvin wave propagation through the outer and inner straits (Fig. 4.7). Coastal boundaries were located along mesh sides where the normal component of velocity is always zero. Due to the irregularities of the coastlines the selection of a particular mesh side to represent part of a coastal boundary is sometimes quite arbitrary. In such cases mesh sides were chosen to ensure conformity between the surface areas of local parts of the model and the actual areas which were determined by planimeter from hydrographic charts.

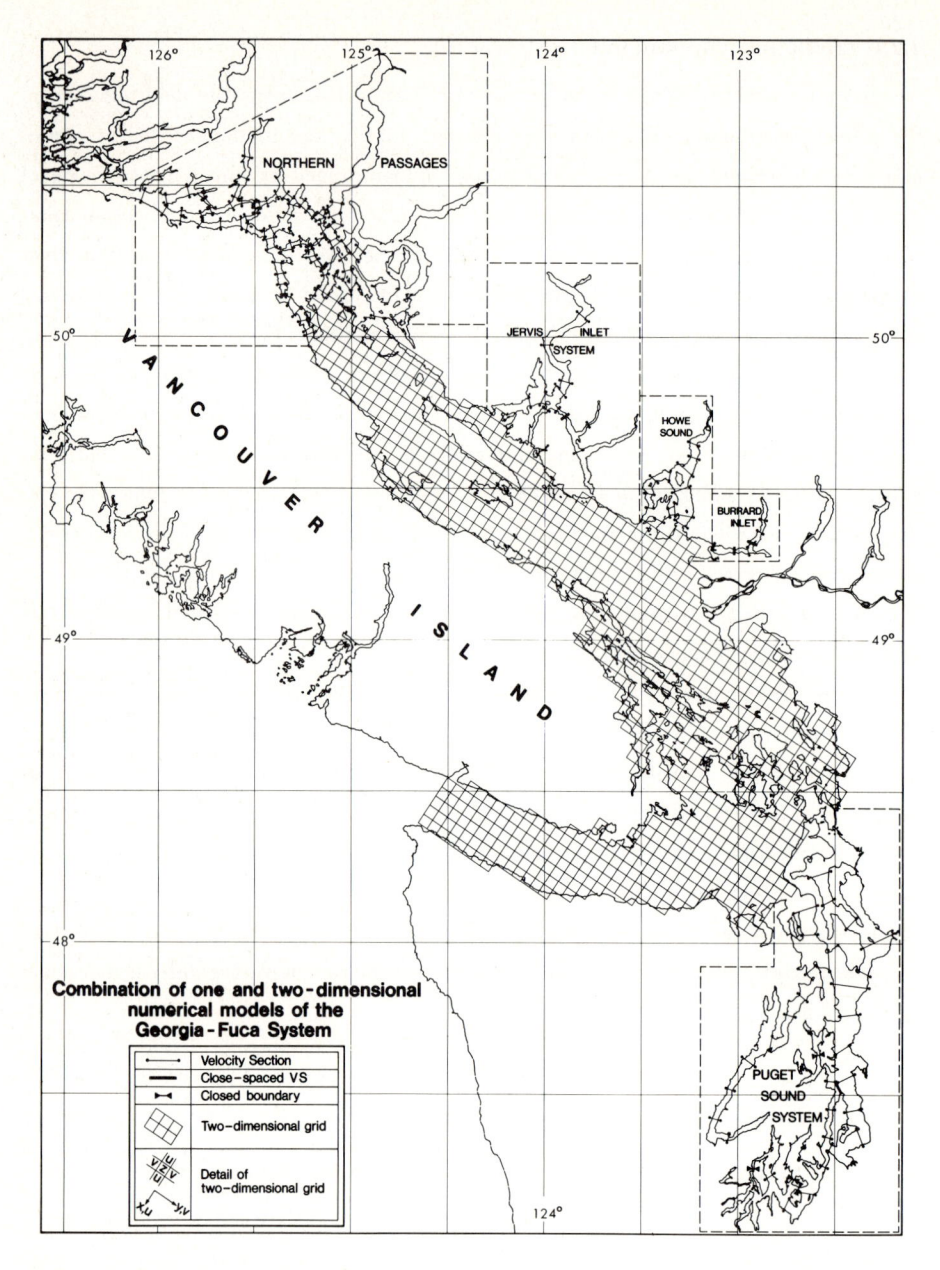

Fig. 4.7 The overall numerical model (GF2) combining one- and two-dimensional schemes.

Depths were specified at elevation points and represented the average depth over a grid square. Channels through the San Juan and Gulf Islands were included in this two-dimensional scheme with the exception of four narrow passes (Active, Porlier, and Thatcher Passes, and Middle Channel, see Fig. 1.8) where strong tidal streams occur.

The individual representations of these passes are discussed in the following sections.

The maximum permissible time step that yielded stable solutions was obtained from the Courant-Friedrichs-Lewy criterion,

$$\Delta t < \frac{\Delta \ell}{\sqrt{2gh_{\max}}}.$$

The resulting time step is 45 s. (This time step is used throughout the overall model though longer time steps were generally employed in the individual adjustments of particular one-dimensional models.)

Ten openings were located in the boundary of the two-dimensional model. The tidal co-oscillation within the system derives primarily from elevations prescribed across the entrance of Juan de Fuca Strait. At the other nine openings, elevations were determined using five one-dimensional models which were joined to the two-dimensional system.

Initially a constant overall frictional coefficient of 0.0025 was used in this model. The final values used were determined only after extensive trials of the extended model. The locations of the tide gauges from which data were obtained to adjust the model for the following schemes (Fig. 4.7) are shown in Fig. 2.1 (p. 46).

4.3.2 The Puget Sound Scheme

Using the first one-dimensional scheme, the Puget Sound system was simulated by seven joined channels. Sections where volume transports were calculated are shown in Fig. 4.7. The distance along the median line between sections was 12 km. Elevations were calculated at points located midway between the velocity sections. Volume transports at the closed end of a side channel were set equal to zero.

Prior to the addition of this model to the two-dimensional scheme the appropriate overall friction coefficient was determined. The M_2 tidal elevations were prescribed at the entrance and Eqs. (4.9) and (4.10) were solved using trial values of the friction coefficients. It was found that good agreement could be obtained between the computed and observed M_2 harmonic constants using a constant coefficient of 0.006 throughout this model.

4.3.3 The Burrard Inlet Scheme

The Burrard Inlet model consisted of a single channel in which volume transports and elevations at ten cross-sections were calculated using the second one-dimensional scheme. The distances between cross-sections were selected to facilitate specification of average widths and depths for the intervening lengths of channel. Two constrictions known as First and Second Narrows, respectively, (see Fig. 1.8) occur in this channel where tidal currents are strong (2–3 m/s).

Initially, M_2 tidal elevations were prescribed at the entrance to the channel. Using an approximate value for the friction coefficient, Eqs. (4.11) and (4.12) were then solved using increasingly detailed schematizations of the Narrows until a solution independent of schematization was obtained. Numerical trials indicate that the dependence of the solution on the schematization is primarily due to the presence of the term $(Q^2/A^3) \cdot (\partial A/\partial s)$ in the momentum equation. Trials were then undertaken to determine the values of frictional coefficients along the channel which gave optimal agreement between observed and computed harmonic constants for the various gauge locations. An additional criterion was available where the phases of the streams through the two Narrows are known from observations. The overall coefficient is 0.006 while values of 0.06 are used at the Narrows.

4.3.4 The Howe Sound Scheme

The Howe Sound model consisted of six joined one-dimensional channels, two of which connect with the Strait of Georgia. Schematically-convenient distances between volume transport sections were employed. Solving Eqs. (4.11) and (4.12) frictional dissipation was adjusted in the same manner as that described for the Burrard Inlet model. It was found that an overall friction coefficient of 0.006 gave good agreement between the observed and computed M_2 tidal harmonic constants for the gauge location #7811 at the closed end of the system (Fig. 2.1). The amplitudes and phases of the tidal streams in the two entrance channels are strongly sensitive to the elevations prescribed, since the openings are in relatively close proximity to each other, and also to the schematization of these channels.

4.3.5 The Jervis Inlet Scheme

Using schematically-convenient distances between volume transport cross-sections, Jervis Inlet was simulated by six joined one-dimensional channels. There are two entrance channels from the Strait of Georgia. Part of the system is separated from the remainder by tidal rapids (Sechelt Rapids). Indicative of the dynamical effect of these rapids, the change in the M_2 tidal amplitude and phase over their length is about 50% and 60°, respectively. Prescribing elevations at the two entrances from local tide gauge data using Eqs. (4.11) and (4.12), the same procedure of optimizing computed values against observations yielded an overall friction coefficient of 0.006. This was increased to 0.06 in the section simulating the Sechelt Rapids.

4.3.6 The Northern Passages Scheme

By far, the most complicated of these one-dimensional models is that which simulates flows entering the northern end of the Strait of Georgia from the open ocean. This system was represented by 25 joined one-dimensional channels and included 94 volume transport sections. Over this region the observed M_2 tidal amplitudes decrease from 115 cm at the single northern boundary opening to a minimum of 62 cm south of

Seymour Narrows, increasing again to 101 cm in the northern Strait of Georgia. Over the same distance the M_2 tidal phase lags increase by 122°. In all, there are ten tidal rapids where, in some instances, the observed streams can exceed 6 m/s.

To determine critical features in schematization and the distribution of friction coefficients, the same M_2 tidal elevations were prescribed at each of the four openings into the Strait of Georgia, based on data from tide gauge #7892. Further M_2 tidal elevations were prescribed at the northern opening using data from gauge #8233. Using estimated friction coefficients, Eqs. (4.11) and (4.12) were solved over the course of repeated trials in which lengths of channel most critical to the flow regime were identified, the density of cross-sections along the channel increased where necessary, and the frictional dissipation adjusted to optimize the agreement between the computed and observed tides. In addition to the tidal data available, the estimated magnitudes and phases of the tidal streams through the more important tidal rapids had been determined from drift-pole measurements. (In particular, analyses of these data, together with local tidal data, by the Liverpool Observatory and Tidal Institute, permit accurate predictions of times of slack water essential for purposes of navigation.) Since the dominant effect of increasing frictional dissipation in one of these narrow passes is to decrease the phase of its tidal streams, observed phase information of this kind assists in adjusting regional friction coefficients. The overall friction coefficient employed in this model was 0.01. Local variations from this value ranged from 0.005 in the deeper parts of Discovery Passage to 0.03 in the smaller channel constrictions.

4.4 SUMMARIZED SEQUENCE OF MODEL DEVELOPMENT

The final version of the model was developed over the course of an extensive series of trials. These will not be discussed in detail but conclusions will be presented insofar as they suggested further modifications and trials.

On the basis of earlier work with the exploratory multi-channel one-dimensional overall model, it appeared reasonable that the amplitudes and phases of the major semi-diurnal and diurnal constituents in the Strait of Georgia could be simulated to within about 5 cm and 5°, respectively, with somewhat greater discrepancies in areas where large spatial rates of change in the harmonic constants occur. References below to unsatisfactory results imply a performance well below these standards. The harmonic constants resulting from computed values were determined by Fourier analyses for trials involving a single constituent or harmonic analysis in the case of mixed tides. All phase lags refer to the phase of the corresponding equilibrium constituent at Greenwich.

Trials were undertaken in five stages. The first stage employed the two-dimensional model with eight boundary openings. The most straightforward, yet dynamically plausible, representation of tides in Juan de Fuca Strait and the Strait of Georgia appeared obtainable from a vertically-integrated two-dimensional model in which Coriolis accelerations were included in the momentum equations. Frictional dissipation was assumed proportional to the square of the computed velocities and elevations were prescribed at boundary openings to simulate the effects of tides in the adjacent ocean and in the mainland inlets.

In this first version the boundary openings included the entrance to Juan de Fuca Strait and seven openings, each the width of a single grid square, representing the main conveying channels into the various inlets. The northern end of the Strait of Georgia was assumed closed except for a single open mesh in the vicinity of gauge #7892 which was intended to allow for tidal fluxes through the northern passages. At least 1 yr of tide gauge data had been obtained in the vicinity of each of these boundary openings.

Assuming an overall friction coefficient of 0.0025, a value commonly applied in work on coastal seas, trials were carried out in which the M_2 and K_1 tidal constituents were individually simulated in the model. The agreement obtained between the harmonic constants deriving from the model and those obtained from observations was excellent.

A further trial was undertaken in which a mixed tide consisting of the sum of the M_2 and K_1 elevations, was prescribed at each boundary opening. These computations showed that no significant non-linear interactions occurred in this version of the model. This appeared consistent with the small amplitudes of the overtides and compound tides that are normally determined in the standard harmonic analysis of 1 yr of tide gauge observations. This was thought to be due primarily to the relatively large depths which prevail throughout most of the system. It will be shown subsequently that significant non-linear interactions do indeed occur.

At this stage of the investigation, there were insufficient data on the harmonic constants of the tidal streams to verify the computed velocity fields. In general, over much of the model these fields showed appreciable rotation of the velocity vectors over a tidal cycle. These rotations appeared greatly influenced by the flows through the various boundary openings.

The magnitudes and phases of the tidal flows through the openings off the Strait of Georgia were, however, quite inconsistent with the anticipations based on results from the earlier exploratory overall one-dimensional model which specifically include inlets associated with these openings. In addition, these flows were superimposed on slowly-fluctuating net flows which, in some cases, were of comparable magnitude with those of tidal period. These irregularities were substantially worse for the K_1 tide compared with that of the M_2. In consequence there was considerable doubt concerning the validity of the computed velocity fields which, since the surface tides in the Strait of Georgia might be substantially held correct by the prescribing of elevations at numerous openings along the mainland boundary, would reflect inadequacies in the dynamical assumptions, frictional adjustment, and finite-difference representation of the topographically-complex region of islands between the Strait of Georgia and Juan de Fuca Strait. Estimated magnitudes of the advective accelerations from the computed velocity fields indicated their dynamical importance in the vicinity of the San Juan Islands. It was decided to extend the model and include these terms in the momentum equations.

In stage two, the two-dimensional with a single one-dimensional scheme and seven boundary openings was used. The one-dimensional model of the Puget Sound system was selected as the first extension of the existing two-dimensional model for a variety of reasons. The dynamical sensitivity of the model would be greatly increased if it were required to correctly proportion incoming wave energy between the Strait of Georgia and the largest of the side channels. It was also desirable that the computed surface tides in the critical region of the San Juan and Gulf Islands should not be favoured by the prescribing of elevations at a nearby boundary opening. Furthermore, the tidal streams in the important area of inner Juan de Fuca Strait are strongly affected by the volume transports through the entrance to the Puget Sound system. It was thus appropriate that the magnitude and phase of these transports be verified.

Trials with single constituent tides (M_2 and K_1) in the combined scheme, using overall friction coefficients of 0.0025 in the two-dimensional and 0.006 in the one-dimensional models confirmed the results previously obtained when each model was operated independently. In particular, the magnitudes and phases of tidal volume exchanges between Puget Sound and Juan de Fuca Strait were in excellent agreement with those obtained when the elevations had been prescribed at that opening.

Illustrative of the results obtained at this stage of the model development, Table 4.1 shows comparisons of the observed M_2 and K_1 harmonic constants with those obtained from the model for representative locations. It is evident that an excellent simulation of the tidal constants in the Strait of Georgia and Puget Sound system had been achieved.

In stage three advective accelerations were included in the two-dimensional momentum equations (4.5) and (4.6) and a further M_2 trial was undertaken. Although no significant overtides were apparent in analyses of the computed elevations, it was found that an increase in mean sea level of 5 cm occurred at all locations in the Puget Sound system. It was also noted that appreciable residual currents were now apparent in regions of the two-dimensional model where large speeds and changes in direction of velocity occurred. There remained, however, the anomalous flows through the various boundary openings in the simulated mainland coast of the Strait of Georgia in stage three. To eliminate these flows and further extend the scope of the model, the one-dimensional Howe Sound, Jervis and Burrard Inlets schemes were now included in the overall model. The northern end of the Strait of Georgia was assumed closed except, as before, for a single boundary opening, in the immediate vicinity of tide gauge #7892.

An initial trial of this version of the model, in which the value of the overall friction coefficient in the two-dimensional model was 0.0025 and the M_2 tidal elevations were prescribed at the two remaining boundary openings showed differences between

Table 4.1 Comparisons of the dominant semi-diurnal (M_2) and diurnal (K_1) amplitudes H (cm) and phases g (°) from observations at representative locations with those obtained from the model when a single constituent tide is prescribed at the seven boundary openings for the early version of the model with one side channel.

Location	Tide gauge number	M_2		K_1	
		H	g	H	g
Port Angeles	7060				
observed		52	75	66	141
calculated		49	68	62	142
Pedder Bay	7080				
observed		34	76	63	149
calculated		33	69	63	149
Finnerty Cove	7140				
observed		45	126	71	157
calculated		42	122	69	156
Friday Harbor	7240				
observed		56	140	76	161
calculated		58	138	76	159
Fulford Harbour	7330				
observed		59	141	76	159
calculated		66	141	77	163
Tumbo Channel	7510				
observed		73	159	81	167
calculated		81	154	83	164
Ferndale	7564				
observed		72	152	80	163
calculated		79	148	83	161
Tsawassen	7590				
observed		82	156	85	164
calculated		86	153	85	164
Winchelsea	7935				
observed		95	161	88	166
calculated		96	159	88	167
Little River	7993				
observed		99	161	90	167
calculated		101	160	90	167
Seattle	7180				
observed		107	140	83	157
calculated		108	141	86	156

observed and computed amplitudes and phases throughout the model which were well below expectations. Further trials in which the distribution of frictional dissipation over the model was varied clearly showed the impossibility of obtaining a satisfactory simulation of the M_2 tide in this manner and that there still remained a fundamental inadequacy in the present formulation of the model. This error now appeared to be associated with the Kelvin wave reflection problem at the northern end of the Strait of Georgia. Since no way existed in which the strong tidal flows through the single fictitious boundary opening at the northern end of the Strait of Georgia could be verified, it was decided to solve the reflection problem by simulating the tides in the complex of passages leading northward through Johnstone Strait to the open sea and in the contiguous mainland inlets. Arrangements were made for current meter observations to verify the magnitude and phase of volume transports through the single northern opening to this further system of one-dimensional channels. Thus, if a viable solution to the reflection problem could be achieved, either a satisfactory simulation would result or further effort could be concentrated on the problem of wave transmission through the San Juan Islands.

In stage four, the two-dimensional scheme with five one-dimensional schemes and two boundary openings was used. Initial trials of the complete model were carried out using an overall friction coefficient of 0.0025 in the two-dimensional model and M_2 tidal elevations prescribed at the two boundary openings. Again, results were obtained which were well below expectation. This problem could not be resolved by adjusting frictional dissipation in the model since improved phases would be accompanied by unimproved amplitudes. Attention was thus directed to the wave transmission problem between Juan de Fuca Strait and the Strait of Georgia.

First, by introducing perturbations of the M_2 tidal flows through the northern passages, it was shown that no excessive changes resulted in the harmonic constants in the amphidromic region of the inner part of Juan de Fuca Strait. Current meter observations over a cross-section of Johnstone Strait near the northern boundary of the model confirmed remarkably well the computed amplitude and phase of the M_2 tidal volume flux. Thus, in the present context, the reflection problem was considered solved.

The basic elements considered in the transmission problem concerned the relative significance of numerical approximations (channel depths, coastal boundaries) and frictional adjustments on tidal propagation through the main (Haro and Rosario Straits) and secondary (Active, Porlier, Thatcher Passes, and Middle Channel) conveying channels to the overall tidal regime. Of the two main channels, the computed tidal volume fluxes through Haro Strait were roughly three times the magnitude of those through Rosario Strait. Accordingly, the next step was to show that the secondary conveying channels did not significantly influence the tidal regime in the Strait of Georgia and the Puget Sound system. This was done by comparing the M_2 tidal harmonic constants obtained from a simulation in which these channels were assumed closed with those deriving from the normal case when they are open.

As previously noted, considerable ambiguity exists in the delineation of the coastal boundaries in the San Juan Island region of the model due to the coarseness of the mesh size. Subject to conservation of surface area, a trial was carried out using a plausible but different arrangement of coastal boundaries. This was found to have a negligible effect on the computed harmonic constants in the Strait of Georgia and Puget Sound. Trials were then carried out to determine the relative effects of adjusting the assigned depths and frictional dissipation in Haro and Rosario Straits. It was in the course of these trials that the key to concomitant improvements in the computed M_2 amplitudes and phases of both the major tidal storage areas of the model, the Strait of Georgia, and the Puget Sound system, was determined. As anticipated, such adjustments in Haro Strait have a much greater influence than when applied to Rosario Strait. This is consistent with the larger volume flow, more complex bottom topography, and greater channel length in the case of Haro Strait.

The dominant effect of increasing frictional dissipation in Haro Strait is to increase the M_2 tidal amplitude in Puget Sound with a corresponding, though smaller, reduction in amplitude in the Strait of Georgia. This is qualitatively consistent with the simple theory of superimposed damped incoming and reflected waves in a basin closed at one end. The important point, however, is that the associated changes in the tidal phase lag are relatively small. If a trial is conducted in which depths are assigned such that there is a contiguous line of meshes corresponding to the maximum depths along Haro Strait and Boundary Pass, the dominant effects are to significantly increase the M_2 tidal phase lag in the Puget Sound system and moderately reduce the phase lag but significantly increase the amplitude in the Strait of Georgia. These adjustments of friction and depth permitted a simulation of the M_2 tide fully consistent with earlier expectations. The overall coefficient of friction in the two-dimensional model was 0.006. This was increased to 0.06 in Haro Strait.

Using the same configuration of frictional coefficients and depths, a trial was now undertaken in which the K_1 elevations were prescribed at the open boundaries. The observed and computed tidal amplitudes were in good agreement but throughout the Strait of Georgia, the computed phase lags were too small by about 15°. Such results could not be improved by any adjustment of frictional dissipation since any improvement in phase had to be accompanied by an unimprovement in amplitude.

In stage five three possibilities were now considered as contributing, at least in part, to the discrepancy between the observed and computed phases of the K_1 tidal constituent in the Strait of Georgia. First, the optimization of the model to simulate the M_2 tide took no account of direct gravitational forcing. In the present case it appeared possible that such forces, though small, might account for some part of this phase error in the K_1 tidal simulation. Secondly, non-linear interactions between the dominant tidal constituents could also contribute to the error. Although attempts to find evidence for such interactions in the first version of the model have been unsuccessful, a number of major changes have been made. Thus, the frictional dissipation had been greatly increased and the advective acceleration terms were included in the momentum equations. Thirdly, there is evidence of the existence of internal tides in both the Strait of Georgia (Chang *et al.*, 1976) and Juan de Fuca Strait (Crean and Miyake, 1976). Thus, energy might be lost at certain frequencies from the barotropic tides in a manner which could not be represented in the model. This third possiblility would logically introduce the need for a three-dimensional model.

Trials were carried out in which the effect of direct gravitational forcing on the simulation of a K_1 tidal co-oscillation in the model was assessed. Since the modelled sea is of limited extent, the time-varying components of the K_1 constituent tide-generating force were assumed independent of spatial variation over the sea. Normally these force components are expressed in spherical polar co-ordinates (Doodson and Warburg, 1941; Proudman, 1953). The components were thus transformed to a Cartesian co-ordinate system, then rotated to conform with the axes of the model to give the gradients of $\overline{\zeta}$ to be introduced into the finite-difference formulations of Eqs. (4.2) and (4.3). Appropriate corrections were made for the earth's crustal displacements due to the tide-generating forces. The K_1 tidal elevations were prescribed on the open boundaries in the usual manner. These trials showed that the introduction of direct gravitational forcing produced no significant change (< 1 cm and $< 1°$) in the harmonic constants of the K_1 tidal constituent.

Attention was now directed to the possibility of non-linear interactions. To obtain the best possible simulation of mixed tides in the system, the elevations prescribed at each boundary opening were the sum of the full 61 tidal constituents normally available from the standard 1-yr analyses by the Marine Environmental Data Service. Using the same configuration of depths and friction which gave optimal results for the M_2 tide, trials of mixed tides were undertaken in the model. It became evident that a reduction in frictional dissipation was necessary. However, the important point was that, in contrast to earlier experience with the M_2 constituent, simultaneous improvements in the tide heights and times of high water in both the Strait of Georgia and the Puget Sound systems could now be effected by a straightforward adjustment of frictional dissipation. Using an overall coefficient of 0.003 in the two-dimensional model and 0.03 in Haro Strait, a 1-wk sequence of mixed tides was simulated. The agreement obtained between the computed and observed tidal elevations in the Strait of Georgia and in the Puget Sound system was excellent. Further comparisons of observed and predicted tidal streams in Juan de Fuca Strait and the Strait of Georgia yielded similar accord. Some of these results characterizing the performance of the model have been reported (Crean, 1976).

The quality of the results suggested that the major discrepancy in the phase of the computed K_1 tide component in earlier individual simulations was no longer significant, a possible consequence of non-linear interaction. To determine whether such was the case, a trial was conducted in which the sum of the M_2 and K_1 constituent tides was prescribed on the open boundaries. The results obtained from this trial may be summarized by comparing the observed and computed M_2 and K_1 harmonic constants at Point Atkinson derived, respectively, from individual and combined simulations of these constituents (Table 4.2). The harmonic analyses in the mixed tidal case employed 48 h of computed tidal elevations. This record length is insufficient to separate the K_1 tide from the O_1 diurnal constituent brought about by the interaction (difference frequency) of M_2 and K_1 constituents. The marked change in phase of the K_1 tide is, however, confirmed by subsequent analyses of elevations obtained from a simulated 30 days of mixed tides. It is evident from these results that, at least in the case of these constituents, a non-linear interaction significant to the successful operation of the model, does indeed occur.

Table 4.2 Amplitudes H (cm) and phases g (°) of the M_2 and K_1 constituents at Point Atkinson (#7795) obtained from the final configuration of the model adjusted to simulate mixed tides when prescribed first individually and then in combination on the open boundaries, and from observations.

Prescribed tide	M_2		K_1	
	H	g	H	g
M_2	99	168	—	—
K_1	—	—	99	148
$M_2 + K_1$	95	164	88	166
Observed	93	158	87	165

It remained to determine whether friction or advective terms in the momentum equations were primarily reponsible for this interaction. Advective acceleration terms were removed from the momentum equations and the trial was continued. The results obtained differed insignificantly from those presented in Table 4.2. It was thus concluded that the primary source of this interaction lay in the friction terms.

4.5 NON-LINEAR TIDAL INTERACTIONS

It is of interest to consider briefly the topic of non-linear tidal constituent interactions and the manner in which the simulation of a major tidal constituent in the model can be affected by the presence or absence of other constituents. The presence of non-linear advective accelerations and bottom friction terms in Eqs. (4.2) and (4.3) enable the local generation of tides within the model, distinct from those prescribed on the open boundaries. These can consist of overtides in which the angular speed is an exact multiple of one of the prescribed tides or compound tides where the angular speed is an exact sum or difference of the speeds of two prescribed constituents. In general, the amplitudes decrease with increasing frequency though more slowly with increasing non-linearity since the interaction term varies as the reciprocal of the sum or difference frequencies. More detailed discussion may be found in Doodson and Warburg (1941) and Dronkers (1964).

For the present studies, the harmonic constants derived from tide gauge data are characterized by small compound tides and overtides. It was thus initially assumed that regional adjustments of frictional dissipation within the model for a single major constituent would be satisfactory for other or summed constituents. This proved not to be the case since model adjustments must be made on the basis of trials in which mixed tides are prescribed on the open boundaries. The primary source of non-linear interactions may reasonably be supposed to occur in the passages between the San Juan and Gulf Islands while appreciable further contributions may derive from the narrow passes located in the system of passages leading from the northern Strait of Georgia to the open sea and also in shallow water where friction is large.

A series of trials was undertaken to assess the effects of non-linear interactions on the major tidal constituents in the model. In these trials mixed tidal elevations, predicted on the basis of the five largest constituents (M_2, S_2, N_2, K_1, and O_1), were prescribed at the two open boundaries of the model over the course of a 30-day simulation. The amplitudes and Greenwich phase lags obtained for these constituents from harmonic analyses of the computed elevations could then be compared with those obtained when a single constituent is prescribed on the open boundaries. The results obtained from these experiments are shown in Table 4.3 for locations representative of the Strait of Georgia (Point Atkinson) and the Puget Sound system (Seattle), respectively. Since both these locations lie inward of the semi-diurnal degenerate amphidromic systems in the inner part of Juan de Fuca Strait, reducing friction should have the effect of increasing both amplitude and Greenwich phase angle for these tides. In the case of the diurnal tides, reducing friction should lead to increasing amplitude and this should be accompanied by a decrease in phase lag.

The damping of the primary semi-diurnal M_2 tide is only moderately affected by the absence of the other constituents. This accords with the conclusion of le Prévost (1976). Secondary constituents are, however, strongly affected by the presence of a primary one. In the case of the semi-diurnal tides, S_2 and N_2, large reductions in amplitude result from the presence of the other constituents, both in the Strait of Georgia and in the Puget Sound. The effect on phase is less pronounced. For diurnal tides, the dominant effect of running the constituent alone, and thus reducing effec-

Table 4.3 Modelled amplitudes H (cm) and Greenwich phases g (°) of the five largest tidal constituents (M_2, S_2, N_2, K_1, and O_1) when prescribed in combination or individually at the model boundaries.

Location	Tides prescribed at openings	M_2		S_2		N_2		K_1		O_1	
		H	g	H	g	H	g	H	g	H	g
Point Atkinson	summed	93.5	162.1	25.4	181.2	19.2	137.2	89.8	165.2	48.2	149.0
	individually prescribed	99.2	164.9	31.7	197.5	29.3	127.6	91.3	148.8	50.2	118.8
	$M_2 + K_1$	96.2	163.0	—	—	—	—	89.8	165.6	—	—
	observed	91.8	159.3	22.9	179.9	18.4	135.4	86.2	165.8	48.3	151.7
Seattle	summed	99.8	143.1	23.8	162.7	18.6	117.0	83.6	158.0	42.8	140.1
	individually prescribed	103.1	145.9	39.4	185.2	33.7	113.0	89.1	142.0	49.7	112.9
	$M_2 + K_1$	103.1	144.1	—	—	—	—	81.6	156.9	—	—
	observed	106.6	139.1	26.3	156.9	23.8	111.0	83.6	156.0	45.9	142.6

tive friction, is to significantly reduce the phase lag both in the Strait of Georgia and Puget Sound. This is particularly evident in the case of the O_1 tide which is also a compound tide derived from the M_2 and K_1 constituents. Reasonable results may be obtained when the summed M_2 and K_1 tides are prescribed on the open boundaries, a minimum requirement for a satisfactory frictional adjustment of the model.

A particular point of interest with respect to non-linear interactions between harmonic constituents in the modelled tide concerns the small L_2 tide. Although it is an astronomical constituent included in the prescribed tidal elevations at the open boundaries of the model, its frequency is also an interaction frequency between the M_2 and N_2 constituents. The modelled L_2 tide displays a classical amphidromic distribution of co-phase and co-range lines centred in Juan de Fuca Strait (Fig. 4.8) near Race Rocks (see Fig. 1.8). This distribution, which is at least qualitatively confirmed by observations, is quite different from that of all the other semi-diurnal constituents which have distributions similar to that of the M_2 tide (Fig. 4.9). The location of the centre of this amphidromic system close to the median line of the channel indicates that it is a consequence of the superposition of Kelvin waves of similar magnitude. It would thus appear that the incoming L_2 wave from the entrance to Juan de Fuca Strait is strongly modified by a non-linear interaction term whose spatial structure is determined by the M_2 and N_2 constituents.

4.6 THE SIMULATION OF MIXED TIDES

The primary aim of this study concerns the reliable prediction of mixed tidal streams throughout the modelled region. The final trial of the model thus consisted of tidal simulations over a 30-day period for which extensive current meter data were available to check the computed tidal streams.

In addition to providing a representative sequence of tides, the computed values permit reasonable estimates of the harmonic constants of the major tidal constituents to be made. Unfortunately, however, the synodic period (183 days) of the K_1 and P_1 constituents is considerably longer than the computed records and these constants must be inferred from the amplitude and phase relations of these constituents determined from longer periods of gauge data for the appropriate locations (Dronkers, 1964). In principle, the run could have been extended to provide data sequences of sufficient length. The high computational cost did not appear to warrant this within the context of the study.

The period selected for this trial extended from March 12 to April 12 (Julian days 72-100) 1973, when the most comprehensive set of simultaneous data yet available were obtained using moored current meters. The elevations prescribed on the open boundaries of the model were predicted on the basis of 61 harmonic constituents. The

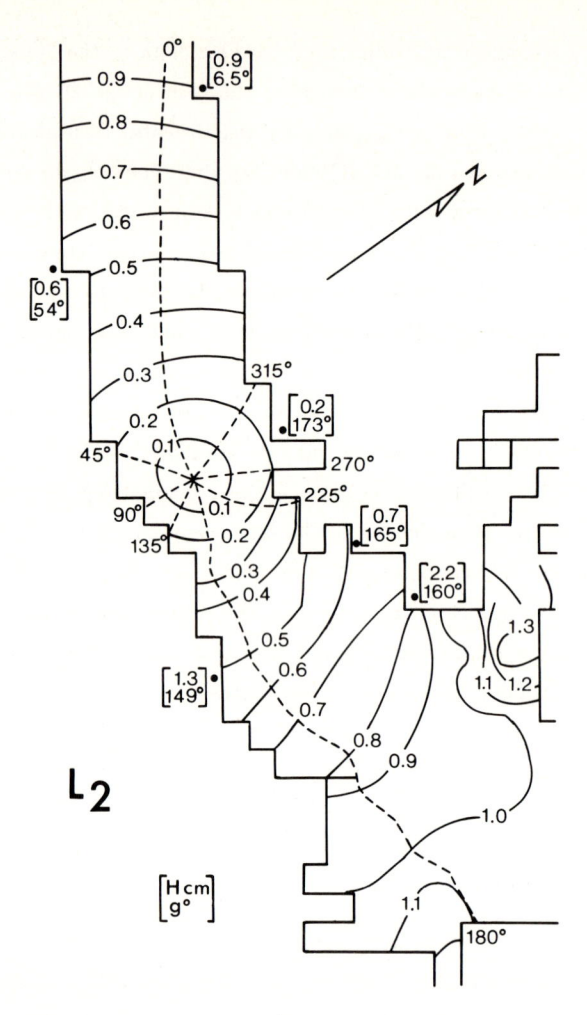

Fig. 4.8 Co-amplitude in centimetres (solid lines) and co-phase (dashed lines) in degrees for the tidal constituent L_2.

friction coefficients and depths conformed to those used in the earlier 1-wk trial of mixed tides.

The co-amplitude and co-phase charts for the major semi-diurnal M_2 and K_1 tides obtained from the modelled values are shown in Figs. 4.9 and 4.10, respectively. A degenerate semi-diurnal amphidrome critically located in the inner part of Juan de Fuca Strait at the entrance to the Puget Sound system is evident. For both constituents the character of the tide in the Strait of Georgia is essentially that of a standing wave. Large spatial changes in values of harmonic constants over the passages leading from the northern Strait of Georgia to the open sea are consistent with the occurrence there of very high tidal stream velocities.

Comparisons of the tidal elevations predicted on the basis of observed harmonic

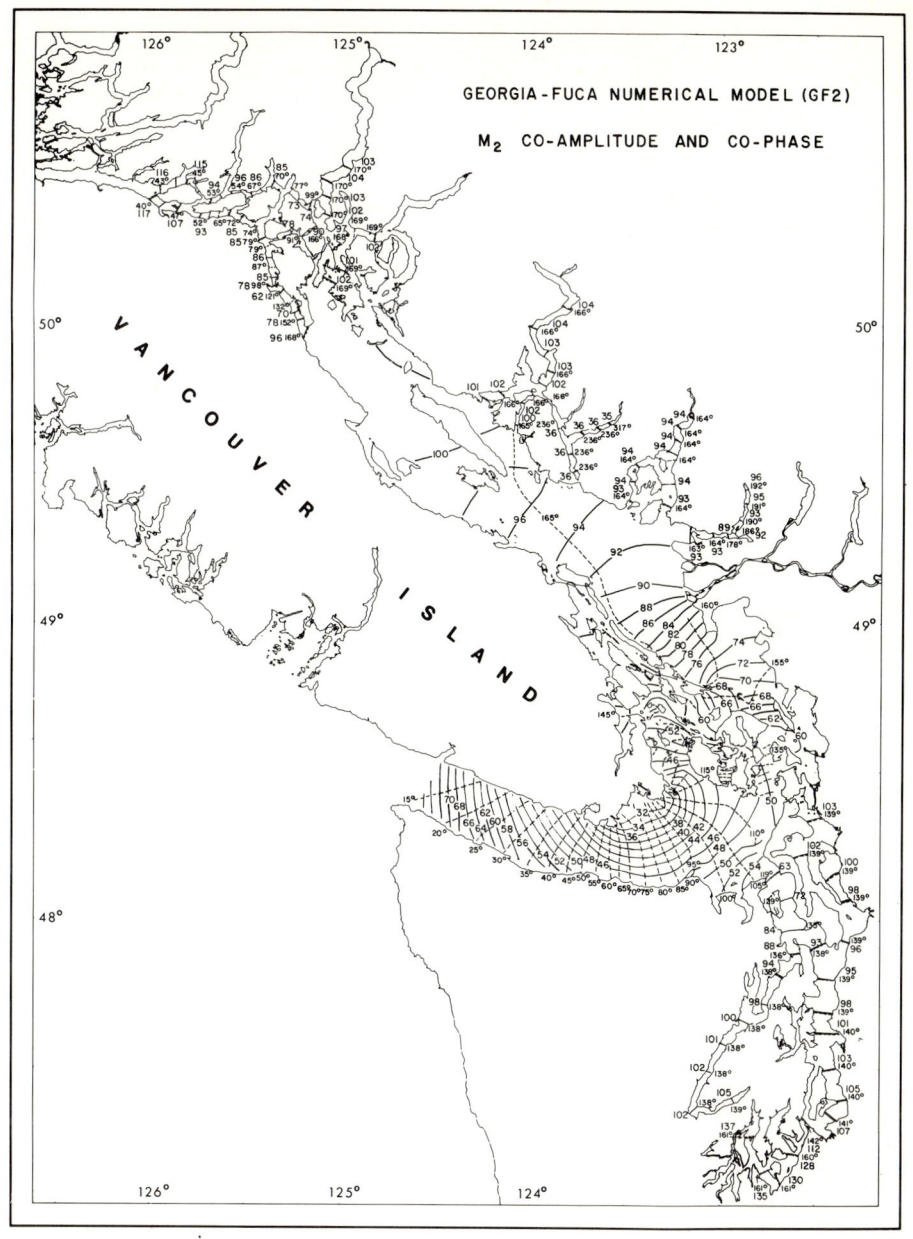

Fig. 4.9 M_2 tidal co-amplitude and co-phase lines computed by the numerical model GF2.

constants with those computed by the model for some typical locations are shown in Fig. 4.11. The excellent agreement obtained at Little River (#7993) and Point Atkinson (#7795) is representative of the performance of the model throughout the Strait of Georgia. The results shown for Fulford Harbour (#7330) are further illustrative of a similar standard of performance in the complex region of the San Juan Islands while

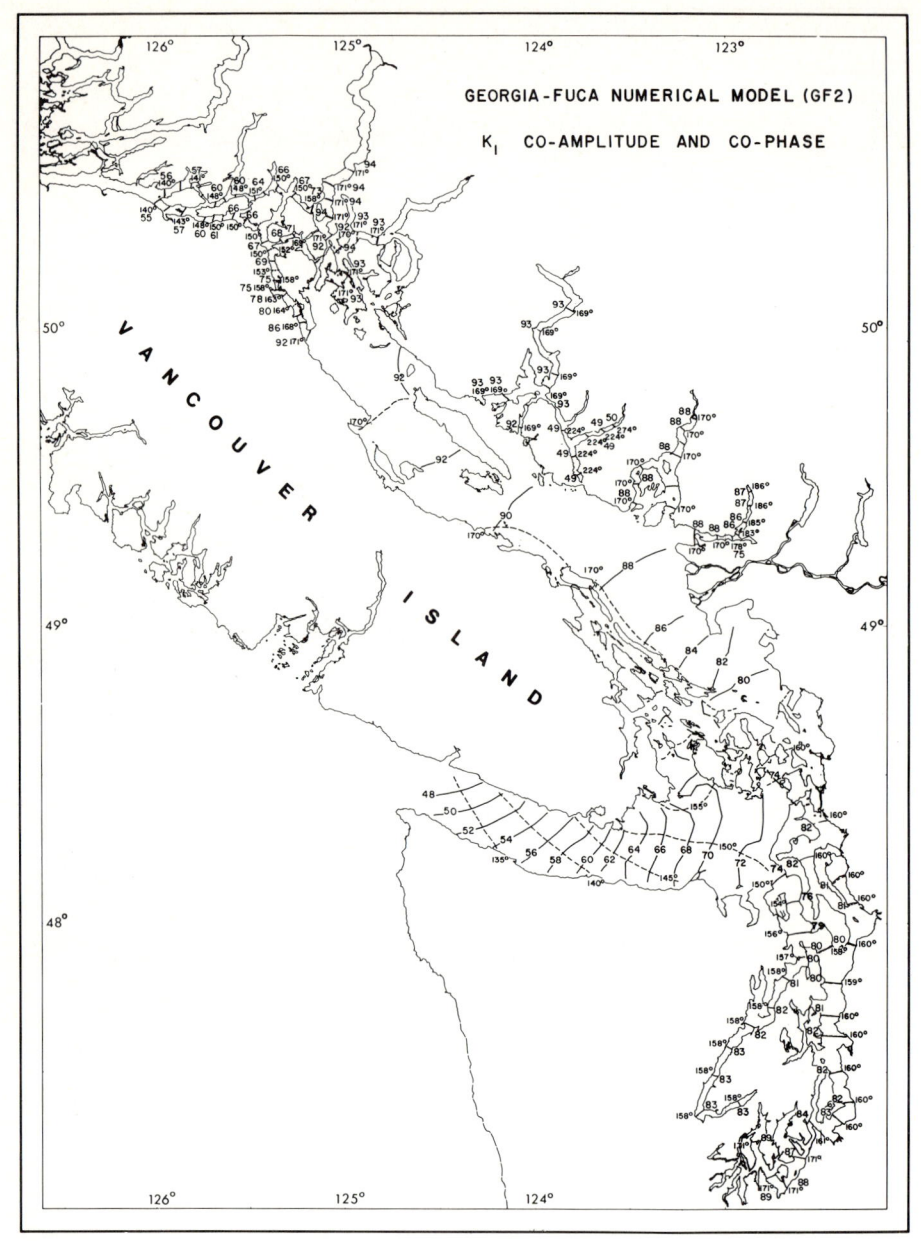

Fig. 4.10 K_1 tidal co-amplitude and co-phase lines computed by the numerical model GF2.

those obtained for Seattle (#7180) are typical for the Puget Sound system. The results at Chatham Point (#8180) are representative of the part of the modelled northern passages which lies seaward of the rapids.

Considering now the tidal current measurements in Juan de Fuca Strait, Fig. 4.12 shows comparisons of the observed tidal streams (band-pass filtered) for Station 113

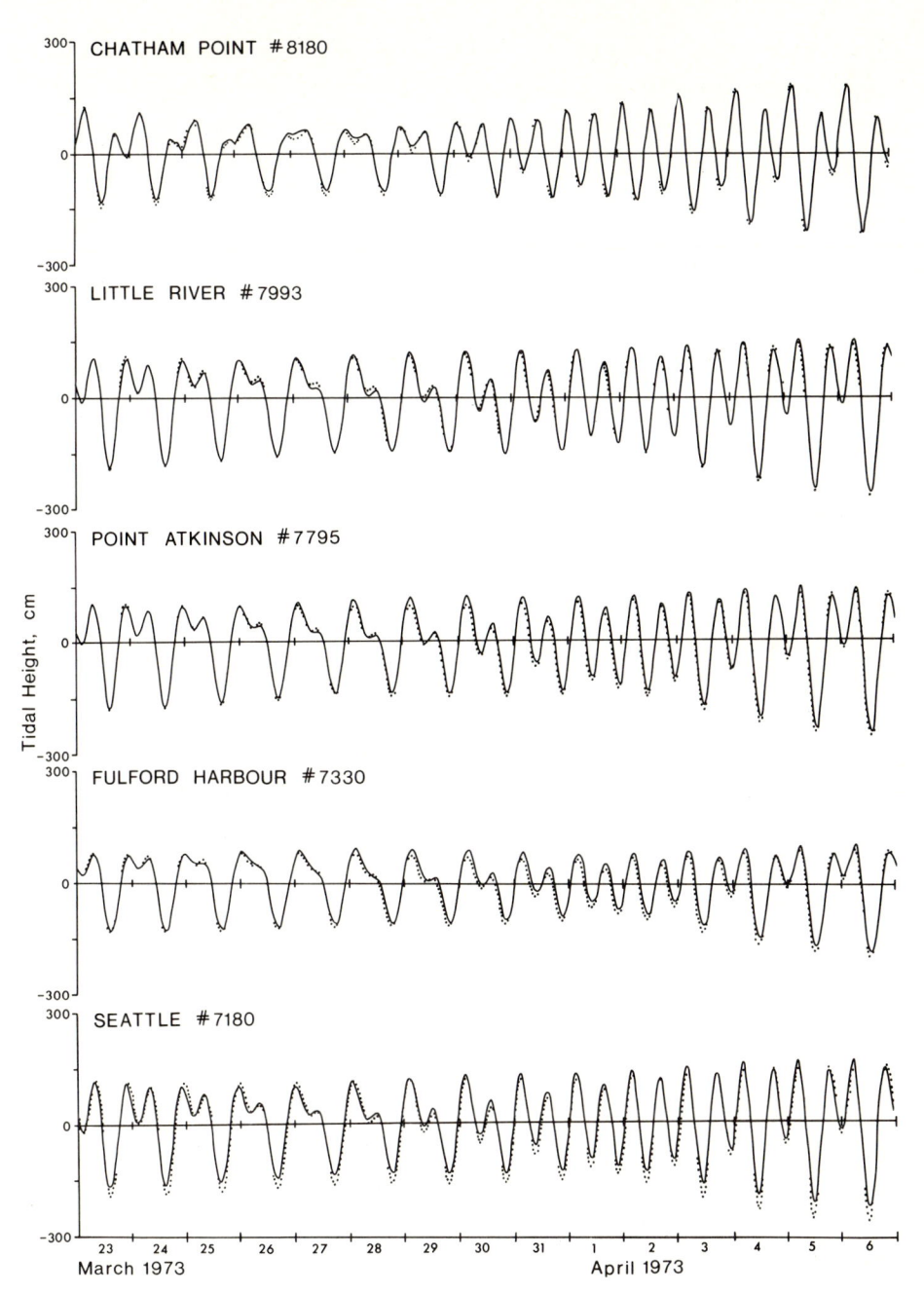

Fig. 4.11 Modelled (dotted lines) and predicted (solid lines) tidal elevations for four typical locations over a representative 15-day sequence of mixed tides (61 constituents). The locations are indicated by gauge numbers in Fig. 2.1.

(Fig. 2.1, p. 46) at depths 15, 50, 100, and 160 m with the corresponding vertically-averaged velocities computed by the model. (It will be noted that the abbreviated current meter record at Station 113 obtained near the sea bottom has been supplemented by data from the neighbouring Station 112. The profiles of tidal currents and the corresponding vertically-averaged velocities plotted from these data were essentially similar at Stations 112 and 113.) Over the period included by these velocity records marked residual flows were occurring in Juan de Fuca Strait associated with the movement of low salinity near-surface waters seaward and a compensatory intrusion of higher salinity water at depth (Fig. 1.11, p. 21) To facilitate comparison with the results from the model, the non-tidal parts of the original records were eliminated by subtracting the residuals remaining after the tidal part of the signal had been suppressed by an $A_{24}^2 A_{25}/24^2 \cdot 25$ filter (Godin, 1972). The current meter records were resolved into components conforming with the U and V directions of the model grid. In this part of Juan de Fuca Strait the U and V components may be regarded as the cross-channel and longitudinal components of velocity, respectively.

The excellent agreement shown by the observed and computed longitudinal components of velocity was typical of all the records obtained over this cross-section. No significant differences in phase occur, even near the bottom (157-160 m) where it might reasonably have been anticipated that the influence of friction could lead to a somewhat earlier turning of the streams. At 15, 50, and 100 m, the ranges contained by observed velocity extremes tend to be larger than those given by the model. Near the bottom, however, the values given by the model are larger than those observed. This accords well with the vertical distribution of velocity normally obtained in open channel flow despite the presence of vertical stratification. Further improvement in the comparisons between these velocities could be obtained by inferring depth-dependent velocities from an empirically-determined and frictionally-induced profile and the depth-mean tidal stream velocities obtained from the model (Dronkers, 1964).

The transverse U components of velocity are subject to additional modes of oscillation not present in the barotropic model. As already noted on p. 17, the stratification in Juan de Fuca Strait, which resembles that of a deep partially-mixed estuary, changed persistently over the course of the observations. At 100 m depth the observed and computed cross-channel velocities agree well with regard to phase. There appears to be, however, a marked increase in the range of observed velocities when the diurnal inequality is minimal. It is of interest to note that over days 79-81 these increased ranges at 100 m were accompanied by an approximate phase reversal of the velocities at 15 m. A concomitant sequence of CTD casts in the vicinity of the line of current meters over this period showed changes in the depths of isopycnals of roughly semi-diurnal period and displacements of up to 40 m.

Over the same period, a single current meter was moored at a depth of 18 m on the median line of the channel south of Race Rocks (Station 001, Fig. 2.1). Due primarily to the smaller cross-sectional area of Juan de Fuca Strait, the velocities are

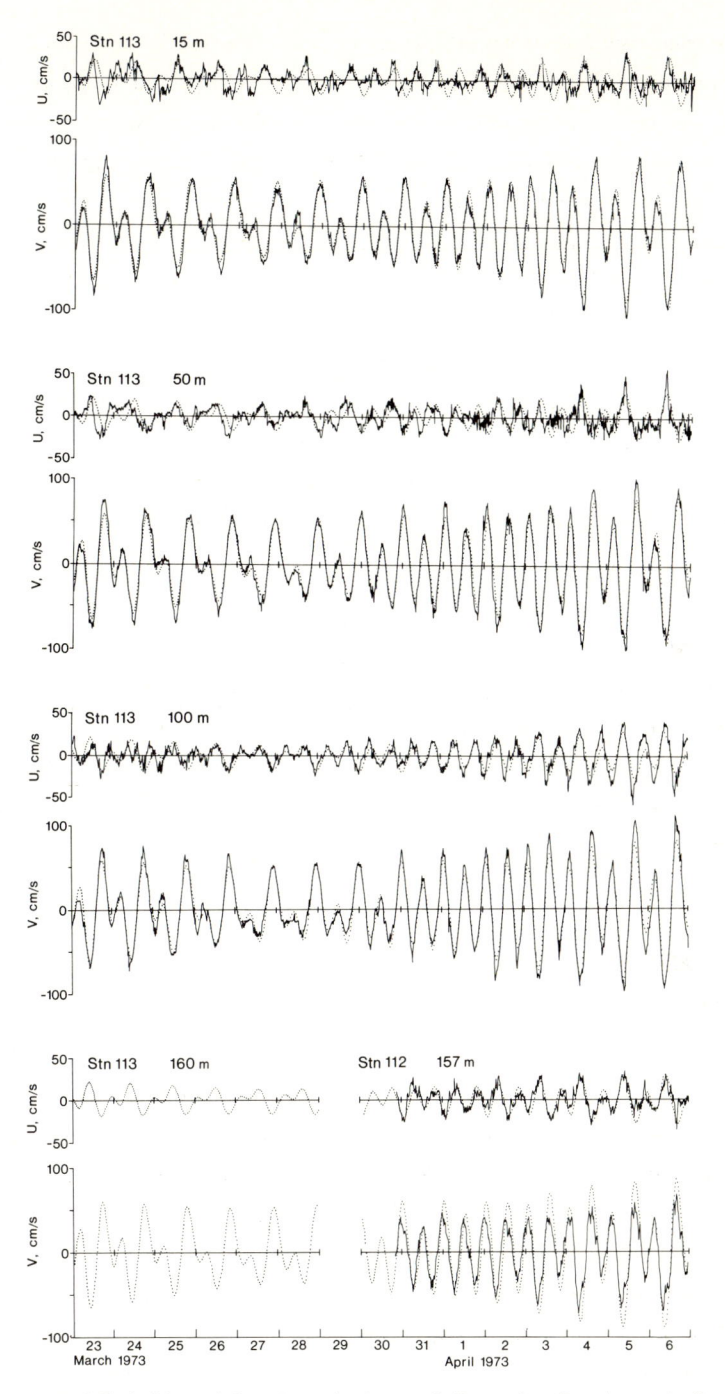

Fig. 4.12 Modelled (dotted lines) and observed (irregular lines) vertically-averaged velocity components at four different depths at Station 113 on line 11 (Fig. 2.1) (from Huggett *et al.*, 1976b).

considerably larger than at the previous location. Comparisons of the observed and computed velocities are shown in Fig. 4.13. The computed U and V components do not conform to the longitudinal and cross-channel directions but rather to north-south and east-west axes. In general, the agreement between the phases of the observed and computed velocities is excellent. As anticipated, the observed velocity ranges are larger than the computed vertically-averaged velocities. Again, the use of an empirically-determined correction generally descriptive of open channel flow would substantially improve the agreement.

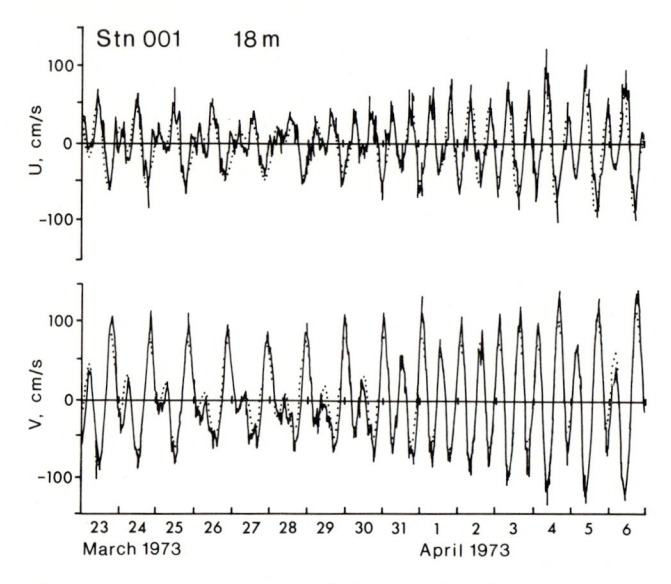

Fig. 4.13 Comparison of the modelled (dotted lines) and observed (solid lines) vertically-averaged velocity components at a depth of 18 m on the median line of the channel south of Race Rocks (Station 001, Fig. 2.1).

A further array of Aanderaa current meters (line 1, Fig. 2.1), was moored over a cross-section of Johnstone Strait near the northern boundary of the model where an average velocity over a cross-section is calculated. This computed velocity was corrected for the slightly different cross-sectional area of the location where the current meters were moored. The computed velocities along this channel indicate a negligible change in the phase of the tidal streams between the two locations. Similar to the case of the Juan de Fuca Strait section, marked residual seaward flows were observed near the sea surface accompanied by compensatory inward flows near the bottom. Density stratification in this channel is small. Comparisons of the observed tidal streams at depths 15, 75, and 225 m for Station 003 centrally located in Johnstone Strait are shown in Fig. 4.14. At all three depths the phasing of the observed streams is in excellent agreement with that obtained from the model. Excellent agreement is also obtained between the observed and computed velocity ranges at 15 and 75 m. At 225 m, the observed velocity ranges are appreciably larger than those computed. Similar results were obtained from other current meters in the array. It has been suggested that a longitudinal internal seiche motion associated with the weak density stratification and excited by the surface tide may be responsible for this anomalous vertical distribution of velocity (Thomson and Hugget, 1980). Subject to this reservation, however, it would appear that velocities computed at the northern boundary of the model are in good accord with observations.

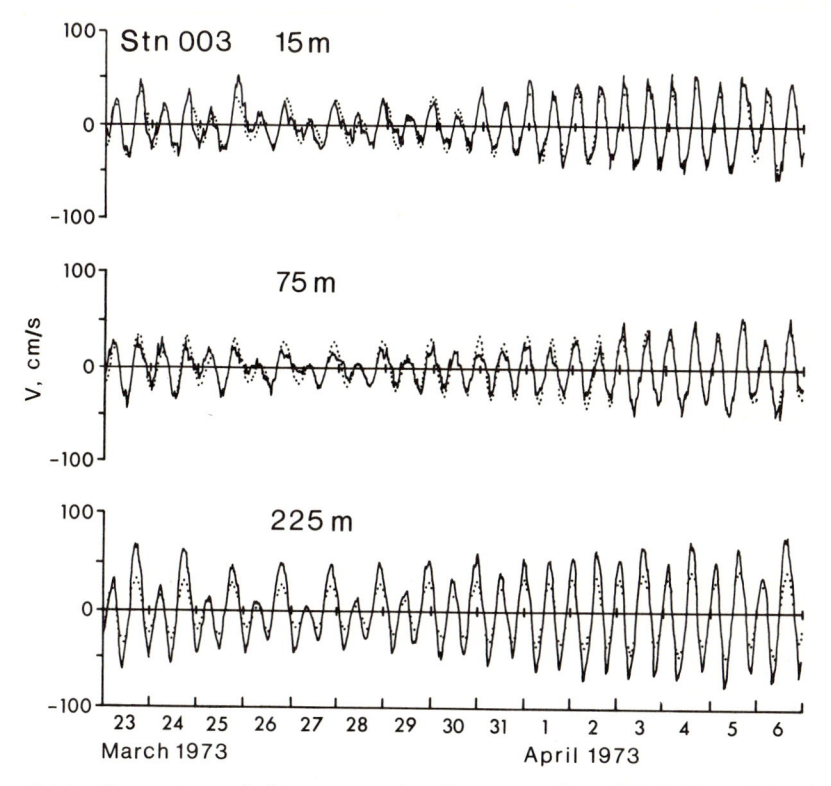

Fig. 4.14 Comparison of the cross-sectionally averaged modelled (dotted lines) and observed (solid lines) tidal velocity components at three different depths at a central location in Johnstone Strait for a representative 15-day period (from Huggett et al., 1976b).

Indicative of the capacity of the model to predict the fast tidal streams in one of the narrow passes, Fig. 4.15 shows the tidal velocities computed by the model for Seymour Narrows with those predicted by the Canadian Hydrographic Service. Again, the observed and computed values are in good agreement. It will be noted that the semi-diurnal character of these oscillations is much more pronounced than in the case of the streams in, for example, Juan de Fuca Strait (Fig. 4.12) where the streams have a pronounced diurnal inequality similar to that of the surface tides. This enhanced semi-diurnal character, which is also characteristic of fast tidal flows in other passes in the model, is readily attributable to the rapid increase in values of non-linear terms with increasing tidal range and larger velocities for such locations.

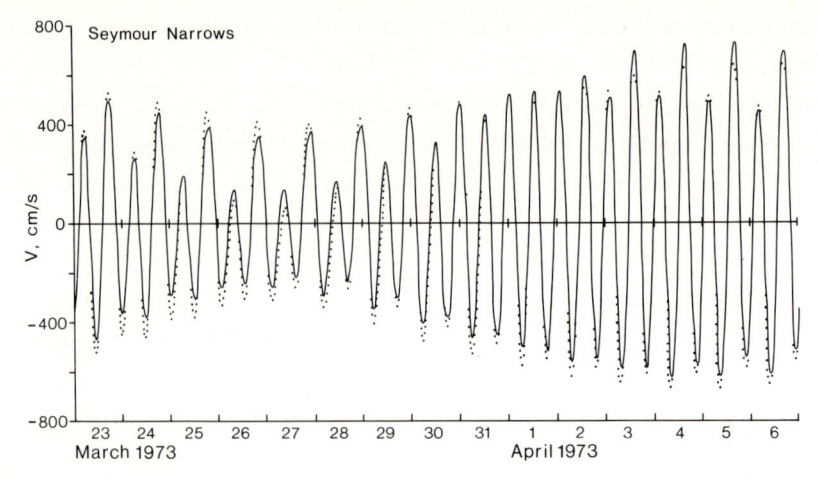

Fig. 4.15 Comparison of the cross-sectionally averaged tidal velocities in Seymour Narrows from the model (dotted line) with those predicted (solid line) on the basis of harmonic constants for a representative 15-day period.

A final point of interest concerns the actual detailed distribution of surface elevations at high water in the northern Strait of Georgia where the tide which has propagated through Juan de Fuca Strait and the Strait of Georgia encounters the smaller, though dynamically significant, tide that enters through the northern opening. This is illustrated in Fig. 4.16 where the elevation maxima are clearly confined to the northeastern corner of the Strait and its associated inlets.

In referring to the early one-dimensional numerical model trials, it was noted that a realistic anticipation of performance for a barotropic numerical model of the system would involve simulating the tidal harmonic constants to within 5 cm and 5° in the Strait of Georgia and Puget Sound. It is thus of interest to compare the constants deriving from harmonic analyses of the computed elevations over a 30-day period for some typical locations with those observed (Table 4.4). In these analyses, the separation of pairs of constituents having synodic periods in excess of 30 days has been done by inference based on known amplitude and phase relationships from 1 yr of tide gauge observations at the corresponding locations.

When harmonic analyses are applied to 30-day records obtained from the tide gauges and to the same period of computed elevations for corresponding locations in the model, excellent agreement is obtained between the respective harmonic constants for the six main constituents. This implies that the 'contamination' of these constituents by others which cannot be separated out on the basis of this record length is essentially similar in both the model and in nature.

In the case of the Strait of Georgia (Little River and Point Atkinson) and Puget Sound (Seattle) agreement has been obtained within 6 cm and 6°. At Fulford Harbour there is a phase discrepancy between the observed and calculated phases of the M_2 tide

Table 4.4 Comparisons of amplitudes H (cm) and phases g (°) of the six dominant constituents obtained from observations at four typical locations with those derived from harmonic analyses of 30 days of computed elevations in the model. The K_1 and P_1 constituents have been inferred using amplitude and phase relationships based on 1 yr of observations.

Location	Tide gauge number	M_2		S_2		N_2		K_1		O_1		P_1	
		H	g	H	g	H	g	H	g	H	g	H	g
Little River	7993												
observed		99	161	25	182	22	138	90	167	49	152	29	166
calculated		101	167	26	184	22	140	91	171	51	151	30	172
Point Atkinson	7795												
observed		93	158	23	179	19	135	87	165	47	151	27	165
calculated		92	164	23	181	20	137	86	170	49	150	29	172
Fulford Harbour	7330												
observed		59	141	14	157	13	116	76	159	42	146	24	158
calculated		59	150	14	163	13	119	76	165	45	146	25	166
Seattle	7180												
observed		107	140	26	158	21	113	83	157	46	144	25	155
calculated		101	140	24	158	21	117	82	161	44	140	25	159

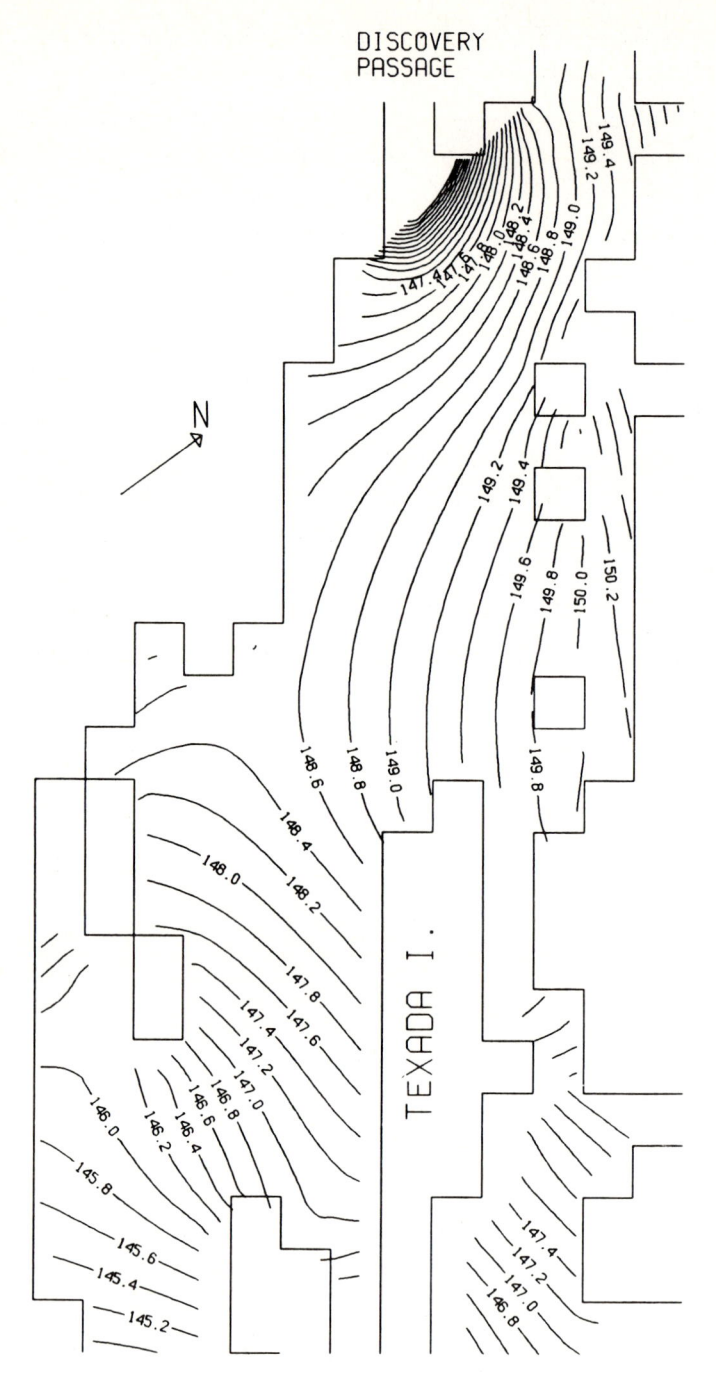

Fig. 4.16 Distribution of surface elevations (cm) at high water in the northern Strait of Georgia.

amounting to 9°. This is, however, a channel where the tidal phase changes rapidly with distance.

Based on the results of sensitivity trials described earlier in stage four of the development of the model, the discrepancies between the computed and observed M_2 tidal amplitudes and phases in the Strait of Georgia and Puget Sound could probably be reduced by manipulating the frictional dissipation and depths in Haro Strait. It appeared, however, that further adjustments of the model were scarcely warranted and that a finer grid schematization of the San Juan and Gulf Island region was required.

4.7 VOLUME TRANSPORTS

A frequent practical question concerns tidal volume transports. The time-varying volume fluxes over a month of 'representative tides' have been calculated for major geographic divisions of the model (Fig. 4.17). Also shown are the energy transports across these sections to be discussed later. At cross-section a, the volume flux entering Juan de Fuca Strait is determined. Volume fluxes illustrate the proportioning of the incoming tidal complement between the cross-sections Puget Sound (b) and the passages leading into the Strait of Georgia, Rosario Strait (c), Middle Channel (d), and Haro Strait (e). The net tidal flux entering the Strait of Georgia from the northern passages is determined over cross-section f. The volume fluxes appropriate to these various cross-sections are shown in Fig. 4.18.

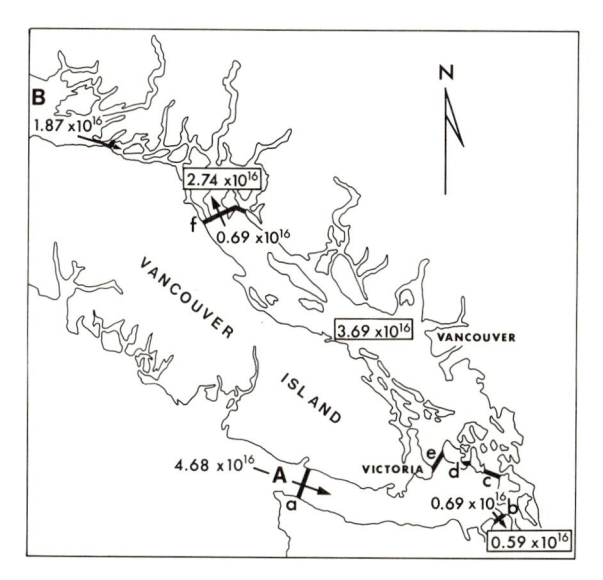

Fig. 4.17 Location of cross-sections where modelled volume transports were calculated. Net energy fluxes (erg/s) and directions through cross-sections and dissipations (boxed values) are indicated over regions between sections and coastlines. The values are obtained from a 30-day model run employing 61 tidal harmonic constituents.

The sign convention for the direction of flow determined by the model GF2 is illustrated in Table 4.5. The transport across the middle of the Strait of Georgia is

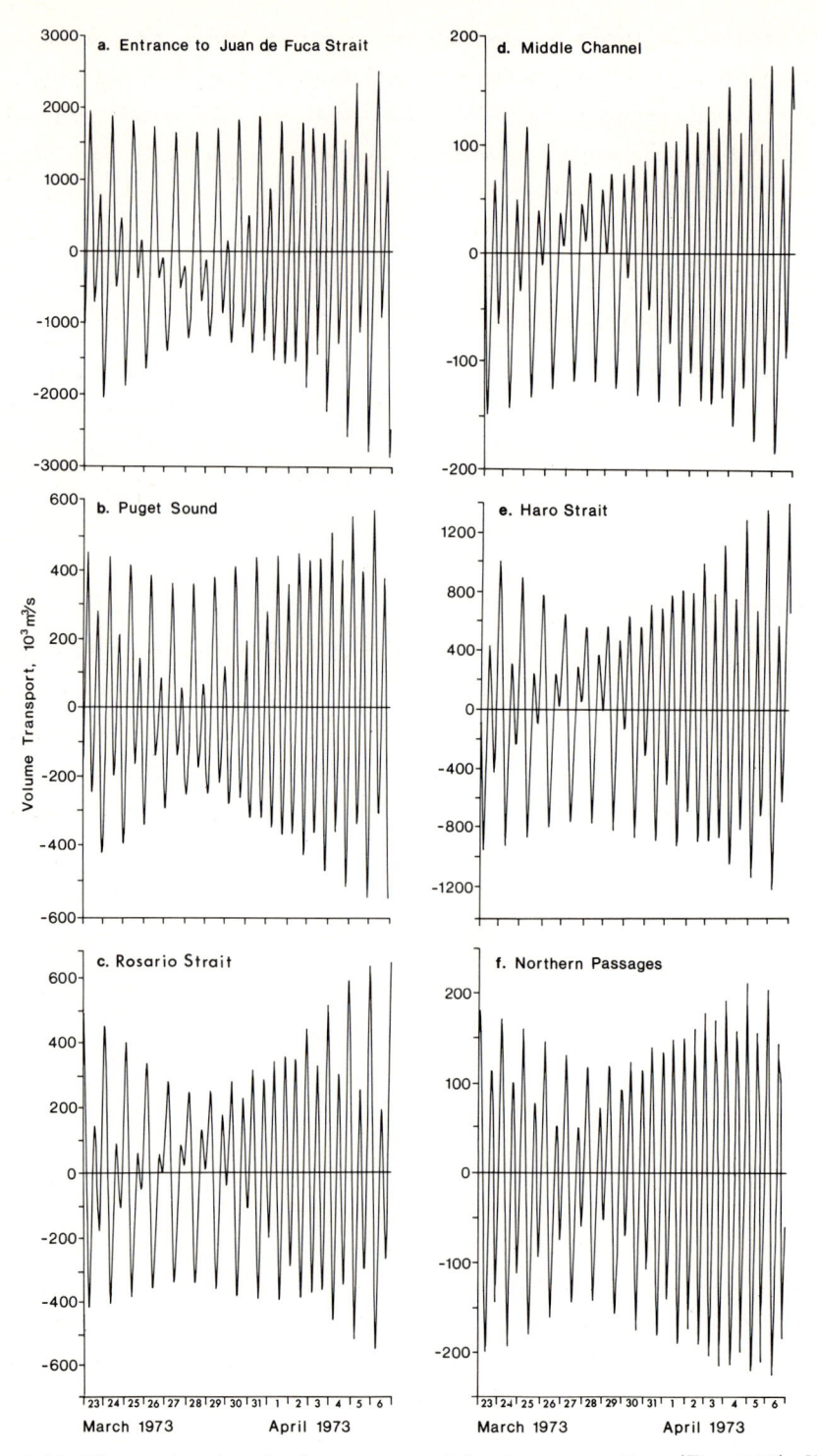

Fig. 4.18 Time series plot of volume transport for six cross-sections (Fig. 4.17). Note differences in scale.

about one third the transport at the entrance to Juan de Fuca Strait.

Table 4.5 Sections of the Georgia/Fuca system used in the calculation of volume transports. Flow towards a destination is indicated by a positive $(+)$ sign and flow away from is shown by a negative $(-)$ sign.

Origin	Sign Convention	Destination
Juan de Fuca Strait entrance, a	+	Strait of Georgia
Puget Sound, b	+	Puget Sound
Rosario Strait (southern end), c	+	Strait of Georgia
Middle Channel, d	−	Strait of Georgia
Haro Strait (southern end), e	−	Strait of Georgia
Northern passages, f	+	Strait of Georgia

4.8 ENERGY CALCULATIONS

Energy conservation requires that the net flux of energy into the system be balanced by frictional dissipation. Following formulation of the energy conservation equation applicable to the numerical schemes, finite-difference formulations employed in computing the energy balance for the combined system of one- and two-dimensional models are described. Any contribution by direct tidal forcing is omitted and will be discussed separately in Section 4.8.2.

The exact energy conservation equation (derived from the original equations of motion) is

$$\frac{\partial}{\partial t}\left[\left(\frac{u^2+v^2}{2}\right)(\zeta+h)+\frac{g(\zeta^2-h^2)}{2}\right]=-C(u^2+v^2)^{3/2}-\nabla\cdot\left[(\zeta+h)\left(\frac{u^2+v^2}{2}+g\zeta\right)\vec{u}\right].$$

The level of zero potential energy is taken as the undisturbed free surface ($\zeta = 0$). The quantity in the divergence term on the right hand side may be re-written as

$$\vec{u}\,(\zeta+h)\left(\frac{u^2+v^2}{2}\right)+\vec{u}\,(\zeta+h)g\zeta=\underbrace{\vec{u}\,(\zeta+h)\left(\frac{u^2}{2}+\frac{v^2}{2}\right)}_{A}+\underbrace{\vec{u}\,g\left(\frac{\zeta^2-h^2}{2}\right)}_{B}+\underbrace{\frac{\vec{u}\,g(\zeta+h)^2}{2}}_{C}$$

where terms A, B, and C are the transport of kinetic energy, the transport of potential energy, and the rate of work done on the open boundary, respectively. Thus, there are two ways of viewing the flux at the boundary. The left hand side of the above equation expresses it as the transport represented by $u(\zeta+h)$ of the mechanical energy per water particle, $(u^2/2+v^2/2+g\zeta)$. The right hand side represents the flux as the advection of kinetic plus potential energy for the entire water column, plus the rate of work on the boundary. Thus, the energy equation may be written as

$$\frac{\partial E}{\partial t}=-F-\nabla\cdot\vec{Q}$$

where E is the total energy of a fluid column, F is the rate of frictional dissipation, and $\vec{Q}$ is the flux of potential and kinetic energies through its boundaries. Integrating over the modelled region,

$$\frac{\partial}{\partial t}\int E\,dA = -\int F\,dA - \int \vec{Q}\cdot\hat{n}\,ds$$

where $\hat{n}$ is an outward-pointing unit normal vector and ds an element of length along the bounding curve.

Values of kinetic energy, potential energy, and dissipation rate were calculated for each mesh. Thus, the kinetic energy term is given by the centred finite-difference equivalent of

$$\left(\frac{u^2}{2}+\frac{v^2}{2}\right)(\zeta+h)\Delta^2,$$

the potential energy term by

$$\frac{g\zeta^2\Delta^2}{2},$$

and the dissipation rate by

$$C(u^2+v^2)^{3/2}\Delta^2.$$

The factor Δ^2 which is the area of a mesh, converts energy densities into the actual energy for that mesh grid. The finite-difference representations are

$$\mathrm{KE}_i = \frac{1}{2}\left(\frac{u_i^2+u_{i-1}^2}{2}+\frac{v_i^2+v_{i-n}^2}{2}\right)(h_i+\zeta_i)\Delta^2,$$

$$\mathrm{PE}_i = \frac{g\zeta_i^2\Delta^2}{2},$$

$$\mathrm{DISS}_i = \mathrm{FCON}_i\left(\frac{u_i^2+u_{i-1}^2}{2}+\frac{v_i^2+v_{i-n}^2}{2}\right)^{3/2}\Delta^2.$$

There is a small inconsistency. FCON_i, the friction coefficient, is considered to be defined at the centre of a mesh. In the two-dimensional model, it entered into calculations of only u_i and v_i. Here it is used in combination with u_i, u_{i-1}, v_i, v_{i-n}. This has significance when it is realized that FCON_i was larger in the Gulf and San Juan Island passages than elsewhere in the system. Certain modifications were necessary when the flow through a mesh side occupied less than the entire side in the various passes. It was assumed that the contribution to the kinetic energy was $(u_i^2 A/4)\Delta$ where A is the cross-sectional area of the passage opening and is used instead of $(h+\zeta)\Delta$. For all one-dimensional channels other than Puget Sound, the computation employed an implicit method (Dronkers, 1969) (Fig. 4.19) where the variables are

u_i = area-averaged velocity = Q_i/A_i,
Q_i = discharge through the ith section,
ζ_i = elevation of the water surface,
A_i = cross-sectional area,
H_i = average depth over the channel segment Δs_i,

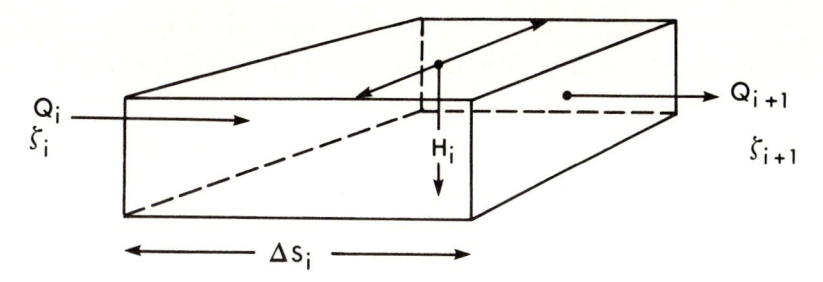

Fig. 4.19 Computational grid for the Dronkers (1969) one-dimensional scheme.

S_i = surface area,
Δs_i = length of channel segment.

The finite-difference forms of the energy equation terms are

$$\text{PE}_i = \frac{g}{2} S_i \left(\frac{\zeta_i + \zeta_{i+1}}{2} \right)^2,$$

$$\text{KE}_i = \frac{1}{2} \left[\frac{u_i A_i + u_{i+1} A_{i+1}}{A_i + A_{i+1}} \right]^2 \left(\frac{A_i + A_{i+1}}{2} \right) \Delta s_i,$$

$$\text{DISS}_i = \text{FK}_i \left[\left(\frac{u_i A_i + u_{i+1} A_{i+1}}{A_i + A_{i+1}} \right)^2 \right] \frac{3}{2} \Delta s_i \left(\frac{A_i + A_{i+1}}{2} \right) \frac{1}{H_i}.$$

For Puget Sound, a scheme similar to the two-dimensional one is used. At a triple junction it is assumed that each inflow contributes one third of the kinetic energy. The energy flux at an open boundary in the two-dimensional region is calculated according to the following expression,

$$u_i \left(\frac{\text{KE}_i}{\Delta^2 (\zeta_i + \zeta_i)} + g\zeta_i \right) (\zeta_i + h_i) \Delta.$$

For a boundary opening into a one-dimensional channel, the corresponding expression is

$$u_i \left(\frac{u_i^2}{2} + g\zeta_i \right) A_i.$$

Figure 4.17 shows the average values of fluxes and dissipation for the various regions of the model. The overall rate of energy dissipation is 7.02×10^{16} erg/s and the overall flux across the open boundaries is 6.55×10^{16} erg/s. These values should be equal but the agreement is considered reasonable in view of the approximations in the finite-difference form of the energy equation. Thus, for example, the frictional dissipation at a point was computed using velocities at four grid points whereas the friction term in the original equations involved velocities from 12 grid points. A particularly interesting feature is the net energy flux from the Strait of Georgia into the northern passages system.

Variations with time of terms in the energy conservation equation are conveniently discussed with reference to the time series of elevations at Point Atkinson (see Fig. 4.11). Figure 4.20a–g shows the potential, kinetic, mechanical total energies, the flux of energy at the open boundaries, both as computed and low-pass filtered, respectively, the energy dissipated over the interior of the system, and the difference between the computed flux of energy and dissipated energy. During periods of large tidal ranges and consequently upon the large diurnal inequality, there is approximately one daily peak in the potential energy time series (Fig. 4.20a). Each of these large peaks is preceded and followed by large peaks in the kinetic energy (Fig. 4.20b). This results in a large daily broad peak in the total energy time series (Fig. 4.20c) which is readily explicable on the basis of the tide at Point Atkinson. The difference between mean sea level and low low water is frequently much greater than the difference between high water and mean sea level. Higher low water is often close to mean sea level. Thus the extreme low waters are the source of the big peaks in potential energy (due to the term involving $\zeta^2 - h^2$ in the energy conservation equation) and associated with each low there is a large ebb and a large flood. The flux at the entrance to Juan de Fuca Strait (Fig. 4.20d) is predominantly positive with a slowly varying mean (Fig. 4.20e). Considering next the dissipation, the anticipated correlation with kinetic energy is evident. It is possible to estimate the temporal rate of change of total barotropic tidal energy in the system (dE/dt) (Fig. 4.20g) in two ways. The signal $E(t)$ can be numerically differentiated with respect to time or the difference between the flux and the frictional dissipation ($Q_T - F_T$) at the boundary can be calculated. The agreement between these two curves is excellent and provides a good check on numerical approximations employed.

4.8.1 Alternative Estimates for the Energy Balance

It is of interest to compare the estimates of energy flux obtained from the numerical model with those estimated from harmonic constants (also from the numerical model). The rate of work done by a single tidal constituent on an open boundary can be calculated by the formula

$$\overline{gh u \zeta} = \frac{\rho g h u_0 \zeta_0 \cos(g_z - g_u)}{2}$$

where u_0 and ζ_0 are the amplitudes of the velocity and elevation and g_u and g_z are their respective Greenwich phases.

The flux estimates for four cross-sections are shown in Table 4.6. The agreement appears reasonable since a 1-month record should include the synodic periods for constituents considered. Furthermore, most of the flux is due to the quantity $\overline{gh\zeta u}$, the pressure-velocity work term and the advection of the potential energy part of the energy transport (*i.e.*, the advection of kinetic energy is considered negligible).

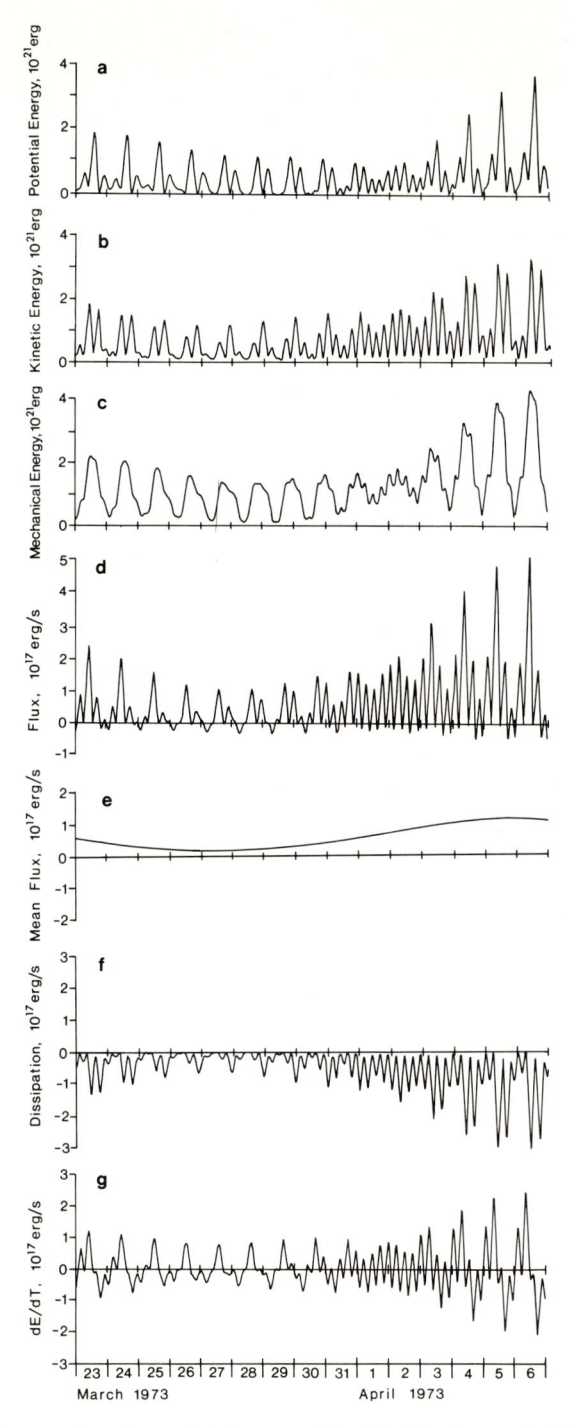

Fig. 4.20 Time series plots of (a) potential, (b) kinetic, and (c) total mechanical energies in the GF2 model. Also shown are the fluxes at the open boundary, both computed (d) and low-pass filtered (e) and the rate of energy dissipation (f). Time series plots of the rate of change of mechanical energy and the difference between energy flux at open boundaries and dissipation are superimposed for comparison (g).

Table 4.6 Energy fluxes for dominant tidal constituents and total mean flux determined directly from modelled values at the boundary openings, the entrance to the Puget Sound system, and between the Strait of Georgia and the northern passages. Fluxes are in 10^{15} erg/s.

Location	M_2	S_2	N_2	K_1	O_1	P_1	(constituents) Total	time series Computed
Juan de Fuca Strait (positive indicates flow into the Strait)	32.10	2.50	2.00	11.90	3.30	1.20	53.0	46.8
Puget Sound (positive indicates flow into the Sound)	4.72	0.41	0.26	0.72	0.16	0.02	6.3	5.9
Northern passages/Georgia connection (negative indicates flow into the passages)	−4.71	−0.31	−0.17	−1.94	−0.60	−0.20	−7.9	−6.9
Pacific entrance to northern passages (positive indicates flow into the passages)	15.56	1.42	0.68	−0.12	0.21	−0.01	17.3	18.7

Some interesting points can be noted from the calculations. The flux into Puget Sound is dominated by the M_2 tide while the K_1 tide assumes a minor role in comparison to its relative importance at the entrance to Juan de Fuca Strait. Also, the agreement in sign between the two methods of calculation for the flux at the northern passages/Strait of Georgia junction is reassuring. Finally, the flux at the Pacific entrance to the northern passages is almost entirely semi-diurnal.

4.8.2 Work Done by or Against the Equilibrium Tide

In the development of GF2 it was determined that the effect of equilibrium tidal forcing was negligible. A generalization of this conclusion is presented in this section where the direct contribution of the energy from the tide-generating potential is calculated and found to contribute little to the overall energy balance.

The equilibrium force may be included in the equations of motion resulting in (for the x component of momentum)

$$\frac{\partial u}{\partial t} + u\frac{\partial u}{\partial x} + g\frac{\partial}{\partial x}(\zeta - \zeta_{\text{eq}}) + \frac{ku|u|}{h+\zeta} = 0$$

where ζ_{eq} is the equilibrium tidal displacement of the water surface valid for stationary gravitational attraction. Thus, in the absence of velocities $u, v = 0$, the solution to the above equation is $\zeta = \zeta_{\text{eq}}$. The water surface forms the equilibrium ellipsoid. The energy equation is modified to become

$$\frac{\partial}{\partial t}(\text{KE} + \text{PE}) = -F - \nabla \cdot Q + g(\zeta + h)\vec{u} \cdot \nabla \zeta_{\text{eq}}$$

where the last term represents the rate of work done by or against the equilibrium tide EW. This term is integrated over the entire two-dimensional grid giving

$$EW = \int\int g(\zeta + h)\,\vec{u} \cdot \nabla \zeta_{\text{eq}}\, dx\, dy.$$

A more convenient expression results from integrating the above expression by parts,

$$EW = \int g(\zeta + h)\,\vec{u} \cdot \hat{n}\zeta_{\text{eq}}\, ds - \int\int g\zeta_{\text{eq}}\nabla \cdot (\vec{u}\,(\zeta + h))dx\, dy$$

where $\hat{n}$ is an outward-directed unit vector and ds is an element of length along the open boundary. However,

$$\nabla \cdot (\vec{u}\,(\zeta + h)) = -\frac{\partial \zeta}{\partial t},$$

hence,

$$EW = \int_s g(\zeta + h)\,\vec{u} \cdot \hat{n}\zeta_{\text{eq}}\, ds + \int_{\text{area}} g\zeta_{\text{eq}}\frac{\partial \zeta}{\partial t}\, dx\, dy.$$

A satisfactory representation of energy fluxes may be obtained by adding up the contribution from individual constituents. For the equilibrium tide, only the M_2 and K_1 constituents are considered and the work associated with each constituent is calculated separately. By a proper choice of time origin, the relevant input functions can be written as

$$\zeta_{eq} = \zeta_0 \cos(\omega t),$$
$$\zeta = A \cos(\omega t - g + pL - \omega S)$$
$$= A \cos(\omega t - \kappa_e),$$
$$u = U \cos(\omega t - \kappa_v),$$

where κ is used to express the tidal phase (i.e., the phase between the observed and equilibrium tides at the place of observation). Therefore, the first term of the equilibrium tide work (denoted the flux term) for a single constituent is

$$\sum_i \tfrac{1}{2} U_i \zeta_0 g h_i \Delta \cos \kappa_i v_i$$

where i is summed over all open boundary meshes and

h_i = water depth,
U_i = velocity amplitude,
κ_{vi} = local phase difference for velocity (calculated for a chosen g_i so that flow out of the grid is positive),
κ_{ei} = local phase difference for elevation.

The term representing the area integral (denoted the local generation term) becomes

$$\sum_i \tfrac{1}{2} g \zeta_i \omega \zeta_0 \sin \kappa_{ei} \Delta^2$$

where ζ_i is the elevation amplitudes. The results obtained from these calculations are shown in Table 4.7. The net dissipation, by comparison is 36.9×10^{15} erg/s.

Table 4.7 M_2 and K_1 net energy fluxes obtained from the fluxes subtracted from the local generation ($\times 10^{15}$ erg/s). Amplitude is in centimetres.

Tidal Constituent	Equilibrium Amplitude	Local Generation	Flux	Net Flux
M_2	10.5	3.81	−3.77	0.04
K_1	14.2	1.76	−1.77	−0.01

There are two points of interest. First, net the equilibrium work is a very small part of the energy balance. Secondly, the generation within the system is not small. However, the energy generated is almost totally propagated out of the system. This has interesting implications for models of the tide in the Pacific Ocean. Although the Georgia/Fuca system is a net dissipator of energy, it is also a source of energy from the forcing potential of the equilibrium tide.

5 THE FINE GRID MODEL: GF3

5.1 INTRODUCTION

The most important region where strong tidal streams occur includes Juan de Fuca Strait, the southern Strait of Georgia, and the connecting passages among the San Juan Islands. A 2-km mesh vertically-integrated model of this part of the system was constructed and operated by prescribing on its open boundaries interpolated elevations obtained from the earlier overall model GF2. It was thus possible to obviate the major difficulties due to proximate boundary openings that were encountered in earlier work.

5.2 DESCRIPTION OF THE MODEL

The scheme used in the fine grid model is shown in Fig. 5.1 where each square corresponds to the four sides of a mesh as illustrated in Fig. 4.1 (p. 72). Coastlines were simulated by mesh sides where the normal component of velocity was set equal to zero. Similarly, a narrow island or land barrier was depicted by a single closed mesh side within the interior of the sea. In some instances ambiguities existed as to which side of a mesh best represented a part of the coastline. Such ambiguities were resolved by ensuring that regional surface areas in the model accorded with those obtained from hydrographic charts.

The average depth over a mesh was specified at each elevation point within the interior of the sea. In regions where banks might be uncovered at low water, notably near the mouth of the Fraser River, an arbitrary depth of 5 m was assigned. This was considered adequate at this stage of development of the model. Any realistic numerical simulation of flows over the shallow banks near the Fraser River mouth would also have to take account of the river discharge and the motion of the salt wedge. A number of additional passes were included within the modelled region such as those leading from the region of the Gulf Islands into the Strait of Georgia. In some cases these were simulated by one-dimensional channels inserted into the two-dimensional scheme. Earlier work with the overall tidal model showed that the comparatively small flows through these passes are not critical to the proper simulation of the tidal streams in the main conveying channels of the system although they do possess considerable local importance. Detailed adjustments of the flows through these passes will be undertaken when pertinent field observations, presently planned, are completed.

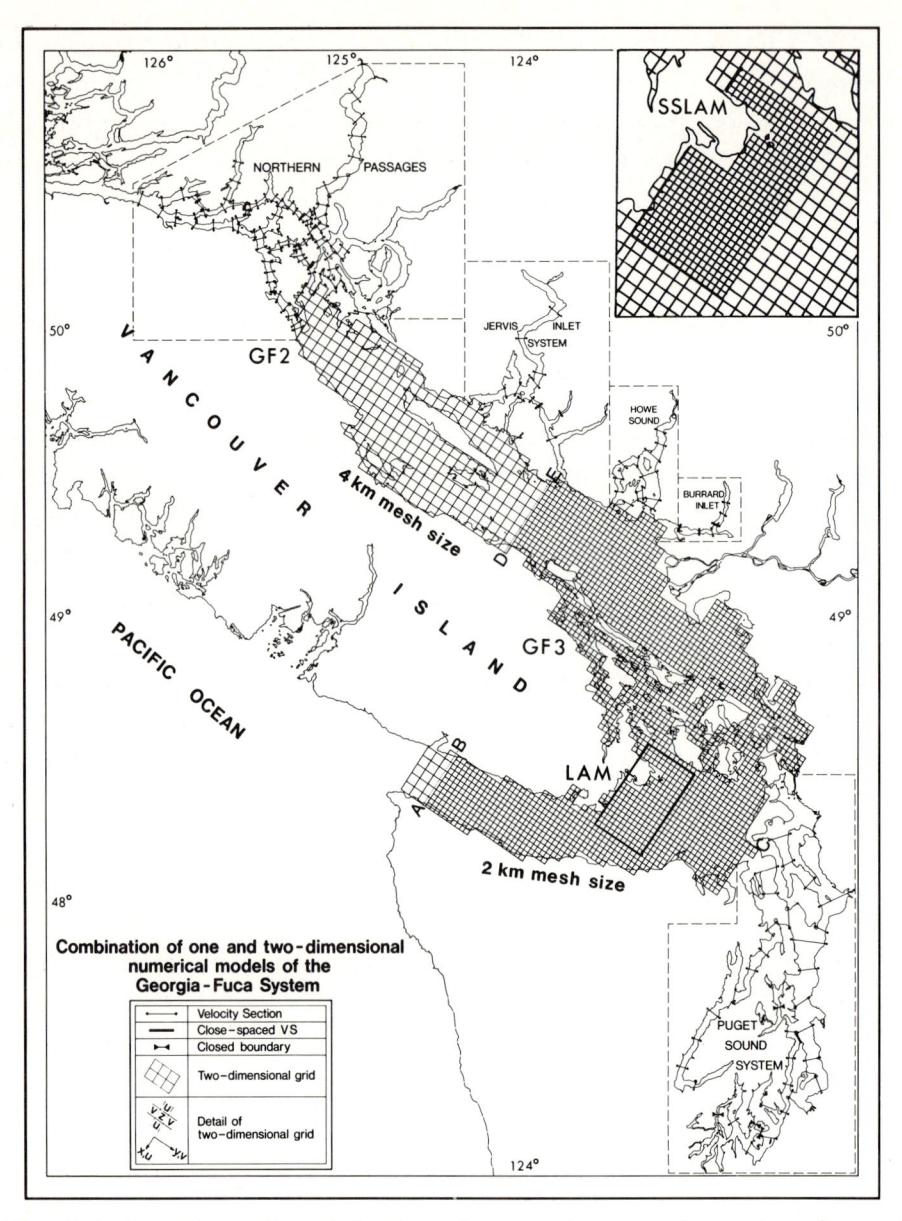

Fig. 5.1 The 2-km grid used for Juan de Fuca Strait and the southern Strait of Georgia. The 1-km mesh inset indicates the location of a finer grid test model (SSLAM) (Chapter 6) in a highly non-linear region of strong tidal streams, subsequently used to study the applicability of boundary conditions from an overall model.

The value of the friction coefficient used in the Strait of Georgia and Juan de Fuca Strait was 0.003. In the region of the channels between the San Juan and Gulf Islands this value was increased to 0.03. These values accord with those used in the overall model (GF2) described in Chapter 4.

When the model was operated in the absence of lateral stresses small grid-scale fluctuations in the directions of velocity vectors (noodling) occurred in regions where the advective accelerations are important. The value of the horizontal eddy coefficient used to eliminate this feature of the model was 10^6 cm^2/s which, in view of the mesh size and time step, implied a relatively small degree of lateral averaging (Kuipers and Vreugdenhil, 1973). This compares to values from 10^6–10^8 cm^2/s used in large-scale ocean models. The problem of numerically evaluating second derivatives near coastal boundaries was avoided by setting the coefficient equal to zero in such circumstances. The implied assumption is that generally the bottom stress acting on the water in a relatively shallow mesh contiguous to the coast is much greater than the lateral stress. The maximum permissible time step yielding stable solutions, obtained from the Courant-Friedrichs-Lewy criterion, is 23 s.

Elevations were prescribed along the three open boundaries of the model shown by lines AB, C, and DE in Fig. 5.1, by interpolation of values from the overall model (GF2) which employed a 4-km mesh. The boundary conditions, as in the case of GF2, were predicted tides based on 61 harmonic constituents. Earlier attempts to operate part of the 4-km mesh model by prescribing elevations from tide gauge data at these locations failed because the model was insufficiently sensitive to permit proper adjustments to the frictional dissipation by optimizing computed and observed elevations around the interior of the modelled sea. Furthermore, the tide gauges at either end of a long open boundary were unable to provide adequate resolution of the slope of the water surface along that boundary required to balance the Coriolis acceleration off the tidal streams moving normal to the open boundary. To overcome these difficulties the distribution of frictional coefficients over the interior of the model and the elevations of the water surface presented at open boundaries were taken from GF2. Subsequent comparisons of the velocity vector fields obtained at comparable instants from the GF2 and GF3 models showed excellent agreement except where modified by the changed representation of local coastal boundaries.

5.3 SIMULATION OF MIXED TIDES

Illustrative of the performance of the model, some comparisons of computed elevations with those predicted on the basis of 61 harmonic tidal constituents over a 7-day period are now presented for the representative locations around the interior of the modelled sea. At Ferndale (#7564) and Tumbo Channel (#7510) (see Fig. 2.1) the good agreement is typical of the results obtained throughout that part of the Strait of Georgia included within the model (Fig. 5.2). The most difficult area of the model in which to obtain good agreement is Haro Strait and the degenerate semi-diurnal amphidromic region in the inner part of Juan de Fuca Strait. These are areas characterized by rapid changes in the spatial distribution of harmonic constants (Figs. 4.8

and 4.9). A comparison of the predicted elevations with those computed by the model for Fulford Harbour (#7330) shows good agreement (Fig. 5.2) although the curves from the model show a small phase lead throughout the record and a modest discrepancy in the tidal ranges exists. More detailed adjustments of frictional dissipation in this part of the model, including that in the passes leading into the Strait of Georgia from the Gulf Islands region, could probably improve the comparison. Such adjustments are presently considered of secondary importance to the successful reproduction of the tidal streams in the main conveying channels of the system and cannot be attempted until the acquisition of further field data. The Pedder Bay gauge (#7080) is located near the semi-diurnal amplitude minimum. The agreement obtained is good although there is evidence of more accentuated higher tidal harmonics in the values from the model.

The primary object of this numerical study concerned the reliable prediction of barotropic tidal streams. As before, the interval of time selected for this trial of the model coincided with the positioning of an extensive array of current meters over a cross-section of Juan de Fuca Strait (line 11, Fig. 2.1). Comparisons representative of the agreement between the vertically-averaged velocities predicted by the model and measured tidal velocities yielded results essentially the same to those shown in Fig. 4.12 (p. 99) and hence are not shown here. The smaller measured cross-channel components of velocity show much more variablility than those directed longitudinally. Occasional intervals of good agreement between the values predicted by the model and those observed are thought to conform essentially to barotropic tidal oscillations while the more irregular motions are associated with the intermittent excitation of semi-diurnal internal modes. Despite the increased resolution afforded to transverse barotropic oscillations in the model GF3, the computed cross-channel components of velocity are essentially identical to those of the coarse grid model GF2.

Comparisons were also made between the elevations and velocities obtained from the overall coarse grid (GF2) and fine grid (GF3) models. No significant differences were noted between the results obtained from each model except in those parts of the velocity fields which are strongly influenced by local coastal boundaries.

Figure 5.3 shows the tidal elevations and the longitudinal component of the tidal streams, obtained from the model for the period 1200 h, 15 March to 1600 h, 16 March, 1973, at a central location in the Strait of Georgia off the mouth of the Fraser River (C, see Fig. 1.1 inset). The range between higher high water and lower low water conforms approximately to the range contained between mean higher high water and mean lower low water at this location. Corresponding vector plots of the flood and ebb velocity fields when the streams are maximal are shown in Figs. 5.4 and 5.5, respectively. Throughout most of Juan de Fuca Strait and the Strait of Georgia the tidal streams are essentially rectilinear. In the inner part of Juan de Fuca Strait some rotatory character to the streams occurs.

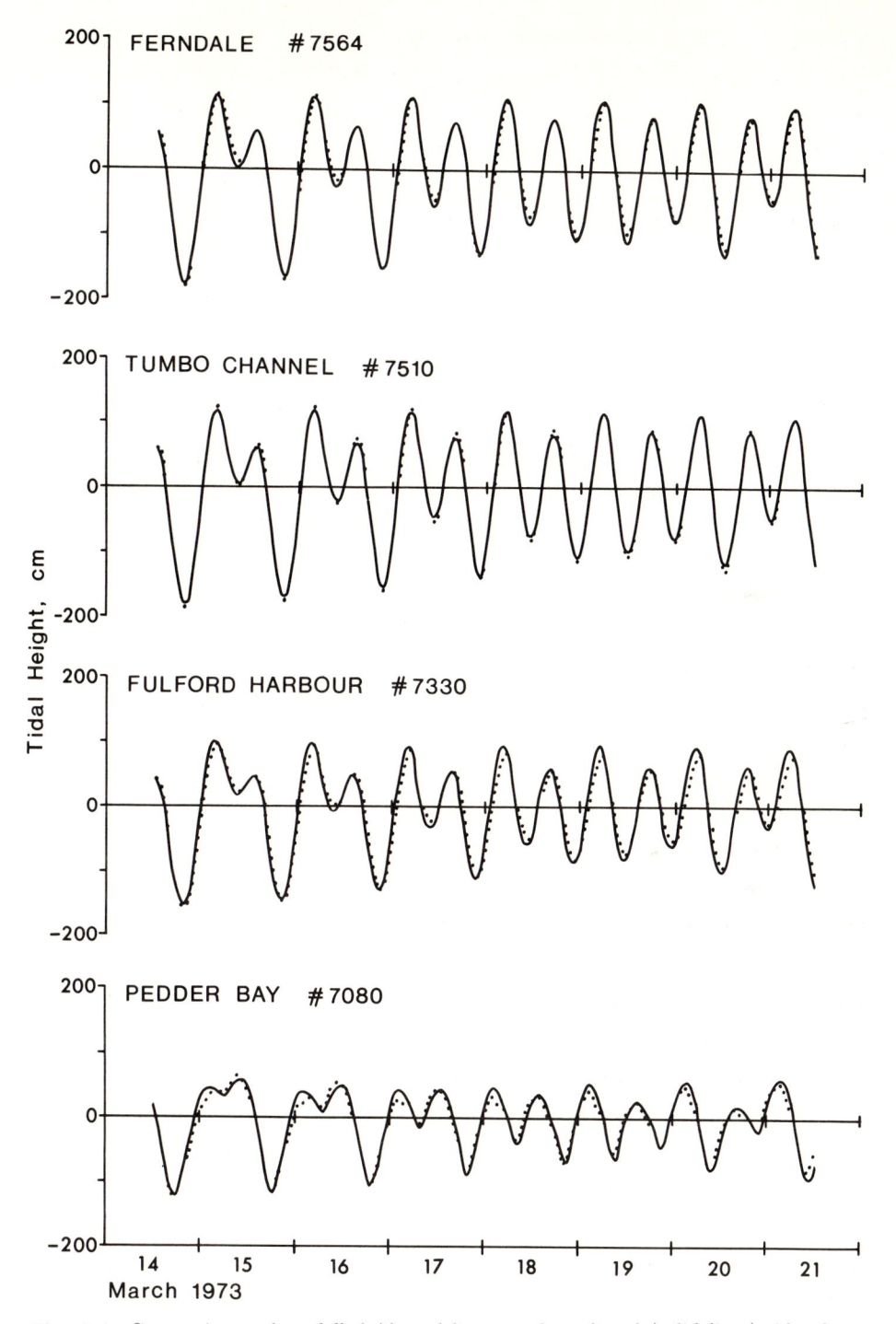

Fig. 5.2 Comparisons of modelled (dotted lines) and predicted (solid lines) tides from the GF3 model for representative locations.

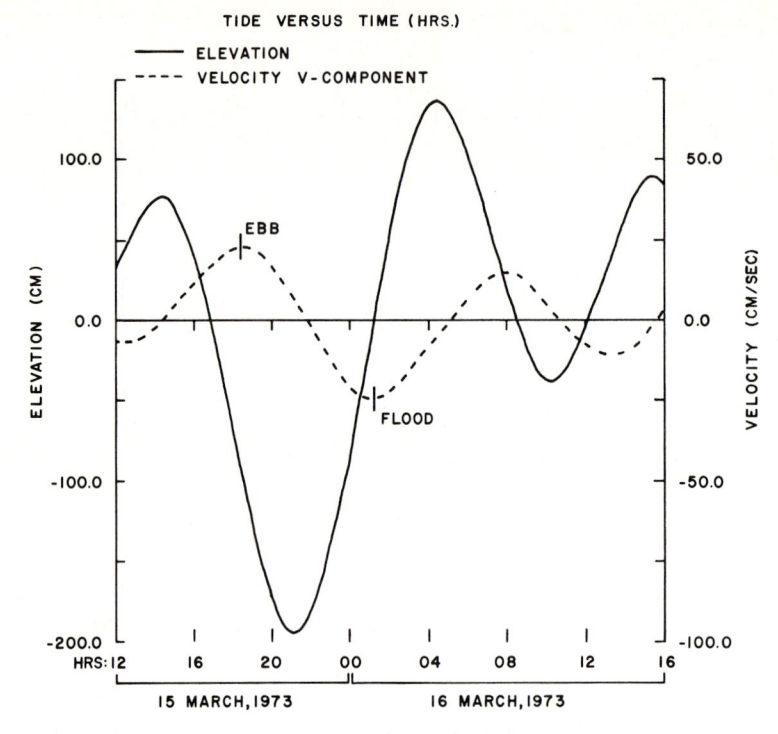

Fig. 5.3 Tidal elevation and the (dominant) longitudinal component of the tidal stream at a central location in the Strait of Georgia off the mouth of the Fraser River.

The velocity vectors suggest some interesting aspects of the tidal flows near coastal features off the southernmost part of Vancouver Island where large speeds and marked changes in flow direction occur. The formation of back eddies changing in location from ebb to flood is clearly evident. Examination of a time sequence of such vector plots indicates that the eddy begins to form when the stream at a particular location has attained sufficient magnitude. The eddy then increases in spatial extent until the stream turns. A further point of interest concerns the difference between strength and direction of the ebb and flood streams in Haro Strait which again emphasizes the importance of local topography. These flow features lie along the primary route followed by water from the Fraser River moving seaward through Juan de Fuca Strait.

A number of interesting dynamical inferences may be made from slopes of the water surface as illustrated by three-dimensional plots such as those shown in Fig. 5.6. In crude terms, all land has been removed above a base plane corresponding to the lowest elevation of the water surface computed in the modelled sea at the particular instant in question. Thus, islands are represented as 'holes' in the sea surface. The intersection of two lines in the sea area corresponds to the centre of a mesh in the model where elevations are calculated while a single line indicates the elevations along a channel represented in the model by an isolated sequence of contiguous meshes. The

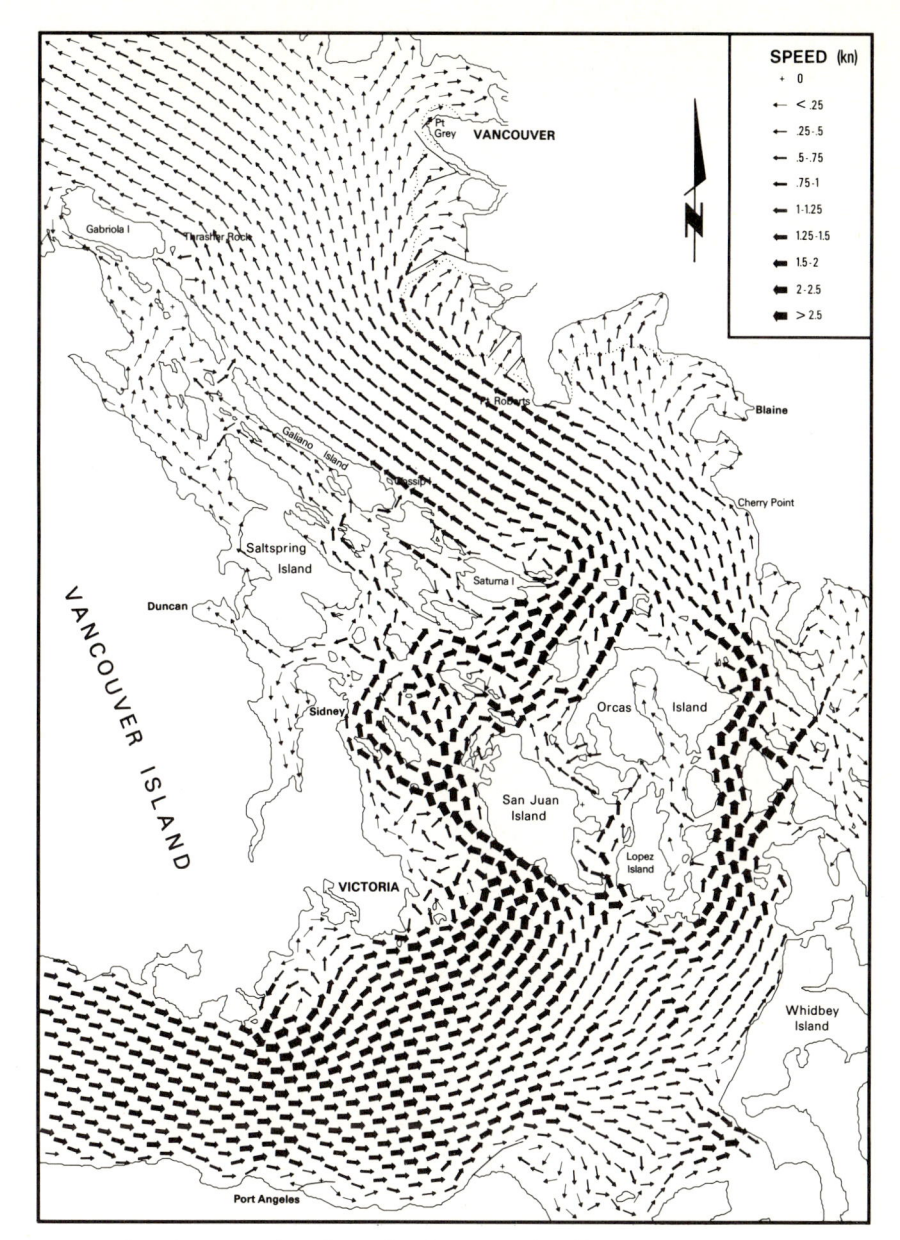

Fig. 5.4 Sample page taken from the Current Atlas (Canadian Hydrographic Service, 1983) showing flood velocity vectors obtained from the model GF3.

fields of elevation coincide in time with the flood and ebb velocity fields shown in Figs. 5.4 and 5.5, respectively.

As noted above, the proper proportioning of tidal elevations and times of high water between the Strait of Georgia and the Puget Sound system required the introduction of high frictional dissipation in the region of the San Juan Islands. The marked slopes of the water surface through this region are clearly evident. Elevations in the area

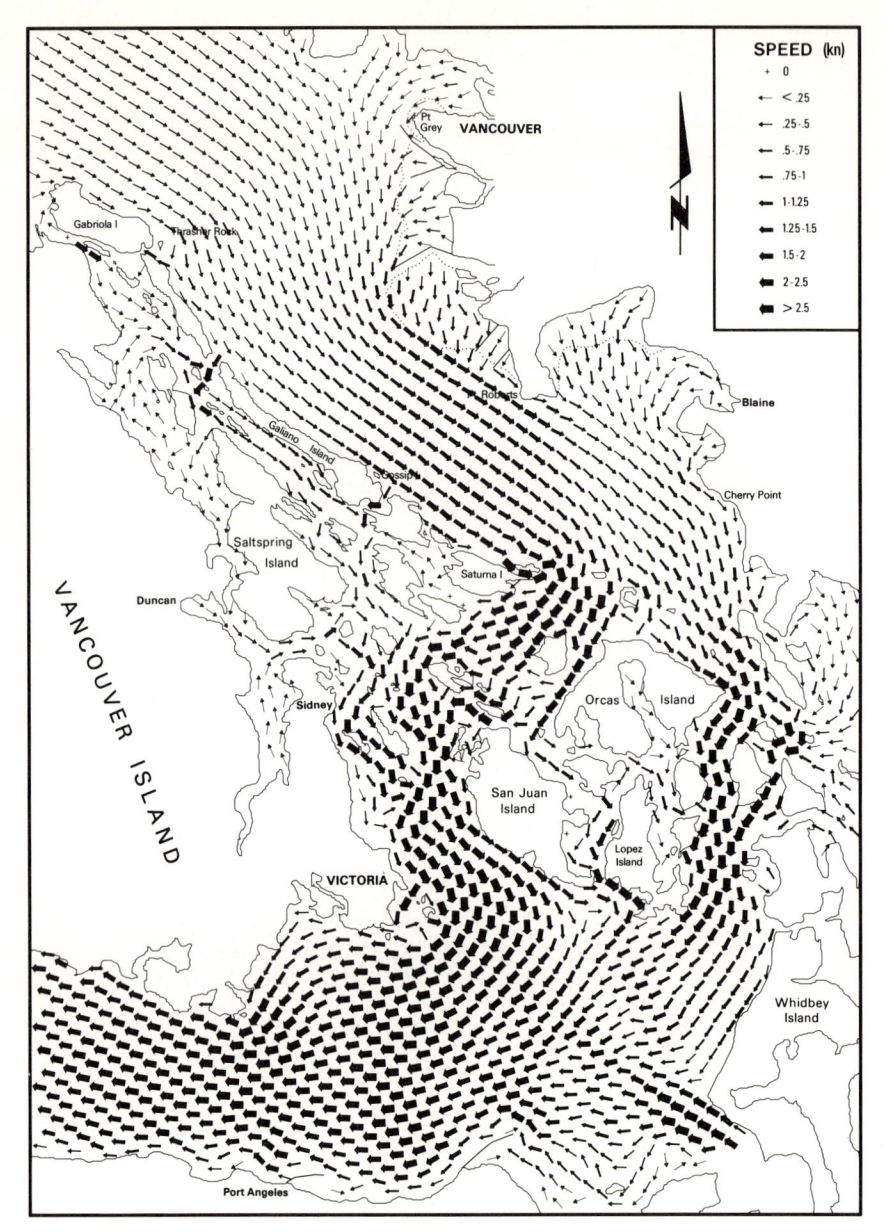

Fig. 5.5 Sample page taken from the Current Atlas (Canadian Hydrographic Service, 1983) showing ebb velocity vectors obtained from the model GF3.

contained between Vancouver Island and the Gulf Islands are similar to those in the northern part of Haro Strait and can differ markedly from those in the Strait of Georgia.

In the Strait of Georgia, cross-channel slopes of the water surface which balance the Coriolis accelerations are due to the tidal streams moving parallel to the major axis of the Strait. In the vicinity of the Fraser River mouth local slopes are associated with the motion of water off or onto the shallow banks. This is indicated by the M_2

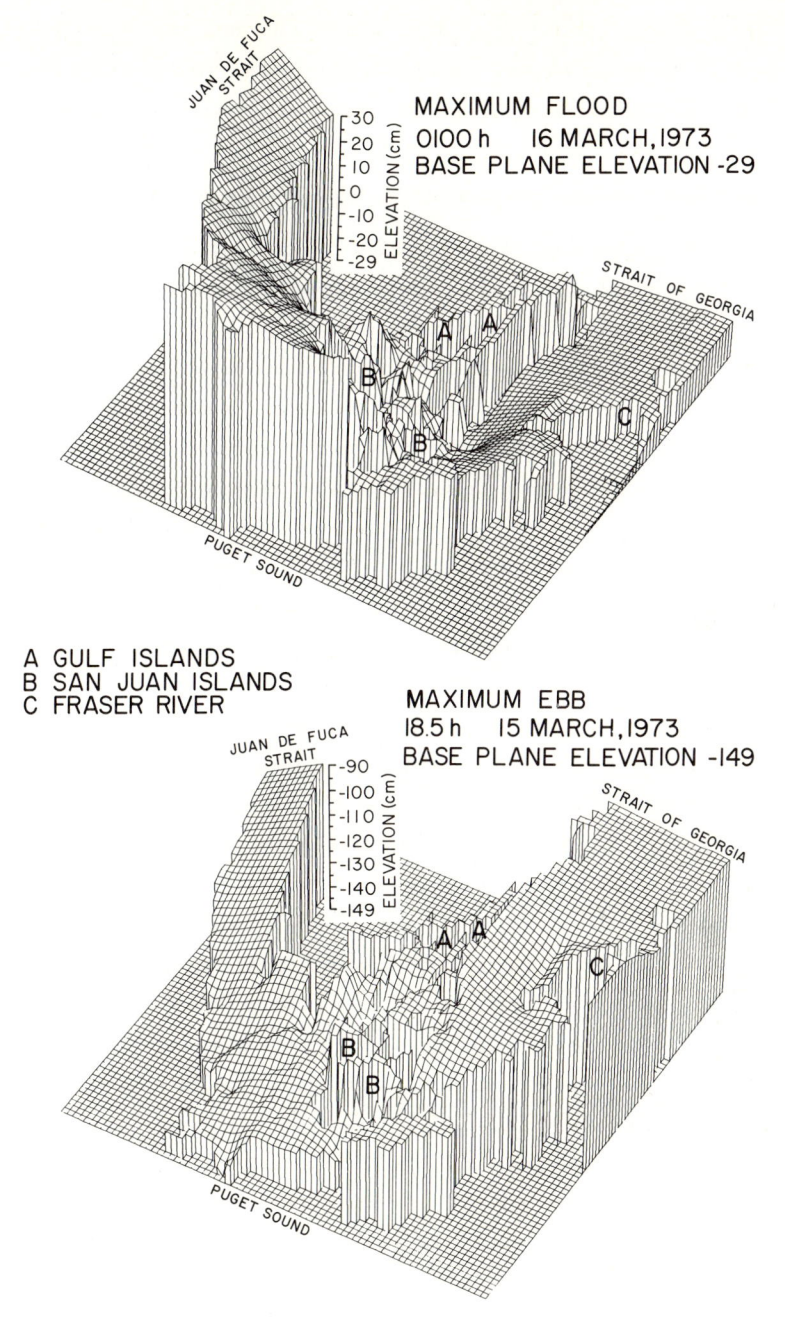

Fig. 5.6 Three-dimensional diagrams showing the shape of the water surface during maximum flood and ebb at the same instant as the plots of flood (Fig. 5.4) and ebb (Fig. 5.5) velocity fields, respectively.

and K_1 tidal current ellipses shown in Chapter 7. A further feature of these diagrams concerns secondary undulations of the water surface in the inner part of Juan de Fuca

Strait. These are more strongly in evidence when the streams are running full than at slack water. At present it appears that they may be associated with depressions in the water surface of some 10–20 cm at the centre of the large eddies near coastal features.

5.4 APPLICATIONS OF RESULTS FROM GF2 AND GF3

An immediate application of the numerical model results was found in the preparation of a tidal current atlas in which the computed flood and ebb tidal velocity fields were presented at hourly intervals over representative tidal ranges (see, *e.g.*, Figs. 5.4 and 5.5). Provided reservation is made for flow features unresolved by the necessarily crude coastline of the model, good predictions of the tidal streams for the main conveying channels of the system were obtained. Of special interest is the growth and decay of large-scale eddies that occur in particular parts of the system. Significant departures of the near-surface velocities from the vertically-averaged tidal velocities occur near the mouth of the Fraser River during freshet. Fields of interpolated velocities obtained from extensive drogue tracks near the river mouth were also included in the atlas. A further model (GF4) has been developed specifically to determine motion in this upper layer and is discussed in Chapter 11.

If the velocity fields have been computed by the model for a sufficient length of time, tidal harmonic constants may be determined for each velocity point. Such constants may then be employed to predict the velocities at any point within the modelled region for times not included with the original sequence of computed tides. Comparisons of velocities computed directly by the model with those predicted using constants from an earlier run of the model for a typical location in Juan de Fuca Strait are shown in Fig. 5.7. Also included within the figure are predictions with and without the use of amplitude and phase relationships known from long tide gauge records to separate those constituents not otherwise separable using the 1 month of velocities computed by the model. Clearly, the use of inference greatly improves the modelled harmonic constants. The availability of such constants makes it possible for a shipboard computer, using a standard tidal prediction program and stored constants, to predict the tidal currents in the vicinity of the vessel, as well as course alterations required to maintain a desired track in the presence of transverse tidal stream. Additional refinement includes the components of surface drift and wind drag on the vessel's superstructure, using wind force data obtained from the ship's anemometer. In the case of search and rescue operations the net drift of a floating object can be estimated, continuously updated, and presented on a small video screen. Uncertainties of initial position could be included to give patterns of probability of the object's location. A further refinement will be the inclusion of relevant hydrographic chart data.

A question arises as to the number of constituents required to provide an adequate representation of the tidal streams for practical purposes. Figure 5.8 shows locations in

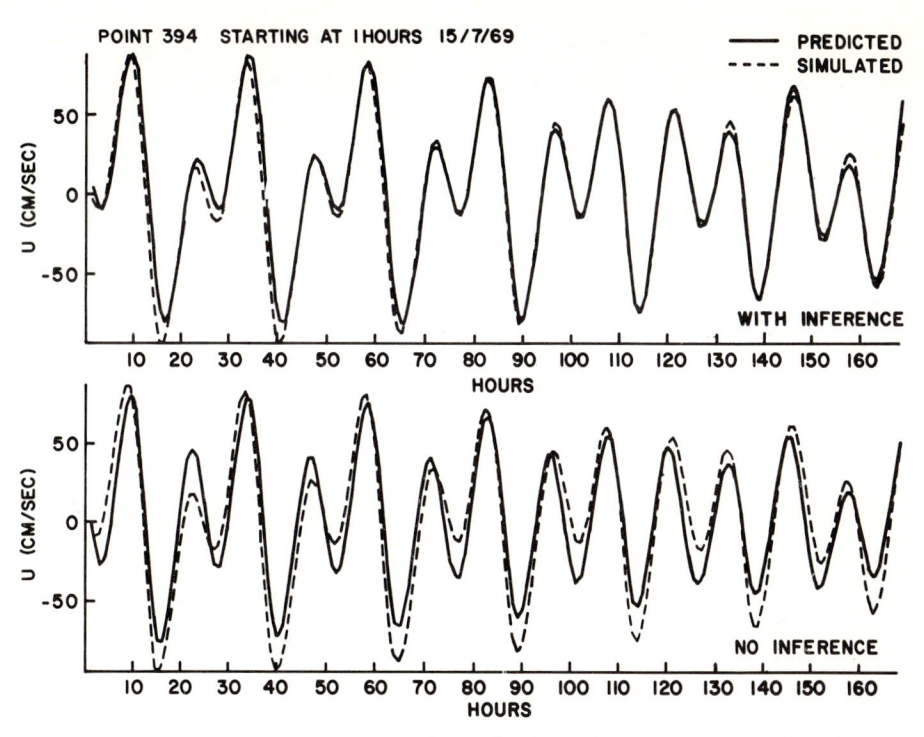

Fig. 5.7 Comparison of predicted and simulated tidal currents at a typical location in Juan de Fuca Strait illustrating the effects of inference.

Juan de Fuca Strait, in the highly non-linear region of Haro Strait, and in the Strait of Georgia where the tidal constituents have been determined. Figures 5.9 and 5.10 show the amplitudes of velocity constituents which are arranged in descending order of magnitude. Also shown is the cumulative amplitude. Representative of Juan de Fuca Strait (4118) and the Strait of Georgia (4438), the V component is shown in Fig. 5.9a and b. Six constituents account for some 80% of the tidal stream velocity. On the other hand, in southern Haro Strait, some 15 constituents are needed to account for a comparable proportion of the tidal stream (Fig. 5.10a and b). Of particular interest is the increased importance of the MO_3 and MK_3 constituents which have frequencies corresponding to interaction frequencies of the M_2 and K_1 constituents.

The primary shortcoming of such a current atlas concerns inadequate resolution of the tidal velocities near coastal features. It would appear that finer grid models can be satisfactorily operated from overall models provided the increased resolution does not lead to significant changes in the dependent variables at the open boundary between the two schemes. The use of GF2 data to prescribe boundary conditions for GF3 is an example of such a telescoping grid strategy. A question arises as to whether or not this can be done for a region of strong tidal velocities undergoing rapid spatial changes. This problem is considered in Chapter 6 in connection with the limited area model

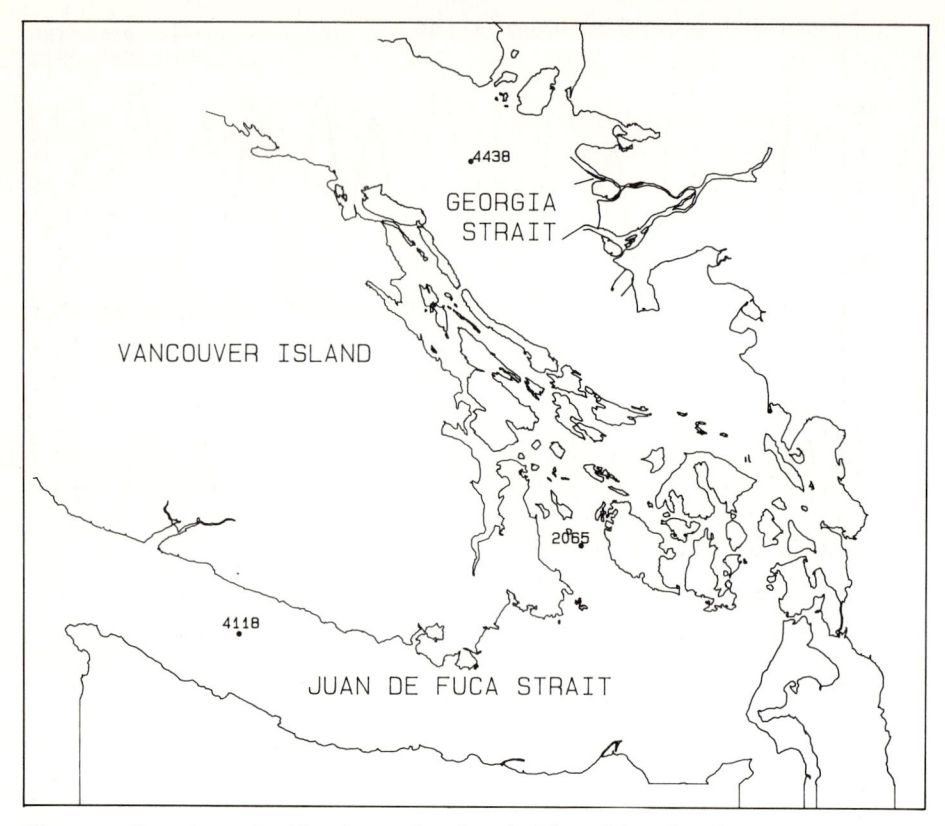

Fig. 5.8 Locations of grid points referred to in Figs. 5.9 and 5.10.

(LAM) and the small-scale area model (SSLAM).

A particularly interesting use of the elevations and velocities computed from the GF3 model is to determine the motion of freshwater which may move southward into the strong tidal mixing area of the San Juan Islands or northward into the Strait of Georgia where the tidal flows are relatively weak. In the latter case, it may eventually move seaward through Johnstone Strait or the accumulated discharge over a lengthy period may be returned southward by the action of winds. The problem is complex. The thickness of the plume is commensurate with the tidal range at the river mouth. There exist extensive areas of drying banks. The intrusion of the salt wedge into the Fraser River is related to the state of the tide and to variations in the strength of the river discharge (Fig. 1.16, p. 27). There is also the problem of simulating propagating shallow fronts between river water and sea-water or the remains of preceding 'plumes' in the Strait.

The simplest approach that holds promise of simulating the dominant aspects of the shallow upper layer motions in the Strait of Georgia, and which utilizes data from the barotropic tidal model GF3, is the vertically-integrated upper layer model GF4

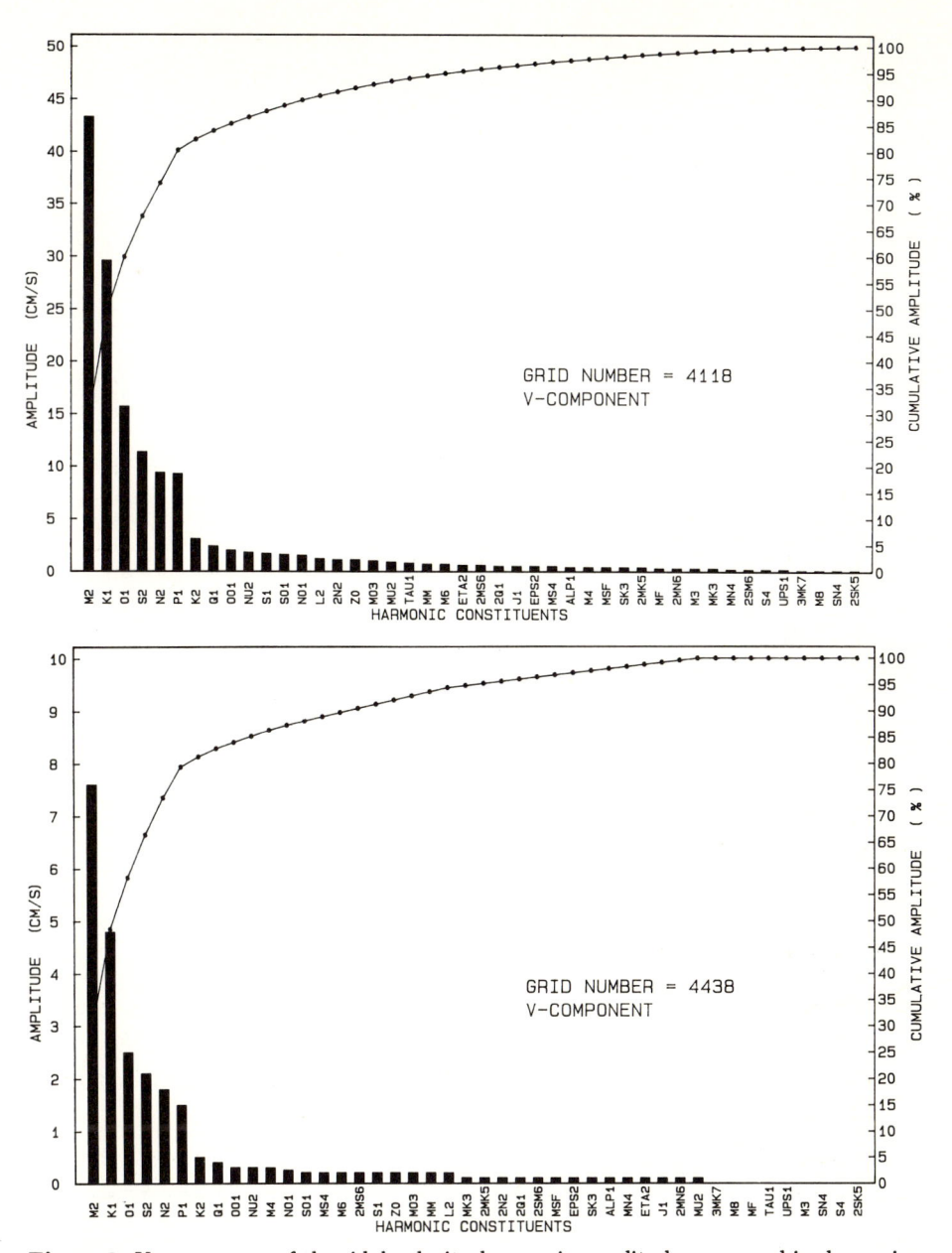

Fig. 5.9 *V* component of the tidal velocity harmonic amplitudes arranged in decreasing order of magnitude in (a) Juan de Fuca Strait (4118) and (b) the Strait of Georgia (4438).

which is described in Chapter 11.

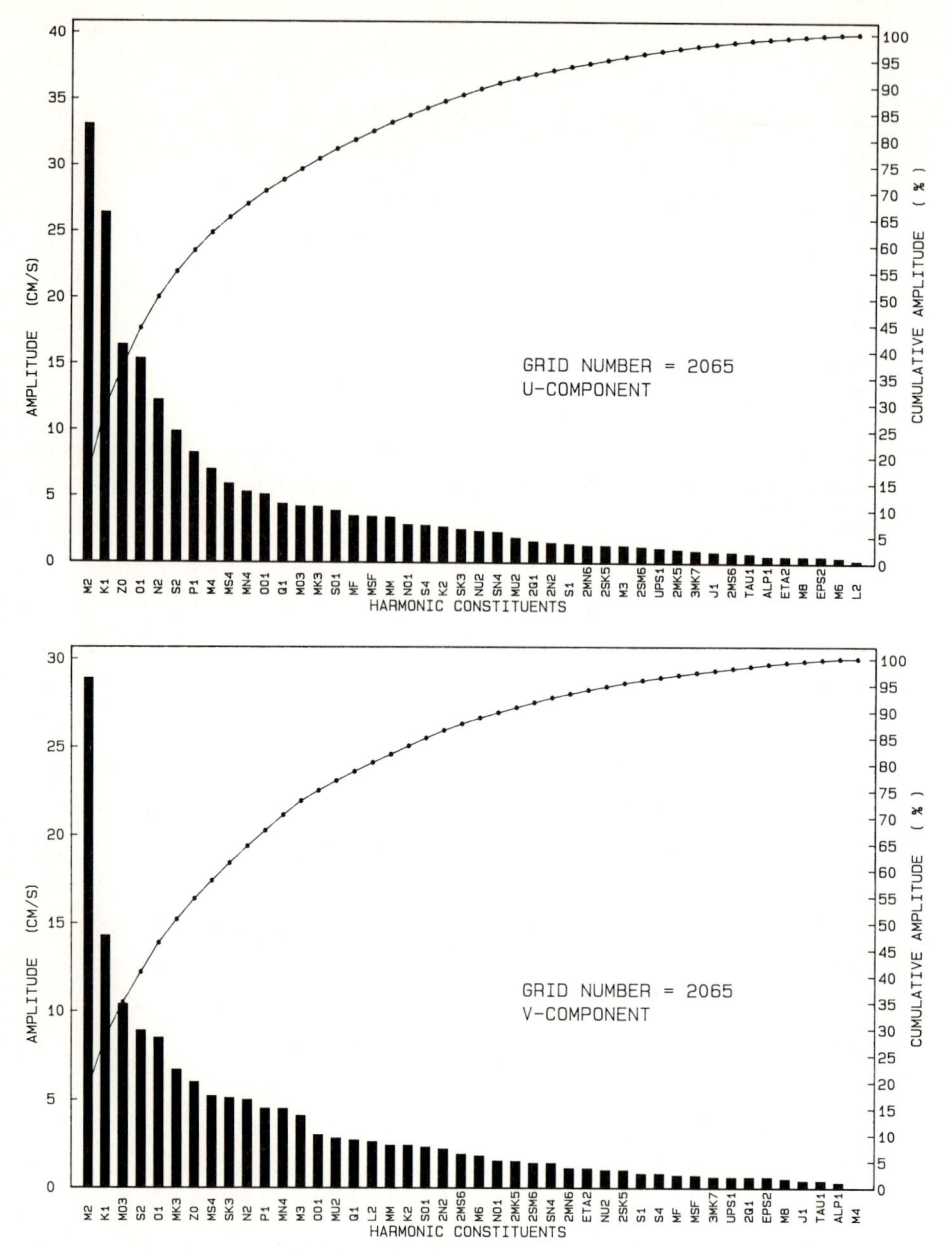

Fig. 5.10 (a) U and (b) V components of the tidal harmonic amplitudes arranged in decreasing order of magnitude in southern Haro Strait (2065).

6 THE LIMITED AREA MODEL (LAM) AND SMALL-SCALE LIMITED AREA MODEL (SSLAM)

6.1 LOCAL AREA MODELLING

A common problem concerns details of tidal currents and residual circulation in the vicinity of coastal engineering installations such as outfalls, docks, and breakwaters. Such detail is not provided by numerical grid schemes that have been described in Chapters 4 and 5. It was noted that satisfactory boundary conditions could, however, be obtained for the fine grid (2-km mesh) partial model (GF3) from the overall coarse grid (4-km mesh) model GF2 of the overall system. The primary open boundaries (AB and DE, Fig. 5.1, p. 116) were located at the entrance to Juan de Fuca Strait and across the deep basin of the Strait of Georgia, regions where the field observations are relatively small. The question arises as to the efficacy of this procedure in a region where swift tidal streams, coastal features, and changes in depth lead to large field accelerations and accompanying strong residual circulation. Although a comprehensive study would require appropriate measurements using moored current meters, it is possible to demonstrate that highly plausible results can be obtained. The test region selected is located near Victoria in the inner part of Juan de Fuca Strait (Fig. 5.1) where, as shown in Chapter 7, large tidal eddies abound.

It is of interest to consider briefly the magnitude of terms occurring in the momentum equations (4.1–4.3) of the model GF2 for this region, particularly the relative importance of the advective accelerations. If these terms are large, it is necessary to include not only elevations but also the u and v velocity components at open boundaries of the limited area model (LAM). Figure 6.1a and b shows the relative magnitude of terms in the momentum equations (4.2) and (4.3) for the limited area selected. These terms have been normalized such that the Coriolis term equals unity on a representative flooding tide. In order from top to bottom in each mesh, these terms are the local time derivative, bottom friction, advective acceleration, and pressure gradient due to surface slope. The u terms apply to the computation of u at the right hand side of the mesh and v to the top side of the mesh. In general, the sum of the four ratios is close to -1 in each grid. Small deviations indicate some inadequacy in the calculation of the various terms. The magnitude of all terms varies rapidly in space and indicates that the GF2 results are unlikely to provide adequate boundary conditions for a small-scale limited area model (SSLAM). Thus, this 'telescoping' procedure should not be too abrupt. It is evident from Fig. 6.1a and b that the advective accelerations are very important.

An initial trial, LAM, concerned the redetermination of the solution for that part of the original (GF3) grid contained in the limited area using the earlier recorded values to prescribe conditions on the open boundaries. In the interests of manageable data storage and convenience, these elevations and velocities (rounded off to one decimal place in c.g.s. units) had been stored at 15-min intervals. The velocities were determined prior to storage, from the volume transports computed by the momentum equations. The work was then repeated using SSLAM.

A basic assumption in applications of this type of procedure is that the open boundaries with the overall model shall be located sufficiently far away from the region of interest in order to be unaffected by local changes in topography of SSLAM.

6.2 THE LIMITED AREA MODEL (LAM)

To ascertain the effects of temporal interpolation, the two trials were carried out in which a cubic spline (employing values at four time levels for the point in question) was fitted to the prescribed boundary values recorded at hourly and 15-min intervals, respectively. It was found that the hourly values resulted in a net loss of fluid from LAM although conservation of fluid within the LAM algorithm itself was correct to one part in 10^{11}. Temporal interpolation of the 15-min boundary elevations and velocities proved entirely satisfactory.

It will be recalled that the solution of the momentum equations at an interior velocity point requires values of the surrounding velocities at the earlier time step. The computational configuration for a u point has been shown earlier in Fig. 4.1 (p. 72) and alternative ways are given (Fig. 6.2) in which the requisite boundary velocities and/or elevations (enclosed in boxes) could possibly be supplied. Other variables shown are computed in the conventional manner.

The first alternative method, (a), prescribes only the u and v velocity components. In the second method, (b), these velocities and also boundary elevations are prescribed. Although the increment of elevation entering into the barotropic pressure gradient term in the momentum equations (4.5) and (4.6) has already been determined from application of the continuity equation for that time step, such an increment can readily be determined from the prescribed u and v values alone. Thus, both methods should be equivalent. It was found that (b) substantially reduces the error in values computed at an interior u point adjacent to the boundary. If the error is defined as the difference between the value of the u component computed at that first point by the overall model and the corresponding value obtained from LAM, the time series of the errors (Fig. 6.3) shows an appreciable reduction when the u, v, and ζ boundary values are prescribed and when the spline fits to data recorded at 15-min intervals are used, from

Fig. 6.1a Relative magnitude of terms in the x-directed momentum equation at time step 53120, a flood tide. Within each grid element the terms are normalized so that the Coriolis term equals $+1$. (Ordered from top to bottom are time derivative, advective acceleration, bottom friction, and surface slope acceleration.)

N ↗

VICTORIA

	0.40 0.50 1.67 -3.69	-0.22 1.73 -2.90 0.40	-0.88 0.97 -2.77 1.56	-2.00 0.53 0.24 0.24					0.56 0.06 -1.64 -0.19
0.01 0.25 0.46 -1.81	0.74 0.46 -0.41 -1.81	1.13 0.65 -1.22 -1.58	1.02 0.72 -0.56 -2.22	0.69 0.49 -0.40 -1.99					0.09 -0.01 -0.46 -0.73
0.77 0.10 -0.32 -1.52	0.47 0.21 0.01 -1.70	0.78 0.34 0.09 -2.25	0.85 0.47 0.08 -2.34	0.43 0.38 0.70 -2.39	-0.05 1.20 2.03 -4.27	-1.42 4.58 1.07 -4.76	-14.3 76.2 -36.0 -24.6	3.27 -0.51 5.61 -9.93	0.09 -0.03 0.71 -1.75
-0.13 0.16 0.52 -1.59	0.24 0.28 0.43 -1.97	0.44 0.38 0.25 -2.10	0.59 0.46 0.17 -2.29	0.77 0.55 0.70 -3.07	0.35 0.97 0.56 -2.91	-0.07 3.73 2.96 -8.03	11.5 -49.8 -35.1 74.9	-1.61 -5.38 11.10 -4.66	-0.99 -0.19 1.92 -1.62
0.37 0.21 0.33 -1.91	0.51 0.31 0.14 -1.97	0.54 0.36 -0.39 -1.50	0.59 0.38 -0.48 -1.45	0.27 0.55 -0.10 -1.67	0.33 0.81 -0.23 -1.91	1.92 2.98 0.74 -6.53	-8.03 -13.00 0.04 19.50	-0.93 -1.82 2.25 -0.63	-0.64 -0.20 0.56 -0.72

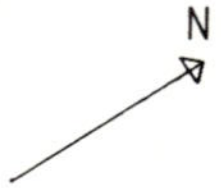

	−0.89 −0.22 0.49 −0.39	−0.42 −0.34 0.80 −1.06	−0.20 −0.04 0.44 −1.22	−0.29 −0.06 −0.29 −0.36	0.80 −0.25 −3.43 1.88			−3.81 0.04 −2.05 4.76	0.43 0.25 −1.54 −0.12
	−0.26 −0.31 0.32 −0.69	−0.66 −0.18 0.85 −1.01	−0.48 −0.10 0.16 −0.61	−0.15 −0.10 −1.15 0.28	−0.10 −1.16 −2.58 3.20			−1.41 0.07 1.76 −0.59	3.18 0.49 −0.96 −4.46
	−1.00 −0.21 0.79 −0.58	−0.68 −0.16 0.03 −0.22	−0.53 −0.21 −0.28 0.06	−0.33 −0.22 −0.73 0.23	−0.21 −0.85 −0.05 0.17	−0.19 −0.42 1.56 −1.93	0.01 −0.02 0.77 −1.78	−0.65 −0.09 0.73 −0.93	−0.10 0.14 0.99 −1.92	5.48 0.51 2.66 −10.40
	−0.76 −0.22 0.04 −0.04	−0.58 −0.25 −0.26 0.07	−0.57 −0.26 −0.10 −0.06	−0.10 −0.29 −0.49 −0.14	−0.56 −0.74 1.08 −0.72	−0.59 −0.12 1.31 −1.64	−0.04 0.03 0.68 −1.61	0.07 0.17 0.65 −1.91	0.11 0.19 0.32 −1.59	2.99 0.21 2.00 −6.39

VICTORIA

N ↗

Fig. 6.1b Relative magnitude of terms in the y-directed momentum equation at time step 53120, a flood tide. Within each grid element the terms are all normalized so that the Coriolis term equals +1. (Ordered from top to bottom are time derivative, bottom friction, advective acceleration, and surface slope acceleration.)

the original overall model. In summary, it was shown that the original solution could be recovered satisfactorily in this manner.

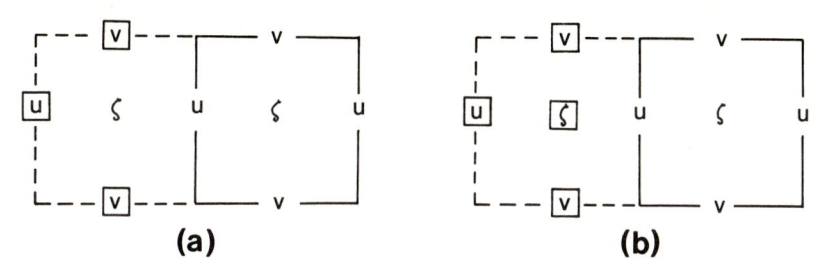

Fig. 6.2 Arrangement of variables in the vicinity of a u point near an open boundary showing alternative arrangements of prescribed values (in boxes).

Since both elevations and component velocities are prescribed on the open boundaries of LAM, a question arises as to whether or not the system is in overdetermined in this region, that is, are the values prescribed in excess of those required to determine a unique solution over the interior of the modelled region. Two considerations indicate that this is not the case. First, the computations in this limited area model merely constitute a repetition of the earlier computations in the overall model. Secondly, the use of such boundary conditions is consistent with theoretical notions based on the method of characteristics (Richardson, 1964). Through each point in the solution space (x, z, t) there exists an infinite family of bicharacteristics. Thus, there is no particular restriction on the number of dependent variables and their observations prescribed on the open boundary. On the other hand, a certain minimal amount of data must be prescribed on the open boundary in order to ensure a correct response over the interior of the modelled region. The amount of boundary data required is primarily determined by the finite-difference formulation of the governing equations.

6.3 THE SMALL-SCALE LIMITED AREA MODEL (SSLAM)

The primary object of these tests is the satisfactory operation of a limited area model but with the employment of a smaller scale grid capable of resolving more detailed features of the local flow field. To the problem of temporal interpolation of boundary values there is now added that of spatial interpolation. The mesh scale selected was 1 km, the depths of the mesh sides constituting the open boundary of SSLAM conforming exactly to those at corresponding locations in the original overall model. Interior depths were reschematized from hydrographic charts. The primary problem concerned the spatial interpolation. In general, any simple spatial interpolation of a boundary velocity will not conserve momentum flux (velocity squared) over the course of the integration. Following unsuccessful trials using velocities, the method

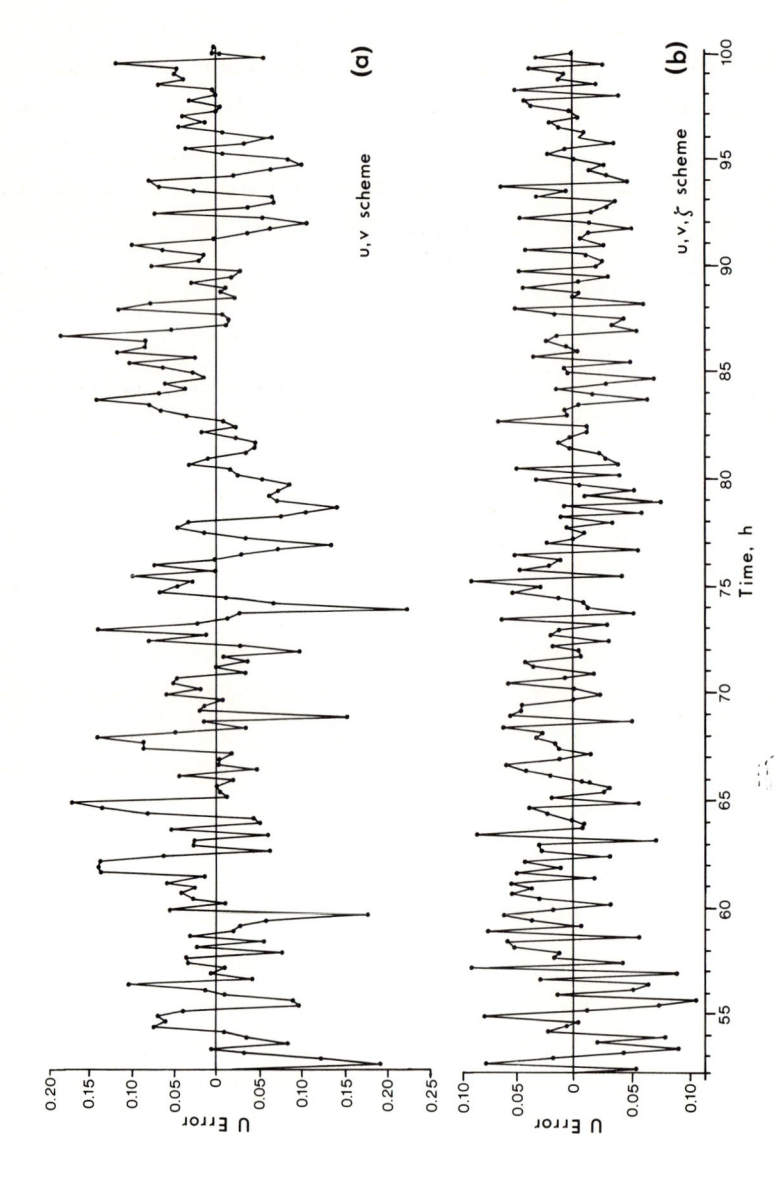

Fig. 6.3 Time sequence of errors (difference between original value from the overall computation and that obtained from LAM) for representative interior u point adjacent to the boundary of LAM when (a) velocity points u, v are prescribed (see Fig. 6.1a) and when (b) velocity and elevation are prescribed (Fig. 6.1b).

employed was to avoid interpolation and to precribe the original transports (temporally interpolated from earlier recorded 15-min data from the overall model). This proved entirely satisfactory and no problems were encountered in the course of the 1-wk tidal computations used in these tests. It is worth noting that, by implication, such transports which include the effect of the averaged surface elevations on either side of the point where the transport is prescribed, include a contribution from elevation points outside the u and v transport points constituting the open boundary of SSLAM.

Representative of the more detailed resolution of the velocity field obtained from SSLAM, Fig. 6.4 compares distributions of vectors in both the fine grid (GF3) model and SSLAM. In the first comparison (Fig. 6.4a and b) the velocities at 0400 h, 15 March 1973 show the greatly increased detail while maintaining a basic conformity for a time when the major eddies off Race Rocks and the southern end of Haro Strait are strongly developed. A second example is shown in Fig. 6.4c and d for 2100 h on the same date for a different stage of eddy development. It will be noted that despite the location of an eddy or partial eddy on the open boundaries, the more detailed resolution confirms the original eddy in the coarser grid model. Thus, for example, the eddy in the upper left hand corner of Fig. 6.4d (approximately west of Race Rocks) is barely perceptible at the coarser scale (Fig. 6.4c).

These results indicate that for a complex region of strong and highly non-linear tidal flows, volume transports recorded at 15-min intervals from an earlier overall coarse grid model computation constitute fully satisfactory boundary conditions for a local finer resolution model of that region.

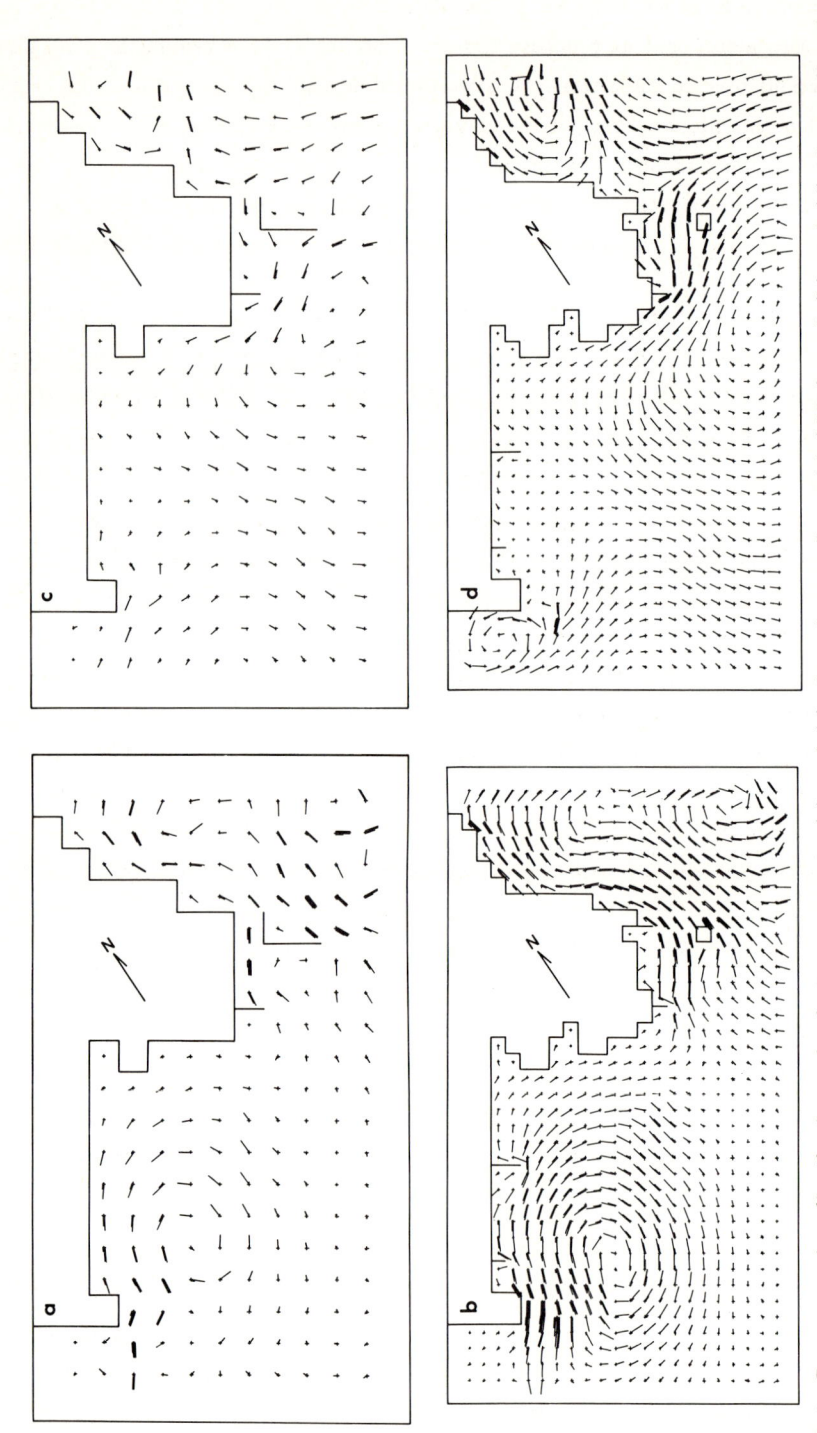

Fig. 6.4 Representative distributions of velocity vectors in (a) and (c) the 2-km mesh model (GF3), and (b) and (d) the small-scale area model (SSLAM) illustrating more detailed resolution of tidal eddies obtained even though close to intersecting the open boundaries.

7 THE OVERALL FINE GRID MODEL: GF7

7.1 INTRODUCTION

The earlier studies using the vertically-integrated barotropic tidal numerical models GF2 and GF3 emphasized the importance of developing an overall fine grid (2 km) model of the system, GF7. Such a model provides adequate resolution of the horizontal residual circulation derived from tidal topographic interactions while at the same time minimizing any spurious net flows associated with the proximity of boundary openings. A further advantage of such a scheme concerns its facile conversion to a fully three-dimensional operation. In various respects this model employs much cruder approximations to a number of narrow passages which were simulated in considerable detail by one-dimensional schemes in the two-dimensional model GF2.

In view of these approximations, it has not been attempted to achieve an accuracy of calibration of GF7 comparable to that obtained from the model GF2 itself. The level of calibration actually achieved is considered satisfactory for the provision of barotropic tidal velocities both for navigational use and for baroclinic calculations in the major conveying channels of the system. An important application of elevations and velocities computed by this model will be described in Chapter 11 in connection with the buoyant-spreading upper layer model of the shallow brackish layer in the Strait of Georgia associated predominantly with discharge from the Fraser River. In this chapter, since the model employs the same equations and is essentially an extension of the earlier model GF3, comment on the numerical scheme is confined to features particular to this overall scheme. Representative comparisons between observed and computed velocities are presented. A detailed discussion of the barotropic tidal residual circulation obtained from this model is compared with the earlier residual circulations of GF2 and GF3.

7.2 NUMERICAL GRID SCHEME

The numerical model GF7 grid layout is shown in Fig. 7.1. In view of the very large grid array required for a geographical representation, the actual computations employ a much smaller array into which dismembered segments of the geographical array are packed (Fig. 7.2). The letters in this array denote corresponding junction points. In certain instances, deep fjords opening from the main conveying channels possess depths more than double the maximum depth occurring in these conveying

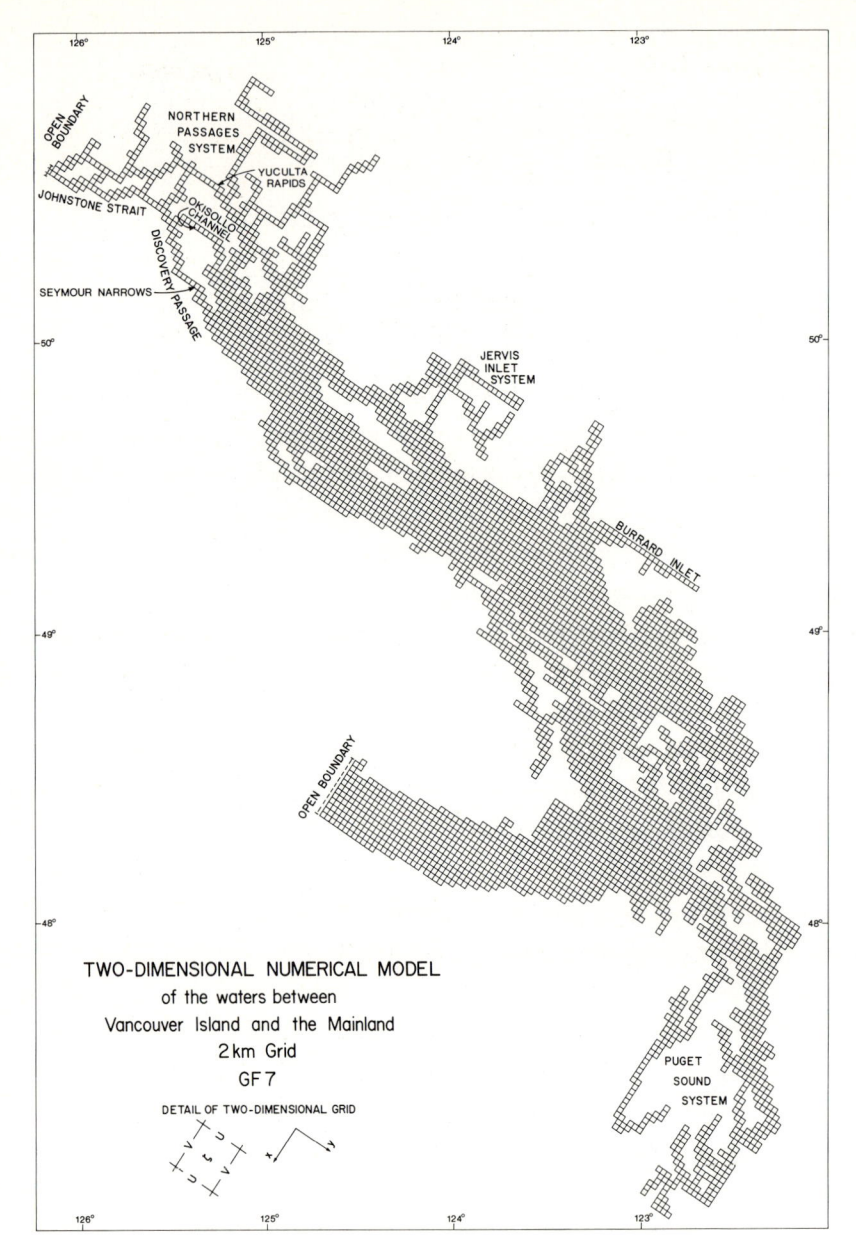

Fig. 7.1 Geographical layout of the numerical model GF7.

passages. To avoid a prohibitive increase in computer costs, length and depth have been adjusted to give an approximately correct dynamical response.

Where narrow passes possess width much less than 2 km, fractional widths have been introduced to give approximately the correct depth and cross-sectional area. In certain instances the dynamically-significant constrictions in the northern passages actually involve local channel divisions and multiple rapids. In the model these are repre-

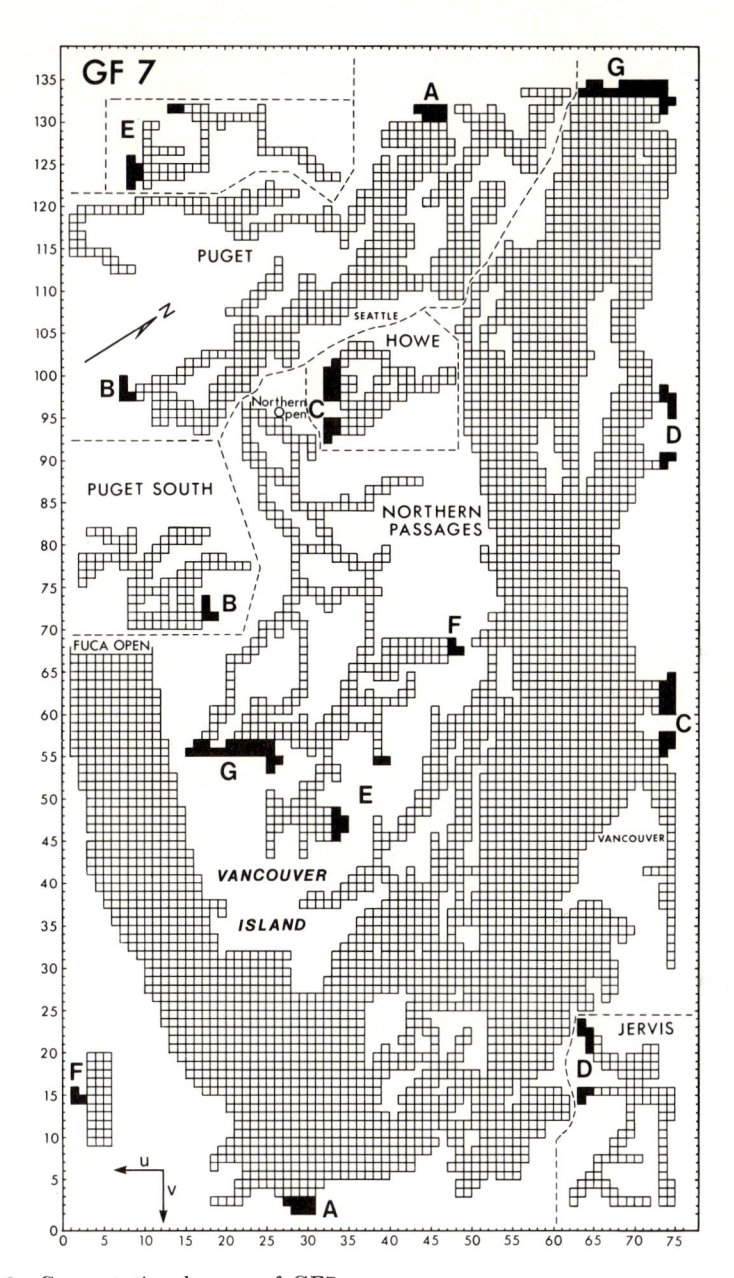

Fig. 7.2 Computational array of GF7.

sented by three reduced width meshes referred to as Seymour Narrows, Okisollo Channel, and Yuculta Rapids (see Fig. 1.30).

7.3 SENSITIVITY TRIALS

A series of 24 trials were carried out to determine the effects of regional adjustments in coefficients of bottom friction and the effects of the approximate schematization at critical locations. To reduce the amount of computation required to provide records of length suitable for the separation of several major tidal constituents, these studies were restricted to the combined M_2 and K_1 tides, thus allowing for the major non-linear tidal constituent interactions found in earlier studies.

Coefficients of bottom friction were varied regionally to approximate the distribution of dissipation determined in the adjustments of the earlier model GF2. The distribution of computed amplitudes and phases of the two tidal constituents, M_2 and K_1, proved to be highly sensitive to such adjustments. It is of interest to note that if an artificial barrier is introduced to block off the effects of the northern passages in the Strait of Georgia, the tidal amplitudes in the latter are reduced approximately by a factor of one half. This should not, of course, be construed as representing such an actual physical intervention in the prototype since the boundary conditions employed in the experiment would no longer apply. These results serve to emphasize, however, influence of these northern passages in the overall tidal propagation in the system. Representative of the results obtained in the most successful of these trials, Fig. 7.3 shows comparisons of computed elevations with those predicted on the basis of harmonic constants derived from analysis of observations at typical locations.

The distribution of frictional coefficients finally adopted is as follows. The overall friction coefficient was 0.003. Exceptions included the Puget Sound system and Rosario Strait (0.006); Haro Strait (0.03); Active, Porlier, Gabriola Passes, and Dodd Narrows of the Gulf Islands (0.03); Seymour Narrows (0.006), Okisollo Channel (0.03), and Yuculta Rapids (0.03) of the northern passages. These coefficients, other than for locations now differently schematized, conform essentially to those employed in the earlier model GF2.

7.4 SIMULATION OF MIXED TIDES

Using the above geographical distribution of friction coefficients, a 30-day simulation in which 61 tidal harmonic constituents were employed to prescribe elevations on the open boundary, was undertaken. Illustrative of this model's performance, harmonic constants for the major tidal constituents (M_2, S_2, N_2, K_1, O_1, and P_1) obtained by harmonic analysis (Foreman, 1977) of the computed tidal elevations for representative locations are shown in Table 7.1. Inference relations based on observed tidal elevations were used to separate the P_1 and K_1 constituents.

Although the agreement between the observed and modelled values falls somewhat short of that obtained from the earlier model GF2, these results should be fully satisfactory for practical applications and for studies of baroclinic calculations with a planned three-dimensional development of this scheme (GF8).

Figure 7.4 shows good agreement between the computed and predicted (based on observations) tidal elevations for a representative 2-wk period at locations indicative of the tidal regimes at Point Atkinson (Strait of Georgia), Victoria (inner Juan de Fuca

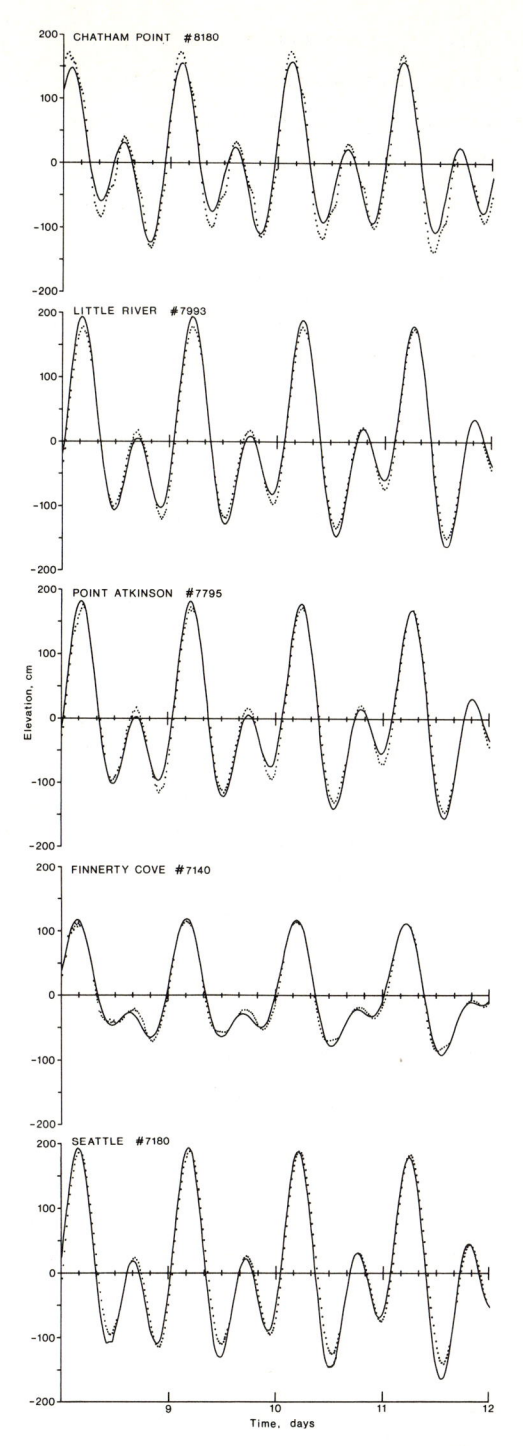

Fig. 7.3 Observed (predicted on the basis of observations) (solid lines) and computed (dotted lines) elevations for the mixed (M_2 and K_1) tides only, at representative locations throughout the system.

Table 7.1 Comparisons of M_2, S_2, N_2, K_1, O_1, and P_1 constituents with amplitudes H (cm) and phases g (°) taken from observations at representative locations and compared with those obtained from the model for a trial of 30 days.

Location	Tide gauge number	M_2		S_2		N_2		K_1		O_1		P_1	
		H	g	H	g	H	g	H	g	H	g	H	g
JUAN DE FUCA STRAIT													
Sooke	7020												
observed		44	52	13	66	9	32	57	146	35	135	17	144
calculated		45	42	13	62	12	20	57	145	36	133	18	144
Port Angeles	7060												
observed		50	77	14	86	12	57	65	137	39	131	—	—
calculated		53	67	15	86	9	63	64	143	39	130	20	142
Victoria	7120												
observed		37	87	10	94	9	63	63	149	37	137	20	148
calculated		37	80	11	98	11	47	62	152	38	138	19	151
PUGET SOUND													
Port Townsend	7160												
observed		65	118	19	129	16	146	67	148	44	139	—	—
calculated		64	127	25	156	24	168	72	160	41	143	22	158
Everett	—												
observed		103	139	25	156	21	113	82	157	45	143	23	155
calculated		98	145	24	168	23	121	78	167	43	149	24	165
Meadowdale	—												
observed		101	142	28	153	23	121	73	155	45	145	—	—
calculated		94	145	23	168	24	111	77	167	43	148	24	165
Seattle	7180												
observed		106	139	26	157	24	111	84	156	46	143	26	154
calculated		100	145	25	168	19	116	78	167	43	149	24	165
Tacoma	—												
observed		114	140	27	159	22	113	83	157	46	142	26	152
calculated		107	146	27	169	20	117	80	167	44	145	25	165

Table 7.1 Continued

Location	Tide gauge number	M_2		S_2		N_2		K_1		O_1		P_1	
		H	g	H	g	H	g	H	g	H	g	H	g
Olympia													
observed		145	156	34	182	27	134	88	164	48	152	26	165
calculated		135	155	34	180	25	130	86	172	47	154	27	171
SAN JUAN ISLAND PASSAGES													
Finnerty Cove	7140												
observed		45	125	11	138	10	101	71	157	41	144	22	156
calculated		47	121	13	139	11	81	70	160	42	145	22	158
Charles Island	7148												
observed		55	105	13	112	13	80	71	148	39	134	22	147
calculated		52	104	15	122	13	67	69	153	41	138	21	152
Reservation Bay	7196												
observed		57	118	16	127	16	94	65	148	41	138	—	—
calculated		57	110	16	127	13	72	72	153	43	138	22	152
STRAIT OF GEORGIA													
Ferndale	7564												
observed		72	152	17	170	14	130	80	163	45	149	24	167
calculated		71	153	19	172	14	122	77	172	44	155	24	171
Tumbo Channel	7510												
observed		73	159	18	180	16	137	81	167	46	152	25	164
calculated		74	161	20	181	15	130	78	180	45	158	24	174
Point Atkinson	7795												
observed		92	158	23	180	18	135	86	165	48	152	27	163
calculated		91	159	25	179	18	128	83	175	47	158	26	173
Waddington	8069												
observed		101	166	26	188	23	146	93	170	51	154	31	170
calculated		91	163	25	184	18	132	85	177	48	159	26	175

Table 7.1 Continued

Location	Tide gauge number	M_2		S_2		N_2		K_1		O_1		P_1	
		H	g	H	g	H	g	H	g	H	g	H	g
Winchelsea	7935												
observed		95	161	24	182	21	138	88	166	48	152	27	167
calculated		92	160	25	180	18	130	83	176	47	158	26	174
Little River	7993												
observed		99	161	24	182	22	138	90	166	49	152	29	166
calculated		94	161	26	181	19	130	84	176	48	159	26	174
NORTHERN PASSAGES													
Chatham Point	8180												
observed		90	73	29	87	21	49	65	150	37	138	20	146
calculated		107	78	37	95	26	50	68	144	40	134	21	143
Duncan Bay	8087												
observed		62	137	16	146	14	117	78	83	46	153	26	167
calculated		63	146	18	159	13	116	167	174	53	160	26	172
Campbell River	8074												
observed		81	148	21	163	17	123	84	165	50	151	26	162
calculated		77	154	21	172	16	123	82	175	47	158	25	173
Surge Narrows	8045												
observed		98	163	25	184	19	132	91	167	48	156	29	167
calculated		83	163	22	183	17	132	83	177	47	159	26	175
Big Bay	8060												
observed		75	143	22	180	24	145	49	128	50	155	29	172
calculated		87	164	23	184	18	132	50	129	48	159	26	175
Yorke Island	8233												
observed		117	40	39	61	26	21	56	140	32	130	17	138
calculated		117	40	40	62	27	21	56	137	34	130	17	135

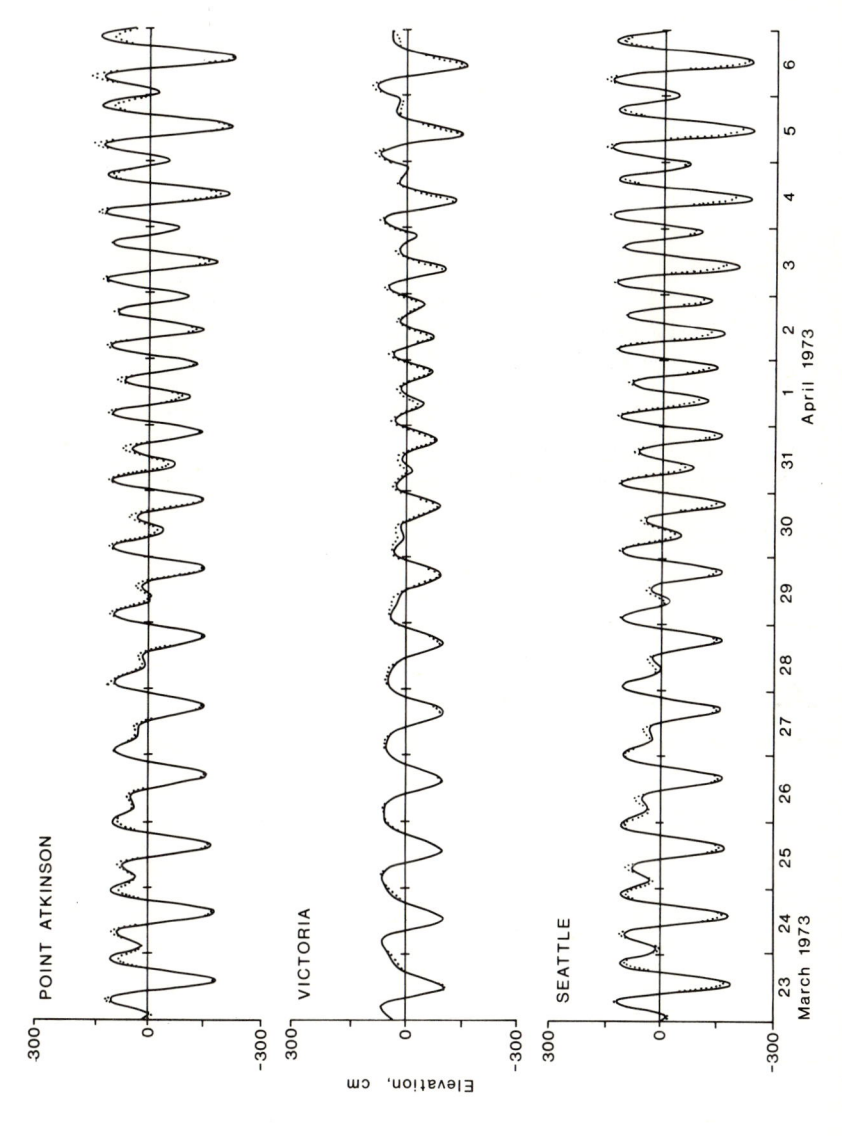

Fig. 7.4 Observed (predicted on the basis of observations) and computed elevations for the mixed tides (61 constituents) for a representative 2-wk period at Point Atkinson, Victoria, and Seattle.

Strait), and Seattle (Puget Sound). This simulation was for one month (12 March to 10 April 1973) using 61 tidal harmonic constituents. Figure 7.5 shows 2 wk of representative comparisons of observed and computed (GF7) velocities from line 11 in Juan de Fuca Strait. The agreement between the observed and computed longtitudinal components of velocity is satisfactory. Comparison between observed and computed cross-channel components is more irregular due to baroclinic influences not represented in the model.

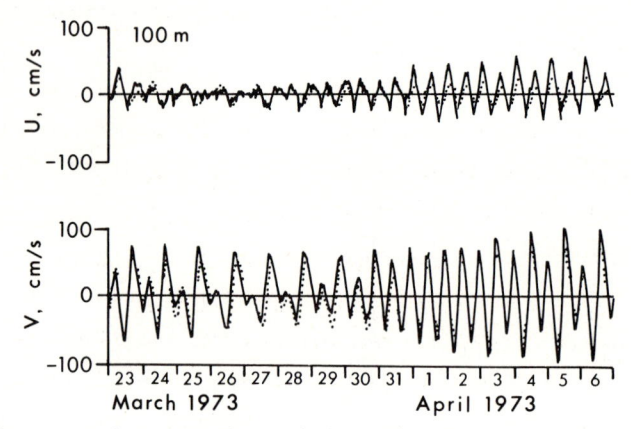

Fig. 7.5 Representative comparisons of observed and computed GF7 velocities from line 11 (Fig. 2.1) in Juan de Fuca Strait. U denotes the cross-channel component and V denotes the longitudinal component.

7.5 HORIZONTAL TIDAL RESIDUAL CIRCULATION

Simulated mixed tidal co-oscillations using the vertically-integrated homogeneous fluid models GF2, GF3, GF7, and the three-dimensional model GF6, all show well developed fields of horizontal residual circulation derived from tidal topographic interactions and friction in regions where the streams are strong. The theoretical nature of such interactions is a complex topic involving the extensive variety of ways in which the oscillating streams transfer vorticity to the mean field of motion (Zimmerman, 1981), due to the presence of many geomorphological features, through non-linear terms in the governing equations and the conditions imposed at bottom or lateral boundaries.

In the present approach, it will be illustrated how the governing equations generate vorticity for a simple case. Horizontal fields of tidal residual circulation for models GF2, GF3, and GF7 will then be discussed. Such circulation for the cases of both constant and varying density in GF6 will be presented later. It will be shown that field observations illustrating the formation of a tide-induced transient eddy and the associated mean flow obtained from arrays of moored current meters, confirm both the eddy and the associated residual circulation.

In the present system, the modelled eddies and associated residual circulations derive essentially from the advective accelerations in the momentum equations. Comparison of results obtained with and without these terms, respectively, show that their omission obviates any significant eddies and residual circulation. Eddy formation by these terms and the distribution of vorticity that results can readily be demonstrated for the very simple case of steady state flow in a flat-bottomed channel containing a thin partial barrier orthogonal to its major axis. The equations and their finite-difference forms are the same as those for model GF2. The Coriolis parameter f is set equal to zero, the horizontal eddy viscosity is 10^6 cm^2/s, the coefficient of bottom friction is equal to 0.003, and the mesh width is 2 km. A constant uniform elevation (50 cm) is prescribed along the elevation points that constitute the upstream boundary. At the downstream boundary, the elevation is assumed to be related to the velocity at the previous time step using Bernoulli's equation.

The relative vorticities resulting from inclusion of the advective accelerations in the two-dimensional momentum equations may be determined from the velocity field by computing the difference between the appropriate velocity gradients $(\partial v/\partial x - \partial u/\partial y)$ at each mesh corner. A steady state flow field is rapidly established. (It should be noted at this point that the centre of each mesh is indicated by a cross. A line emanating from the cross denotes the velocity vector averaged for each mesh centre. To prevent excessive overlapping, the vector is represented by a series of parallel lines which may be summed to give the appropriate speed.) In the absence of advective terms in the momentum equations there is flow around the barrier and weak vorticity is generated (Fig. 7.6a and b). When the advective terms are included, a well developed eddy results in the lee of the barrier along with considerable vorticity (Fig. 7.6c and d). It is of interest to note that a weak grid-scale vorticity forms upstream from the narrow passages due to the transfer of energy by the advective terms down to the minimum wavelength resolvable by the grid $2\Delta x$. Accumulation of such energy at this wavelength is prevented through dissipation by the horizontal eddy viscosity.

Extending these notions to the eddies formed in the models, the effect of oscillating streams is to produce eddies alternately in the lee of the barrier with respect to the flood or ebb stream. This gives rise to a residual circulation pattern resembling a 'figure 8' pattern. A further common configuration evolves with the formation of two back eddies near to, and flanking the source as an emergent jet enters a broad channel. In general, a small eddy starts to form some time after the onset of a particular tidal stream, reaching its maximum geographical extent as the tide begins to turn. The reverse stream then starts to accelerate over the part of the eddy that favours its own direction. The formation of such eddies will be dependent not only on topography but also on the strength and phase of its generating stream. The eddies resulting from the mixed character of the tides and varied topography of inner Juan de Fuca Strait give

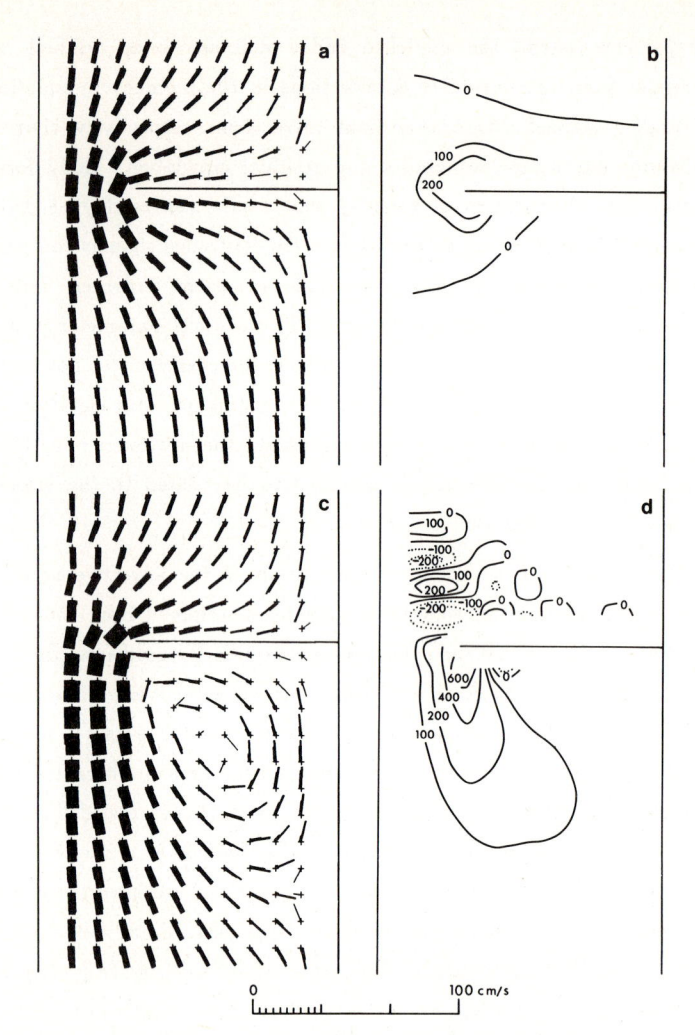

Fig. 7.6 Field of velocity vectors resulting from the insertion of a thin barrier into the steady state flow field in a non-rotating flat-bottomed rectangular channel. (a) Velocity vectors, (b) vorticity when the advective terms are omitted, (c) and (d) with the terms included.

rise to a remarkably complex pattern, the primary features of which are reproduced in the models. The growth and decay of the eddies in this region can be followed in the hourly tidal current charts (employing data from the model GF3) for representative tides (Canadian Hydrographic Service, 1983).

The residual tidal velocity vector fields obtained from the models employ the zero frequency component of u- and v-directed velocities averaged at the centre of each mesh, from harmonic analyses of 31 days of computed tidal velocities. Figures 7.7 and 7.8 show residual tidal current vector plots for the 4-km mesh model GF2 and the 2-km grid model GF3, respectively. No horizontal eddy viscosity was employed in the former model. Although the exceedingly complex pattern of strong residual currents is poorly resolved in GF2, the results are of interest when compared with those obtained from the higher resolution (2-km mesh) models GF7 (Fig. 7.9) and GF3, and subsequently with the three-dimensional 4-km mesh model GF6. The major eddies contributing to the overall residual circulation are at least qualitatively present. It has been estimated that about five mesh widths are required for the adequate numerical resolution of an eddy (Kuipers and Vreugdenhil, 1973).

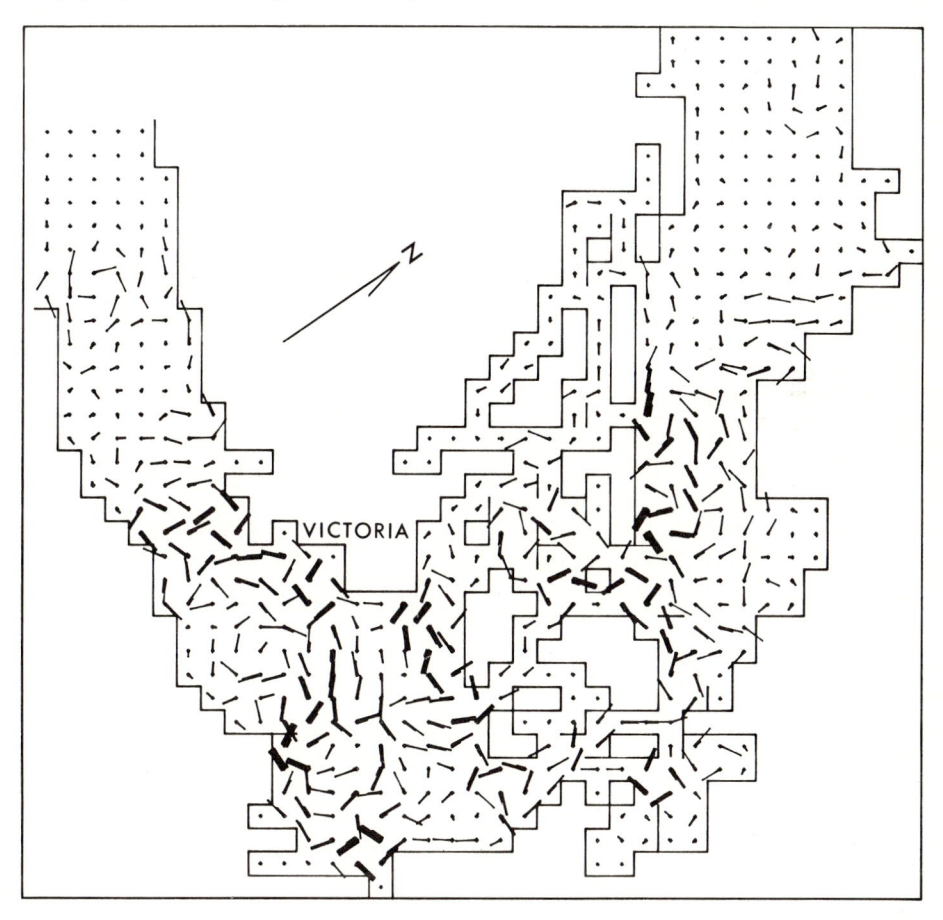

Fig. 7.7 Residual current vectors obtained from harmonic analyses of 1 month of tidal velocities computed by the coarse-grid model GF2 when predicted tidal elevations based on 61 constituents are prescribed on the open boundaries.

The 2-km mesh models GF3 and GF7 would appear to provide adequate resolution of the major eddies that dominate the overall pattern of tidal residual circulation. A comparison of the residual velocity vectors obtained from the model GF3 (Fig. 7.8) with those obtained from the overall model GF7 (Fig. 7.9), although showing the same general features, serves to illustrate a spurious net flow originating at the Puget Sound boundary opening and moving seaward through Juan de Fuca Strait in the case of GF3. For GF7 the pattern of residual circulation, though highly complex, is essentially closed

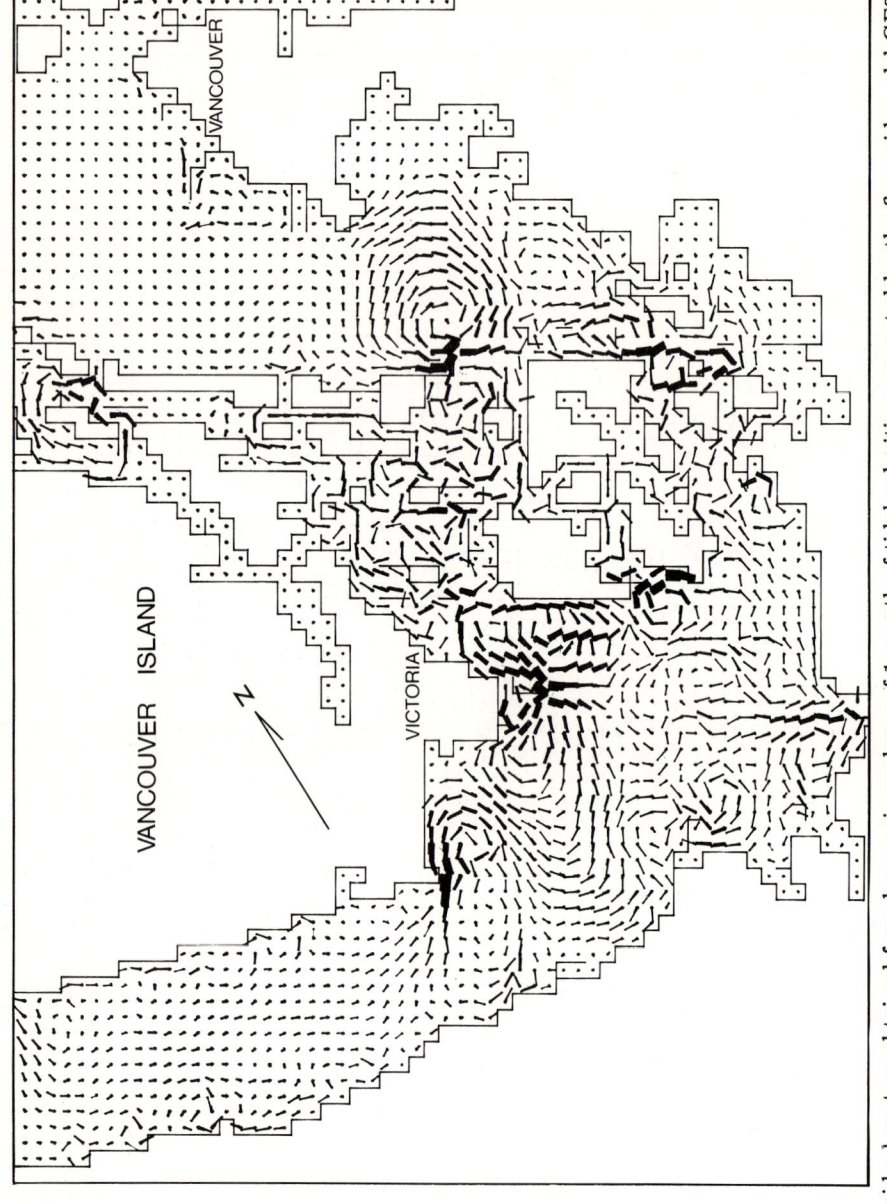

Fig. 7.8 Residual vectors obtained from harmonic analyses of 1 month of tidal velocities computed by the fine grid model GF3 when interpolated elevations obtained from the coarse grid model GF2 are prescribed on the open boundaries.

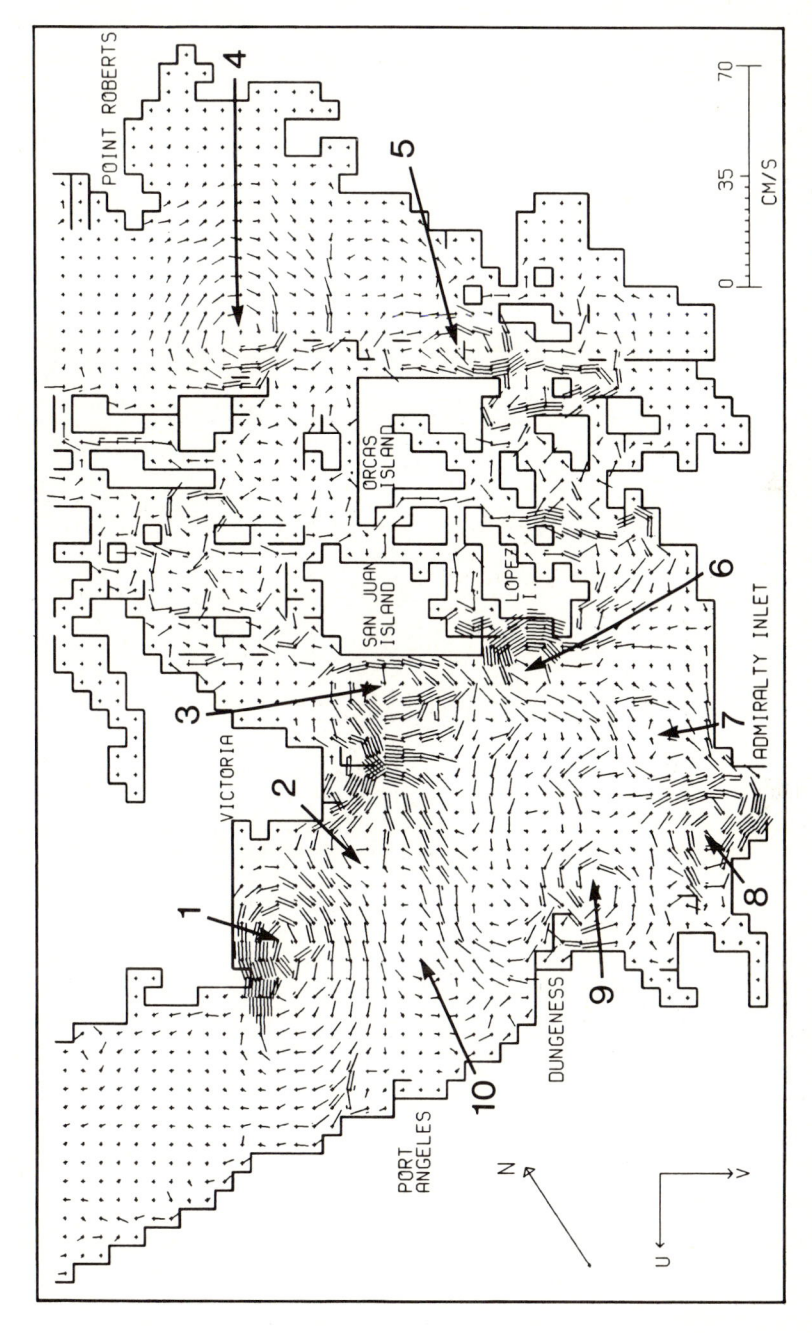

Fig. 7.9 Residual vectors obtained from harmonic analyses of 1 month of tidal velocities computed by the fine grid model GF7 when interpolated elevations obtained from the coarse grid model GF2 are prescribed on the open boundaries. Arrows denote the major eddies that have been predicted by the model GF7.

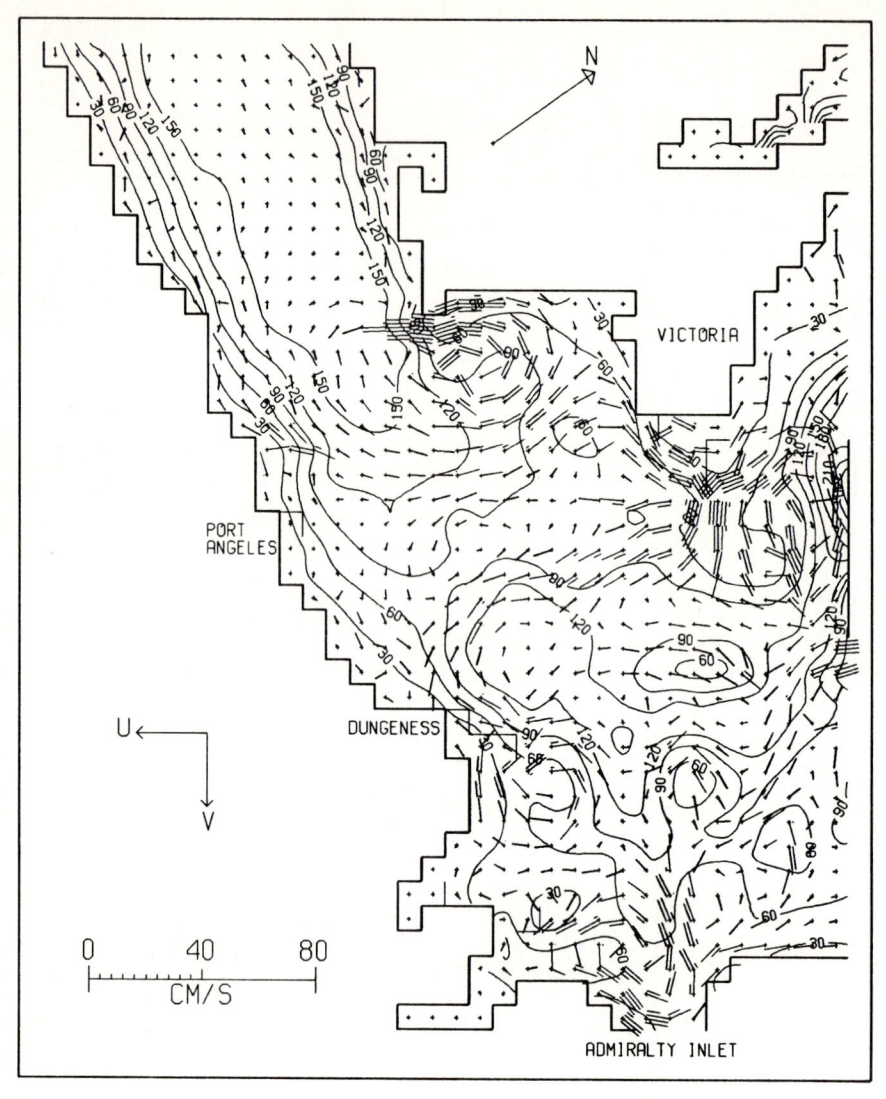

Fig. 7.10a Residual current vectors obtained from harmonic analyses of 1 month of tidal velocites computed by the fine grid model GF7 when interpolated elevations obtained from the coarse grid model GF2 are prescribed on the open boundaries for Juan de Fuca Strait. Bottom topography (m) is also shown.

in character. This is consistent with the results of earlier test trials involving a simple configuration as shown in Chapter 5. The general overall pattern of residual circulation obtained in the southern Strait of Georgia, San Juan Island passages, and Juan de Fuca Strait is shown in Fig. 7.9 where the residual eddies are numbered 1 to 10. A more detailed representation, including the bottom topography which is intimately related in the generation of such circulation, is shown in Fig. 7.10a and b. In a number of instances the eddies dominating this field of tidal residual circulation may be associated

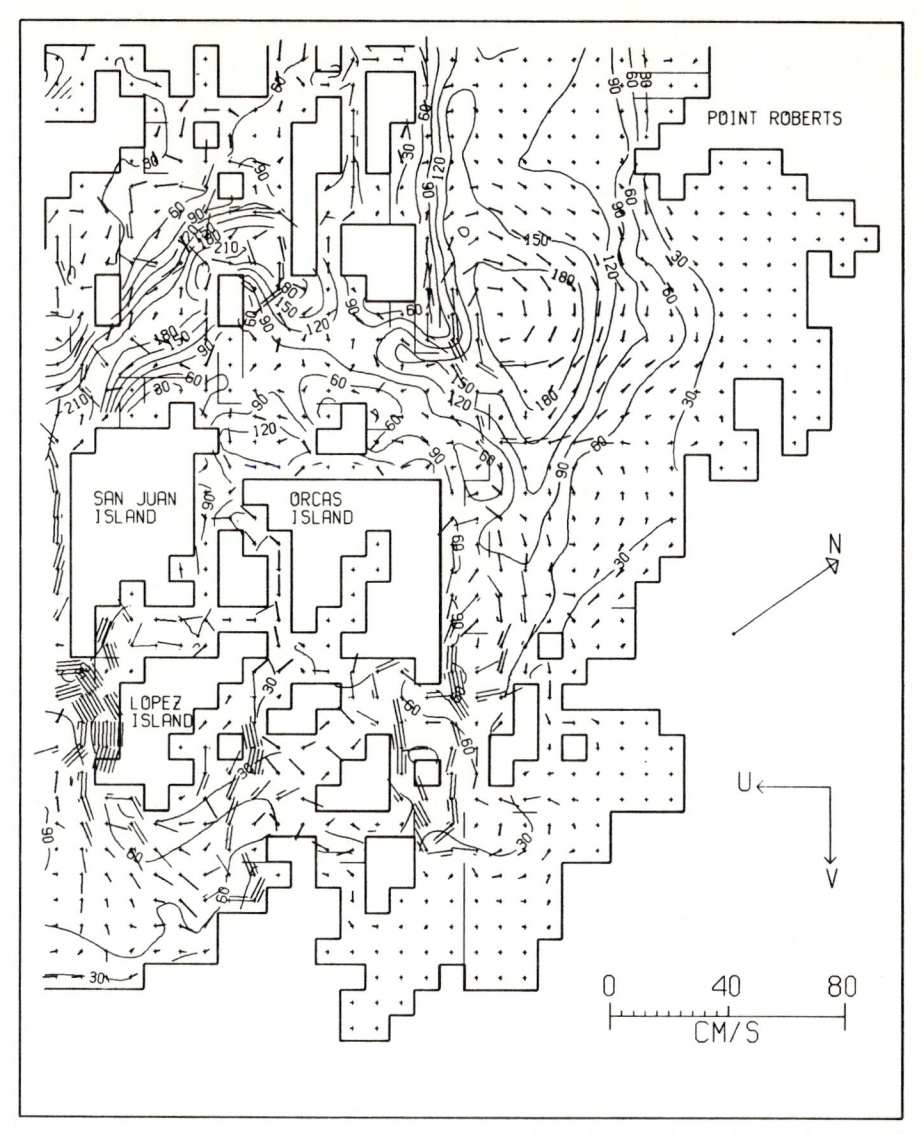

Fig. 7.10b Residual current vectors obtained from harmonic analyses of 1 month of tidal velocities computed by the fine grid model GF7 when interpolated elevations obtained from the coarse grid model GF2 are prescribed on the open boundaries for the Strait of Georgia. Bottom topography (m) is also shown.

with eddies formed by the tidal streams at some particular stage of the tidal cycle. Thus the residual eddies 1, 3, 4, 5, and 9 likely derive from eddies clearly evident at the later stage of a flooding tide as illustrated in Fig. 7.11. The poorly resolved eddy numbered 2, south of Victoria and eddy 8 near the mouth of Puget Sound reflect the influence of an ebbing tide (Fig. 7.12). Eddy 10 would appear to be a consequence of the oppositely-directed cross-strait currents associated with the major eddies 1 and 3. Eddy 6 is doubtful because of its dependence on the crudity of schematization in the

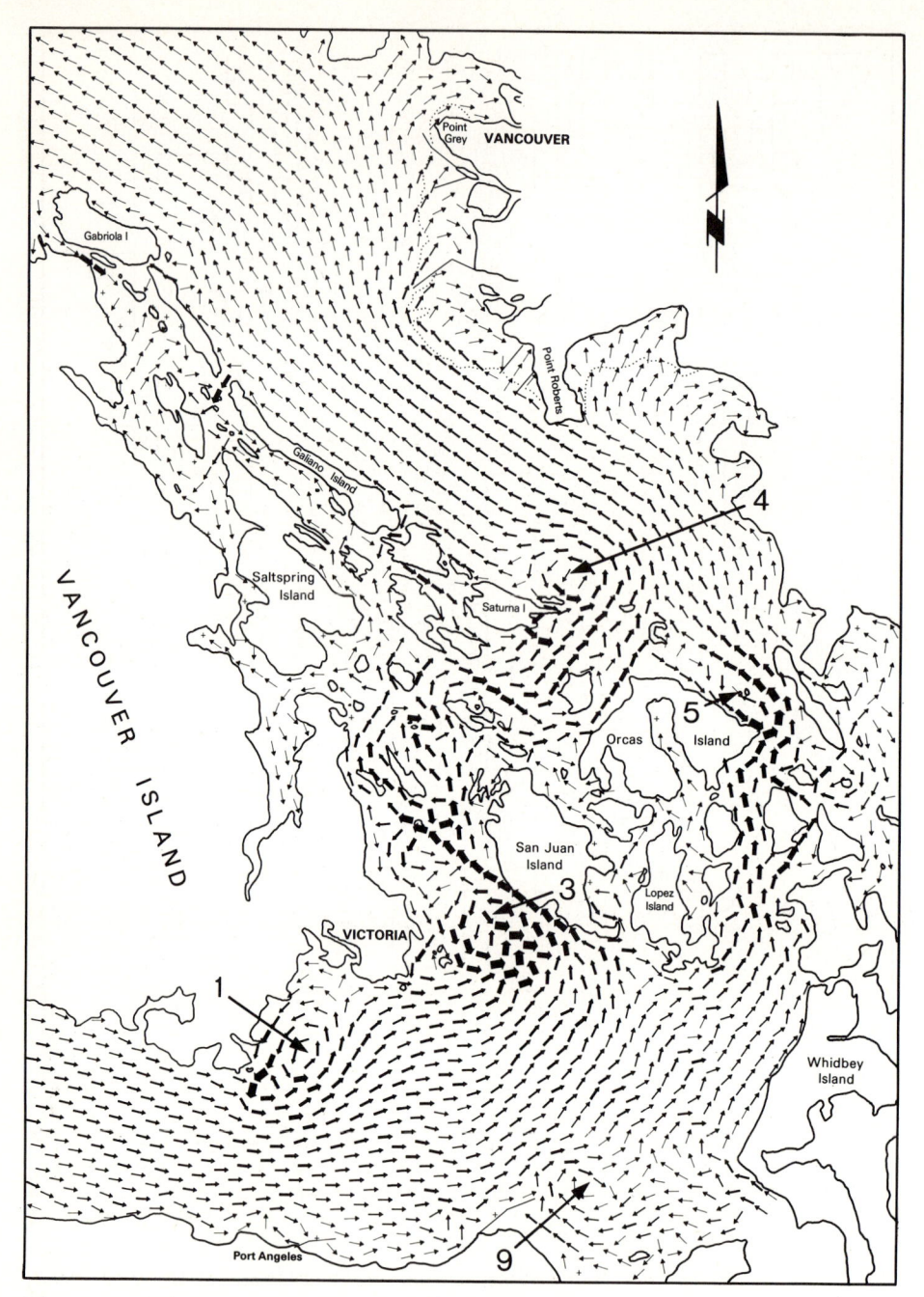

Fig. 7.11 Tidal currents at the end of a flood tide. Shown are the areas where eddies 1, 3, 4, 5, and 9 have been predicted by the model GF7.

area of strong tidal streams entering and leaving Middle Channel.

An interesting feature of these modelled tidal eddies concerns the local depression of the water surface of the centre of the major eddies 1, 3, and 4 of order 10–15 cm.

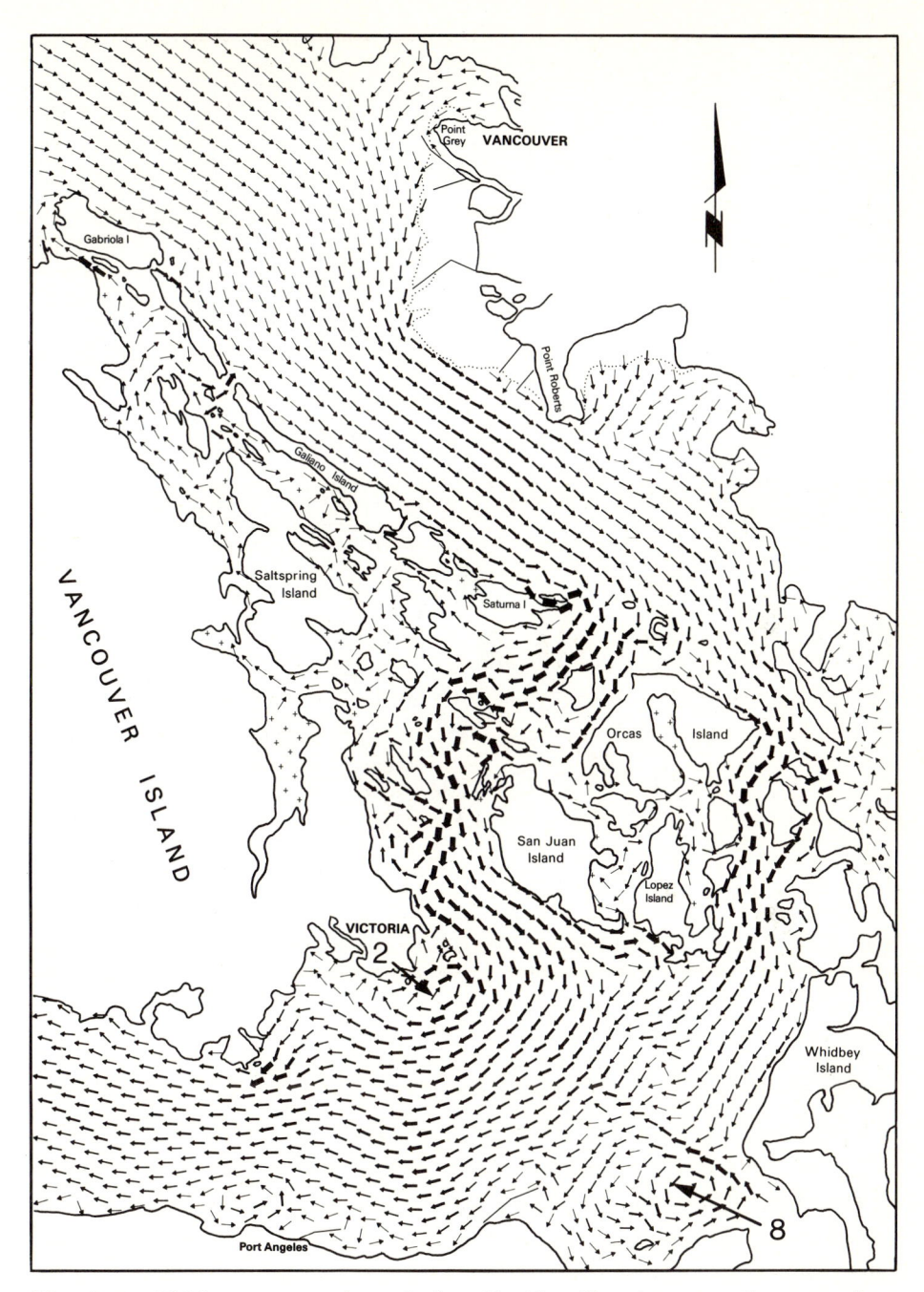

Fig. 7.12 Tidal currents at the end of a ebb tide. Also shown are the areas where eddies 2 and 8 have been predicted by GF7.

This is illustrated by the distribution of sea surface elevation at a time when the tidal streams are flooding strongly (Fig. 7.13).

One other major eddy predicted by the model is formed by the flood tidal jet

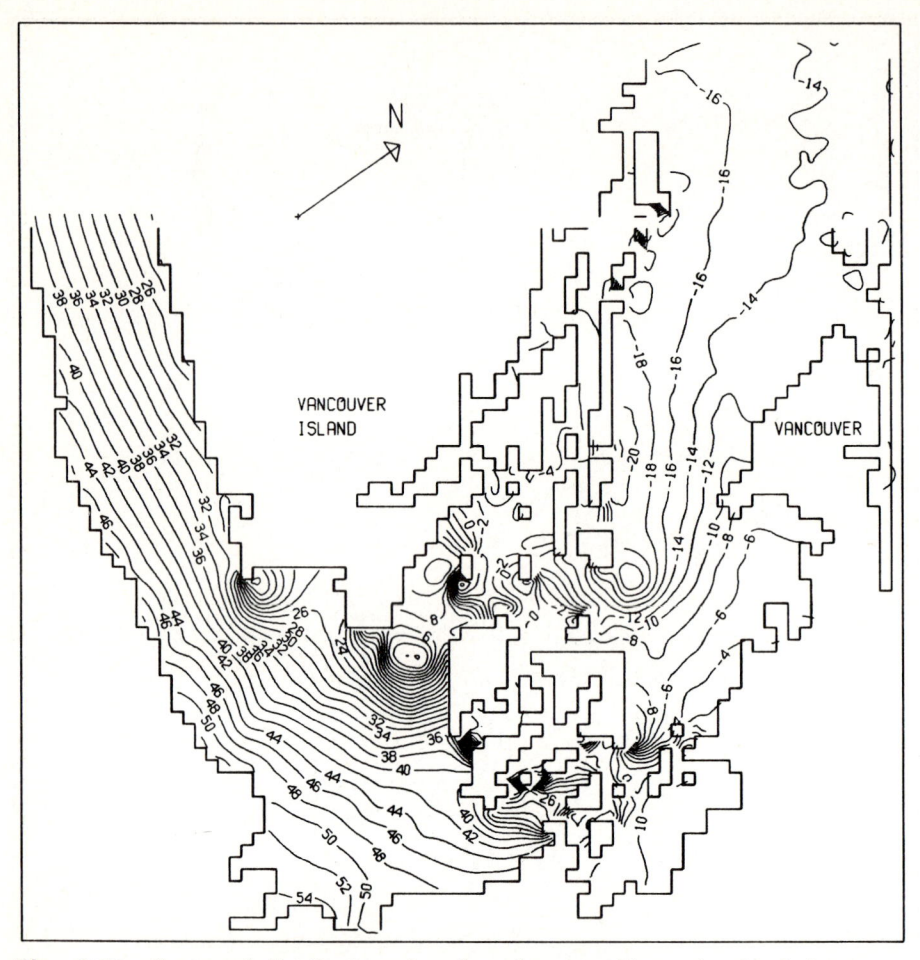

Fig. 7.13 Contoured distribution of surface elevations illustrating local depressions (cm) of the sea surface near the centres of the major eddies 1, 3, and 4 on a flooding tide.

entering the northern Strait of Georgia from the southern end of Discovery Passage (Fig. 7.14). There is a strong net southerly flow along the Vancouver Island shore and slower return flow in the adjacent deeper water. This flow is of particular interest in connection with baroclinic circulation deriving from the ebb tidal flow of strongly vertically-stratified water from the northern Strait of Georgia into Discovery Passage and the reverse flood flow of mixed water at an appropriate depth of neutral buoyancy.

The question arises as to whether the various features of the tidal residual circulation predicted by these vertically-integrated models accord with what actually happens in nature. Any such verification will require that these purely tidal features of the flow field be sufficiently strong to stand out against the fluctuations associated with the estuarine and wind-driven circulation or disturbances from the open ocean. The existence of the flood tidal eddies that account for eddies 1, 3, and 4 in the residual

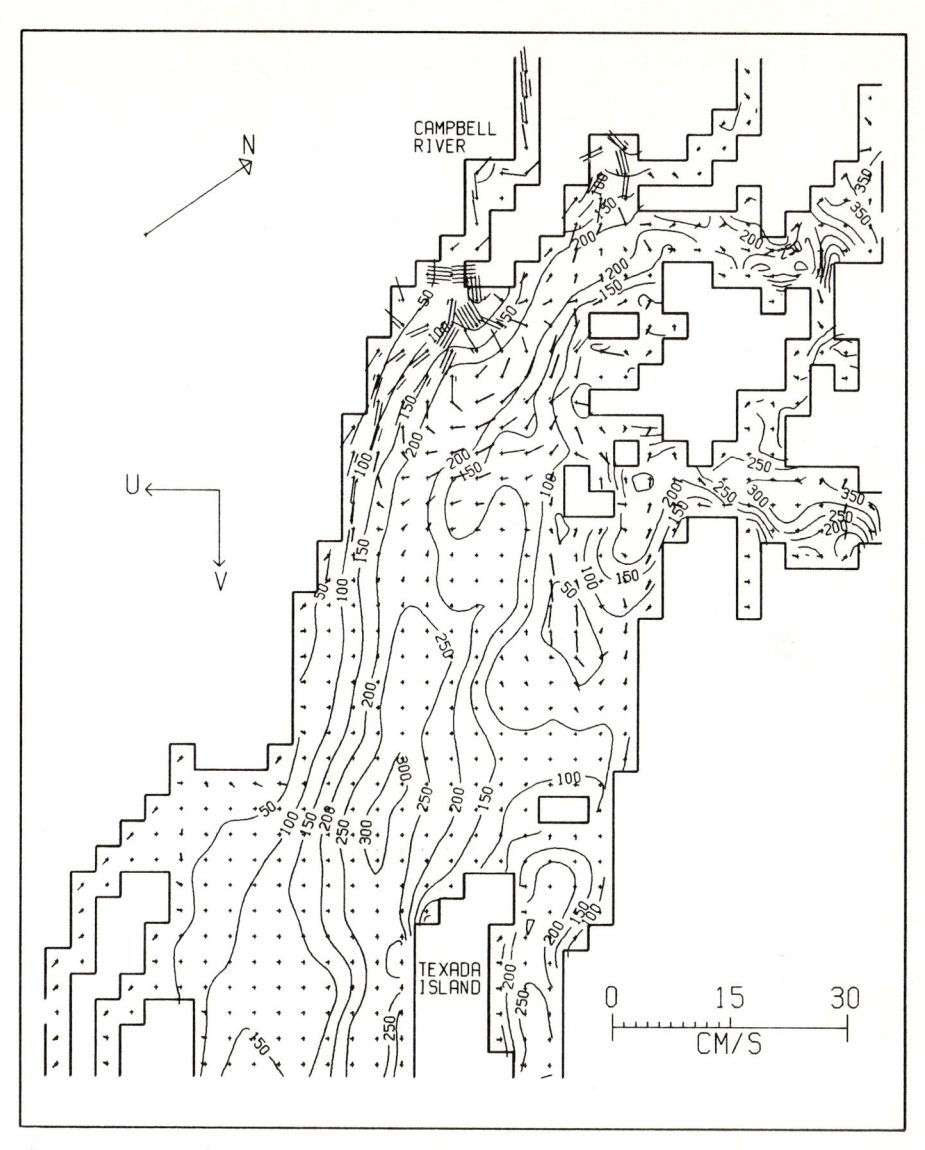

Fig. 7.14 Residual current vectors obtained from harmonic analyses of 1 month of tidal velocities computed by the fine grid model GF7 when interpolated elevations obtained from the coarse grid model GF2 are prescribed on the open boundaries for the northern Strait of Georgia. Bottom topography (m) is included.

circulation (Fig. 7.9) has been verified by tracking free floating surface current followers. Figure 7.15 shows the velocity vectors inferred from such tracks for the later part of a flooding tide at the location of eddy 3. Also shown is the comparable tidal velocity vector field obtained from model GF3. There exists good agreement between the observed and modelled directions and speeds.

The residuals derived from analyses of 30 days of velocities computed by the model GF3 (Fig. 7.8) and those inferred from moored current meters are shown in Fig. 7.16

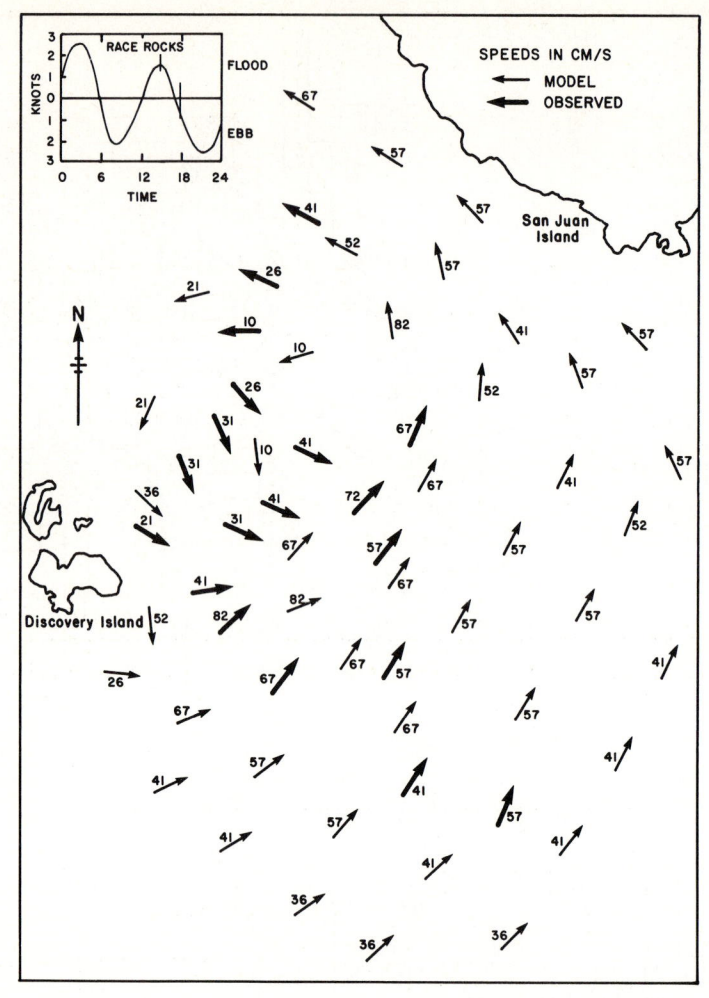

Fig. 7.15 Unit vectors showing directions and speeds of the flood tidal stream at the southern end of Haro Strait obtained from the model and from drogue tracks. This tidal eddy is associated with the residual tidal eddy 3 in Fig. 7.9.

for the general area of eddies 1, 2, and 3. Confirmation of the basic residual circulation obtained from the model with that derived from the moored current meters is shown in Fig. 7.17 for eddy 4. The same convention with respect to presentation has been employed as in Fig. 7.11. Where two or three current meters occupy different depths at the same location, the results have been averaged. Again, it will be noted that there is generally good agreement between the modelled speeds and directions.

From these results it may be concluded that there exists a strong and complex pattern of residual circulation induced by tidal topographic interactions in the southern part of the system and that the major features of that pattern may be inferred from the vertically-integrated 2-km mesh models GF3 and GF7. It should be noted, however,

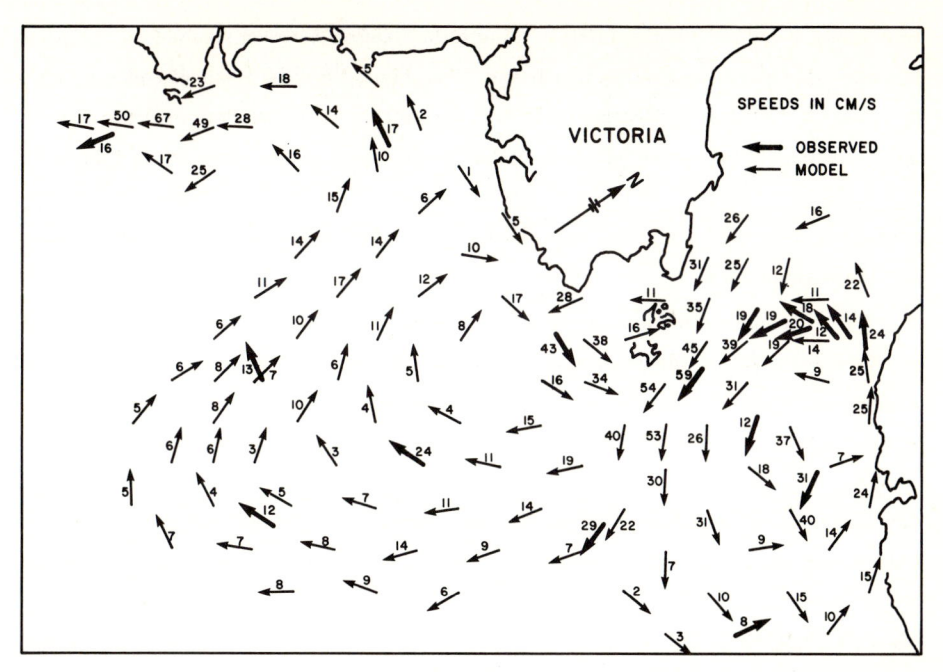

Fig. 7.16 Comparison of residual tidal velocities obtained from moored current meters in a region extending from Haro Strait to Race Rocks with those obtained from the model.

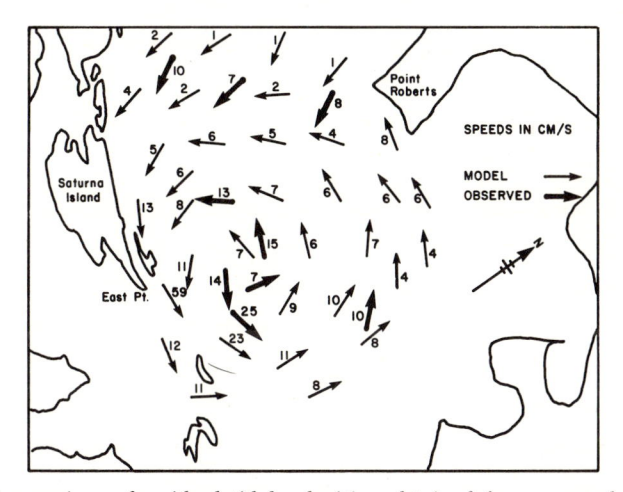

Fig. 7.17 Comparison of residual tidal velocities obtained from moored current meter records in the southen end of the Strait of Georgia with those obtained from the model (eddy 4 in Fig. 7.9).

that some modifications may result when vertical residual circulation is taken into account. This will be referred to in Chapter 13 in connection with the three-dimensional model GF6.

The pattern of tidal residual circulation described above persists on a geological time scale. It is of particular interest to note the relation between residual currents and bottom contours in Figs. 7.10a and b and 7.14. Thus, the tidal scour of soft sediments in the southern Strait of Georgia could well account for the large bottom depression. It might reasonably be speculated that coastal features such as the sandspits of Ediz Hook and Dungeness (see Fig. 1.8) derive partly from the return flows of residual circulation, associated with eddies 1 and 3, and partly from wave action (Thomson, 1981), along the Washington coast.

The phenomenon of rotating tidal streams can derive from a variety of causes such as the direct effect of tide-generating forces, the combination of a progressive wave with a standing oscillation, the proximity to shelving coasts, and the effect of the earth's rotation (Doodson and Warburg, 1941). For a strongly dissipative, topographically complex rotating system, the attribution of the rotating character of the streams to any cause or combination of causes for any freshwater location within the modelled fluid would request extensive analyses of the relative magnitudes and phasing of terms in the governing equations. It would, however, appear probable that the single largest contributing effect in the system is that of the earth's rotation. It was noted in Chapter 3 that an exploratory solution to the problem of Kelvin wave reflection in a frictionless rectangular gulf closed at one end and having dimensions approximating those of Juan de Fuca Strait and the Strait of Georgia, indicated that significant rotating character in the tidal streams should be confined to a limited region in the vicinity of the reflecting barrier. This derives from the presence of Poincaré waves which are confined to an area close to the reflecting barrier in a narrow deep gulf. A very approximate calculation using a 20-km width and 200-m depth indicates a decay distance from the barrier of 6–7 km for M_2 and K_1 tides.

The actual topography of the 2-km mesh model GF7 may be expected to introduce partial reflections where marked changes in the widths and depths of the major conveying channels occur. In general, the major constrictions are associated with the San Juan Island passages between Juan de Fuca Strait and the Strait of Georgia and the channels flanking Lasqueti and Texada Islands (see Fig. 1.30) in the northern part of the latter strait. With widths of the broad channels approaching these constrictions, at least of order 20 km, it may reasonably be supposed that the model will resolve the first and second cross-channel Poincaré modes. Resolution of the third mode would be minimal and there should be little energy associated with this mode.

Figures 7.18a–d and 7.19a–d show the distribution of the tidal current ellipses (in the interests of clarity for alternate meshes only) for M_2 and K_1 in the major conveying channels of the system. It should be noted that the velocity scales vary in these diagrams to facilitate presentation of the elliptical character of the streams. Away from the reflection and close to lateral boundaries, ellipticity diminishes and local tidal flows in the vicinity of a boundary tend to parallel that boundary. In the model

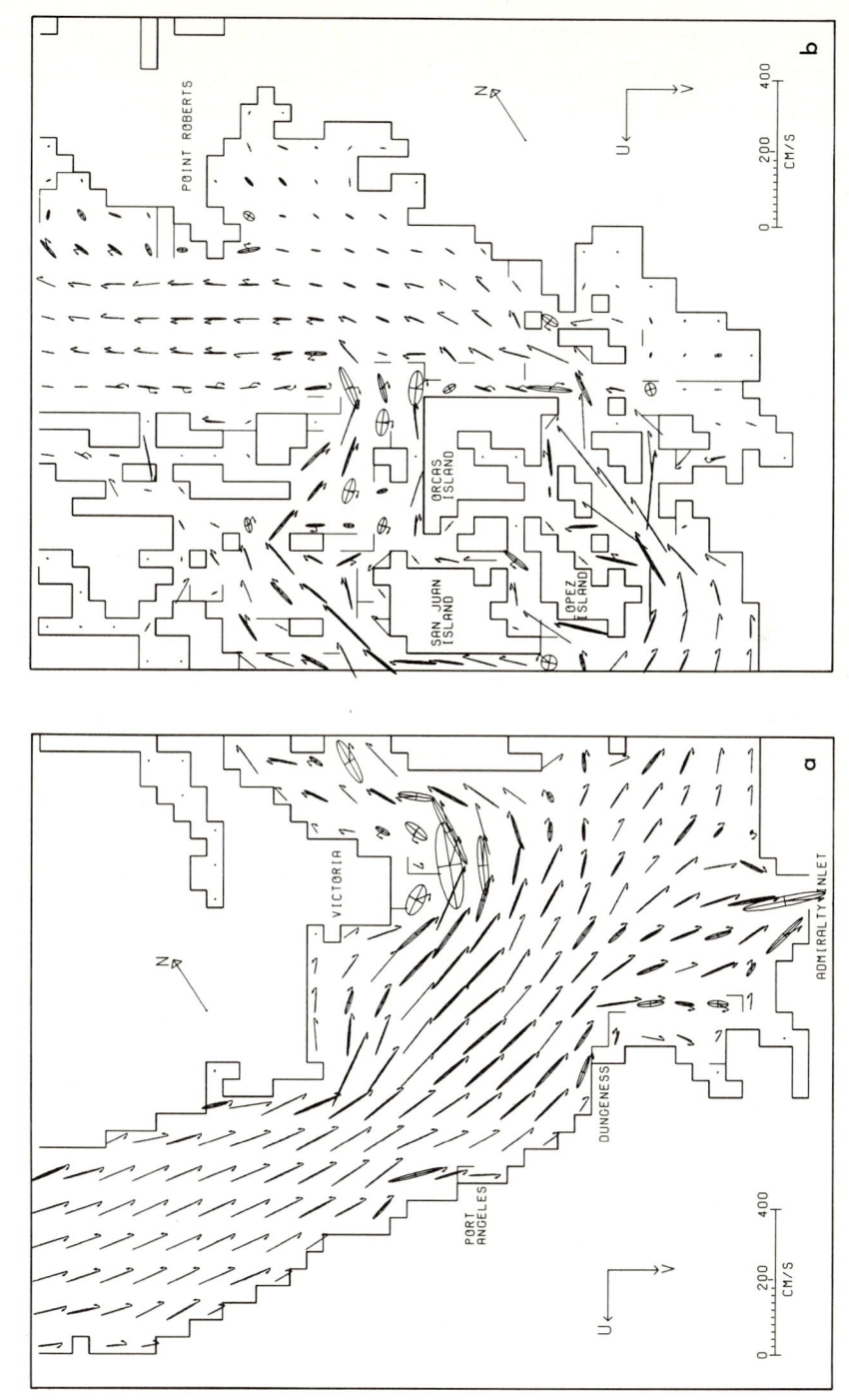

Fig. 7.18 M₂ tidal current ellipses for (a) Juan de Fuca Strait and (b) southern Strait of Georgia.

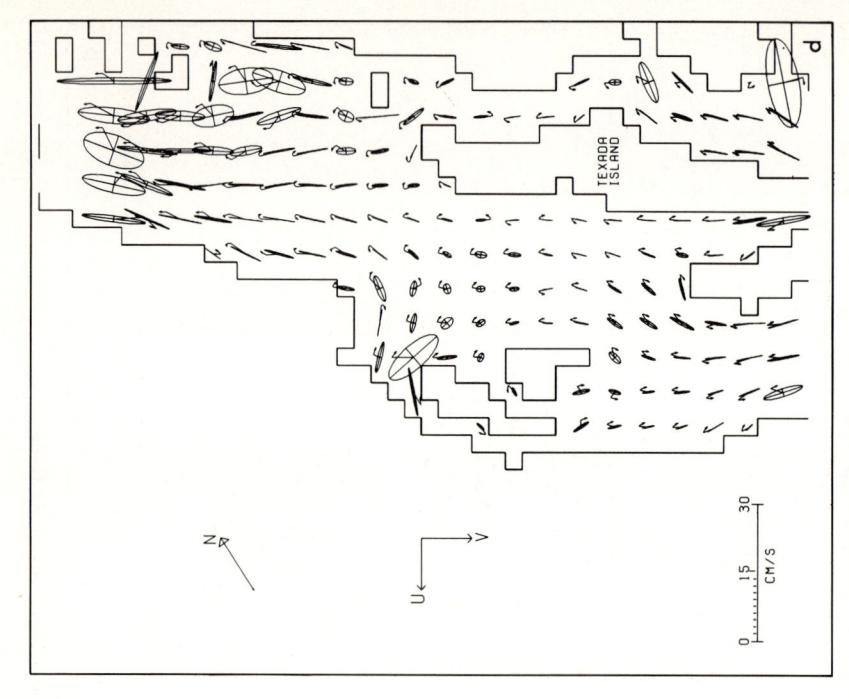

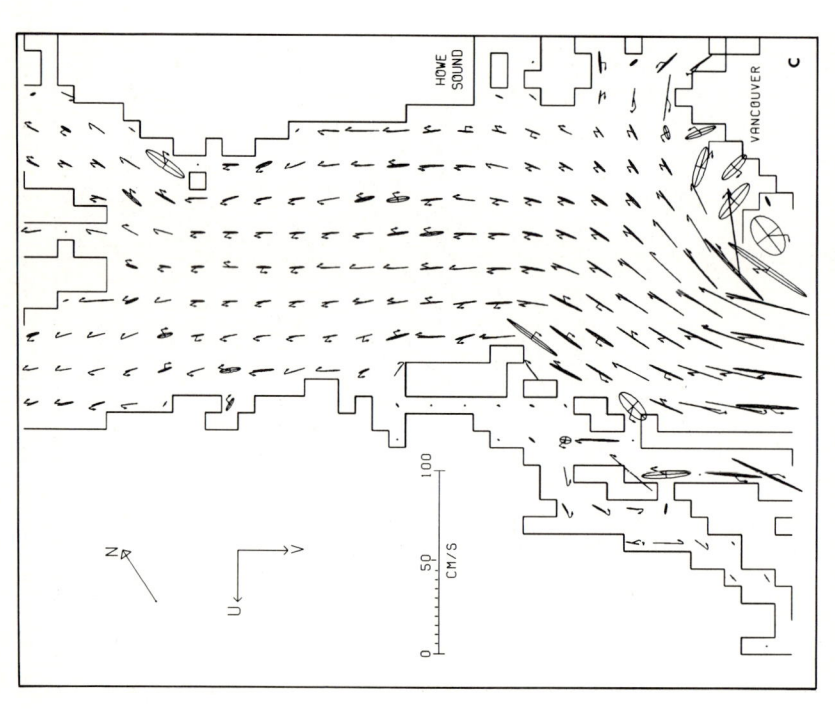

Fig. 7.18 M_2 tidal current ellipses for (c) central Strait of Georgia and (d) northern Strait of Georgia.

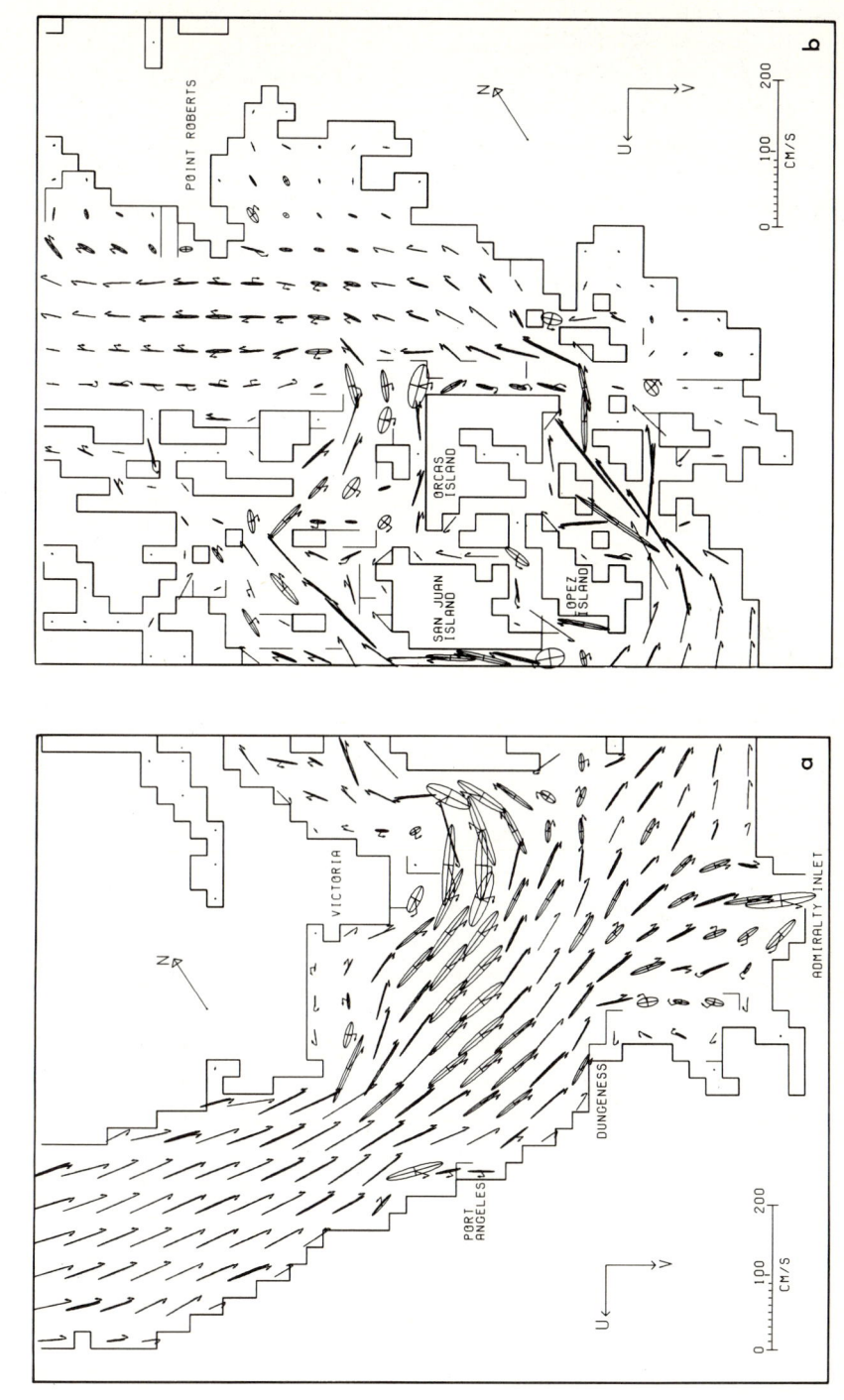

Fig. 7.19 K_1 tidal current ellipses for (a) Juan de Fuca Strait and (b) southern Strait of Georgia.

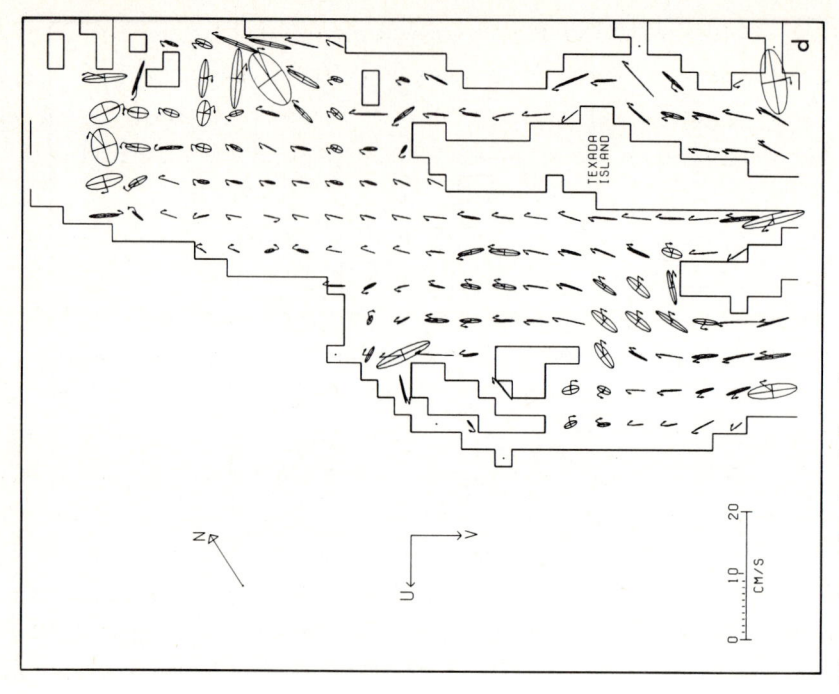

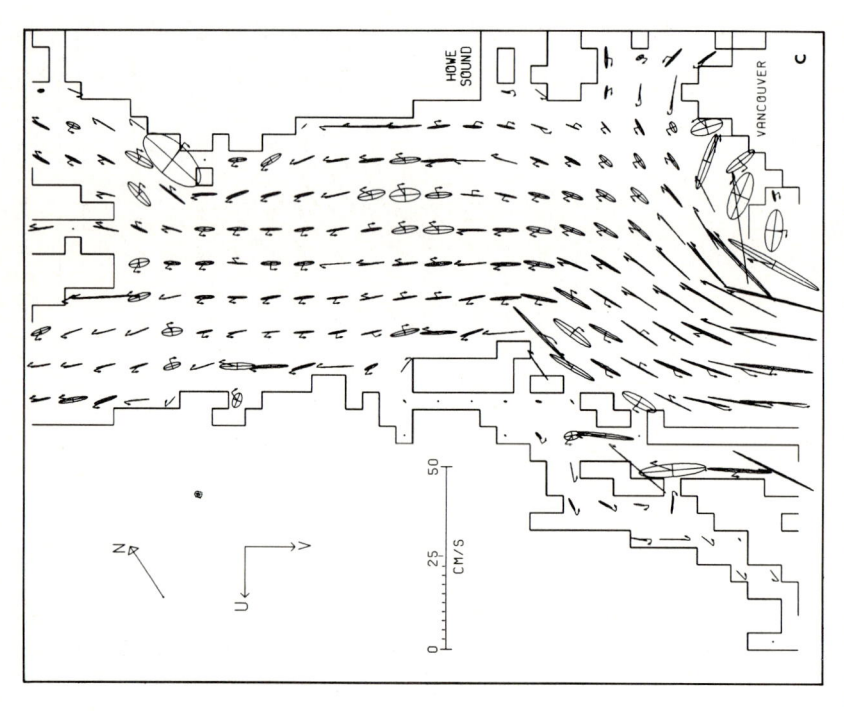

Fig. 7.19 K_1 tidal current ellipses for (c) central Strait of Georgia and (d) northern Strait of Georgia.

significant ellipticity in Juan de Fuca Strait is associated with the approaches to Haro Strait. In the Strait of Georgia, pronounced ellipticity is associated with the area of Boundary Pass, the region immediately south of Lasqueti and Texada Islands, and the approaches to the complex of passages opening off the northern Strait of Georgia. A further area where pronounced rotatory character is evident is the shallow banks adjoining the Fraser River delta. It may reasonably be supposed that this is associated with the effect of the shallow banks and to the effects of friction.

The direction of the flooding streams corresponds to the major axis of each ellipse directed towards the northern Strait of Georgia or the closed end of local embayments. Throughout, the Greenwich phase of the M_2 tidal streams is of order 40–90°. If a line is superimposed upon each ellipse through its centre intersecting the ellipse at the time of local high water, the angle between the major axis and that line denotes the phase difference between local high water and the M_2 tidal stream maximum. At the outer end of the line denoting the stream direction at the time of high water, there is a tick mark indicating the direction of rotation and hence the direction of positive increase in phase angle between the occurrence of the flood stream maximum and that of local high water. It can readily be shown that the direction of energy flux coincides with the direction of the tidal stream vector at the time of local high water. This flux is related to the averaged product $\overline{u\zeta}$ of tidal elevation and stream velocity over a tidal cycle. Thus, for some point within the modelled system having a local tidal elevation given by

$$\zeta = \zeta_0 \cos(\omega t - g_c)$$

and assuming that the stream ellipse at that point having its major axis coinciding with the x axis, the tidal stream velocity components are given by

$$U = U_0 \cos(\omega t - g_c),$$
$$V = V_0 \sin(\omega t - g_c)$$

where the respective velocity amplitudes and Greenwich phase are U_0, V_0, and g_c and the elevation amplitude is ζ_0 and phase is g_e. At the time of high water $\omega t = g_e$ and hence the velocity components will be

$$U = U_0 \cos(g_e - g_c),$$
$$V = V_0 \sin(g_e - g_c).$$

Thus, the direction of the energy flux coincides with the direction of the tidal stream at the time of local high water. It should, however, be noted that this is strictly applicable when considering a single tidal harmonic in a simple system and that its application in the present context is approximate.

In the region of the M_2 degenerate amphidrome in the inner part of Juan de Fuca Strait, the co-phase lines are roughly orthogonal to the median line (Fig. 4.9, p. 95) and have a range of values similar to those of the M_2 flood tidal streams. There is, accordingly, a clear indication of a flux of energy from the open boundary towards the region of high frictional dissipation in the general area of the San Juan Islands. In general, there is a flux of energy directed towards the northern end of the Strait of Georgia. A point of interest concerns the energy flux in the vicinity of tidal eddies such as the southern end of Haro Strait. In the Strait of Georgia near Boundary Pass, the shallow banks around the mouth of the Fraser River, and at the northern opening leading into Discovery Passage, the direction of the energy flux is essentially the same as that of the velocity vectors that describe the eddy.

Comparison of M_2 and K_1 tidal current ellipses indicates substantial reduction in current speeds for the K_1 tide, but otherwise, a similar distribution in elliptical character occurs throughout the system. A point of interest concerns the current ellipses in the central Strait of Georgia immediately northwest of the entrance to the Howe Sound system (Fig. 7.19c) which are associated with a shallow bank (see Fig. 1.2). These ellipses are also present for the M_2 tide (Fig. 7.18c).

8 NORMAL MODES

8.1 INTRODUCTION

Not uncommonly, free or forced oscillations of natural water bodies may be excited by meteorological or seismic events. In the first instance these oscillations are called storm surges and in the latter case they are referred to as tsunamis. Applications of the barotropic models (GF2 and GF7) to determine the response of the system to instances of meteorological and seismic forcing are described in Chapters 9 and 10, respectively. Of particular significance in discussing these phenomena is the determination of the barotropic normal modes which indicate frequencies at which significant levels of energy can accumulate. In this chapter, the response of the model GF2 to an input signal comprising a range of frequencies covering the principal normal modes likely to occur in the system is described. Subsequent data on normal mode frequencies that may be of practical assistance in the interpretation of tide gauge records is then provided. Literature available on the theory of normal modes may be found in Murty (1984).

8.2 GRAVITATIONAL AND ROTATIONAL NORMAL MODES

Hough (1898) distinguished between natural oscillations of the first and second classes on the following basis. If σ is the frequency of oscillation and ω is the frequency of rotation, then oscillations of the first class are those for which $\sigma \to \sigma_0 (\neq 0)$ as $\omega \to 0$ and oscillations of the second class are those for which $\sigma \to O(\omega)$ as $\omega \to 0$. Bjerknes et al. (1934) distinguished between these two types of oscillation by means of the ratio $\sigma/2\omega$. Gravity modes (oscillations of the first class) are those for which $\sigma/2\omega \geq 1$. Elastoid-inertia modes (oscillations of the second class) are those for which $\sigma/2\omega \leq 1$. For the gravity modes, gravity appears in the frequency equation. In the case of the rotational (elastoid-inertia) modes, the frequency for a given mode is a function mainly of the ratio of the depth of the liquid to the radius of the container and gravity does not play an important role in the frequency equation. Here this discussion is restricted to gravity modes. In the mathematical analysis this restriction is imposed by introducing the approximations of the shallow water theory called the quasi-static approximation (Bjerknes et al., 1934). This means that in the vertical direction, equilibrium exists not only before the motion but also during the motion, with vertical accelerations of the liquid being considered negligible compared to that of gravity.

In studying tidal motions on a rotating earth Kelvin (1879) considered a shallow layer of water in a circular flat-bottomed cylinder and assumed the quasi-static approximation to the pressure field. Kelvin considered small rotations and neglected the curvature of the free surface due to rotation. If σ is the frequency of the rotating mode, σ_0 is the frequency of the non-rotating mode, and ω is the rotation frequency, then the result obtained is

$$\sigma^2 = \sigma_0^2 + 4\omega^2.$$

This result shows that rotation increases the frequency and thus increases the restoring tendency of the system when disturbed. However, if the curvature of the free surface is taken into account, this is not always true, especially for the higher modes.

8.3 METHOD OF DETERMINATION

For water bodies with a simple topography, the normal modes of oscillation may readily be determined by analytic methods (Murty, 1984). For more complex systems realistic results may be obtained with the aid of numerical models that have reasonable finite-difference approximations to the actual topography. In this case, the system may be forced by introducing some disturbance as an initial condition and determining the amplitudes of the decaying oscillations over a suitable range of frequencies at representative locations over the interior of the model. This requires some suitable radiation boundary condition to be prescribed at the open boundaries of the model. An alternative procedure is to prescribe at the open boundaries a forcing function consisting of summed sinusoids of unit amplitude and possessing a suitable range of frequencies. The model response will show large amplification or gain for locations associated with the excitation of a normal or natural mode.

8.4 OSCILLATORY RESPONSE OF THE SYSTEM

An initial experiment providing an estimate of the frequency of the fundamental gravitational normal mode was carried out using the original simplified version of the one-dimensional model GF1 (which includes Juan de Fuca Strait and the Strait of Georgia as well as the Puget Sound) in which the northern end of the Strait of Georgia was assumed closed. Oscillations having unit amplitude ranging in period from 1–24 h were prescribed at the entrance to Juan de Fuca Strait. The resulting distribution of gain at Point Atkinson (representative of the Strait of Georgia for this frequency range) as a function of frequency is shown in Fig. 8.1. The period at which maximum gain occurs is 19.2 h which constitutes a rough approximation to the period (16 h as indicated by the higher resolution model GF2) of the fundamental gravitational mode of the system with the entrance to the northern passages closed.

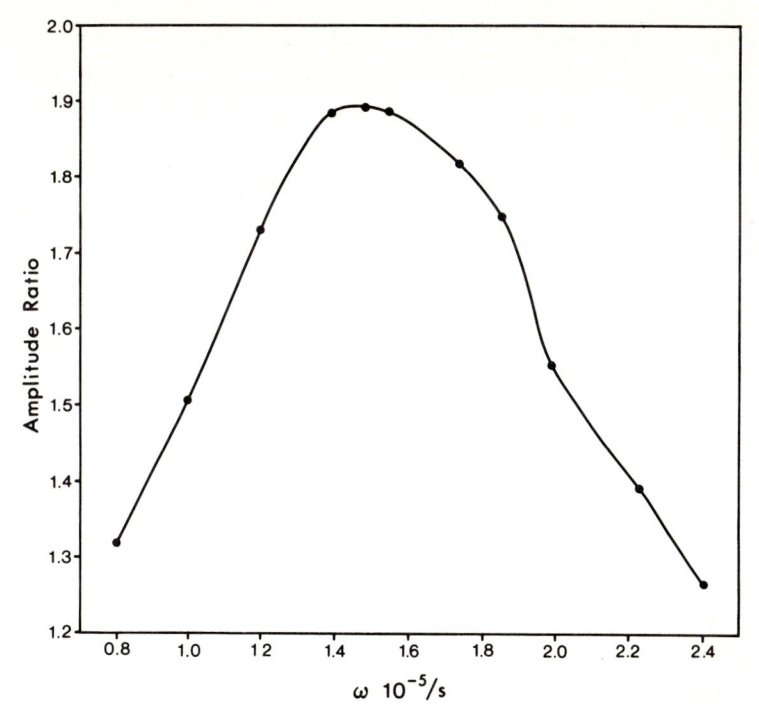

Fig. 8.1 Amplification ratio versus frequency for oscillations having unit amplitude ranging in period from 1–14 h prescribed at the entrance to Juan de Fuca Strait in a short version of the one-dimensional model GF1.

A numerical experiment essentially of the same nature as above was carried out using the combined system of one- and two-dimensional models, GF2. Given the overall length and approximate phase speed of propagating free surface disturbances in the system, the 4-km mesh should provide adequate resolution of modes of period 1 h or greater. It will be recalled that a typical free surface wave speed in this system is about 100 km/h. In this application of the GF2 model, in addition to the higher resolution and inclusion of the earth's rotation, an input function was applied at openings A and B (Fig. 4.7, p. 80) consisting of summed sinusoids, each of unit amplitude and having periods of 1–24 h in hourly increments. The resulting gain, expressed as a function of frequency for locations (1–13) representative of the major conveying channels, is shown in Fig. 8.2. It will be noted that large peaks at about an hourly period occur in Juan de Fuca Strait. If it is assumed that the inner end of the Strait is closed and a rough estimate is made of its free surface wave speed, its overall length would suggest a natural period of oscillation of roughly this order. The distribution of peaks in the overall diagram suggests that there is a substantial measure of decoupling between Juan de Fuca Strait and the Strait of Georgia at the shorter periods. It has been observed that a tsunami propagating into the system (with a period of about an hour) was substantially decimated in its passage through the San Juan Islands (Murty, 1977).

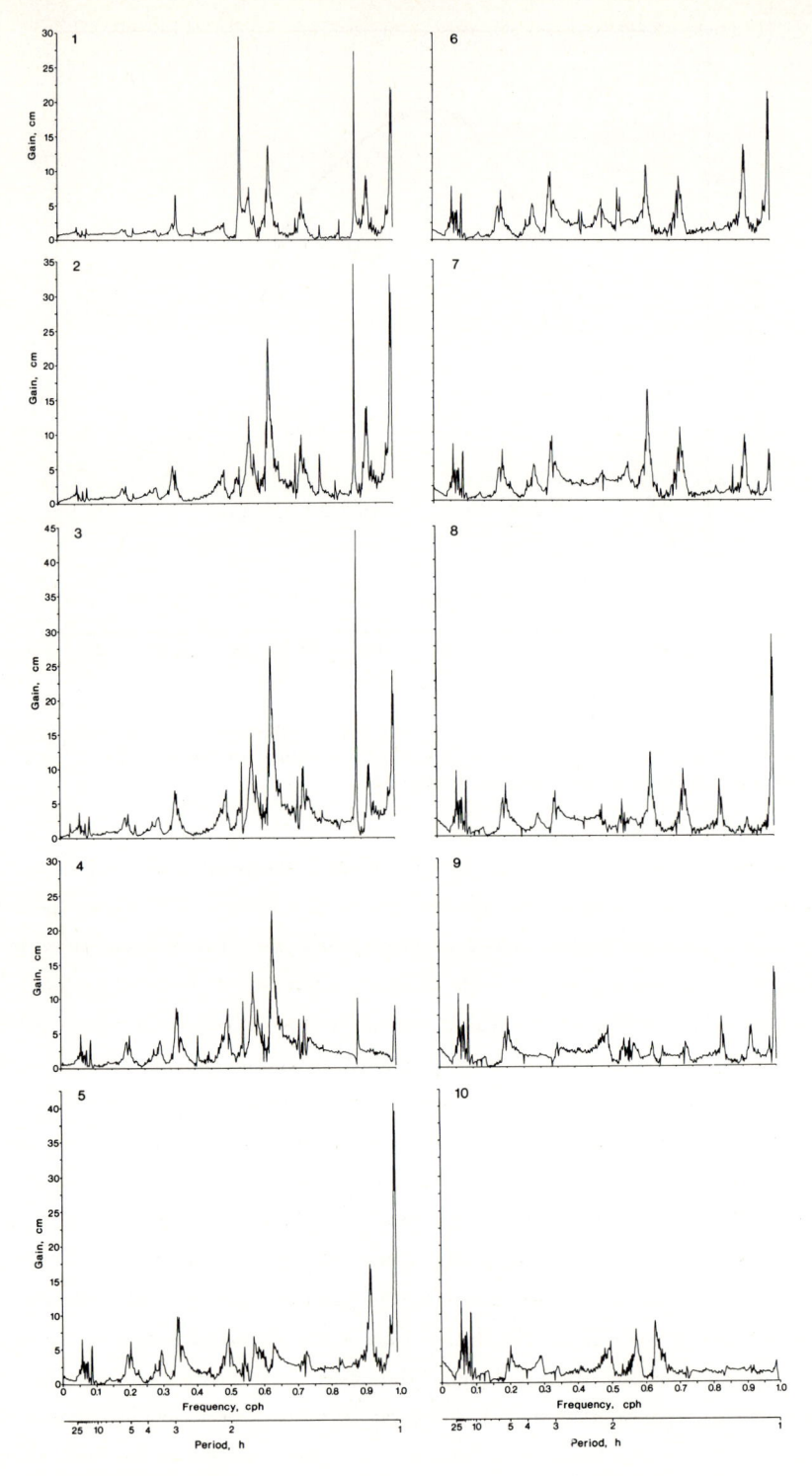

Fig. 8.2 Gain versus frequency curves at 13 locations in the GF2 grid. The last panel shows grid locations.

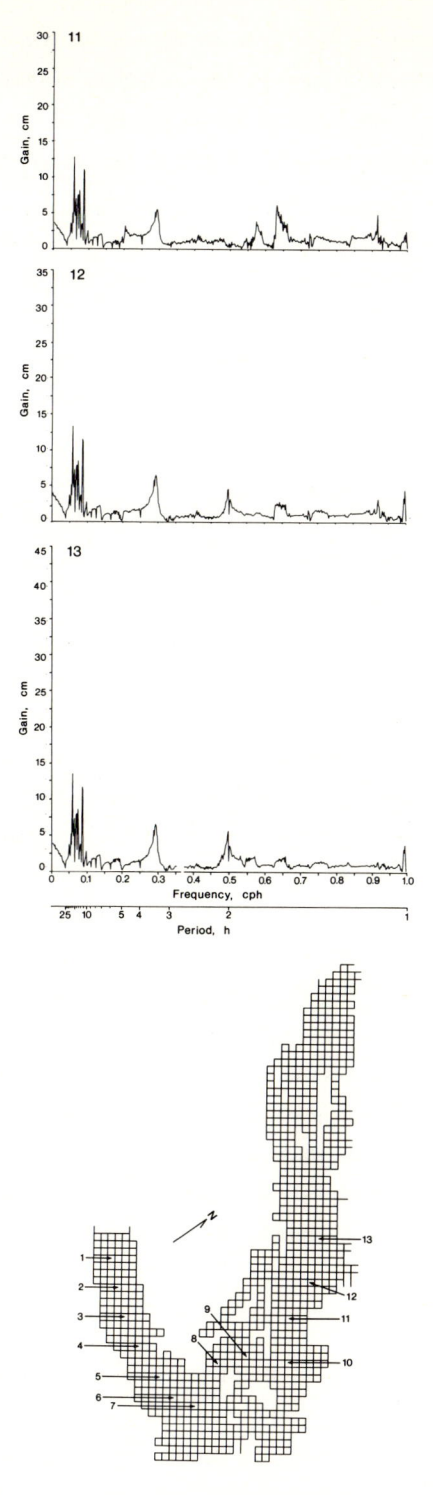

Fig. 8.2 Continued.

The reasons why this tsunami from Juan de Fuca Strait did not enter the Strait of Georgia with any appreciable energy are the decoupling between the Straits and the strong dissipation through the San Juan Island passages.

With respect to the other peaks, the dominant periods appear to occur at approximately 16, 5, and 2 h. Although this system possesses a highly complex topography, it is of interest to note that, when simulated by the GF2 model, the ratios of these smaller periods to the fundamentals (16 h) accord well with theoretical anticipation for a rectangular bay (Murty, 1977) where it is shown that the shorter periodicities confirm to $\frac{1}{3}, \frac{1}{5}, \frac{1}{7} \ldots$ of that of the fundamental mode.

Figures 8.3, 8.4, and 8.5 show the contoured distribtions of gain over the interior of the system for 16, 5, and 2 h, respectively. For the first case, the gain increases monotonically from the mouth of Juan de Fuca Strait to the northern end of the Strait of Georgia. In the case of the 5-h period two regions of peak gain occur, one at the northern end of the Strait of Georgia and the other in the inner part of Juan de Fuca Strait. For the 2-h period peaks in gain are located at the northern and southern extremes of the Strait of Georgia and approximately half way in from the mouth of Juan de Fuca Strait. For the normal mode with the 16-h period, the maximum gain is about 8.0 near Vancouver and 8.5 in the northern Strait of Georgia. The gain decreases to 6.5 in the San Juan Island passages and decreases monotonically from 5.0 in the inner part of Juan de Fuca Strait to 1.0 at the western end. For the normal mode with the 5-h period the gain is about 3.0 in the northern Strait of Georgia, decreasing to about 0.5 north of Vancouver. The gain increases again at the eastern end of Juan de Fuca Strait but decreases to 1.0 at the mouth of Juan de Fuca Strait. The mode with the 2-h period has a more complicated structure. The gain is about 1.8 at the northern end of the Strait of Georgia, decreasing to 1.4 north of Vancouver. In Howe Sound the gain is 2.0, increasing to 3.0 in the San Juan Island passages. In Juan de Fuca Strait the gain is 1.0 at both ends and about 2.6 in the middle.

The distribution of the gain is consistent with the results shown in Figs. 8.3–8.5. However, the distribution of gain shown in Fig. 8.2 reveals that there might be another normal mode around the 3-h period. An application of these normal mode estimates to storm surges in the Strait of Georgia will be discussed in Chapter 9.

8.5 INTERNAL MODES

It is not proposed to discuss at length possible internal modes occurring in this system, a complex topic requiring further extensive investigation. It is, however, of interest to provide some estimate of the internal modes and associated wave speeds occurring in the Strait of Georgia. A rough estimate of the time required for the system to return to equilibrium following some excitation may be obtained from the times taken for a surface or internal disturbance to travel the overall length of the

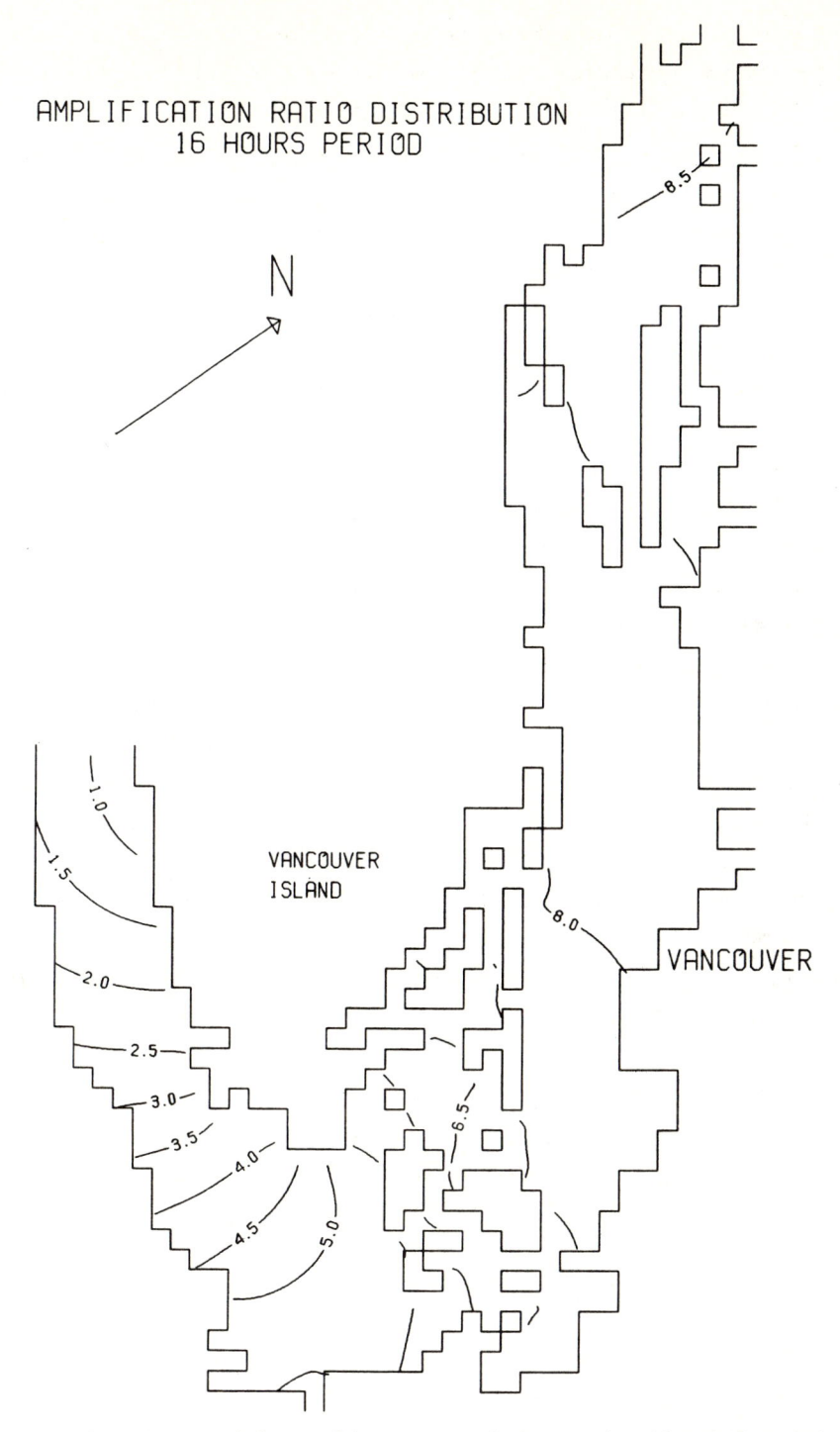

Fig. 8.3 Distribution of the amplification ratio for the mode with a 16-h period.

Strait of Georgia. Figure 8.6 shows representative vertical profiles of density from the

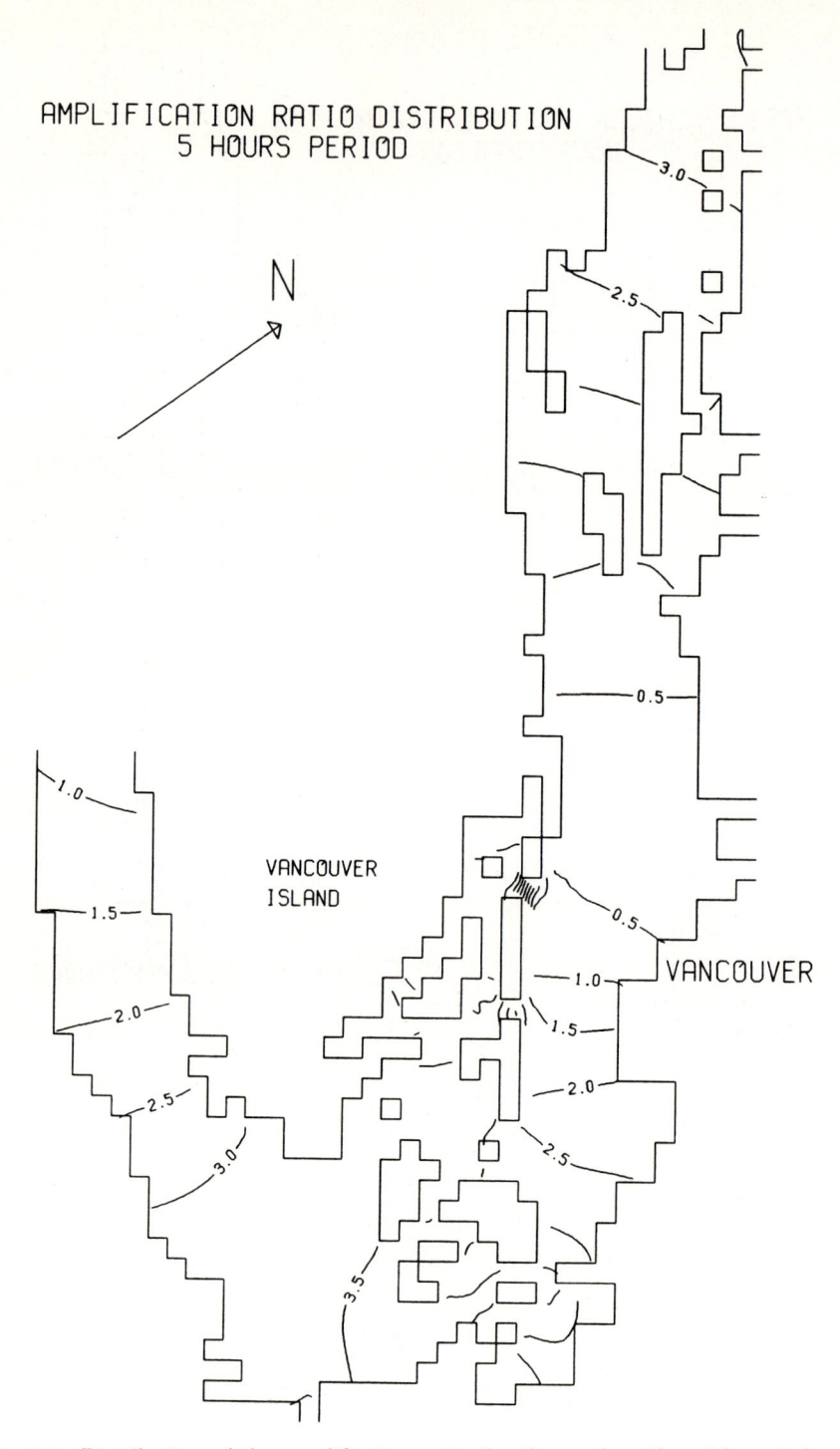

Fig. 8.4 Distribution of the amplification ratio for the mode with a 5-h period.

deep central basin of the Strait of Georgia and associated distribution of Brünt-Väisälä

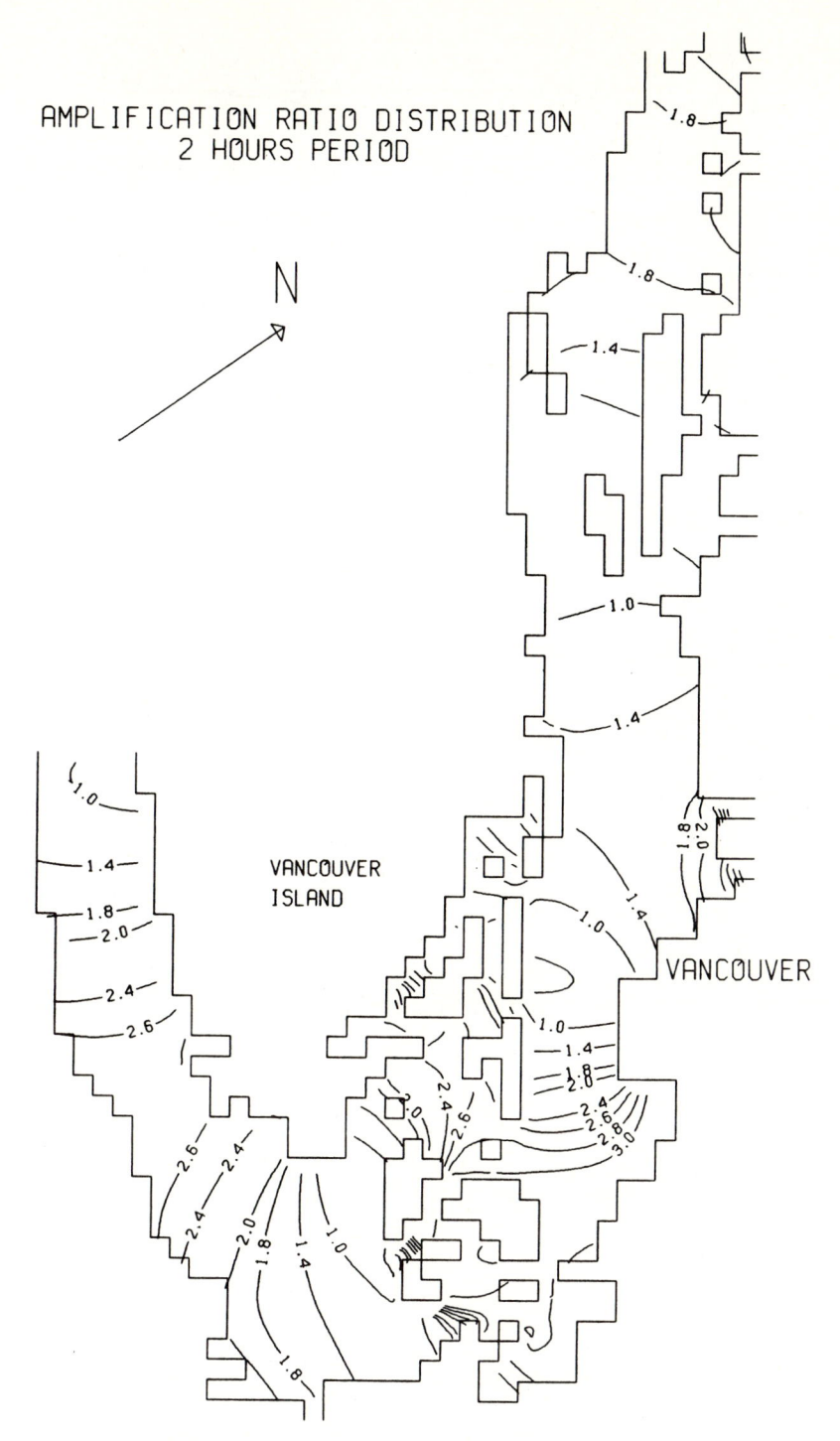

Fig. 8.5 Distribution of the amplification ratio for the mode with a 2-h period.

frequencies. The associated depth distribution of vertical displacements in the first three

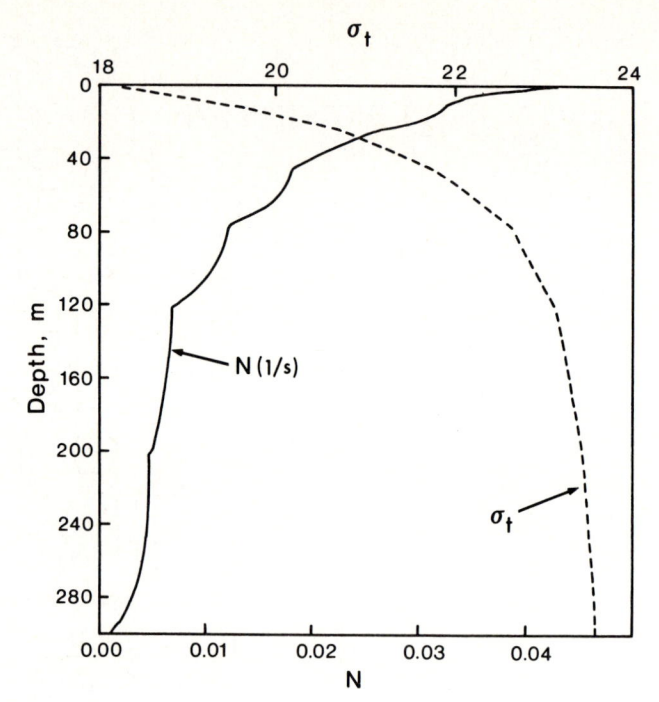

Fig. 8.6 Typical vertical profile of density in the central part of the Strait of Georgia, and the associated Brünt-Väisälä frequency.

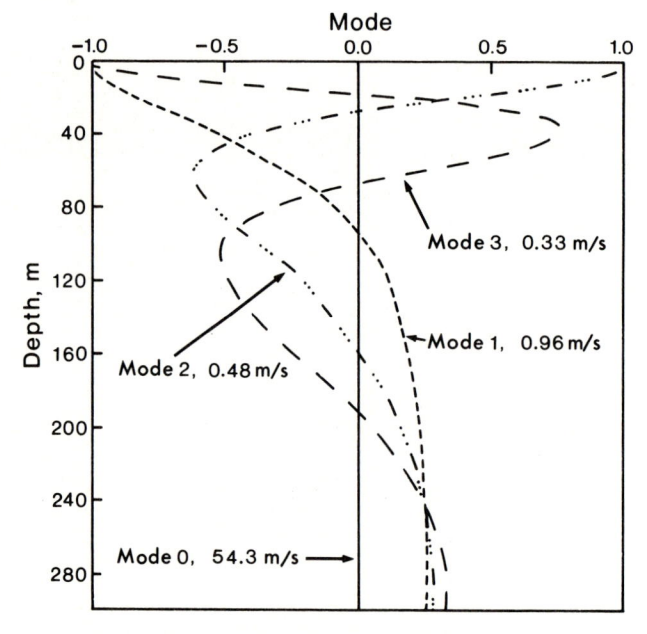

Fig. 8.7 Vertical distribution of the internal modes.

internal modes of oscillation are shown in Fig. 8.7.

Wave speeds for the zeroth and the first, second, and third internal modes are 54.3, 0.96, 0.48, and 0.33 m/s, respectively. Thus, the corresponding relaxation times are about 0.5, 2.7, 5.3, and 7.7 days, respectively. It is of interest to note that the mode with a 0.5-day period could be an internal M_2 tide having significant cross-channel velocities. The three other modes could represent internal Kelvin waves possessing insignificant cross-channel velocities. In practical terms the higher modes decay more rapidly than the fundamental mode and a reasonable estimate of the internal relaxation time for present purposes is of order 5 days.

9 STORM SURGES

9.1 INTRODUCTION

Storm surges constitute one of the world's foremost natural hazards. In general, they derive from the action of wind stresses and pressure gradients associated with the movement of strong weather systems over exposed water bodies. On entering shallower coastal waters, the surge can build up to lethal proportions.

Although the deep and relatively protected coastal seaways of British Columbia and Washington present storm surge hazards much inferior to those in some other regions of the world (Murty, 1984), unusually high sea levels can lead to extensive waterfront damage. A typical example of this occurred in December 1982 when an exceptionally deep extra-tropical cyclone in the North Pacific followed a northeasterly course onto the British Columbia coast (Fig. 9.1). The track of the low pressure centre across the Pacific Ocean for the period 0600 z, 14 December to 1800 z, 16 December, 1982 is shown by the dashed line.

In this study, the barotropic numerical model GF2, consisting of joined one- and two-dimensional numerical schemes simulating the waters between Vancouver Island and the mainland, was employed to determine the response of the system to the raised sea levels on its open boundaries and the effects of winds and atmospheric pressure gradients within the system.

9.2 MODEL DEVELOPMENT CONSIDERATIONS

It is necessary at the outset to establish a surface to which water elevations, whether observed, predicted, or modelled, are referred. Neglecting the relatively small effects of Bernoulli-type accelerations, this is taken to be a surface of constant gravitational potential coinciding with long term mean sea levels at the various gauging locations employed in this study. For modelled water elevations, the aforementioned surface is approximated by the same surface passing through long term mean sea levels at the northern and southern boundary openings of the model (Fig. 4.7, p. 80). Although this surface is assumed to be fixed with respect to the earth, the observed and modelled vertical displacements are derived from gauges which themselves are undergoing vertical displacements due to earth tides. These latter are assumed to be uniform over the study region.

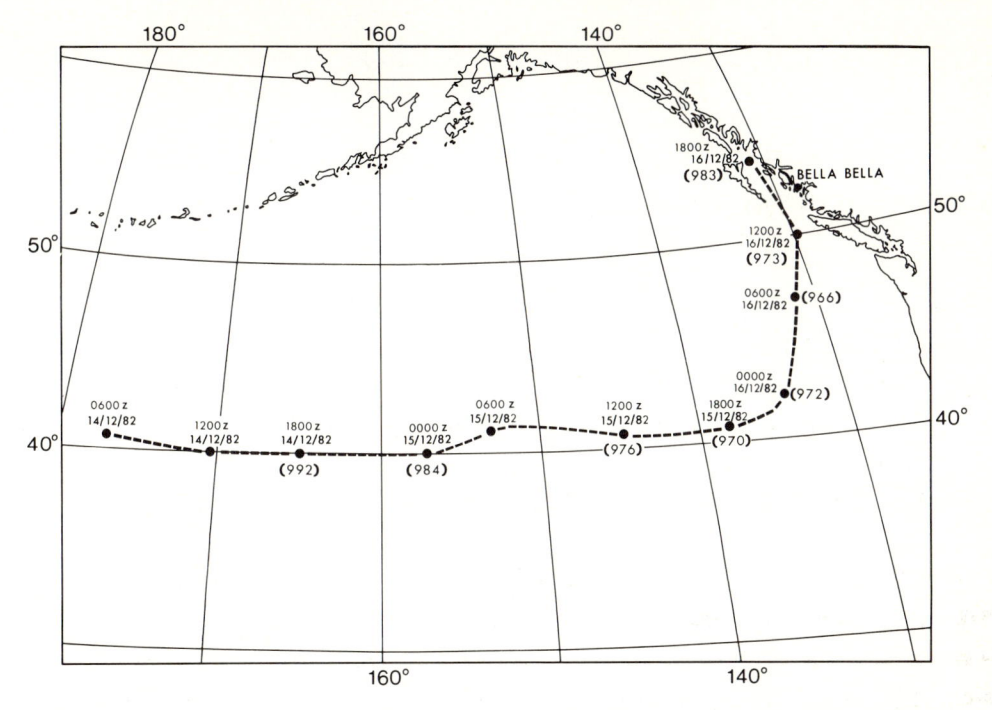

Fig. 9.1 Track of the extra-tropical cyclone of December 1982. Black dots denote locations of the centre of the low at 6-h intervals. Time (GMT), date, and pressure (mb) at the storm centre are also shown.

Vertical displacements of the water surface other than those of tidal origin are now briefly considered. These can be derived from changes in atmospheric pressure (inverse barometer effect), effects of winds, either locally or over the adjacent ocean, freshwater runoff, and density changes in the water column (steric effect). In this present barotropic simulation, the two latter effects are not explicitly included. The observed surge will thus include any contribution from shifts in the density distribution within the system associated with the passage of the disturbance. Such a contribution will not be present in the values computed by the model. Subsequent studies for the baroclinic models GF4, GF5, and GF6 suggest that some further work might be warranted in this connection.

In general, over this whole region monthly mean sea levels are high in winter and low in summer. A smaller secondary sea level maximum, associated with the Fraser River freshet and subsequent dilution, occurs in the Strait of Georgia. In the present context, it is of interest to consider regional variations of the 1982 annual mean sea levels from the long term means and those of the December 1982 monthly means from the 1982 annual means (Table 9.1). The 1982 annual mean sea levels show an increase over the long term means from 2–10 cm. The December 1982 monthly mean sea levels show increased sea levels over the 1982 annual mean from 14–22 cm. Thus,

Table 9.1 Regional variation of the 1982 annual and December 1982 monthly mean sea levels (m) from long term average values for the tide gauge locations used in this study.

Location	Tide gauge number	Long term mean*	1982 Annual mean	(Annual − Long term)	1982 Monthly mean	(Monthly − Annual)
Port Renfrew	8525		2.05		2.27	0.22
Neah Bay	8512	1.97	1.99	0.02	2.20	0.21
Victoria	7120	1.88	1.98	0.10	2.13	0.15
Seattle	7180	4.39	4.48	0.09	4.64	0.16
Fulford Harbour	7330	2.26	2.32	0.06	2.49	0.17
Ferndale	7564	3.54	3.59	0.05	3.76	0.17
Point Atkinson	7795	3.10	3.16	0.06	3.31	0.15
Little River	7993	3.26	3.29	0.03	3.47	0.18
Bella Bella	8976	2.84	2.90	0.06	3.09	0.14
Mean				0.06		0.17
S.D.				0.03		0.02

* Relative to gauge datum

surge elevations considered below will contain these annual as well as monthly mean contributions.

In the present context, the surge elevations are defined as the difference between observed elevations and predicted elevations using long term annual mean sea levels and tidal harmonic constants. Two of the small long-period constituents, the solar annual (SA) and the solar semi-annual (SSA) tend to have much larger amplitudes and spatially more variable Greenwich phases than would reasonably be anticipated on the basis of equilibrium tidal theory. It is further known that the amplitudes of these constituents tend to be small compared to the meteorologically dominated contribution at these frequencies. For present purposes, the SA and SSA constants have been omitted from the list of tidal harmonic constants employed in predictions used to define surge levels.

In the ensuing discussion it is helpful, in the interests of clarity, to restrict the term 'predicted' to elevations determined on the basis of astronomical data and the tidal harmonic constants and use the term 'modelled' for tidal or surge elevations taken from the numerical model. For a discussion of the model see Chapter 4.

An initial model run simulating tides in the region for the same period as that in which the surge occurred was undertaken. Predicted tidal elevations based on harmonic constants but omitting the SA and SSA terms were prescribed at the boundary openings of the model. This calculation was repeated but with prescribed surge elevations at the open boundaries. In the case of Juan de Fuca Strait, the prescribed elevations at Port Renfrew and Neah Bay (see Fig. 1.1) were obtained by subtracting long term mean sea level from the observed elevations of the water surface. For the northern boundary opening, comparable gauge data were not available and it was necessary to use an approximation. An estimate of the surge elevations obtained by subtracting predicted tides from observed water surface elevations at Bella Bella (#8976), the nearest tide gauge in operation at that time (Fig. 9.1). The resulting surge elevations were then added to tidal elevations predicted on the basis of harmonic constants to provide the elevations prescribed at the northern boundary opening of the model. The modelled surge is defined as the difference between the modelled elevations when surge elevations are prescribed at its open boundaries and the modelled tidal elevations for the same location when predicted tides only are prescribed at the boundary openings.

9.3 DISCUSSION OF RESULTS

Figure 9.2 shows the tide at representative locations in the system. Comparison of the modelled and predicted tides shows that the numerical model provides a good tidal simulation over the period in question and illustrates tidal elevations which are subtracted from the modelled surge elevations to yield the modelled surge.

Hourly surge levels in the vicinity of the model boundaries (Port Renfrew and Bella Bella) and the displacements of the water surface due to the inverted barometer effect

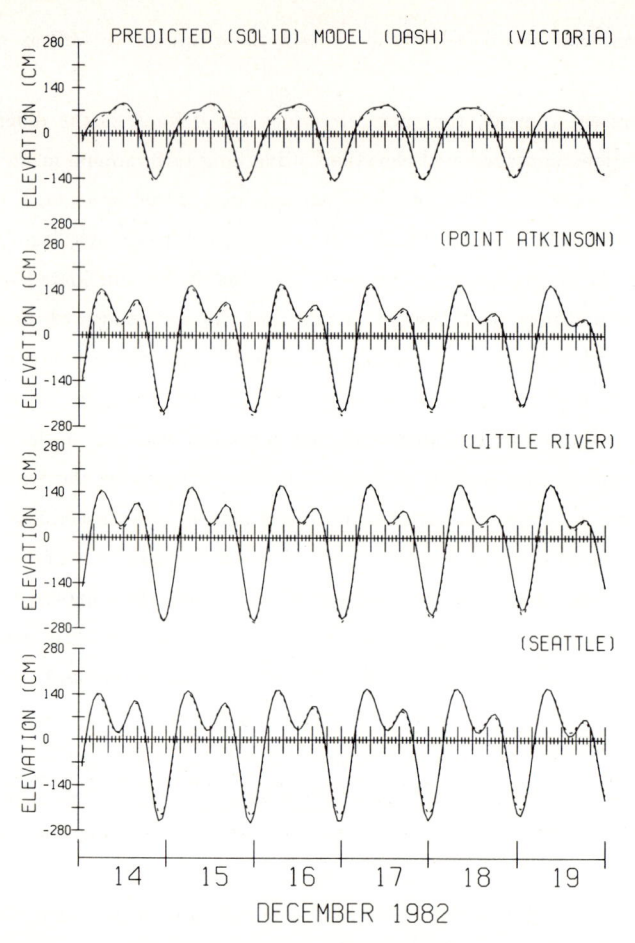

Fig. 9.2 Predicted (solid line) and modelled (dashed line) tides at Victoria, Point Atkinson, Little River, and Seattle for the period 14–19 December 1982.

on the open coast (Tofino) and in the interior (Vancouver) of the modelled region are shown in Fig. 9.3. The surge elevations at Port Renfrew show a smaller initial peak at 1700 z on 14 December followed by a second strong peak at 0000 z, 16 December and a third large peak at 2000 z, 18 December. It will be shown later that these three peaks cohere markedly with particular low pressure minima in the weather system as it moves onto the coast. The nearest operating tide gauge to the northern boundary of the model during this disturbance was located at Bella Bella, some 300 km north of Port Renfrew. At this location three peaks also occur although much less pronounced.

Sea level elevations at Tofino and Vancouver due to the inverted barometer effect were determined by subtracting hourly sea level atmospheric pressures from long term mean atmospheric pressure. The equivalence of a decrease of 1 mb to a 1 cm rise in sea level was assumed. The inverted barometer effect accounts for roughly one third to one half of the observed surge elevation at Port Renfrew and also displays the three

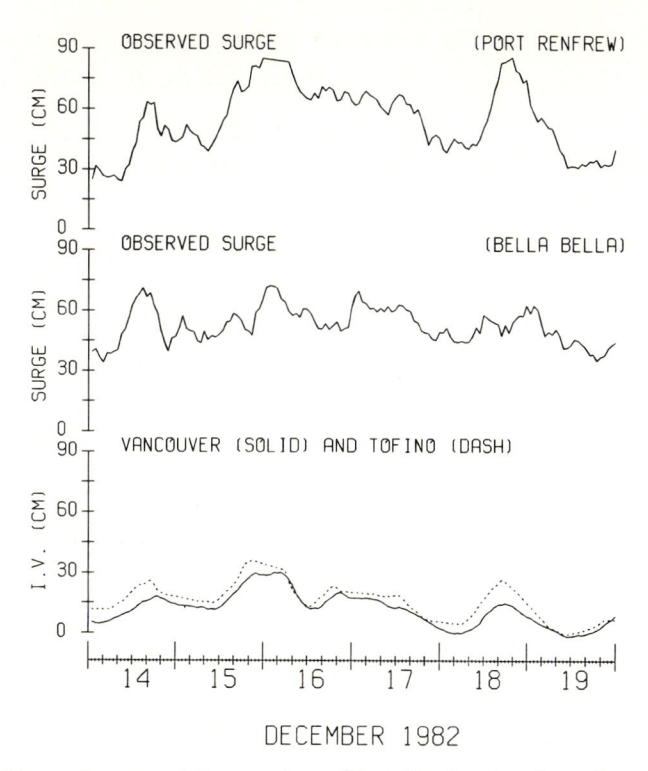

Fig. 9.3 Observed surge at the southern (Port Renfrew) and northern (Bella Bella) boundary openings. Also shown are the displacements of the water surface due to the inverted barometer effect using observed atmospheric pressures on the open coast (Tofino) and in the interior (Vancouver) of the modelled region.

distinctive peaks referred to above. Thus, the peaks are entirely consistent with the commonly occurring sequence of events associated with the movement of an intense low pressure system onto a coastline. The initial peak is due to dropping pressure as winds undergo an orographic increase when the leading edge of the system encounters the coast. This encounter with coastal mountains brings about an alteration in the path of the system which roughly parallels the coast. Following this recurvature, the system extracts further energy from the water and intensifies for a second time leading to the second peak. The third peak derives from the curvature of the coastline (Fig. 9.1). In this case most of the system lies over the land and again an orographic increase occurs in the wind field with an accompanying decrease in the pressure field. This is followed by an effective translation of the entire system onto land and its ultimate dissipation. The actual response of the sea level itself is thus determined predominantly by both the winds and the inverted barometer effect. It will be noted that the sea level increase due to the atmospheric pressure effect is persistently higher at Tofino than at Vancouver mainly due to the deflection of the low pressure system by the coastal mountains. The uncorrected modelled surge elevations in the interior of the system thus assume at any time a uniform atmospheric pressure over the system equal to that exerted at the open boundary. The model results must thus be corrected for the increase in weight of the air column over the interior of the system compared to that of the open boundary. Accordingly, in order to obtain the modelled surge elevations, the difference between the Tofino and Vancouver sea level elevations is subtracted from the computed elevations when observed tide and surge elevations are prescribed on the open boundaries.

The observed and modelled surge elevations for the same four representative locations within the system are shown in Fig. 9.4. The agreement is considered satisfactory in view of the fact that in this model the surge is represented as a solely barotropic phenomenon. The discrepancies between the observed and modelled values are of the same order as the effects produced by winds acting over a fully three-dimensional baroclinic model (GF6) of the region, as will be discussed in Chapter 13.

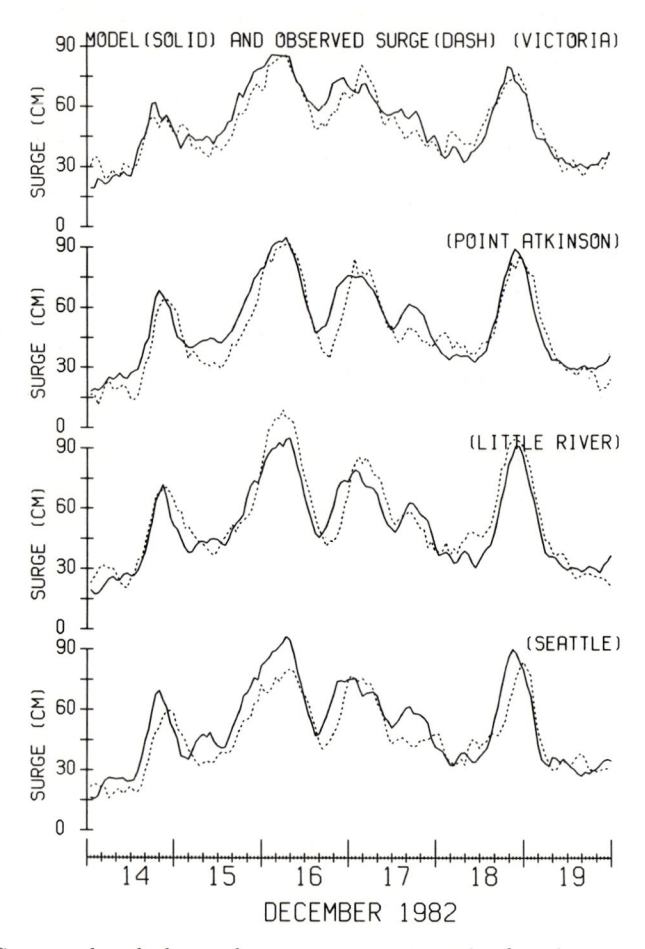

Fig. 9.4 Computed and observed surges at representative locations.

The surge incoming from the mouth of Juan de Fuca Strait undergoes relatively little amplification within the system, the largest value occurring in the northern Strait of Georgia (Little River, about 30%). The phase lag between the occurrence of surge peaks in the Strait of Georgia and that on the boundary is readily discernible from the figure and is of order 3 h. This is reasonably consistent with the time taken for a disturbance to propagate from the mouth of Juan de Fuca Strait. An effective surface wave speed may be estimated from the overall length of the system from the mouth of Juan de Fuca Strait to the northern end of the Strait of Georgia and from the period (16.1 h) of the first normal barotropic mode, as described on p. 172. This yields an effective speed of propagation of a surface disturbance through the system of close to 100 km/h. Thus it should take 2.5 h for the surge to travel from the southern boundary opening to Point Atkinson. Whereas the observed surge at Port Renfrew is characterized by three distinct peaks, for the modelled surges at Point Atkinson, Little River, and Seattle, the second of the peaks is followed by two smaller peaks of diminishing displacement. The time interval between these peaks (about 16 h) corresponds to that of the first barotropic normal mode in the system. Thus, it would appear that the highest peak of this surge event was associated with the excitation of the system's first normal mode.

A question arises as to the direct contribution to surge elevations made by winds over the interior of the modelled system. Using the analyses of the Canadian Meteorological Centre (Dorval, Quebec) and the Pacific Weather Centre (Vancouver, British Columbia), the sea level atmospheric pressure field and the observed winds provided a consistent evolving sequence of surface wind stresses that were determined and applied over the surface of the modelled sea. The observed wind speeds and directions at Sand Heads for this time (Fig. 9.5) are used in illustration.

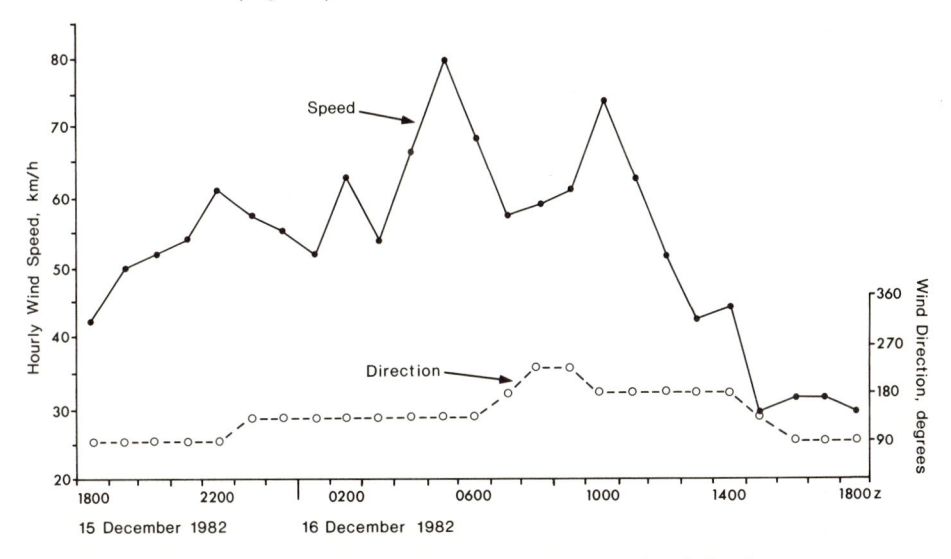

Fig. 9.5 Mean hourly wind speed and wind direction at Sand Heads.

The observed and modelled surges, the latter with and without, respectively, the application of winds, are shown in Fig. 9.6 for the major peaks of the surge in the early hours of 16 December. The effect of the winds is to produce a small (< 10 cm) increase in the modelled surge elevations at Point Atkinson, Little River, and Seattle.

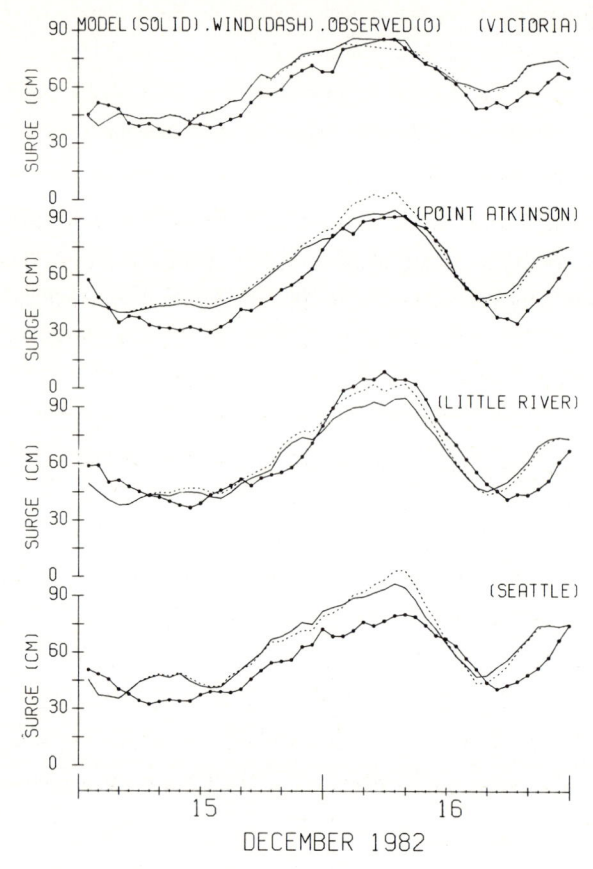

Fig. 9.6 Comparison of the modelled surge without wind (solid line) and with the prescribed observed wind field (dashed line), respectively, with the observed surge (dots) for representative locations about the time of the major surge peak in the early hours of 16 December 1982.

9.4 CONCLUSIONS

Despite the absence of baroclinic effects, the model approximates the basic features of the storm surge occurring from 14–16 December 1982 in the North Pacific Ocean. Roughly one third of the surge elevations is due to the inverted barometer effect, the remainder primarily to the surge propagating into the system from the open ocean through Juan de Fuca Strait. In the process, the surge undergoes little amplification, the largest increase (about 30%) occurring in the northern Strait of Georgia. The effect of local winds is to produce only a small (<10 cm) increase in the surge heights.

10 TSUNAMIS

10.1 INTRODUCTION

Tsunamis are oscillations of the water level either due to earthquakes in the ocean or due to underwater volcanic explosions or submarine slides. Large underwater nuclear explosions also can give rise to tsunamis. In principle, tsunamis fall into the category of long gravity waves which are in the same class as tides and storm surges. For the theory of tsunami generation and propagation, see Murty (1977).

Most tsunamis on the globe occur in the Pacific Ocean, due mainly to the seismic (and to a lesser extent, volcanic) belt that girdles the Pacific basin. In principle, tsunamis can occur in the Strait of Georgia system (which includes Juan de Fuca Strait and the Puget Sound) from three different source areas: (1) tsunamis generated from earthquakes in Alaska and the Aleutian Islands, (2) earthquakes on the coasts of British Columbia, Washington, and Oregon, (3) earthquakes in the Strait of Georgia system. Other possibilities include earthquakes elsewhere in the Pacific Ocean, such as Kamchatka, Japan, Chile, and Peru. However, it is expected that significant tsunamis can occur in the Strait of Georgia system mainly from earthquakes in the system itself and to a lesser extent, from earthquakes on the Washington or Oregon coasts, in which case some tsunami energy can enter the system through the mouth of Juan de Fuca Strait.

10.2 SIMULATION OF THE TSUNAMI OF 23 JUNE 1946

On 23 June 1946 an earthquake with a magnitude of about 7.3 occurred on Vancouver Island (Fig. 10.1). The epicentre was close to the eastern shore at 49.76° N and 125.34° W and the surface magnitude was 7.2 ± 0.1. Associated with the earthquake was a vertical displacement of up to 3 m on land and on the ocean bottom (Rogers and Hasegawa, 1978). The resulting tsunami killed one person and caused some damage along the coastline. Rogers and Hasegawa suggested that there was no tsunami associated with this earthquake other than some seiche action resulting from landslides following the earthquake. However, from hydrodynamical considerations it is not possible not to have a tsunami when there is a vertical displacement of the sea bottom by as much as 3 m. From 1940–1946 due to the second World War and its economical effects, tide gauges in the system were not regularly maintained and operated satisfactorily. Hence, it is not surprising that the Vancouver tide gauge did not register even a small tsunami. Very little tsunami energy would have passed through the Gulf and San Juan Island passages to be registered on the Victoria gauge.

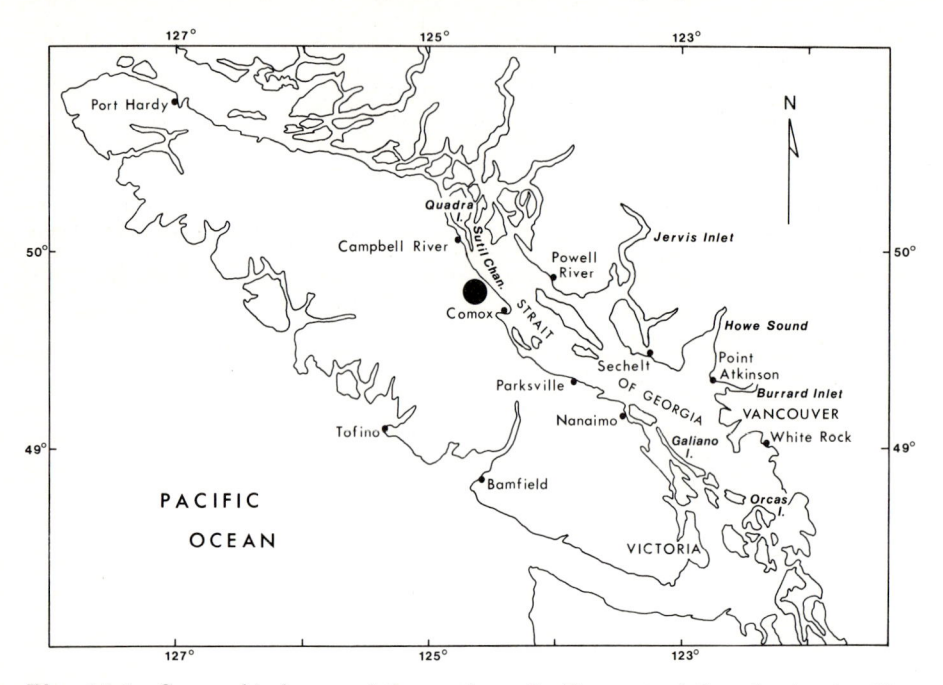

Fig. 10.1 Geographical map of the southern Pacific coast of Canada showing Vancouver Island and the Strait of Georgia. Epicentre of the 23 June 1946 earthquake is shown by the large solid circle.

In this study the water level disturbances that occurred following this earthquake were numerically simulated. As input into the numerical model, the ground motion (Fig. 9c of Rogers and Hasegawa, 1978) was used in which vertical motion of up to 3 m, mostly on land but partly in the Strait of Georgia, occurred.

The x axis of a right handed Cartesian co-ordinate system was taken along the length of the Strait of Georgia and the y axis was taken along its width with the origin located at the northwest corner. The grid size (Fig. 10.2) in both directions was 2.62 km. There are a total of 91 grid points in the x direction and 36 grid points in the y direction. Although this model is similar to GF2 and GF3, some differences do exist as will be seen below.

The partly linearized shallow water equations are (Henry 1982; Murty, 1977, 1984)

$$\eta_t = -(du)_x - (dv)_y. \tag{10.1}$$

$$u_t = -g\eta_x + fv - F^u + G^u, \tag{10.2}$$

$$v_t = -g\eta_y - fu - F^v + G^v, \tag{10.3}$$

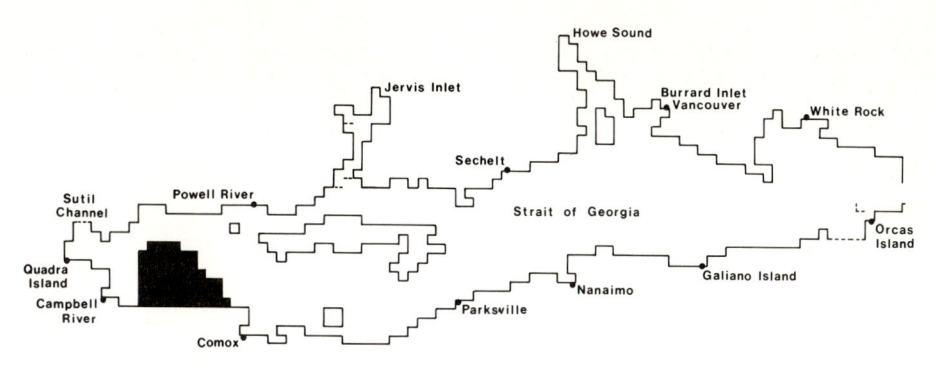

Fig. 10.2 Grid used for numerical simulation of the tsunami. Locations are shown where the computed water level time series is given in subsequent figures. The black area denotes bottom deformation due to the earthquake.

where subscripts denote differentiation and

$\eta(x, y, t)$ = elevation of the water surface above mean level,
$u(x, y, t)$ = depth-averaged velocity in the x direction,
$v(x, y, t)$ = depth-averaged velocity in the y direction,
$d(x, y)$ = mean water depth,
x, y = Cartesian co-ordinates in the horizontal plane,
f = Coriolis coefficient (assumed constant),
t = time,
F^u, F^v = friction terms.

A quadratic bottom friction of the following form used is

$$F^u = \frac{ku(u^2 + v^2)^{1/2}}{d} \quad \text{and} \quad F^v = \frac{kv(u^2 + v^2)^{1/2}}{d} \tag{10.4}$$

where k is a dimensionless bottom friction coefficient. The terms G^u and G^v in Eqs. (10.1–10.3) represent forcing terms such as surface wind stress and equilibrium tide gradient which may vary with x, y, t. For storm surge computations, these terms represent the meteorological forcing terms and for tidal computations, they represent the tidal potential. For the present problem of tsunami generation and propagation, these terms are ignored and the forcing comes in the form of an increasing water level directly above the bottom (of the estuary) displacement.

A simple Richardson grid was chosen as the basis for the finite-difference scheme on the grounds that it minimizes storage and permits particularly simple representation of coastlines. At interior points of the grid, Eqs. (10.1–10.3) are represented by the following finite-difference forms,

$$\frac{n'_{i,j} - n_{i,j}}{\Delta t} = -\frac{(d_{i,j} + d_{i+1,j})u_{i+1,j} - (d_{i-1,j} + d_{i,j})u_{i,j}}{2\Delta x}$$
$$- \frac{(d_{i,j} + d_{i,j+1})v_{i,j+1} - (d_{i,j-1} + d_{i,j})v_{i,j}}{2\Delta y}, \tag{10.5}$$

$$\frac{u'_{i,j} - u_{i,j}}{\Delta t} = -g\frac{\eta'_{i,j} - \eta'_{i-1,j}}{\Delta x} + f\tilde{v}_{i,j} - F^u_{i,j} + G^u_{i,j}, \qquad (10.6)$$

$$\frac{v'_{i,j} - v_{i,j}}{\Delta t} = -g\frac{\eta'_{i,j} - \eta'_{i,j-1}}{\Delta y} - f\tilde{u}'_{i,j} - F^v_{i,j} + G^v_{i,j}. \qquad (10.7)$$

where

Δt = time step,

$\Delta x, \Delta y$ = grid interval sizes in the x and y directions, respectively,

$d_{i,j}$ = mean water depth at elevation points $\eta_{i,j}$.

$$\tilde{u}_{i,j} = \tfrac{1}{4}u_{i,j-1} + u_{i+1,j-1} + u_{i,j} + u_{i+1,j}, \qquad (10.8)$$

$$\tilde{v}_{i,j} = \tfrac{1}{4}v_{i-1,j} + v_{i,j} + v_{i-1,j+1} + v_{i,j+1}. \qquad (10.9)$$

Primes indicate variables updated during the current time step. Unprimed varibles are those evaluated at the previous step. The use of old (unprimed) values of v in the Coriolis term in Eq. (10.6) and new (primed) values of u in the corresponding term in Eq. (10.7) is necessary for stability. Fortunately, it also eliminates the need to store any but the most recently updated values of each variable provided that the equations are applied in the order given, i.e., at each time step all the $\eta_{i,j}$ are updated, then all the $u_{i,j}$, and finally all the $v_{i,j}$. The same stability and storage conclusions apply if variables are evaluated in the order η, v, and u using old values of u in the v equation and new values of v in the u equation. To reduce possible bias, the stepping subroutines evaluate the variables in the order η', u', v' on odd numbered steps and η', v', and u' on the even-numbered steps.

Strictly speaking, Eqs. (10.5–10.7) imply that $u_{i,j}$ and $v_{i,j}$ are evaluated $\Delta t/2$ later than $\eta_{i,j}$ but normally they are regarded as pertaining to the same time level. The distinction is important only when calculating quantities which depend on phase differences between elevation and velocity, e.g., energy flux, and then only when there are relatively few time steps per wave period.

A choice of a time step of 30 s satisfies the following stability criterion,

$$\Delta t \leq \frac{\Delta x \Delta y}{(g d_{\max}(\Delta x^2 + \Delta y^2))^{1/2}} \qquad (10.10)$$

where $d_{\max}$ is the maximum depth in the computational region.

Where there are good grounds for assuming that no waves enter the model from an adjacent water body, it is appropriate to use a radiation condition on the sea boundary between the two. This permits waves reaching the sea boundary from the interior of the model to pass out of the model domain (Henry, 1982).

When choosing the model grid initially, radiating sea boundaries parallel to the x axis of the model should be placed to run through v points on the grid. Similarly, those parallel to the y axis should run through u points. It is assumed that the radiation problem can be treated one-dimensionally at each velocity point on the sea boundary and thus, that the surface elevation and normal velocity at the boundary are related by

$$\text{outward normal velocity} = \left(\frac{g}{d}\right)^{1/2} \times \text{elevation}.$$

Since there are no elevation points actually on the boundary, the nearest interior elevation value is taken instead so that the formulae used in the stepping subroutines for u points on radiating boundaries facing in the positive or negative x direction are, respectively,

$$u_{i,j} = \left(\frac{g}{d_{i-1,j}}\right)^{1/2} \eta_{i-1,j}$$

or

$$u_{i,j} = -\left(\frac{g}{d_{i,j}}\right)^{1/2} \eta_{i,j}.$$

Similarly, at radiating sea boundaries facing in the positive or negative y directions, the formulae used are, respectively,

$$v_{i,j} = \left(\frac{g}{d_{i,j-1}}\right)^{1/2} \eta_{i,j-1}$$

or

$$v_{i,j} = \left(\frac{g}{d_{i,j}}\right)^{1/2} \eta_{i,j}.$$

When this type of radiation boundary condition is used the permissible time step must be reduced by 50%. Hence, in the stability criterion (10.10) the denominator should be multiplied by a factor of 2.

The vertical ground displacement of up to 3 m (Rogers and Hasegawa, 1978) was prescribed as the initial condition. Actually a mound of water 1–3 m high was prescribed initially in the area designated in Rogers' and Hasegawa's paper and the numerical simulation involved propagation of this initial elevation (mound of water). As expected, the tsunami did not propagate with significant amplitudes Strait of Georgia-wide.

Figure 10.3 shows the time series of the computed water levels at various locations along the western and eastern shores of the Strait of Georgia including some islands. Table 10.1 lists the approximate range of the water level associated with the tsunami at these locations. Some areas farther from the initial ground subsidence, associated with the earthquake, have a greater tsunami range than areas closer. This is mainly due to damping of the tsuanmi waves in shallow areas and amplification in some inlets due to resonance.

In particular, the tsuanmi range along the western shores of the Strait of Georgia is examined first. Even though Comox is farther from the epicentre than Campbell River, the tsunami range was greater at Comox. Similarly, even though Parksville is

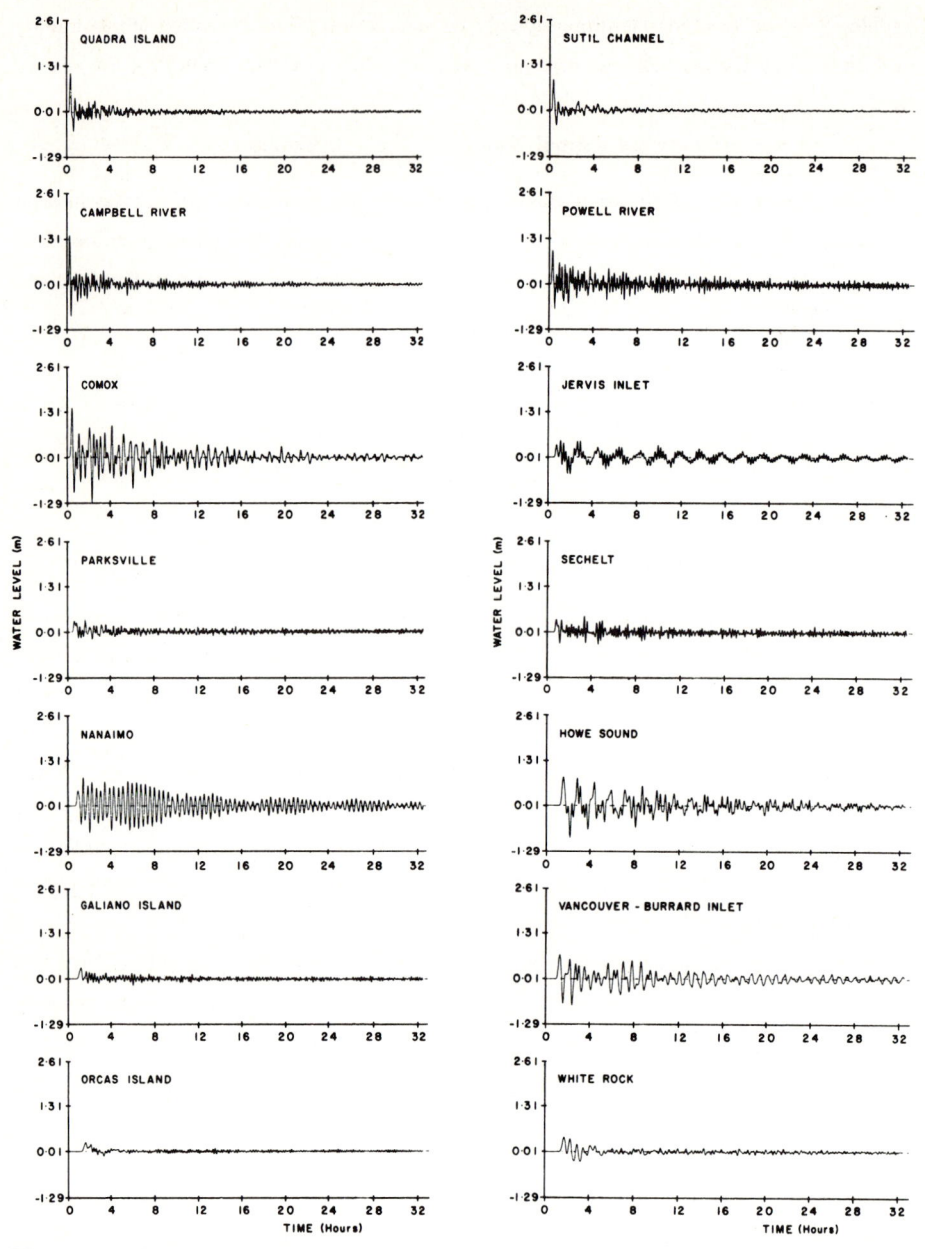

Fig. 10.3 Computed water levels at several locations along the western and eastern shores of the Strait of Georgia resulting from the earthquake of 23 June 1946.

closer than Nanaimo to the epicentre, the tsunami amplitude at Parksville is smaller because the extensive shallows and tidal flats at Parksville slowed the tsuanmi waves significantly. The amplitudes at Galiano and Orcas Islands were also due to damping and the absence of resonance effects.

On the eastern shore of the Strait of Georgia, some amplification of the tsunami

Table 10.1 Approximate tsunami range (m) associated with the 23 June 1946 earthquake at several locations in the Strait of Georgia. The locations are listed from north to south for the western and eastern shores of the Strait, respectively (from Murty and Crean, 1985).

Western shore locations	Tsunami range	Eastern shore locations	Tsunami range
Quadra Island	−0.5 to 1.2	Sutil Channel	−0.4 to 1.0
Campbell River	−0.9 to 1.4	Powell River	−0.8 to 1.0
Comox	−1.3 to 1.4	Jervis Inlet	−0.4 to 0.4
Parksville	−0.2 to 0.3	Sechelt	−0.3 to 0.4
Nanaimo	−0.9 to 0.9	Howe Sound	−0.9 to 0.9
Galiano Island	−0.2 to 0.2	Vancouver	−0.8 to 0.7
Orcas Island	−0.1 to 0.2	White Rock	−0.4 to 0.4

in Howe Sound and Burrard Inlet is noticeable. Also associated with this tusnami, a regular seiche occurred in Jervis Inlet. It would have been better if comparison could be made between these computed water levels and levels recorded on tide gauges. Rogers and Hasegawa (1978) mentioned that the Vancouver and Victoria tide gauges showed no tsunami.

In conclusion, the computed water levels shown in Fig. 10.3 and Table 10.1 are unverified against any observations. It is not claimed here that these water levels necessarily occurred. Rather, it is shown that if the vertical ground motion (Rogers and Hasegawa, 1978) has occurred, a small tsunami would have occurred in the Strait of Georgia.

The tsunami of 23 June 1946 was also simulated using the 2-km grid GF7 model. Figure 10.4a–d shows the tsunami height contours at selected times. Note that in the early stages the tsunami waves have not yet propagated into the southern part of the Strait of Georgia. Figure 10.5 gives the time series of tsunami amplitude at selected locations.

10.3 FUTURE TSUNAMI ESTIMATES

Further modelling studies of tsunamis in the system are in progress in a collaborative effort between Canada (T.S. Murty) and the United States (G.T. Hebenstreit) using the GF7 model. Two types of tsunami sources are dealt with in these studies. In the first instance, earthquakes outside the mouth of Juan de Fuca Strait are considered and a tsunami generation and propagation model will be used which will include the coasts of Washington, Oregon, and part of British Columbia. The output from this model will then be used to provide boundary conditions at the mouth of Juan de Fuca Strait for further tsunami propagation into Juan de Fuca Strait, Puget Sound, and the

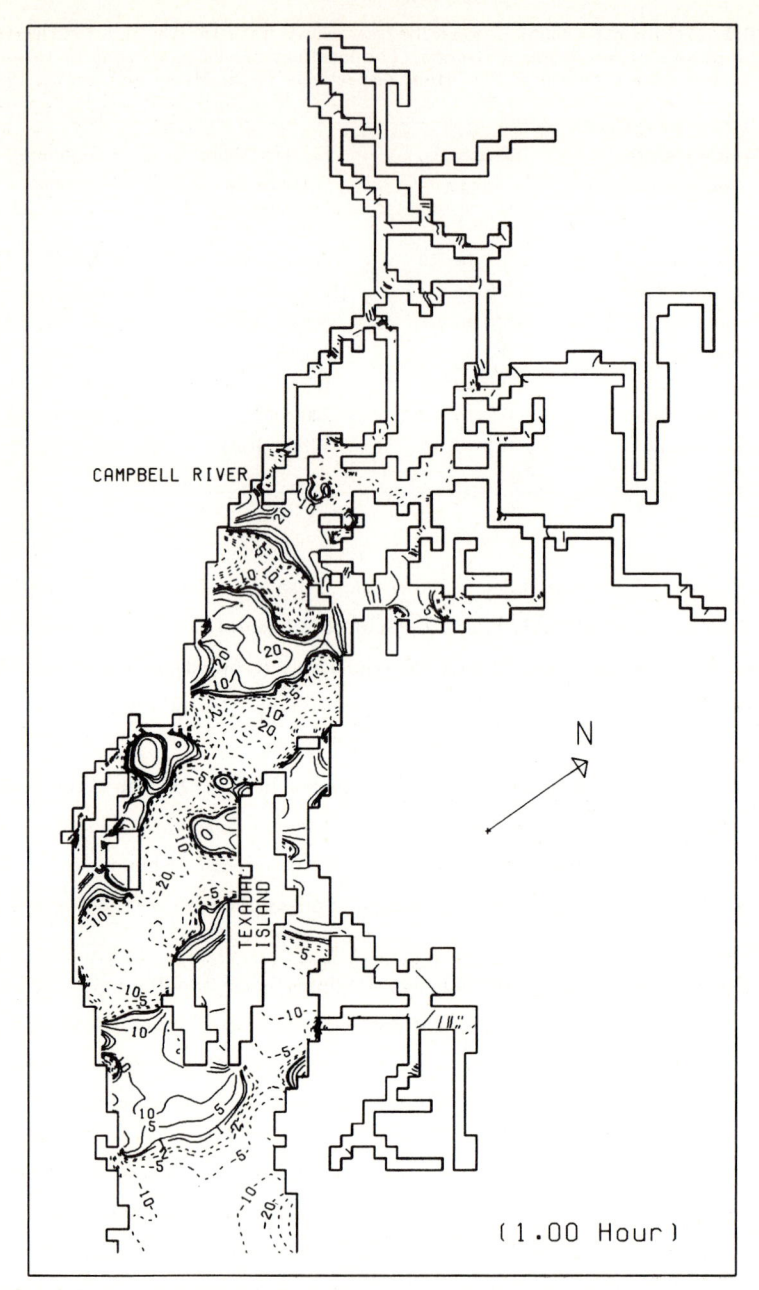

Fig. 10.4a Northern Strait of Georgia at 1 h. Contours are in centimetres.

Strait of Georgia, using the GF7 model. In the second case, earthquake sources inside the system itself will be considered and the tsunami generation and propagation will be handled through the GF7 model. In this case no outer model will be necessary.

Estimates of maximum tsunami amplitudes for a 100-yr and 500-yr tsunami in the Puget Sound, Strait of Georgia, and in the U.S. waters in Juan de Fuca Strait were

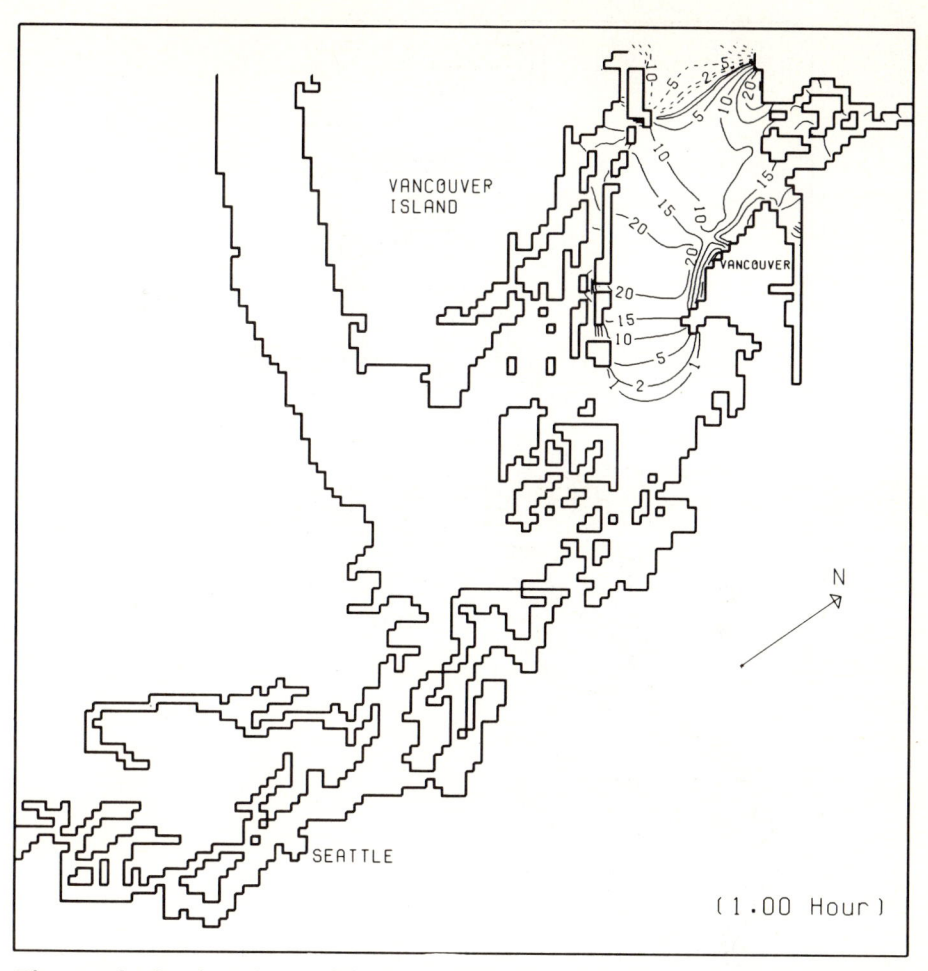

Fig. 10.4b Southern Strait of Georgia and Puget Sound system at 1 h. Contours are in centimetres.

made by Garcia and Houston (1975). They determined the response of the Puget Sound system to an incoming tsunami using a model with a finer grid than that used in propagating the tsunami across the deep ocean. Although tsunamis are generated in several areas along the rim of the Pacific Ocean, only the Aleutian Trench generates tsunamis capable of causing significant run-up in the study areas with sufficient frequency to influence 100- and 500-yr run-up values. (Note that since tsunamis are aperiodic phenomena, the authors do not subscribe to the concept of 100- and 500-yr tsunamis. However, since little work has been done on tsunamis in the Puget Sound area, this work has been included.) Historical evidence, tsunami source characteristics and orientation, and numerical computer programs discussed in Houston and Garcia (1974) were used in the selection of the Aleutian Trench as the sole tsunamigenic area. The hypothetical uplift of the water surface used and the dimensional parameters which can be

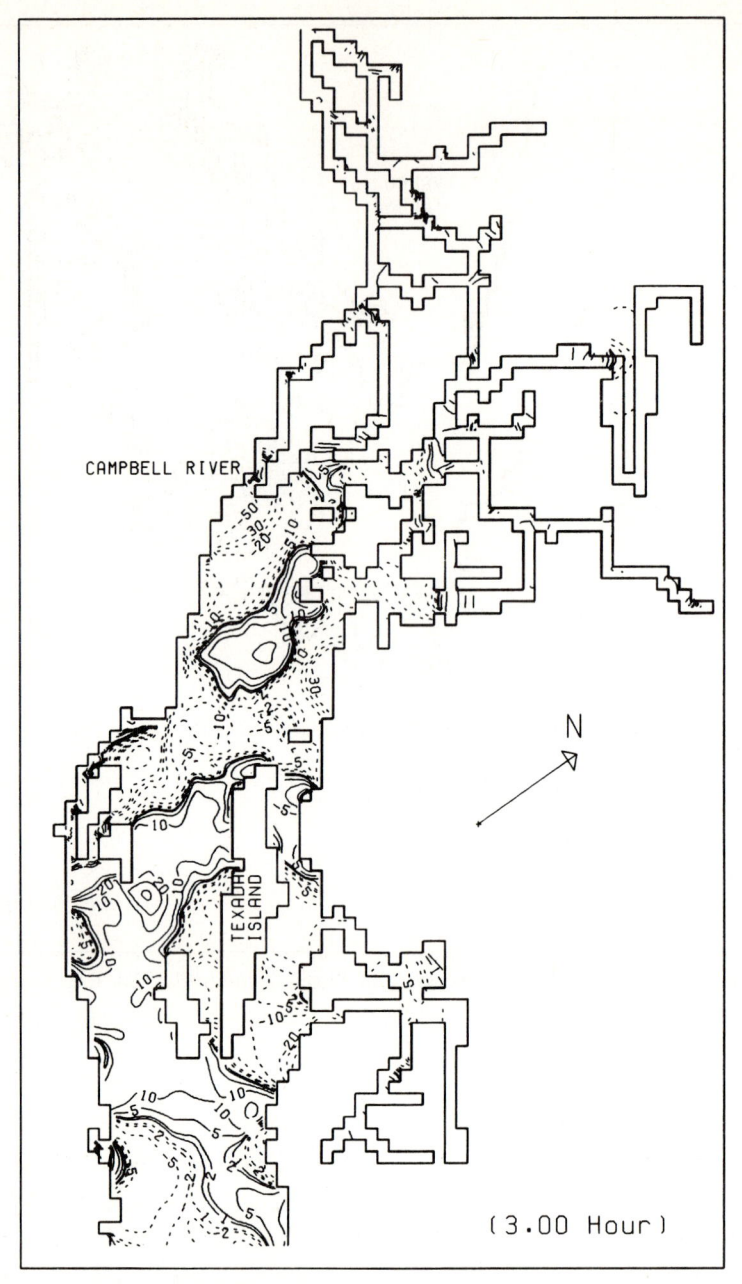

Fig. 10.4c Northern Strait of Georgia at 3 h. Contours are in centimetres.

varied to represent tsunamis of different intensities in the selected tsunamigenic areas were also formulated therein.

The probability of generation of tsunamis of different intensities and the maximum height of the uplift deformation for the standard source were defined for these different intensities by Houston and Garcia (1974). The Aleutian Trench was divided

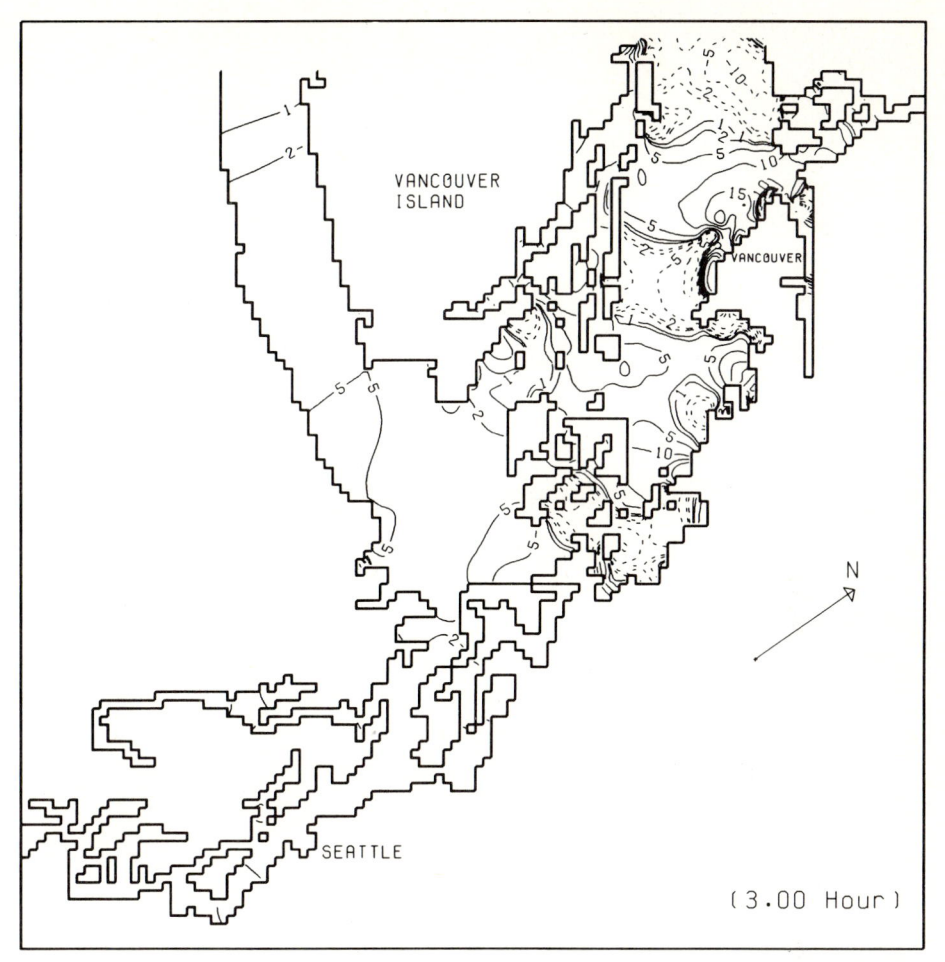

Fig. 10.4d Southern Strait of Georgia and Puget Sound system at 3 h. Contours are in centimetres.

into 12 segments and the wave amplitudes resulting from locating sources of varying intensities in each of the segments were calculated for points near mouths of coastal seas considered. The responses of the water bodies were determined by the numerical and analytical solutions mentioned earlier. Each wave amplitude inside the water body has an associated probability distribution and the sum of wave amplitudes defines a probability distribution from which the cumulative probability distribution for a wave amplitude greater than or equal to a particular value is obtained. Run-up was set equal to wave amplitude at the shore.

A cumulative probability distribution $P(z)$ for run-up at a given site equal to or exceeding a particular value due to the astronomical tide and a tsunami was determined using an approach discussed by Houston and Garcia (1974). This approach makes use

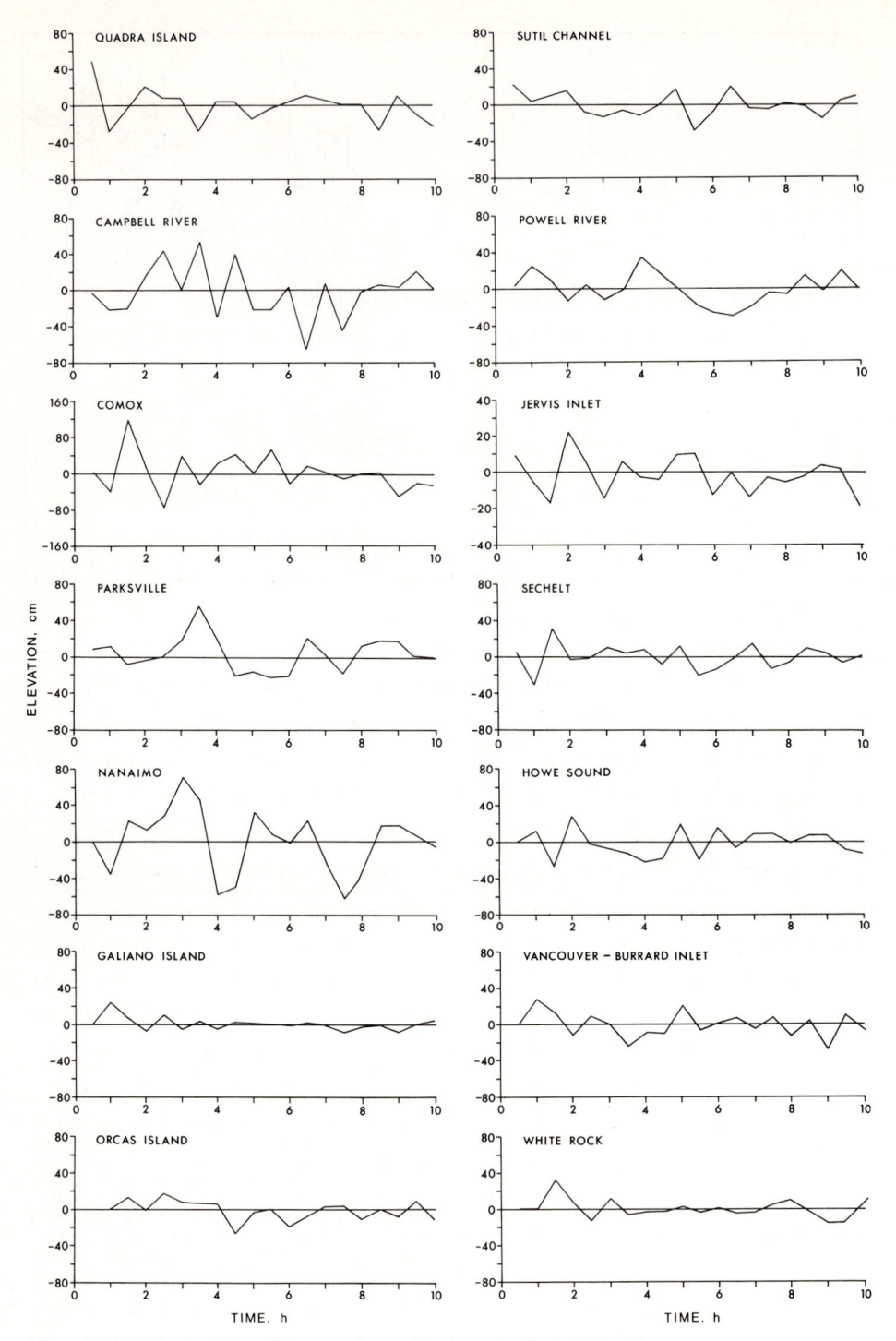

Fig. 10.5 Time series of tsunami amplitudes at selected locations.

of the following relationship,

$$P(z) = \int_{-\infty}^{\infty} f_\beta(\lambda) P_S(z - \lambda) d\lambda \qquad (10.11)$$

where

z = run-up at any time above local mean sea level,
$P(z)$ = cumulative probability distribution for run-up at a given site $\geq z$ due to the maximum wave of the tsuanmi and the astronomical tide,
$f_\beta(\lambda)$ = the probability density for the astronomical tide,
$P_S(z)$ = cumulative probability distribution for run-up at a given site $\geq z$ due only to the maximum wave of a tsunami,

and where

$$f_\beta(z) = -\frac{dP_\beta(z)}{dz} \qquad (10.12)$$

and $P_\beta(z)$ is the probability of run-up at the same location $\geq z$ due only to the astronomical tide, here approximated by a Gaussian distribution (tidal run-up = tidal level). Equation (10.11) was solved numerically by superimposing tsunami wave trains on the tides for a 1-yr period.

The estimated run-up due to a 100- and 500-yr tsunami is given is given by Houston and Garcia (1974) in the form of R_{100} and R_{500}, respectively, where run-up is in feet. The datum for all run-up computations (R_{100} and R_{500}) is mean sea level.

Ground displacements which produce tsunamis with intensities ranging from 2–5 in intensity increments of one half are centred at the midpoint of each of the 12 segments of the Aleutian Trench. An investigation of tsunamigenic sites in the Pacific Ocean indicated that tsunamis generated along the Aleutian or Peru-Chile Trenches only cause significant run-up along the western coast of the United States. Furthermore, it was found that the 100- and 500-yr run-up values for the areas studied by Garcia and Houston (1975) were determined by tsunamigenic sites in the Aleutian Trench alone and were not significantly influenced by tsunamis generated in the Peru–Chile Trench.

A computer program for a finite-difference numerical scheme was used (Houston and Garcia, 1974) to generate a tsunami (using a given ground displacement as input) and to propagate it across the deep ocean. The linearized long-wave equations were the governing equations in the program. This program used a commonly available tape which contained bathymetric data at intervals of one degree of latitude and longitude. The program then interpolated these data to create another bathymetric grid for computational purposes that has 15 bathymetric data points per degree of latitude and longitude. While this procedure was adequate for those regions of the ocean that are relatively featureless, it was not satisfactory for wave propagation across the continental rise and portions of the continental shelf. To avoid this difficulty, the program was modified to include bathymetric data obtained from National Ocean Survey bathymetric charts. These data were smoothed from a minimum of 180 points to 15 points (compatible with the grid used for computation) per degree of latitude and longitude. The program was used to propagate tsunamis to the vicinity of the Juan de Fuca Strait. Local bathymetric irregularities, *e.g.*, shoals and channels, which could not be resolved by the large mesh grid covering part of the Pacific Ocean near the continental shelf, along with non-linear terms and vertical accelerations which were neglected in the linearized long-wave equations, produced effects which had to be determined by calibration. Results of a computer simulation of the generation and propagation of the 1964 Alaska tsunami were compared with tide gauge records obtained near the mouth of Juan de Fuca Strait. Differences in comparison are attributed to these local bathymetric irregularities and the neglected terms in the equations of motion. In this manner, the simulation of the 1964 tsunami was calibrated to reproduce the effects observed at the coastal sea mouth.

Once the tsunami wave amplitudes at the mouth of Juan de Fuca Strait were known, the response of this partially enclosed body of water to the incoming tsunami was determined by using a computerized numerical scheme which employed a fine-mesh spatial grid. Variable bathymetry and Chezy frictional coefficients in the water body were allowed as input to the program. The ratio of tsunami amplitude to water depth in parts of the system was great enough so that non-linear advective terms could be significant and were therefore included in the equations of motion. The period of the tsunami waves entering the Puget Sound was 1.8 h. This period was observed during the 1964 tsunami.

The statistical effect of astronomical tides on tsunami run-up was determined by an analytical solution of the convolution integral (10.11). Tsunamis arriving at Puget Sound have, in the past, exhibited characteristic wave trains consisting of a number of waves of significant amplitude. The statistical effect of astronomical tides on tsunami run-up for such a situation when more than a single maximum tsunami wave is important must be determined through a numerical solution because an analytical solution is intractable. A numerical approach similar to that used by Petrauskas and Borgman (1971) was used to solve this problem. The period and number of significant waves of a tsunami which could be expected at a site were determined by analyzing wave records obtained during the 1964 tsunami. It is assumed here that the number and period of waves observed during the 1964 tsunami are indicative of the respective response of these areas to other high intensity tsunamis. Three waves with a period of 1.8 h were chosen for the tide effect analysis of Puget Sound.

A tsunami with an intensity between 2 and 5 in increments of one half is generated in one of the 12 segments of the Aleutian Trench and arrives at a site on the western coast of the United States with its significant waves having an amplitude which can be determined by the techniques described earlier. The probability of such an event occurring is equal to the probability of a tsunami of some particular intensity being generated somewhere in the Aleutian Trench, given by

$$n(i) = 0.065 e^{-0.7 l i}$$

and multiplied by $\frac{1}{12}$ because it is assumed that earthquakes occur uniformly throughout the length of the segmented trench.

Each of the possible tsunami wave trains of intensity range 2–5 from one of the 12 segments is then superimposed upon the astronomical tide occurring at a site during a year. The maximum tsunami plus tide elevation for the period is assigned a probability equal to $\frac{1}{12}$ multiplied by Eq. (10.12) for a particular intensity i, multiplied by q (360 min/number of minutes per year = 6.85×10^{-4}). The tsunami is superimposed upon the astronomical tide for 360-min intervals for the entire year. By following this procedure for all tsunamis of intensity 2–5 for the 12 segments, a cumulative probability distribution for run-up at a given site equal to or exceeding some value due to the superposition of the tsunami and the astronomical tide was determined. The one in 100-yr and one in 500-yr run-up values at a site were determined from this probability distribution.

By applying the methodology described earlier, 100- and 500-yr run-up values were calculated for the Puget Sound area. The effect of the astronomical tides on run-up varied from location to location. The more pronounced the tidal range, the more significant was the increase in run-up due to the influence of the astronomical tides. For example, tsunami waves in Puget Sound had small amplitudes and run-up values were governed largely by the effect of astronomical tides. Therefore, although waves had larger amplitudes at Port Townsend compared to Seattle, the greater tidal range at Seattle resulted in larger combined run-up values.

The effect of the astronomical tides on run-up was also dependent upon the probability distribution of tsunami wave amplitudes at a location. The tidal contribution to run-up is usually greater for locations protected from tsunamis than for those exposed. This can readily be seen for an analytic solution of Eq. (10.11).

For simplicity, consider only a single source region (e.g., the Aleutian Trench) and let $P_S^1(z)$ be represented by an exponential function

$$P_S^1(z) = A_1 e^{-\alpha_1 z} \tag{10.13}$$

where the superscript $_1$ denotes location 1.

Suppose that tsunamis at a second location occur with the same frequency as that for the first location but always produce twice as great a run-up. Then

$$P_S^1(z) = P_S^2(\hat{z}) \tag{10.14}$$

where

$$\hat{z} = 2z$$

and

$$P_S^2(\hat{z}) = A_2 e^{-\alpha_2 \hat{z}}$$
$$= A_2 e^{-2\alpha_2 z}. \tag{10.15}$$

Therefore,

$$A_1 e^{-\alpha_1 z} = A_2 e^{-2\alpha_2 z} \tag{10.16}$$

where

$$A_1 = A_2 \quad \text{and} \quad \alpha_1 = 2\alpha_2.$$

Houston and Garcia (1974) found that the net effect of the astronomical tide was to produce a $P(z)$ identical with $P_S(z)$ except for a shift of z by an amount $(\sigma^2/2)\alpha$. The value σ^2 is the tidal variance and equals $\sum_{m=1}^{\infty} C_m^2$ where C_m is equal to the mth tidal constituent. To evaluate Eq. (10.11),

$$P_\beta(z) \approx f_\beta(z) = \frac{1}{\sqrt{\pi}\sigma} e^{-z^2/2\sigma^2}.$$

Since σ^2 varies very little between two locations, the effect of the astronomical tide on run-up is approximately twice as large for location 1 as for location 2.

BAROCLINIC MODELS

11 THE UPPER LAYER MODEL: GF4

11.1 INTRODUCTION

Basic features of the surface circulation as determined from field studies have been presented in Chapter 1. It has also been noted that in the area contiguous to the mouth of the Fraser River the dilute upper layer in the Strait of Georgia surmounts a sharp pycnocline and would hence suggest a layer type model with a moveable lower interface. On the other hand, the geographical extent over which such a model is applicable varies not only with the tidally and seasonally modulated river discharge rate but also with the winds. Throughout the remainder of the Strait appreciable surface dilution remains although the vertical gradient of density is weak. Near the river mouth, entrainment of underlying denser water into the fast-flowing turbulent upper layer is important. Near the boundary openings the fast tidal streams bring about both strong turbulent mixing and complex patterns of residual circulation. The extensive area of shallow banks around the river mouth is characterized by landward movements of upper layer water moving over a salt wedge on the flood while on the ebbing tide vertically-mixed water moves off the banks. During a tidal cycle, the layer undergoes vertical tidal displacement over a range comparable with its own thickness.

Against this complexity there remains the fact that the extensive accumulation of data from surface drifters indicates that the surface current patterns conform to a relatively few coherent flow features which respond to the changing tides, winds, and river discharge. This suggested the possibility of developing a numerical model providing theoretical insight into the structure and variability of the basic circulation patterns. Such a model would require considerable innovation with respect to approximating conditions at the underside of the layer and at the open boundaries. It should also have to be able to simulate the tidally-interacting and Coriolis-deflected jet of river water while permitting the situation to be occasionally dominated by surface wind drift. Further considerations would concern the depths of water in the Strait, which greatly exceed those of the thin upper layer, the need for a relatively small horizontal grid scale, and the requirement of keeping computing costs at a level which would permit an extensive program of numerical experiments.

These objectives have been achieved. A vertically-integrated buoyant-spreading upper layer model (GF4) of the Strait of Georgia has been developed in which the barotropic pressure gradients derived from the mixed tidal co-oscillations within the system are obtained from a prior calculation (GF7) or from harmonic constants derived from such a calculation, and then prescribed for each time step at each velocity point within the model. Furthermore, the tidal velocities underlying the layer can be approximated by those obtained from GF7. Thus, a basic assumption is that the pressure gradients below the upper layer are those of the barotropic tide and any baroclinic contributions from the thin upper layer are negligible.

11.2. GOVERNING EQUATIONS

For clarity and brevity, the essential theoretical notions are presented for the two-dimensional case of fluid particle motions in a vertical plane through a two-layer system, the z axis being taken positive upward.

The basic equations governing volume, salt, and momentum conservation in the upper layer are

$$\frac{\partial u}{\partial x} + \frac{\partial w}{\partial z} = 0, \tag{11.1}$$

$$\frac{\partial s}{\partial t} + \frac{\partial (su)}{\partial x} + \frac{\partial (sw)}{\partial z} = 0, \tag{11.2}$$

$$\frac{\partial u}{\partial t} + \frac{\partial u^2}{\partial x} + \frac{\partial (uw)}{\partial z} + \frac{1}{\rho}\left(\frac{\partial p}{\partial x} - \frac{\partial \tau_{x,x}}{\partial x} - \frac{\partial \tau_{x,z}}{\partial z}\right) = 0 \tag{11.3}$$

where, assuming averaging over a time interval suitable for the removal of turbulent fluctuations, the notation is

$u(x, z, t)$ = horizontal component of velocity,
$w(x, z, t)$ = vertical component of velocity,
$p(x, z, t)$ = pressure,
$s(x, z, t)$ = salinity,
$\rho(x, z, t)$ = density = $a + bs$ where a and b are constants which depend on water temperature,
$\tau_{x,x}(x, z, t)$ = horizontal component of stress.
$\tau_{x,z}(x, z, t)$ = vertical component of stress.

These equations are vertically integrated to remove the z dependence. It is assumed at this stage of development that u, s, and ρ in the upper layer are independent of depth. This appears justified in virtue of results from trials of the model. (Significant vertical velocity and salinity gradients are shown by observations to exist in the region around the river mouth. Revision of this assumption is planned for later work.)

Vertical integration of the continuity equation (11.1) requires use of the kinematic conditions applied at the free surface $z = \zeta(x, t)$ and at the permeable interface $z = \eta(x, t)$, respectively, as follows,

$$\frac{D}{Dt}(\zeta - z) = \frac{\partial \zeta}{\partial t} + u(\zeta)\frac{\partial \zeta}{\partial x} - w(\zeta) = 0, \tag{11.4}$$

$$\frac{D}{Dt}(z-\eta) = -\frac{\partial \eta}{\partial t} - u(\eta)\frac{\partial \eta}{\partial x} + w(\eta) = w_+ - w_-. \tag{11.5}$$

In Eq. (11.5) account has been taken of possible fluxes of mass into and out of the upper layer. The physical significance of w_+ and w_- is described in detail later but it should be mentioned at this point that both w_+ and w_- may be expected to vary regionally depending on local properties of the flow.

Vertically integrating the equation of continuity (11.1),

$$\frac{\partial}{\partial x}\int_\eta^\zeta u\, dz - u(\zeta)\frac{\partial \zeta}{\partial x} + u(\eta)\frac{\partial \eta}{\partial x} + w(\zeta) - w(\eta) = 0.$$

Substituting Eqs. (11.4) and (11.5) and setting the layer depth $h = (\zeta - \eta)$ and the horizontal transport $U = uh$,

$$\frac{\partial h}{\partial t} + \frac{\partial U}{\partial x} = w_+ - w_-. \tag{11.6}$$

Vertically integrating Eq. (11.2),

$$\frac{\partial}{\partial t}\int_\eta^\zeta s\, dz - s(\zeta)\frac{\partial \zeta}{\partial t} + s(\eta)\frac{\partial \eta}{\partial t}$$
$$+ \frac{\partial}{\partial x}\int_\eta^\zeta us\, dz - u(\zeta)s(\zeta)\frac{\partial \zeta}{\partial x} + u(\eta)s(\eta)\frac{\partial \eta}{\partial x} + \big(w(\zeta)s(\zeta) - w(\eta)s(\eta)\big) = 0. \tag{11.7}$$

At the interface $z = \eta$, the quantity $w(\eta)s(\eta)$ represents an average over a time period sufficient to remove turbulent fluctuations and has two components. One is associated with motion of the interface denoted by $w_{\text{int}}s_\eta$. The other is due to the flux of salt relative to the interface. The flux associated with entrainment is written as w_+s_0 where s_0 is the salinity of the lower layer; the flux associated with depletion is written as w_-s, where s is the average salinity in the upper layer, assumed appropriate to the negative flux of salt due to depletion. Thus,

$$w(\eta)s(\eta) = w_{\text{int}}s_\eta + w_+s_0 - w_-s, \tag{11.8}$$

or

$$w(\eta)s(\eta) = \left(\frac{\partial \eta}{\partial t} + u_\eta\frac{\partial \eta}{\partial x}\right)s_\eta + w_+s_0 - w_-s. \tag{11.9}$$

Using Eq. (11.9) and the kinematic boundary condition (11.4), Eq. (11.7) simplifies to

$$\frac{\partial}{\partial t}(hs) + \frac{\partial}{\partial x}(us) = w_+s_0 - w_-s \tag{11.10}$$

where it has been assumed that s is uniform in the vertical.

To vertically integrate the momentum equation (11.3) and include the barotropic tidal forcing as described above, it is necessary to give prior consideration to the pressure gradient term. In particular, the fundamental simplifying assumption is that the horizontal pressure gradient in the deep underlying layer is solely determined by a purely barotropic tidal co-oscillation within the system.

Considering initially the pressure at some depth z in the upper layer and assuming that vertical variations in density can be ignored,

$$p = \int_\zeta^z \rho g \, dz' = \rho g (\zeta - z) \tag{11.11}$$

and in the lower layer,

$$p = \int_\eta^\zeta \rho g \, dz + \int_z^\eta \rho_0 g \, dz' = h \rho g + (\eta - z) \rho_0 g \tag{11.12}$$

where ρ_0 is the uniform density of the lower layer.

Vertically integrating the horizontal pressure gradient term in the upper layer momentum equation (11.3) and substituting Eq. (11.11),

$$\int_\eta^\zeta \frac{1}{\rho} \frac{\partial p}{\partial x} \, dz = \frac{g}{\rho} \left[\frac{h^2}{2} \frac{\partial \rho}{\partial x} + \rho h \frac{\partial \zeta}{\partial x} \right]. \tag{11.13}$$

Differentiating Eq. (11.12), the resulting lower layer pressure gradient is set equal to the predetermined pressure gradient for purely barotropic tidal co-oscillations, thereby yielding an equation for the slope of the free surface to be used in the upper layer computations,

$$\frac{\partial \zeta}{\partial x} = \frac{\partial}{\partial x} \left[\left(\frac{\rho_0 - \rho}{\rho_0} \right) h \right] + \frac{\partial \zeta_T}{\partial x} \tag{11.14}$$

where $\zeta_T(x,t)$ is the elevation of the water surface obtained from an earlier barotropic tidal computation using the vertically-integrated model GF7.

Substituting the right hand side of Eq. (11.14) into Eq. (11.13) gives the vertically-integrated pressure gradient for the upper layer in terms of layer thickness and imposed barotropic tidal slopes,

$$\int_\eta^\zeta \frac{1}{\rho} \frac{\partial p}{\partial x} \, dz = \frac{\partial}{\partial x} \left[\frac{g' h^2}{2} \right] + gh \frac{\partial \zeta_T}{\partial x} \tag{11.15}$$

where $g' = g \left(\frac{\rho_0 - \rho}{\rho_0} \right)$.

Since the velocity u and density ρ have been assumed to be vertically uniform in the upper layer, Eq. (11.3) may readily be vertically integrated, taking account of the vertical momentum transports across the interface and incorporating the pressure term (11.15),

$$\frac{\partial U}{\partial t} + \frac{\partial}{\partial x} \left(\frac{U^2}{h} \right) + \frac{\partial}{\partial x} \left(\frac{g' h^2}{2} \right) + gh \left(\frac{\partial \zeta_T}{\partial x} \right)$$
$$+ \frac{k U_R |U_R|}{h^2} - \frac{\rho_a C_D u_w^2}{\rho_0} - u_T w_+ + \frac{U}{h} w_- = 0 \tag{11.16}$$

where $U_R = h(u - u_T)$
and

$U(x,t)$ = vertically-integrated transport in the upper layer,
$g'(x)$ = reduced gravity, defined above,
k = frictional coefficient at the interface,
ρ_a = density of air,
C_D = drag coefficient at the sea surface,
u_w = wind speed in the x direction,
ζ_T = predetermined barotropic tidal elevation,
u_T = predetermined barotropic tidal velocity in the lower layer,
w_+, w_- = positive (entrainment) and negative (depletion) vertical velocity components, respectively,

and where the following substitutions have been made:

$$\tau_{s,x}(\zeta) = \rho_a C_D u_w |u_w|,$$

$$\tau_{b,x}(\eta) = \frac{k U_R |U_R|}{h^2}.$$

Generalizing the above equations to include a lateral dimension and incorporating eddy viscosity terms to parameterize small-scale turbulent processes, the vertically-integrated equations employed in the model are

$$\frac{\partial h}{\partial t} + \frac{\partial U}{\partial x} + \frac{\partial V}{\partial y} = w_+ - w_-,$$

$$\frac{\partial S}{\partial t} + \frac{\partial}{\partial x}\left(\frac{US}{h}\right) + \frac{\partial}{\partial y}\left(\frac{VS}{h}\right) = w_+ s_0 - w_- \frac{S}{h},$$

$$\frac{\partial U}{\partial t} + \frac{\partial}{\partial x}\left(\frac{U^2}{h}\right) + \frac{\partial}{\partial y}\left(\frac{UV}{h}\right) + \frac{\partial}{\partial x}\left(\frac{g'h^2}{2}\right) - fV + \frac{kU_R\sqrt{U_R^2+V_R^2}}{h^2}$$
$$+ \frac{\partial}{\partial x}\left(A\frac{\partial}{\partial x}U\right) + \frac{\partial}{\partial y}\left(A\frac{\partial}{\partial y}U\right) = u_T w_+ - \frac{U}{h}w_- - gh\frac{\partial \zeta_T}{\partial x} + \tau_{s,x},$$

$$\frac{\partial V}{\partial t} + \frac{\partial}{\partial x}\left(\frac{UV}{h}\right) + \frac{\partial}{\partial y}\left(\frac{V^2}{h}\right) + \frac{\partial}{\partial y}\left(\frac{g'h^2}{2}\right) + fU + \frac{kV_R\sqrt{U_R^2+V_R^2}}{h^2}$$
$$+ \frac{\partial}{\partial x}\left(A\frac{\partial}{\partial x}V\right) + \frac{\partial}{\partial y}\left(A\frac{\partial}{\partial y}V\right) = v_T w_+ - \frac{V}{h}w_- - gh\frac{\partial \zeta_T}{\partial y} + \tau_{s,y},$$

where

$U, V(x,y,t)$ = vertically-integrated transports in the upper layer,
$h(x,y,t)$ = layer thickness,
$u, v(x,y,t)$ = vertically-averaged velocities in the upper layer, $= U/h, V/h$,
$S(x,y,t)$ = vertically-integrated salinity,
$s(x,y,t)$ = average salinity in the upper layer, $= S/h$,
$g'(x,y,t)$ = reduced gravity, previously defined,
$\zeta_T(x,y,t)$ = barotropic tidal elevation (from GF7),

u_T, v_T = barotropic tidal streams (from GF7),
$U_R, V_R(x, y, t)$ = relative transports = $U - u_T h$, $V - v_T h$,
f = Coriolis parameter,
A = horizontal eddy viscosity,
$K(u, v, h, \rho)$ = quadratic friction coefficient,
$w_+(u, v, h, \rho)$ = entrainment velocity,
$w_-(u, v, h, \rho)$ = depletion velocity,
$\tau_{s,x}, \tau_{s,y}(x, y, t)$ = wind stress components at the sea surface.

11.3 BOUNDARY CONDITIONS

The implementation of a numerical scheme discussed above presents a variety of special considerations with respect to the conditions to be applied at the lateral open boundaries and the upper and lower interfaces of the modelled plume. The conditions described below were determined over an extensive series of experimental simulations (Stronach, 1977). They permit long trials in which, for a given river flow, stable horizontal distributions of salinity and layer depth are achieved and sustained in at least good qualitative agreement with the spatially and temporally sparse data presently available for such a labile and extensive (7,000 km^2) system. Concomitant horizontal velocity fields evolve stable cyclic behaviour in response to the prescribed tidally-modulated river flow and applied barotropic tidal forcing. These considerations constitute the primary justification of the model.

11.4 FINITE-DIFFERENCE EQUATIONS

The numerical formulation of the governing equations follows the semi-implicit scheme developed by Flather and Heaps (1975) for Eqs. (4.4–4.6). Though satisfactory for tidal calculations, a question arises as to its efficacy in the context of an upper layer model in which the fluid velocity may approach the small amplitude internal gravity wave velocity (internal Froude numbers close to unity). Provided the Courant-Friedrichs-Lewy criterion is satisfied when the free surface wave speed is replaced by an appropriate internal wave speed (see Chapter 4), no difficulty arises. Figure 11.1 shows the computational grid for the upper layer model. This grid is part of the overall 2-km mesh model grid (GF7). Thus, computed elevations and velocities from GF7 are readily transferred to GF4.

Certain spatial averages are defined below for use in the finite-difference equations which follow. The indexing is the same as that used in the earlier two-dimensional models. All fields are considered to be at the same time level. A single spatial mesh index is employed such that adding or subtracting n (the number of columns in the grid) specifies, respectively, the adjacent mesh in the row above or below. The averaging

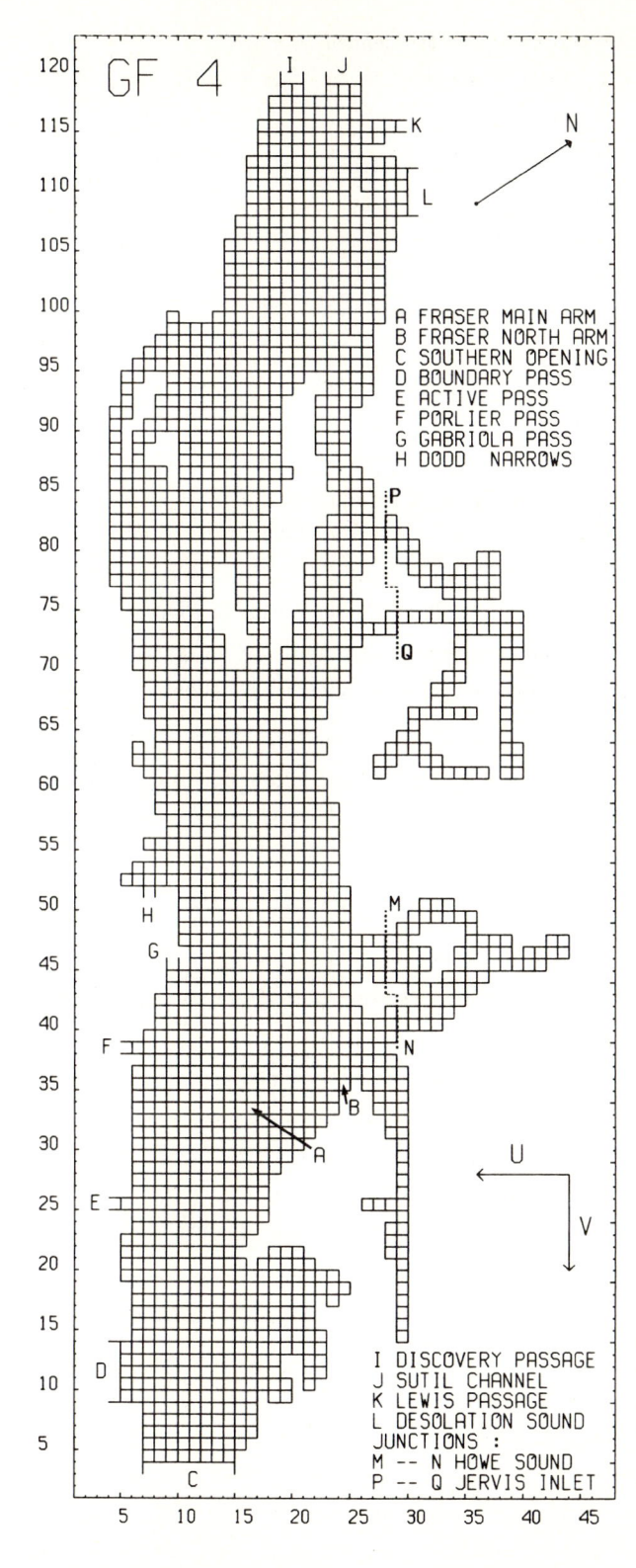

Fig. 11.1 Numerical grid used for the vertically-integrated, buoyant-spreading upper layer model GF4. This grid is identical to the appropriate part of GF7.

A FRASER MAIN ARM
B FRASER NORTH ARM
C SOUTHERN OPENING
D BOUNDARY PASS
E ACTIVE PASS
F PORLIER PASS
G GABRIOLA PASS
H DODD NARROWS
I DISCOVERY PASSAGE
J SUTIL CHANNEL
K LEWIS PASSAGE
L DESOLATION SOUND
JUNCTIONS:
M -- N HOWE SOUND
P -- Q JERVIS INLET

operations for the finite-difference formulations are defined as follows:

$$\overline{U}_i^x = \tfrac{1}{2}(U_i + U_{i+1}),$$
$$\overline{U}_i^y = \tfrac{1}{2}(U_i + U_{i-n}),$$
$$\overline{V}_i^x = \tfrac{1}{2}(V_i + V_{i+1}),$$
$$\overline{V}_i^y = \tfrac{1}{2}(V_i + V_{i-n}),$$
$$\overline{Z}_i^x = \tfrac{1}{2}(Z_i + Z_{i+1}), \qquad (Z \equiv h),$$
$$\overline{Z}_i^y = \tfrac{1}{2}(Z_i + Z_{i-n}),$$
$$\overline{\overline{U}}_i = \tfrac{1}{4}(U_i + U_{i-1} + U_{i+n} + U_{i-n}),$$
$$\overline{\overline{V}}_i = \tfrac{1}{4}(V_i + V_{i+1} + V_{i-n} + V_{i-n+1}),$$
$$\overline{\overline{Z}}_i = \tfrac{1}{4}(Z_i + Z_{i+1} + Z_{i-n} + Z_{i-n+1}),$$
$$\overline{\overline{Z}}_i^x = \tfrac{1}{4}(Z_i + 2Z_{i+1} + Z_{i+2}),$$
$$\overline{\overline{Z}}_i^y = \tfrac{1}{4}(Z_{i-n} + 2Z_i + Z_{i+n}).$$

A slight change in the notation is used in the above definitions in which z replaces h from the continuous partial-differential equation form. With $\Delta \ell$ as the spatial grid size and Δt as the time step, the following set of finite-difference equations is obtained. The continuity equation is

$$[Z_i(t+\Delta t) - Z_i(t)]\frac{1}{\Delta t} = -[U_i(t) - U_{i-1}(t) + V_{i-n}(t) - V_i(t)]\frac{1}{\Delta \ell} + WP_i(t) - WN_i(t) \quad (11.17)$$

and the salt equation is given by

$$[S_i(t+\Delta t) - S_i(t)]\frac{1}{\Delta t} = \left[\frac{U_i(t)}{\overline{Z}_i^x(t)}\overline{S}_i^x(t) - \frac{U_{i-1}(t)}{\overline{Z}_{i-1}^x(t)}\overline{S}_{i-1}^x(t)\right.$$
$$\left. + \frac{V_{i-n}(t)}{\overline{Z}_i^y(t)}\overline{S}_i^y(t) - \frac{V_i(t)}{\overline{Z}_{i+n}^y(t)}\overline{S}_{i+n}^y(t)\right]\frac{1}{\Delta \ell}$$
$$+ WP_i(t)S_0 - WN_i(t)\frac{S_i(t)}{Z_i(t)}. \quad (11.18)$$

For reduced gravity,

$$G_i(t+\Delta t) = 24 - 0.8\frac{S_i(t+\Delta t)}{Z_i(t+\Delta t)}. \quad (11.19)$$

The finite-difference representation of the x-directed momentum equation is

$$[U_i(t+\Delta t) - U_i(t)]\frac{1}{\Delta t} = f\overline{\overline{V}}_i(t)$$
$$- \frac{1}{2\Delta \ell}\left[G_{i+1}(t+\Delta t)(Z_{i+1}(t+\Delta t))^2 - G_i(t+\Delta t)(Z_i(t+\Delta t))^2\right]$$
$$- K\left[\frac{U_i(t) - U_T(t)\overline{Z}_i^x(t)}{\overline{Z}_i^x(t)}\right.$$
$$\left. \times \sqrt{(U_i(t) - U_T(t)\overline{Z}_i^x(t))^2 + (\overline{\overline{V}}_i(t) - V_T(t)\overline{Z}_i^x(t))^2}\right]$$

$$-\frac{1}{\Delta\ell}\left[\frac{(\overline{U}_i^x(t))^2}{\overline{\overline{Z}}_i^x(t)} - \frac{(\overline{U}_{i-1}^x(t))^2}{\overline{\overline{Z}}_i^x(t)}\right]$$

$$-\frac{1}{\Delta\ell}\left[\frac{\overline{U}_i^y(t)\overline{V}_{i-n}^x(t)}{\overline{\overline{Z}}_i} - \frac{\overline{U}_{i+n}^y(t)\overline{V}_i^x(t)}{\overline{\overline{Z}}_{i-n}}\right]$$

$$-gTSX_i(t) + \overline{WP}_i^x(t)U_T(t) - \overline{WN}_i^y(t)\frac{U_i(t)}{\overline{Z}_i^x(t)}$$

$$+\frac{\overline{Z}_i^x}{(\Delta\ell)^2}\left[A_{i+1}\left(\frac{U_{i+1}(t)}{\overline{Z}_{i+1}^x(t)} - \frac{U_i(t)}{\overline{Z}_i^x(t)}\right)\right.$$

$$+ A_{i-1}\left(\frac{U_{i-1}(t)}{\overline{Z}_{i-1}^x(t)} - \frac{U_i(t)}{\overline{Z}_i^x(t)}\right) + A_{i+n}\left(\frac{U_{i+n}(t)}{\overline{Z}_{i+n}^x(t)} - \frac{U_i(t)}{\overline{Z}_i^x(t)}\right)$$

$$\left.+ A_{i-n}\left(\frac{U_{i-n}(t)}{\overline{Z}_{i-n}^x(t)} - \frac{U_i(t)}{\overline{Z}_i^x(t)}\right)\right]. \tag{11.20}$$

For the y-directed momentum,

$$[V_i(t+\Delta t) - V_i(t)]\frac{1}{\Delta t} = -f\overline{\overline{U}}_i(t+\Delta t)$$

$$-\frac{1}{2\Delta\ell}\left[G_i(t+\Delta t)(Z_i(t+\Delta t))^2 - G_{i+n}(t+\Delta t)(Z_{i+n}(t+\Delta t))^2\right]$$

$$- K\left[\frac{V_i(t) - V_T(t)\overline{Z}_{i+n}^y(t)}{(\overline{Z}_{i+1}^y(t))^2}\right.$$

$$\left.\times \sqrt{(V_i(t) - V_T(t)\overline{Z}_{i+n}^y(t))^2 + (\overline{\overline{U}}_i(t) - U_T(t)\overline{Z}_{i+n}^y(t))^2}\right]$$

$$-\frac{1}{\Delta\ell}\left[\frac{\overline{U}_{i+n}^y(t)\overline{V}_i^x(t)}{\overline{\overline{Z}}_{i+n}(t)} - \frac{\overline{U}_{i+n-1}^y(t)\overline{V}_{i-1}^x(t)}{\overline{\overline{Z}}_{i+n-1}(t)}\right]$$

$$-\frac{1}{\Delta\ell}\left[\frac{(\overline{V}_{i+n}^y(t))^2}{\overline{\overline{Z}}_{i+n}^y(t)} - \frac{(\overline{V}_i^y(t))^2}{\overline{\overline{Z}}_i^y(t)}\right]$$

$$+ \overline{WP}_{i+n}^y(t)V_T(t) - \overline{WN}_{i+n}^y(t)\frac{V_i(t)}{\overline{Z}_{i+n}^y(t)} - gTSY_i(t)$$

$$+\frac{\overline{Z}_{i+n}^y(t)}{(\Delta\ell)^2}\left[A_{i+1}\left(\frac{V_{i+1}(t)}{\overline{Z}_{i+n+1}^y(t)} - \frac{V_i(t)}{\overline{Z}_{i+n}^y(t)}\right)\right.$$

$$+ A_{i-n}\left(\frac{V_{i-1}(t)}{\overline{Z}_{i+n-1}^y(t)} - \frac{\overline{V}_i(t)}{\overline{Z}_{i+n}^y(t)}\right) + A_{i+n}\left(\frac{V_{i+n}(t)}{\overline{Z}_{i+2n}^y(t)} - \frac{V_i(t)}{\overline{Z}_{i+n}^y(t)}\right)$$

$$\left.+ A_{i-n}\left(\frac{V_{i-n}(t)}{\overline{Z}_i^y(t)} - \frac{V_i(t)}{\overline{Z}_{i+n}^y(t)}\right)\right] \tag{11.21}$$

where TSX_i and TSY_i are the slopes of the water surface obtained from a barotropic tidal model of the same area and WP_i and WN_i are the entrainment and depletion velocities, respectively.

The ordering in time of these equations is important. In each elevation, the thickness Z_i and salt content S_i are calculated using values of the derivatives of U_i and V_i from the previous time step. Then U_i is calculated using the derivatives of Z_i and S_i (or G_i) from the current time step and the previous values of V_i in the Coriolis term. Finally, V_i is calculated using derivatives of the current Z_i and S_i and the current U_i in the Coriolis term. In the entrainment, depletion, and friction functions, and in the non-linear and eddy viscosity terms, the values of U_i, V_i, and S_i from the previous time step are always used.

11.4.1 Free Surface and Interfacial Boundaries

The wind stress applied at the sea surface is given by

$$\tau_{s,x} = \rho_a C_D U_w \sqrt{U_w^2 + V_w^2},$$
$$\tau_{s,y} = \rho_a C_D V_w \sqrt{U_w^2 + V_w^2}$$

where

U_w, V_w = x and y components of the surface wind, respectively,
ρ_a = density of air = $1.2 \times 10^{-3} \text{g/cm}^3$,
C_D = drag coefficient = $1.5 \times 10^{-3} \text{g/cm}^3$.

At the lower surface, the model requires that allowance be made for various degrees of permeability. These considerations may be incorporated into the kinematic condition applying at the lower interface by writing

$$\frac{\partial \eta}{\partial t} + u(\eta)\frac{\partial \eta}{\partial x} + v(\eta)\frac{\partial \eta}{\partial y} - w(\eta) = -w_+ + w_-$$

where w_+ is an upward and w_- a downward component of vertical velocity. Each is to be empirically related to the computed vertically-averaged horizontal velocities and densities in the upper layer and to the predetermined barotropic tidal velocities and uniform density (assumed known from observation) in the hypothetical lower layer. Thus, w_+ denotes a net transfer rate of fluid upward through the lower boundary surface into the upper layer which, in the absence of horizontal divergences, would result in a net rate of descent of the lower boundary surface relative to the ambient fluid. Similarly, w_- denotes a net transfer rate of fluid downward through the lower boundary surface and a corresponding tendency toward a displacement rate upward of that surface relative to the ambient fluid. On the basis of observation, the practical application of these variables in the numerical scheme provides for entrainment when

$$w_+ > 0, \quad w_- = 0,$$

for mixing when

$$w_+ \simeq w_-,$$

and for layer depletion (thinning) when

$$w_- < 0, \qquad w_+ = 0.$$

In the absence of winds, the first context (entrainment) applies most obviously to the fast-flowing turbulent upper layer in the region near the river mouth. The second (mixing) applies where a thickening of the pycnocline occurs away from the river mouth (particularly near boundary openings where the constricted passages can give rise to strong barotropic tidal streams and large vertical velocity shears); the third where the lower boundary surface tends to slope upward under the influence of low vertical shear and weak upper layer turbulence, ultimately intersecting the free surface.

Further illustrative of the physical meaning of w_-, it is of interest to consider, for the steady state case, the depth of the bounding isohaline surface in the x, z planes. Where this surface slopes upward to the sea surface, net depletion is occurring and a negative vertical velocity is implied. Under these conditions, in the absence of significant entrainment, the continuity equation becomes

$$\frac{\partial(hu)}{\partial x} = w_+ - w_-.$$

If the rate of entrainment is low, w_- dominates and as salinity increases (due to a non-zero w_+), the layer decreases in thickness in the downstream direction ($\partial h/\partial x < 0$), presuming that spatial variation in u is not significant and u is positive.

However, a more appropriate description of the process is to recognize that although water particle velocities are approximately horizontal, the ever present mixing in the upper part of the water column serves to incorporate saline water into the upper layer. This mixing is accomplished by a change in the shape of the salinity profile, with the result that a particular isohaline surface moves closer to the top of the water column. In order to express this concept in the context of a layer equation, the layer is assumed to thin as expressed by non-zero depletion velocity. Unfortunately, the term 'velocity' in the phrase 'depletion velocity' sometimes leads to a more active interpretation (i.e., bulk water particle motions) than is really intended (a parameterization of mixing with an ongoing process of layer redefinition).

Considering now the empirical formulation of expressions relating w_+ and w_- to the computed flow variables, the prime consideration at this stage of development is to approximate the general features of the upper layer described in Chapter 1. Extensive studies have been made of the entrainment process (Turner, 1973). Proceeding from standard parameterizations in the present work, the vertical entrainment velocity has been formulated as

$$w_+ = K_E \sqrt{u_R^2 + v_R^2}$$

where u_R and v_R are the differences between the u and v components of velocity and the known velocities of the underlying water, and

$$K_E = \alpha + \beta F_i^2$$

where α and β are constants determined through numerical experiments and F_i is the internal Froude number. The inclusion of F_i, which is a measure of the ratio of inertial to buoyancy forces, stems from observations (Cordes et al., 1980) which indicate that entrainment is large near the river mouth, decreasing rapidly thereafter, and from early numerical experiments which indicated that the coefficient K_E must vary according to the type of flow.

The primary physical roles of w_- are to provide a transition from vertical entrainment to vertical mixing and to control the layer depth in regions of low buoyancy near the model boundaries. In the context of mixing it should reflect the differential velocity between the upper and lower layers. In the context of layer thickness it should either inhibit surfacing or plunging to the sea bottom of the isohaline interface of the model, eventualities incompatible with existing data (Crean and Ages, 1971). Accordingly, the following formulation is employed.

$$w_- = K_D \sqrt{u_R^2 + v_R^2}$$

where

$$K_D = \gamma(g' - 12)h \quad \text{for} \quad g' > 12$$

and

$$K_D = 0 \quad \text{for} \quad g' < 12$$

where γ is a constant and g' is the buoyancy which equals $g(\rho_0 - \rho)/\rho_0$. For present purposes the number 12 has simply been selected as the central value of the practical range (0–24) of g' in this model.

11.4.2 Lateral Boundaries

Lateral boundaries include coasts, coastal openings, river mouths, and the vertical interface between the vertically-integrated model of the shallow banks near the river mouth and the upper layer model. At mesh sides denoting coastal boundaries, the normal component of velocity is set equal to zero and a free-slip lateral boundary condition is used.

Based on the best available estimates, the daily mean Fraser River discharge is proportioned between the North Arm (15%) and the Main Arm (85%). The flows at both river mouths were presumed to be modulated by the changing tidal elevations in the Strait of Georgia. In order to predict these river flows, a two-term linear prediction filter was applied to time series of measured river mouth velocities (Stronach, 1977) and corresponding tidal elevations at Point Atkinson. This filter was used to modulate the observed daily discharge proportioned between the primary and secondary river mouths.

Salt wedge effects are thus ignored relative to the major entrainment occurring at the river mouths themselves. The boundary condition applied at the Main Arm mouth is complicated by the fact that the direction of the entrant stream is inclined to both grid axes. In order to preserve the entrant direction of flow at the Main Arm mouth the momentum flux was given by $Q^2 \cos \theta$ for the x-directed momentum equation and $Q^2 \sin \theta$ for the y-directed equation where Q is the flow and θ is the angle of the jetty to the x axis. (An alternative scheme, based upon resolving the river mass flux into two components, did not provide a satisfactorily directed entrance stream with respect to the model grid.) Note that these fluxes are introduced at location A in Fig. 11.2.

A demanding problem in the implementation of an upper layer model of this type concerns the numerical conditions to be imposed at the open boundaries. In the model under consideration, such openings are located near regions of swift tidal streams, strong mixing, and tidally-driven recirculation. It is required that the layer, at least in a computational sense, remain a viable entity up to the model boundary and not introduce spurious effects over the interior of the model. As noted by Roache (1972), removal of the computational boundary away from the region of interest (the general area of the river mouth) can assist in attaining agreement of solutions in the latter with the physics to within some acceptable level of tolerance. Fundamentally, however, the location of the boundaries was determined by the desire to carry the model to the extreme limits of its geographic relevance. A model restricted to the immediate river mouth area would not really serve to answer the important question concerning the manner in which Fraser River water enters into the large-scale circulation of the Strait of Georgia.

Two fundamental notions suggest a viable approximation to conditions on the open boundaries. First, it was noted in the course of numerical experiments that the internal Froude number which, as a ratio of inertial to gravitational buoyant-spreading forces, characterizes such a layer flow, underwent little spatial variation in regions distant from the river mouth. In general, the internal Froude number is large near the mouth, thereafter falling rapidly and from then changing very slowly with distance. Secondly, considering the flow along the constant density surfaces associated with a buoyant jet, entrainment tends to be confined to the flow field relatively close to the source, while remote from the source there occurs flow predominantly along the constant density surfaces. These density surfaces are positioned through some balance of advective and turbulent mixing processes. Accordingly, the numerical scheme must accommodate a transition from a process involving the net transfer of fluid volumes possessing the density of the underlying ambient fluid to one in which mixing through compensatory exchanges of fluid volumes can occur but with limited exchange of mass. Near the boundary openings, significant vertical velocity shears can arise between the fluid in

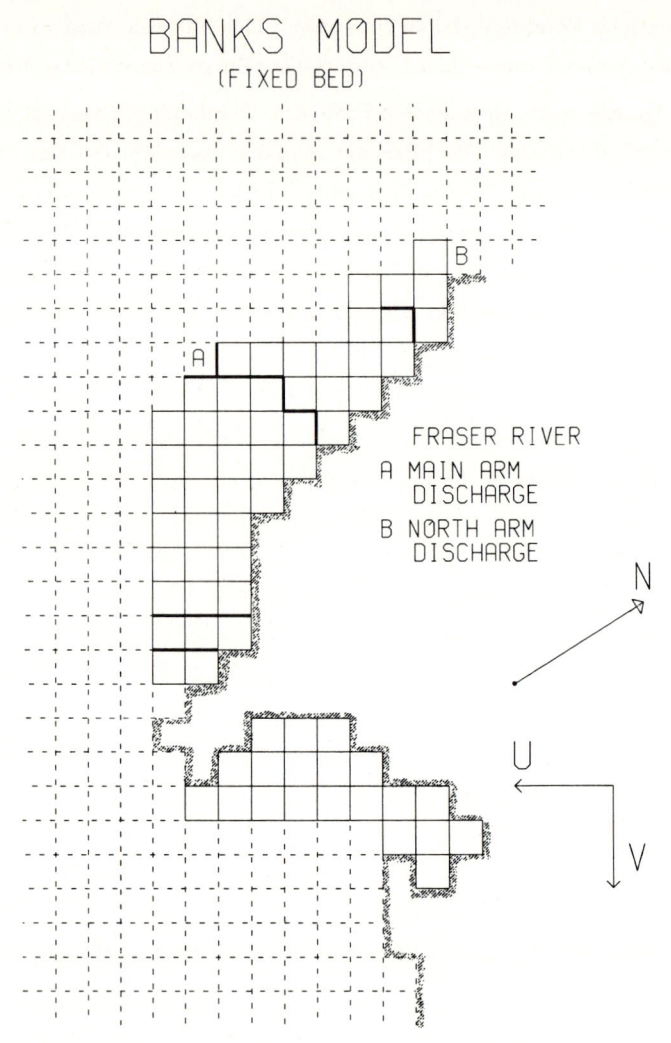

Fig. 11.2 Detail of the numerical grid scheme in the vicinity of the main freshwater sources A and B. Dashed lines denote the upper layer scheme and light solid lines show the vertically-integrated fixed bed model of the shallow banks of the Fraser River delta. The heavy solid lines indicate breakwaters.

the layer which responds to both the prescribed barotropic tidal and buoyant pressure gradients and the assumed predetermined underlying barotropic tidal velocities. This conduces to mixing and deepening of the layer. Thus, an additional process must occur which permits advection out of the modelled layer. The basic sequence of advective and mixing processes over a tidal cycle is not known in detail but presumably involves a flow of layered fluid into the mixing region on an ebb and a return of the mixed fluid on a flood.

Numerical formulation of these notions is now described. The open boundaries are located along mesh sides where the normal component of velocity must be determined from quantities known at the previous time step in the interior of the model. An internal Froude number, employing averaged velocities at the centre of each mesh, is calculated for the interior three meshes along a normal to the boundary and immediately contiguous to the actual boundary mesh side. At the boundary the second derivative of the internal Froude number normal to the boundary is assumed equal to zero. This yields a relationship between the depth of fluid in the mesh immediately inside the boundary and the normal component of velocity through the mesh side that constitutes the open boundary. The sign of the boundary velocity is taken to be the same as that of the first velocity component in from the boundary. Thus, the condition at the boundary assumes a linear change in the ratio of inertial to gravitational buoyant-spreading forces with distance from the boundary. This is a major simplification since the motion in the modelled layer responds not only to a buoyant-spreading pressure gradient but also to the applied predetermined barotropic tidal pressure gradients and wind stress at the sea surface as shown in Eqs. (11.19) and (11.20). Two distinctive situations arise. In one, the buoyancy is relatively high and the net outward flow through the open boundary responding predominantly to the buoyant pressure gradient is merely modulated by the tide. This is the type of situation that exists where the Coriolis-deflected river discharge finds seaward egress in the northeast corner of the model. In the second situation characterized by high energy and low buoyancy the barotropic tidal pressure gradients dominate the horizontal motion in the vicinity of the boundary opening. This is characteristic of the openings in the southeast and northwest corners of the model. It has been shown above, on the basis of observations, that these latter openings constitute sources of higher salinity surface water which intrude into the modelled layer.

In this high buoyancy case, the flood and ebb tides differ significantly. Considering first an ebbing tide, the barotropic tidal and buoyancy pressure gradients act in concert to move fluid towards the open boundary. The boundary condition, requiring linear spatial variation normal to the boundary of the ratio of the squared outflow velocity to the product of buoyancy and thickness, results in an increase in both outflow and thickness. For a flooding tide, the barotropic tidal pressure gradient, now opposing the buoyant-spreading pressure gradient, moves fluid away from the boundary mesh towards the interior of the model. The boundary condition fixing the ratio of inflow velocity to buoyancy pressure gradient, results in a relatively small inflow velocity and a consequent reduction in the depth of fluid in the boundary mesh. At the same time, the predetermined barotropic tidal flood stream, underflowing the thinning layer near the boundary, brings about maximal vertical velocity shear. Both w_+ and w_- (provided the buoyancy is sufficiently low), which are activated primarily by this shear, rise to peak values. Thus, a strong vertical mixing event is simulated in the course of a flood tide.

Depending on the constants of proportionality employed, the net result averaged over several tidal cycles is to form a thickened layer of mixed water adjacent to such open boundaries. Thus, a buoyant-spreading horizontal pressure gradient is formed, dependent primarily on the spatial thickness gradient which moves mixed water into the modelled surface layer. This is opposed by the buoyant-spreading pressure gradient of brackish water from the river mouth. In the course of numerical experiments, it has been found that this formulation of conditions at open boundaries has relatively small effect on the computed field of motion over the interior of the modelled region where the velocity results essentially from forcing by wind, tide, and river discharge. The formulation, deemed satisfactory at the present stage of development, is clearly susceptible to much improvement given a suitable concomitant program of observations.

It is also enlightening to consider the analogy between the upper layer model and the kinematic equations describing river flow (Lighthill, 1957; Whitham, 1974). In the case of river flow, the full hydrodynamic equations require an upstream and a downstream boundary condition. The kinematic approach is to replace the momentum equation and the downstream boundary condition relating flow rate to water depth with a relation derived from the momentum equation. (The latter relation is used in river flow gauging, for example.) The continuity equation is then solved using the remaining upstream flow boundary condition. Whitham (1974, Chapter 10) presents a mathematical justification for this procedure. The analogy with the upper layer model is obvious. The river flow rate is retained and the downstream boundary condition is replaced by the internal Froude number condition. In both cases, a fundamental requirement is that the momentum equation be dominated by the terms other than $\partial u/\partial t$. In the case of the plume, friction, entrainment, momentum flux, Coriolis force, and tidal pressure gradients do in fact dominate the $\partial u/\partial t$ term.

A problem arises in connection with the formulation of an upper layer model with respect to the extensive area of shallow banks in the general vicinity of the river mouth. Extensive field observations indicated that the flooding tide moves a shallow salt wedge onto the banks while the later stages of ebbing tide involve mixing of the water column to near uniformity, presumably due to bottom generated turbulence. The net result is that the shallow banks area can act as a source of mixed water entering the upper layer. This effect can be approximated in the following manner. The part of the vertically-integrated model GF7 which includes the shallow banks is isolated (Fig. 11.2) and the mesh sides corresponding approximately to the edge of the bank are taken as the boundary between a fixed-bed banks model and the 'floating' upper layer model. The fixed-bed banks model equations and schematization have been previously described (GF7).

Prior to discussing the boundary conditions at this lateral interface between the two models, it is desireable to emphasize certain aspects of the upper layer model. It is assumed that the range of thickness occurring over the layer is small with respect to the much deeper lower layer, the depth of which does not enter formally into the the upper layer equations. Furthermore, the barotropic tides are introduced into these equations in the form of pressure gradients. Thus, the sea level datum of the layer model, with respect to the sea bottom, may be chosen at will. The choice is made that the free surface of the upper layer model coincides with the height of the water surface in the meshes immediately seaward of the banks edge as given by the vertically-integrated model GF7 and an appropriate additional elevation corresponding to the increased buoyancy of the upper layer. Thus, in effect, the layer model rises and falls with the vertical tides in the system like the actual river plume in practice. Accordingly, the pressure gradient entering into the equation for a horizontal velocity normal to a mesh side on the boundary between the two models is composed of two parts. One is due to the difference in elevation between the prescribed tide (known from GF7) in the mesh immediately seaward of the banks boundary and the adjoining mesh on the banks immediately inward of the boundary. The other is due to the buoyant-spreading pressure gradient in the upper layer.

There remains the problem of how to relate the layer depth to the bank depth at a boundary mesh side. For a flooding tide two cases may be distinguished. If the layer thickness exceeds the depth of water between the surface and sea bottom in the adjoining bank mesh, the flood stream onto the banks is assumed to consist solely of upper layer water. If the layer thickness is less than the depth on the banks, it is assumed that the part of the water column corresponding to the layer depth advects upper layer water onto the banks. The flow in the lower part of the water column (*i.e.*, through a depth corresponding to the difference between the overall depth of water in the bank mesh and the layer thickness in the adjoining mesh) is assumed to have a salinity corresponding to that of the lower layer. This, in effect, approximates the mixing on the bank of saline water in the wedge and water from the upper layer. On an ebbing tide, it is assumed that all the mixed water from the banks model enters the upper layer.

11.4.3 Initial Conditions

An initial uniform salinity field is prescribed. This salinity is selected as intermediate between that of the deep sea-water in the Strait of Georgia and lower salinity water issuing from the river mouth. The total freshwater content of the upper layer is thus initially less than it should be, and the model is required to flush itself during the start up phase. It is further assumed that the layer is of uniform thickness (4 m) and at rest. Under prescribed tidally-modulated high river discharge (about $10,000$ m^3/s) and from rest, the system requires about 2 wk to attain essentially steady distributions of salinity and layer thickness. Such a time period is consistent with the flushing time of the upper layer.

11.5 RESULTS

Validation of the model's performance presents a number of difficulties. This is partly due to inadequate coverage of such an extended range of rapidly changing physical processes as well as a dearth of suitable wind observations needed to prescribe the wind stresses for the simulation of actual events. The response to tides and the tidally-pulsed river jet, the formation of wind-induced coastal jets, and layer deepening along certain boundaries must be considered as well as the motion deriving from adjustment of the mass field associated with the onset and relaxation of wind events.

It may, however, be readily demonstrated that the model reproduces a wide range of phenomena in at least good qualitative agreement with theoretical anticipation and existing observations. In Sections 11.5.1 and 11.5.2 the response of the model to various combinations of wind and tide is considered. Section 11.5.3 addresses the residual circulation while Section 11.5.4 compares modelled and observed results.

The results presented are for the case of high river discharge ($10,000$ m^3/s), these being of primary interest in that they describe the maximal seasonal departure of surface currents away from the tidal velocities obtained over the greater part of the water column as derived from the vertically-integrated model. Satisfactory results have also been obtained for conditions of low river flow ($2,000$ m^3/s).

11.5.1 River and Tide Interactions in the Central Strait of Georgia

Tides in the system are of the mixed type and change persistently in a pattern that effectively repeats itself every 18.6 yr. The major source of freshwater discharge (the Fraser River) is strongly modulated by the tide and enters a region in the Strait of Georgia where the tidal streams undergo a marked change in direction and a severalfold decrease in amplitude in the region of the river mouth.

It is of interest to first illustrate the rapidly changing complex patterns of flow reproduced by the model in the vicinity of the river mouth over a representative 24-h sequence. Figure 11.3 shows the tidal elevations at Point Atkinson, representative of the tide in the Strait of Georgia off the Fraser River mouth, with times corresponding to the numbering sequence of the vector plots (Fig. 11.4). Thus, hours 1 to 3 show the diminishing flood streams south of the mouth, the Coriolis-deflected river discharge north of the river mouth, and the initiation of an eddy in the vicinity of the freshwater sources A and B. This corresponds to the last 3 h of a weakly flooding tide.

About the time of higher high water, this eddy reaches its maximum geographical extent (20 km) then starts to dissipate as the southward-moving streams on its eastern side increase while those on its western side vanish and (hours 4, 5, 6) as the ebb streams to the south of the river mouth gather momentum. As these latter attain their maxima and start to diminish towards lower low water, the discharge from the river mouth builds to maximum values (hours 7, 8, 9). With the turn of the tide the flood

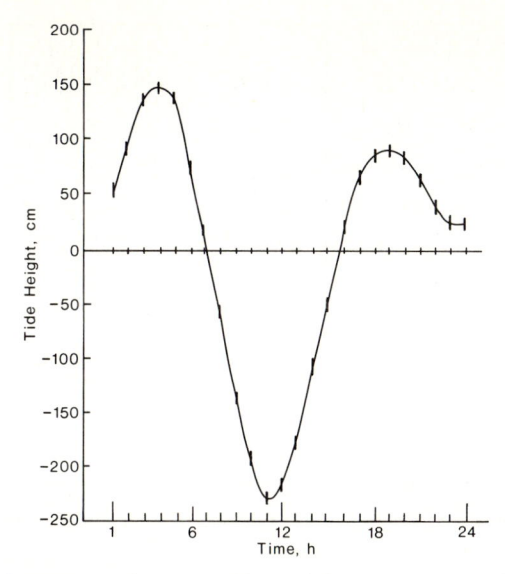

Fig. 11.3 Tidal elevations and times at Point Atkinson corresponding to the numbered sequence in Fig. 11.4.

streams strongly augment the Coriolis-deflected northward river discharge (hours 10, 11, 12).

As the flood streams attain maximum value and start decreasing (hours 13 to 15), an eddy again starts to form north of the main source of freshwater discharge at about the time of lower high water (hours 16 to 18). This eddy grows in extent but with the inception of a weak ebbing tide loses its distinctive character (hours 19 to 21). About the time of higher low water (hours 22 to 24) the river discharge attains a weak secondary maximum and the weak flooding streams again augment the northward Coriolis-deflected river flow.

11.5.2 River, Tide, and Wind Interactions in the Overall Strait of Georgia

To further illustrate the basic features of the overall model performance, fields of salinity, layer thickness, horizontal and relative vertical (*i.e.*, relative to the lower interface) velocities are presented under conditions of high river discharge, ebbing and flooding tides, and a range (± 2.5, ± 5, ± 10 m/s) of wind speeds, directed parallel to the major axis of the model. In these tests the model was operated for a 30-day period from initial conditions of rest. A uniform layer thickness of 4 m and salinity of $20^0/_{00}$ was assumed. The river flow prescribed was 10,000 m^3/s. The last 3 days of this run were repeated but each time with the application of a steady uniform wind field. The plots presented were taken from the large ebb and the large flood preceding and following, respectively, lower low water on the third day. From a practical point of view, the times of high river discharge are of greatest interest since, under these conditions,

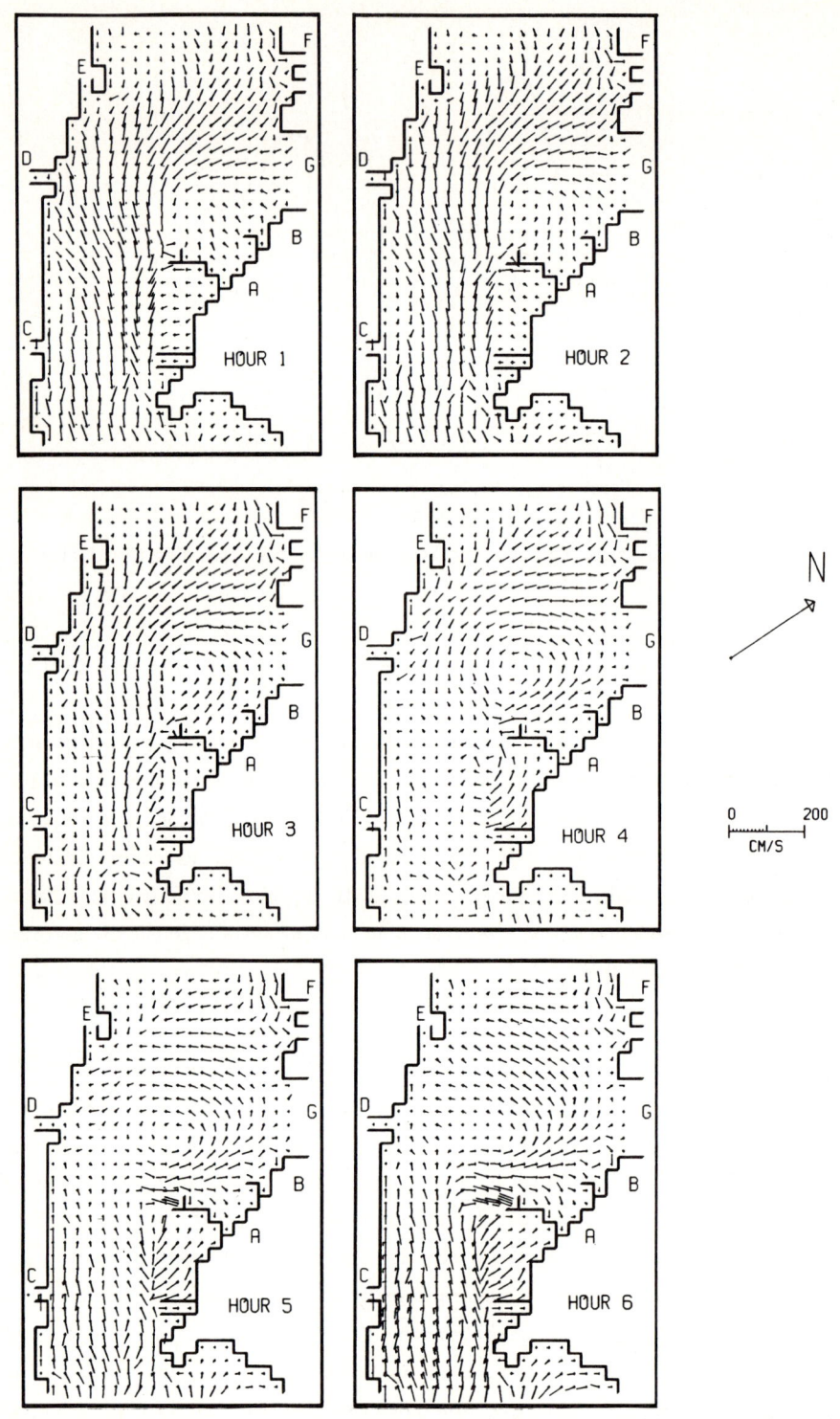

Fig. 11.4 Hourly plots over a 24-h period of velocity vector fields computed by the model GF4 illustrating the interaction of tidally-pulsed river discharge with tides in the Strait of Georgia.

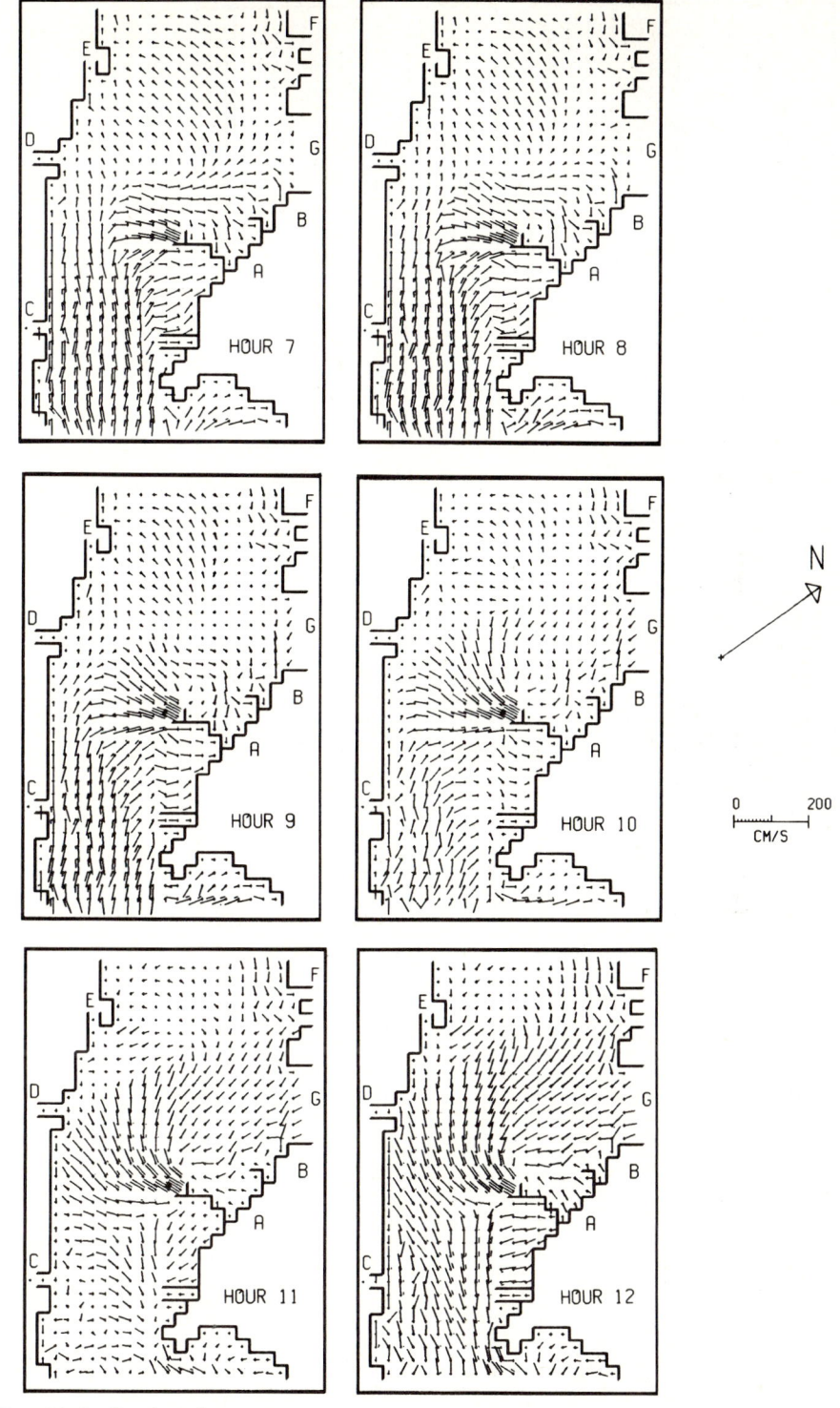

Fig. 11.4 Continued.

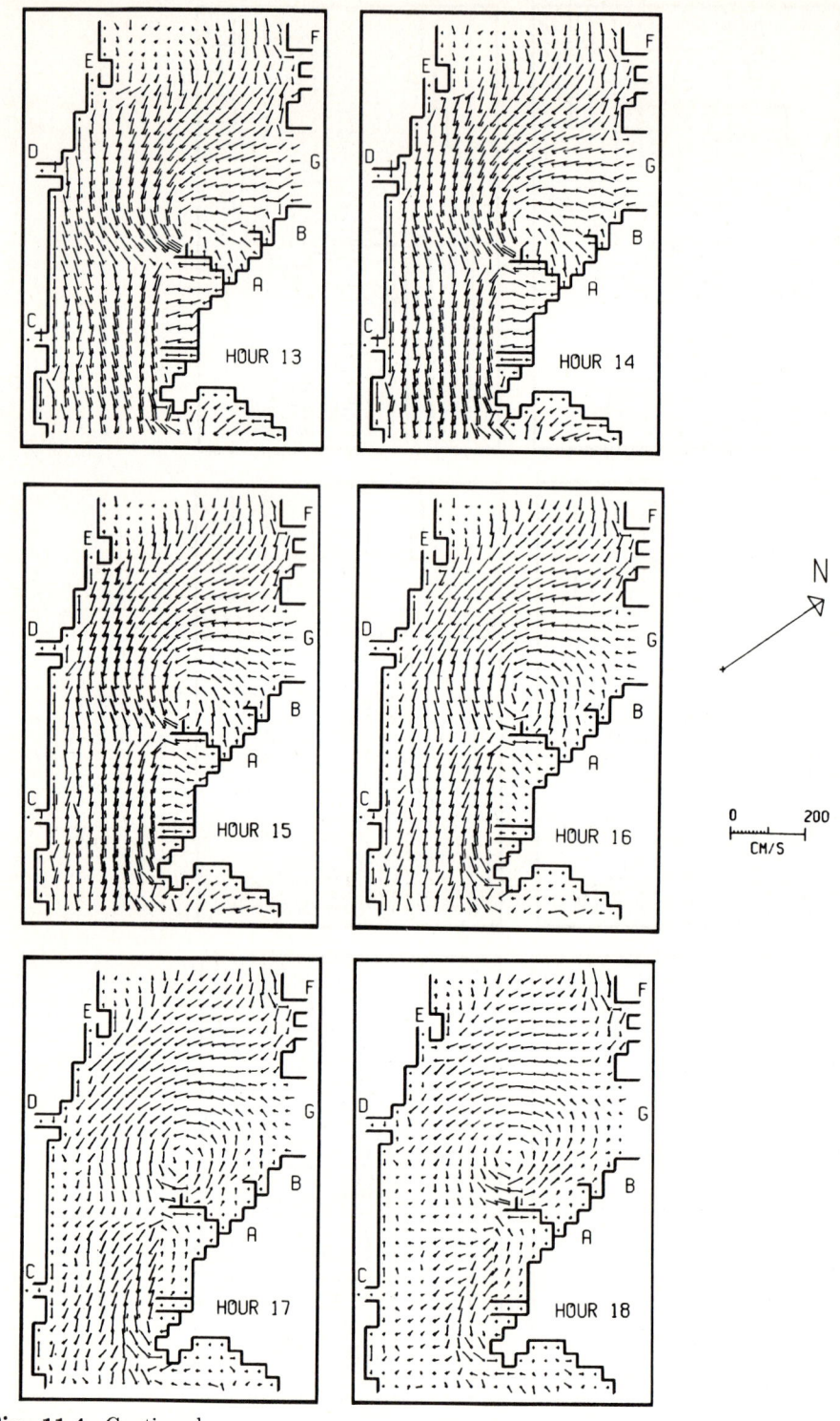

Fig. 11.4 Continued.

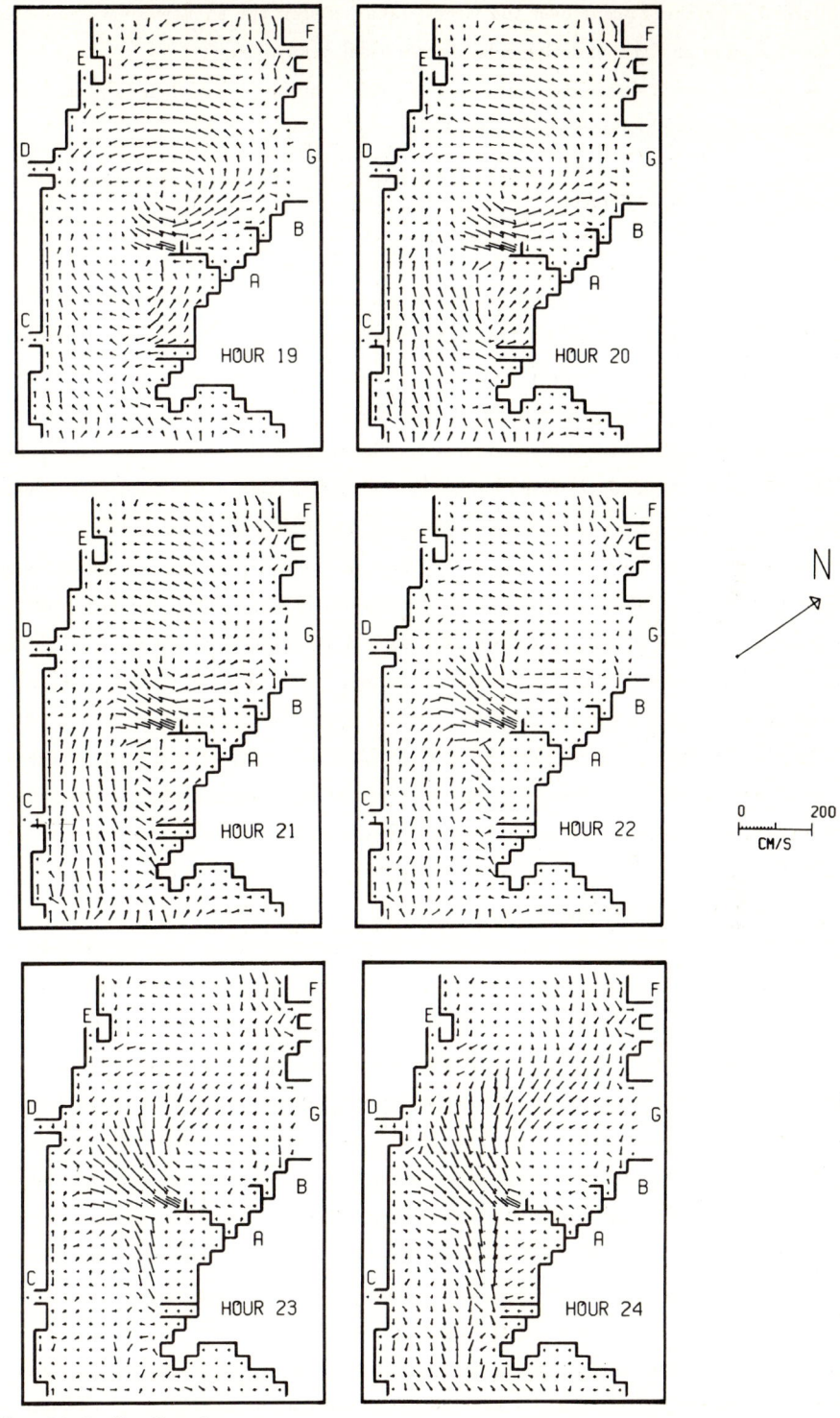

Fig. 11.4 Continued.

the largest differences occur over the largest area between the shallow river-dominated surface currents and the underlying barotropic tidal velocities.

Attention now is directed initially to the model's performance when tides and river discharge are prescribed and winds are absent.

11.5.2.1 No wind forcing

Ebb Tide

The velocity vector field corresponding to an ebbing tide 3 h before lower low water is shown in Fig. 11.5a. There are strong ebb surface currents in the southern Strait of Georgia similar to those obtained from the vertically-integrated model GF7. The river discharge which is steadily increasing at this stage of the tide, maintains its entrant direction rather more than half way across the Strait from the edge of the banks. The ebbing flow from the shallow banks model into the upper layer model accords well with anticipation. based on the overall vertically-integrated fine grid model GF7.

A local area of higher velocities in the central Strait of Georgia off Howe Sound (Fig. 11.1) results from readjustment of the plume thickness after the last pulse of discharge augmented by the prescribed barotropic tidal pressure gradients. The ebbing stream at the southern end of Discovery Passage (I) is also in evidence and is in agreement with observations and results from the vertically-integrated model. A patch of southeasterly flow along the western side of the Strait south of this opening derives from the mixing processes that constitute a source of intermediate salinity water which enters the surface layer. This again is in qualitative agreement with observations. The locally high velocities occurring in Jervis Inlet are associated with a channel constriction and fast flowing tidal streams.

The tendency for the Coriolis-deflected discharge to move in a northwesterly direction along the mainland coast is illustrated by the salinity distribution shown in Fig. 11.5b. This accords with anticipation based on shallow drogue tracks such as those illustrated in Fig. 1.26 (p. 37). Higher salinity water is evident in the northwest and southeast region of the model, consistent with the proximity of strong tidal mixing.

The layer thickness distribution (Fig. 11.5c), consistent with the concomitant salinity distribution, shows the formation of a pool of lower salinity water extending northwestward along the mainland coast. Pools of higher salinity mixed water have formed near the northwestern and southern strongly tidal boundary openings. Local layer thickening in the vicinity of the boundary between the banks and upper layer models is associated with the ebb flow off the banks.

Conditions at the interface are presented using contours of the summed $(w_+ + w_-)$ vertical velocities relative to the lower boundary surface (Fig. 11.5d). Over most of the region this net velocity is small. This is a reflection of the fact that the actual magnitudes of both w_+ and w_- are small. The largest values are positive and occur

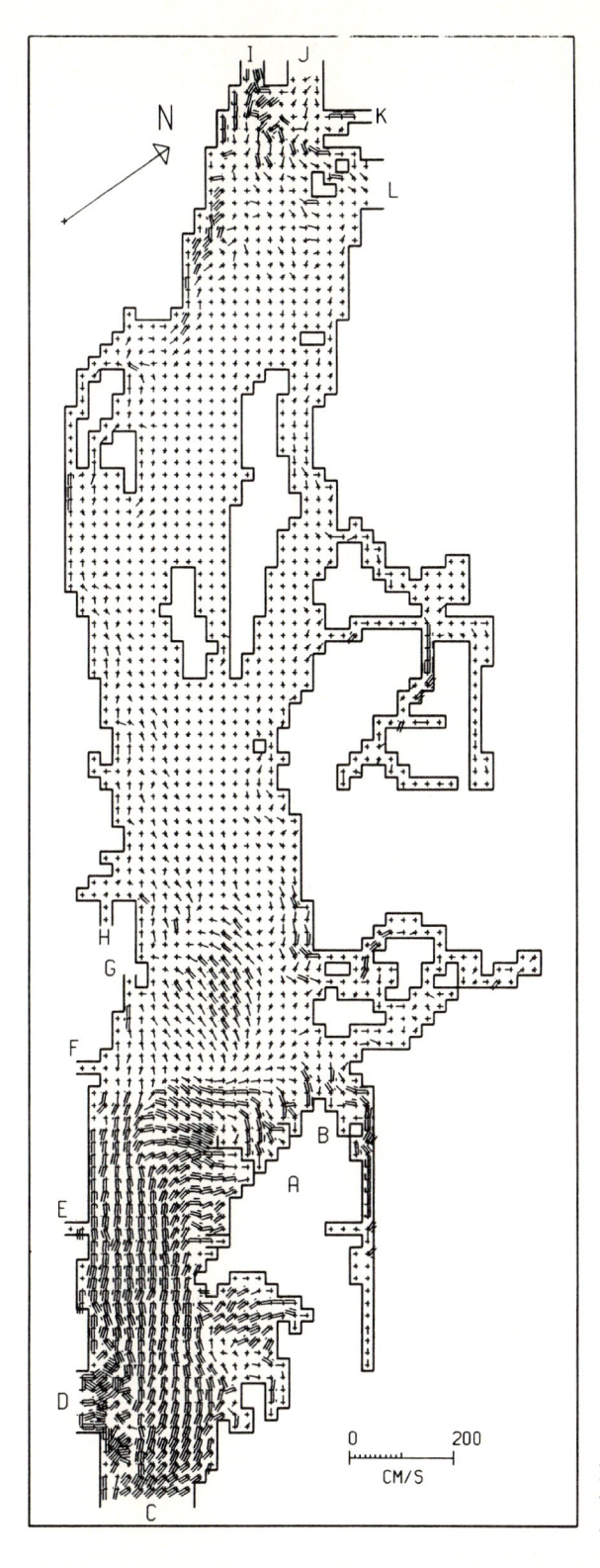

Fig. 11.5a Distributions of velocity (cm/s) for ebb tide and no wind.

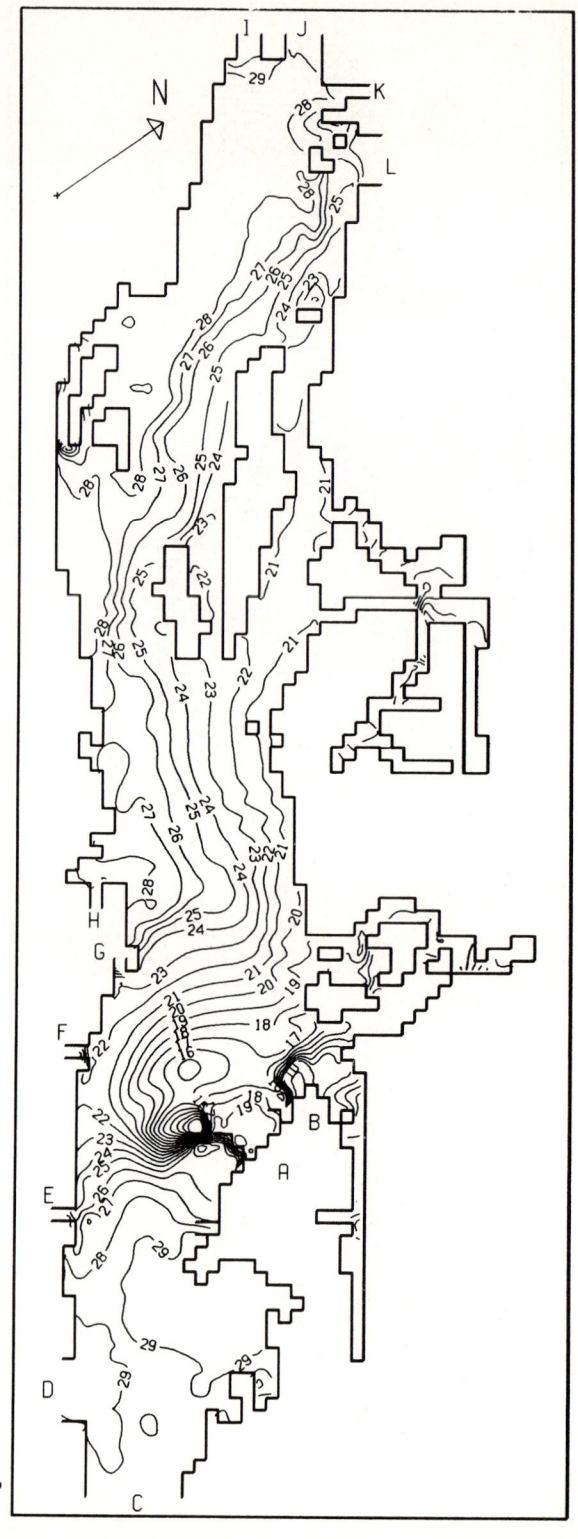

Fig. 11.5b Distributions of salinity (⁰/₀₀) for ebb tide and no wind.

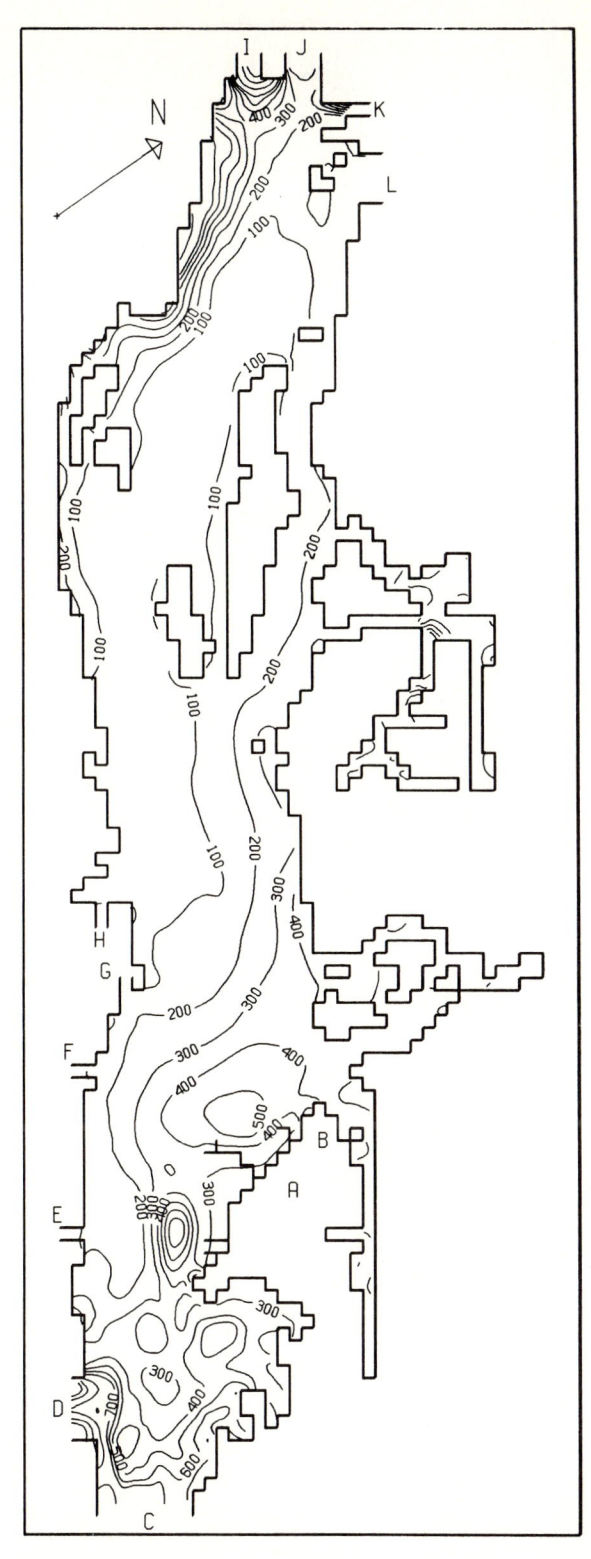

Fig. 11.5c Distributions of layer depth (cm) for ebb tide and no wind.

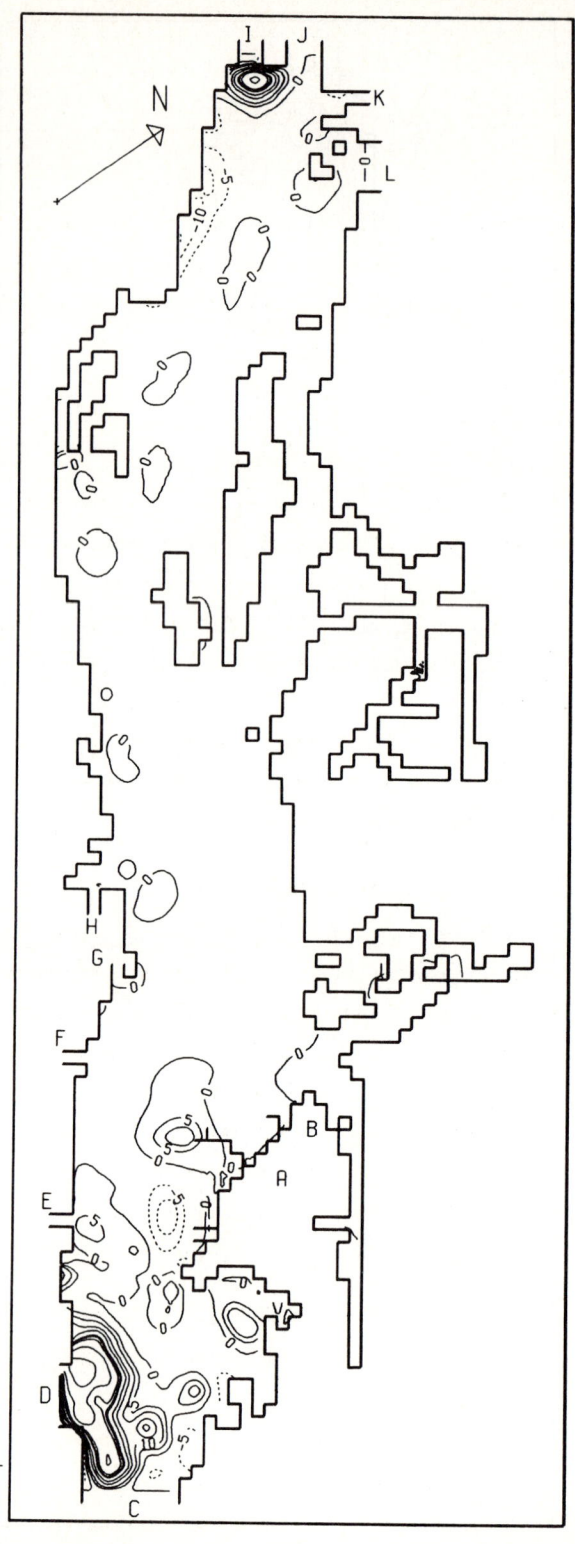

Fig. 11.5d Distributions of relative vertical velocity ($\times 10^3$ cm/s) for ebb tide and no wind.

near the boundary openings where tidal streams are strong and indicate a downward displacement of the interface as well as the formation of mixed water which exits on the ebb tide. Positive velocities near the river mouth indicate entrainment.

Flood

The Fraser River discharge rate rises to a maximum at about lower low water. The flood tidal streams (Fig. 11.6a) act in concert with the Coriolis-deflected freshwater outflow to produce a major movement of water towards, and to the northwest along, the mainland coast. The flood streams south of the river mouth are strongly developed and can be seen moving onto the banks. The tide is flooding at the southern end of Discovery Passage and the southward flow of mixed water from the Passage along the Vancouver Island shore is enhanced.

The salinity distribution (Fig. 11.6b) is essentially similar to that obtained on the ebb tide and is again dominated by the northward movement of lower salinity water along the mainland coast. The thickness distribution (Fig. 11.6c) is also similar to that of the ebbing tide. The flow of mixed water along the Vancouver Island shore south of Discovery Passage is deepened. The irregular pool of mixed water at the southwestern corner of the model is deeper than on the ebb. The net relative vertical velocity distribution (Fig. 11.6d) shows that large values of w_+ and w_-, corresponding to strong vertical mixing, occur in this area. Over most of the modelled region, the values of these quantities remain small.

11.5.2.2 Wind forcing effects

The foregoing representative ebb and flood illustrations of the computed fields of velocity, salinity, thickness, and net vertical velocity relative to the lower interface summarize the basic aspects of the model's performance in the absence of winds. Despite the rather arbitrary nature of the boundary conditions, the major features of the river plume accord well with available observations. The question now arises as to how the upper layer model responds to winds, particularly those parallell to the major axis of the Strait and thus most likely to be of oceanographic significance. The ensuing sections illustrate the performance of the model, again for the flood and ebb conditions of strong river flow as used above, thus enabling direct comparison.

Ebb Tide, Northwest Wind of 2.5 m/s

The velocity field (Fig. 11.7a) in the southern part of the Strait of Georgia is essentially similar to that for the case of no wind. Over much of the northern area of the Strait the wind-induced flow, which is Coriolis-deflected towards the Vancouver Island coast, induces an enhanced southerly flow aided by the ebbing tide. This effect is most pronounced in the central and western areas of the Strait across from Howe Sound. The salinity field (Fig. 11.7b) shows the region of significant dilution confined

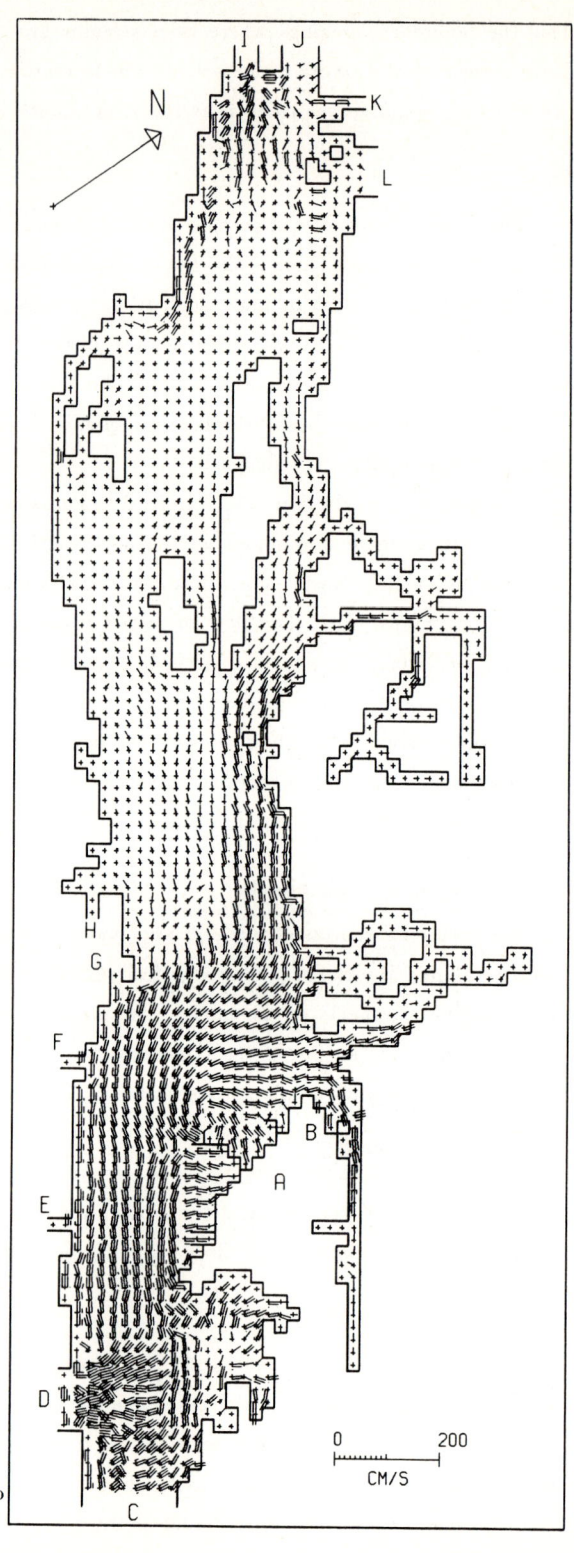

Fig. 11.6a Distributions of velocity (cm/s) for flood tide and no wind.

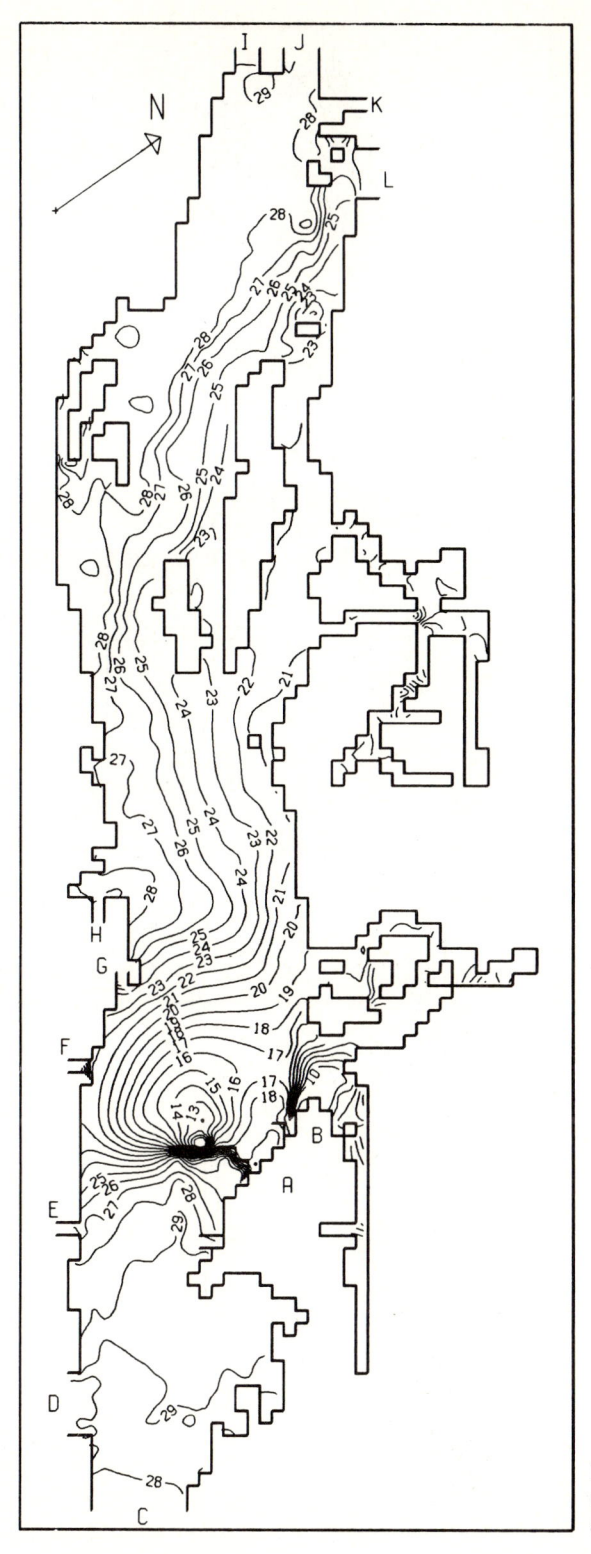

Fig. 11.6b Distributions of salinity (‰) for flood tide and no wind.

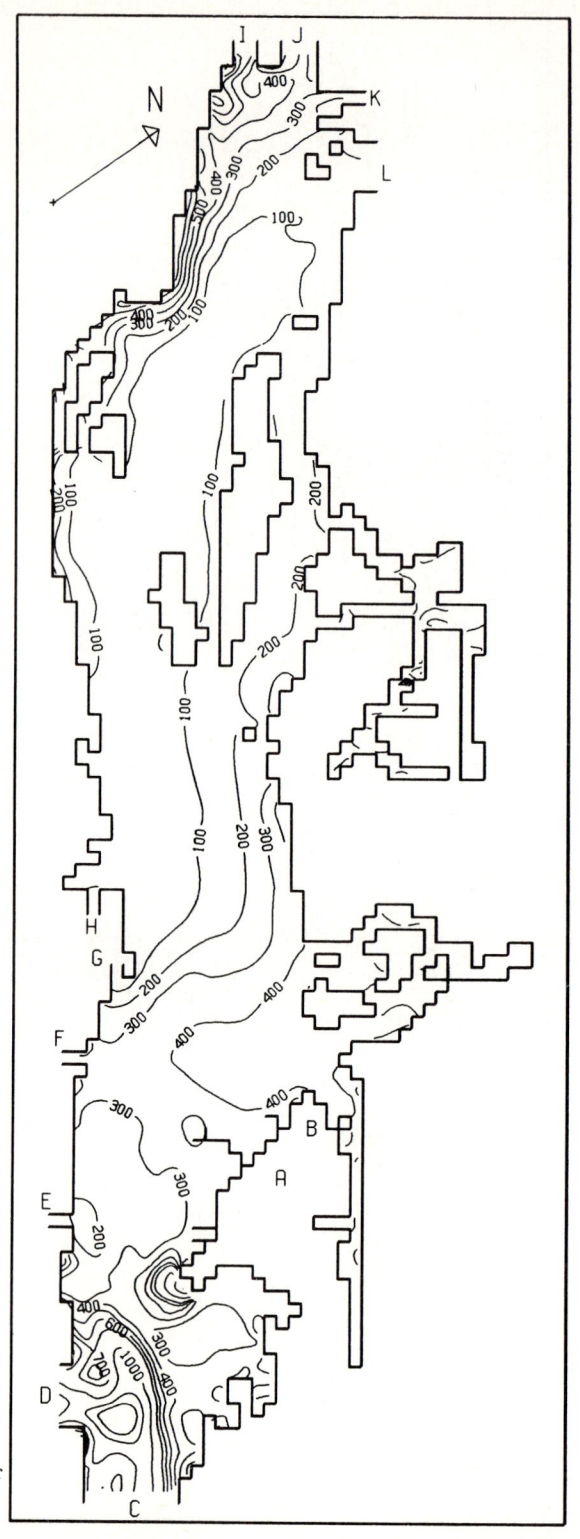

Fig. 11.6c Distributions of layer depth (cm) for flood tide and no wind.

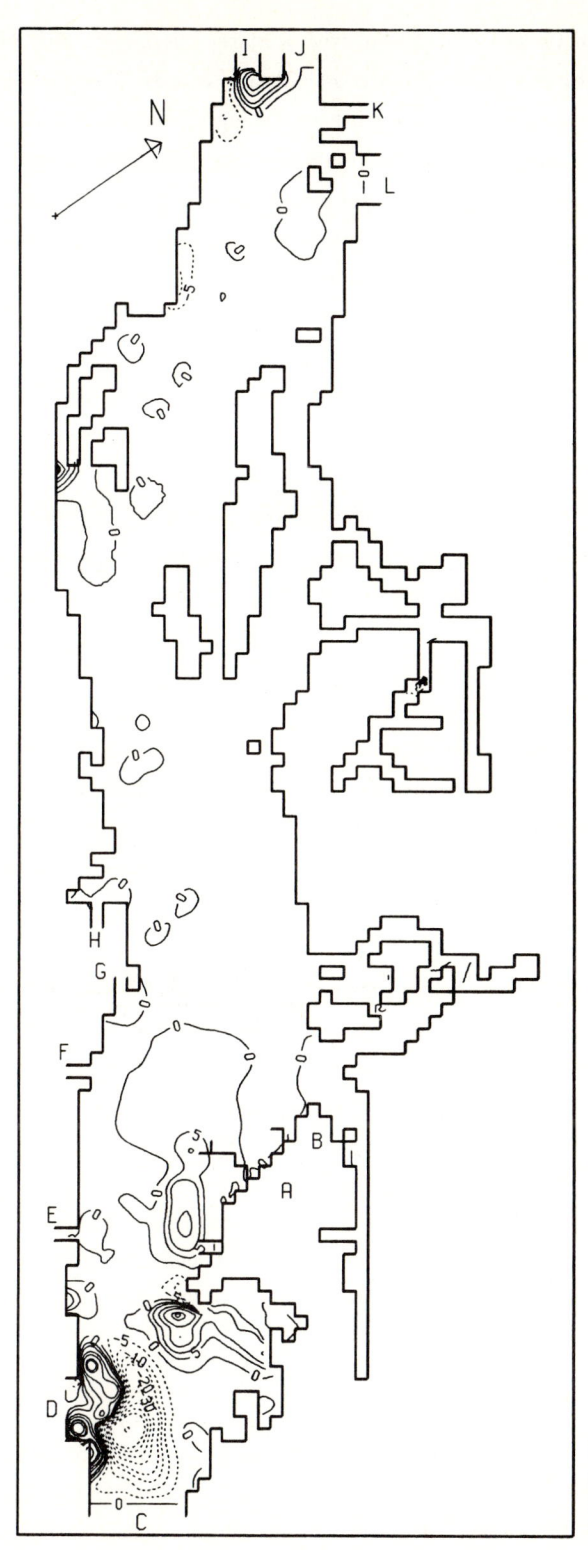

Fig. 11.6d Distributions of relative vertical velocity ($\times 10^3$ cm/s) for flood tide and no wind.

to the mainland coast north of the river mouth but with a marked horizontal gradient across the Strait near the southern ends of Lasqueti and Texada Islands. There is also a higher degree of dilution which extends farther southward along the Vancouver Island shore. The primary feature of interest in the distribution of thickness (Fig. 11.7c) is the appearance of a local deepening of the layer along the coast of Vancouver Island in the northern portion of the Strait. The field of vertical velocity relative to the lower boundary (Fig. 11.7d) does not differ significantly from the no-wind case.

Flood Tide, Northwest Wind of 2.5 m/s

The velocity field (Fig. 11.8a) resembles that of the flood, no-wind case (Fig. 11.6a). The flooding streams offset the formation of a southerly flow along the Vancouver Island shore in the central Strait of Georgia. The salinity and thickness distributions (Fig. 11.8b and c) differ negligibly from that on the ebb tide (Fig. 11.7b and c), evidence of the slow adjustment rate of the density field. Again, the field of vertical relative velocity (Fig. 11.8d) differs marginally from that obtained in the case of the flooding tide in the absence of wind.

Ebb Tide, Southeast Wind of 2.5 m/s

The velocity field (Fig. 11.9a) is similar to that of the ebb and no-wind field, (Fig. 11.5a), for the southern Strait of Georgia. In the central region there is a weak ebb flow in a generally easterly direction but a wind-enhanced local northward flow along the mainland coast. In the northern region the ebbing tide is augmented by the wind stress except for a short patch of southerly-directed flow along the Vancouver Island shore.

The salinity field (Fig. 11.9b) shows a closer confinement of the low salinity water along the mainland coast north of Howe Sound with some increase in layer thickness (Fig. 11.9c). Again, the component of vertical relative velocity is essentially unaltered (Fig. 11.9d).

Flood Tide, Southeast Wind of 2.5 m/s

Comparisons of the computed fields (Fig. 11.10a–d) with those of the corresponding no-wind fields show the only significant change to be a somewhat enhanced flood flow with a bias in the direction toward, and some intensification of the flow along, the mainland coast.

Ebb Tide, Northwest Winds of 5 m/s

Over the entire Strait of Georgia, except for a small region of northward-ebbing stream near the entrance to Discovery Passage, the velocity vectors (Fig. 11.11a) show an ebbing tide much enhanced by the wind with some intensification of flow along the eastern shores of Vancouver and Texada Islands. A patch of low velocities in the central

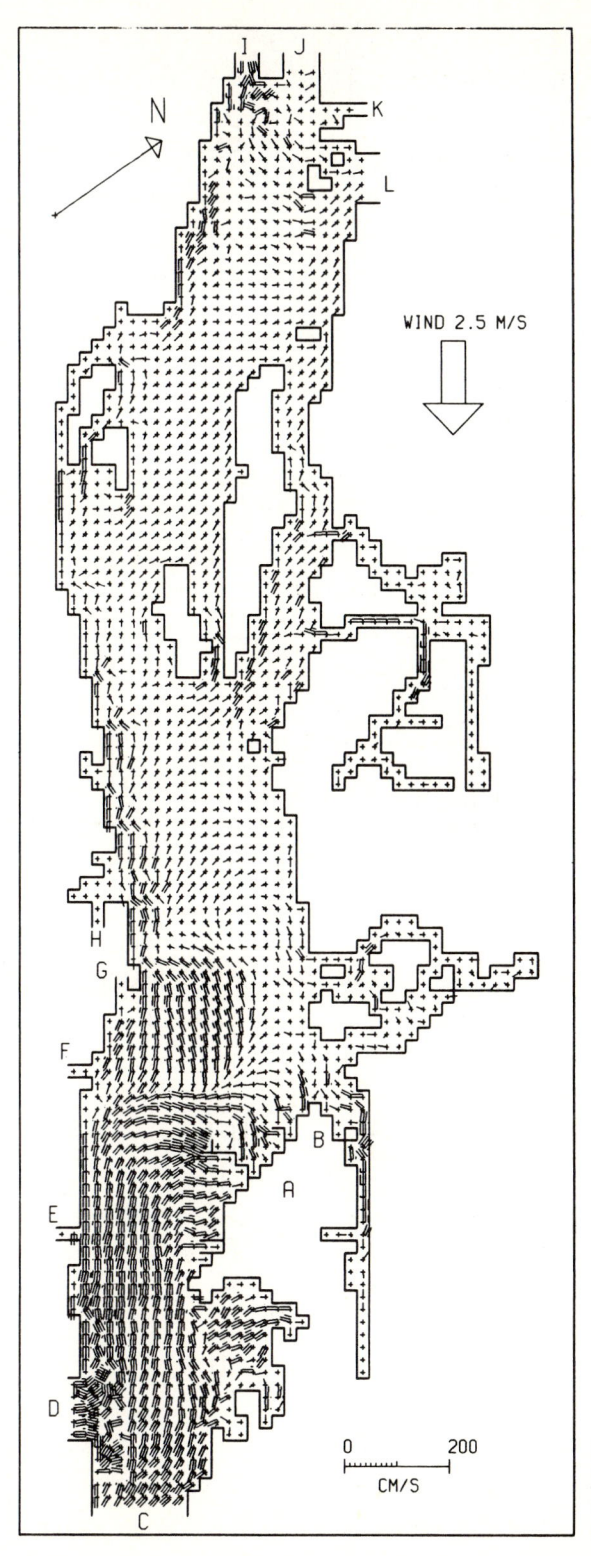

Fig. 11.7a Distributions of velocity (cm/s) for ebb tide and wind of 2.5 m/s from the northwest.

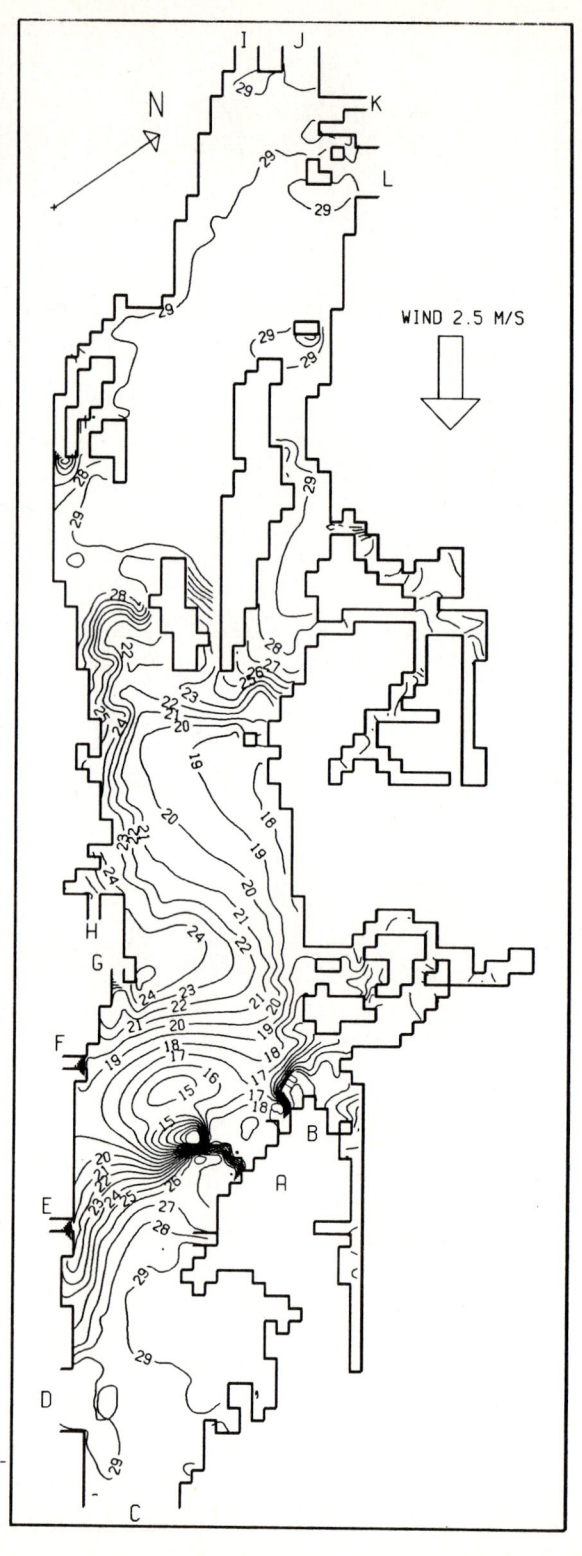

Fig. 11.7b Distributions of salinity ($^0/_{00}$) for ebb tide and wind of 2.5 m/s from the northwest.

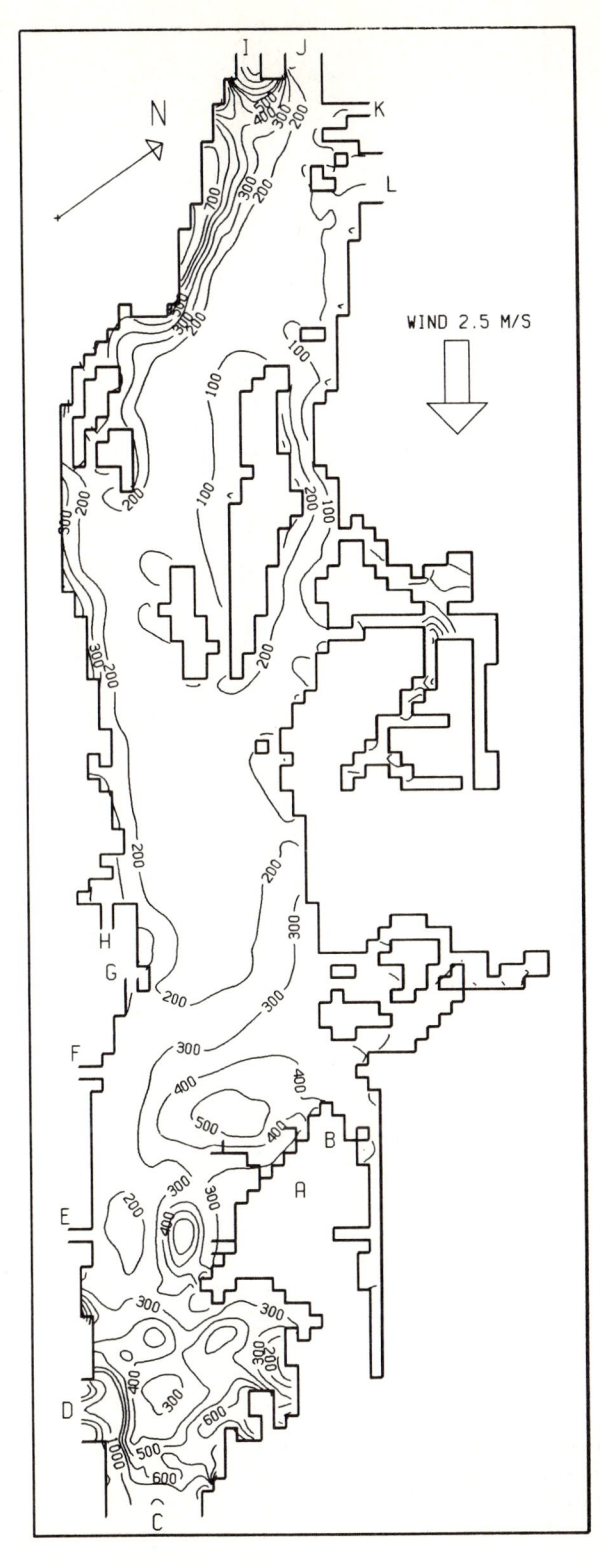

Fig. 11.7c Distributions of layer depth (cm) for ebb tide and wind of 2.5 m/s from the northwest.

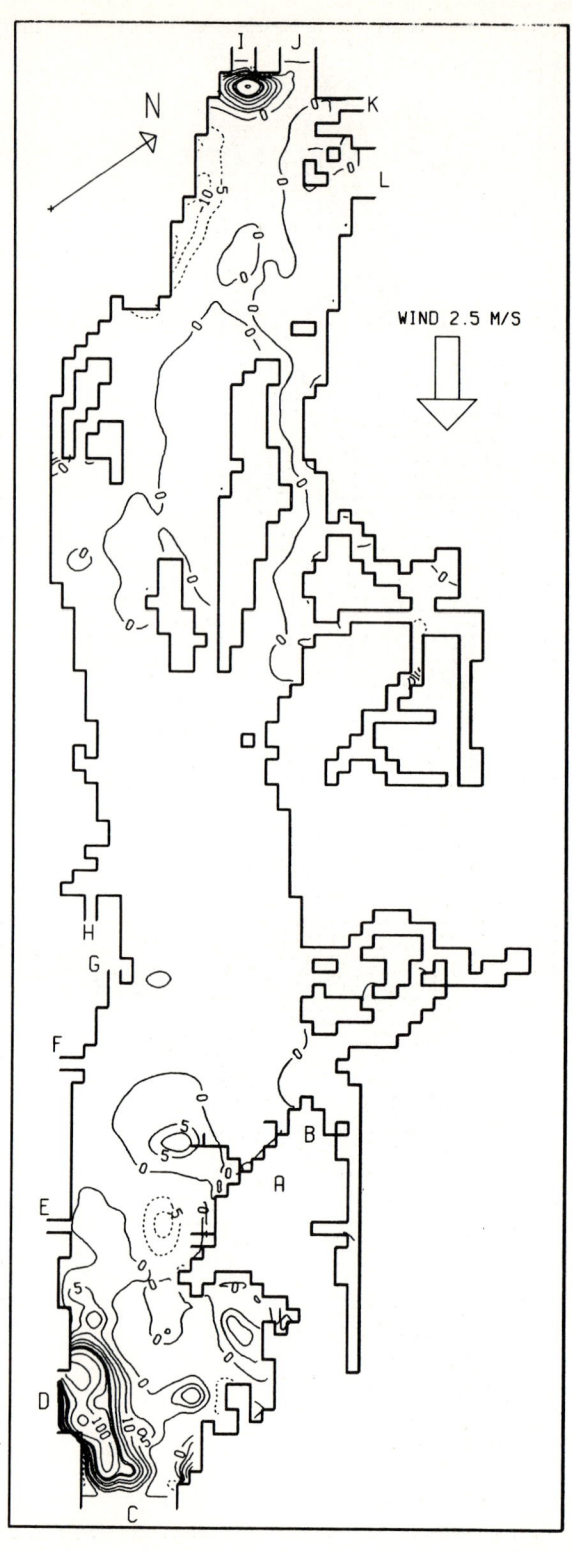

Fig. 11.7d Distributions of relative vertical velocity ($\times 10^3$ cm/s) for wind of 2.5 m/s from the northwest.

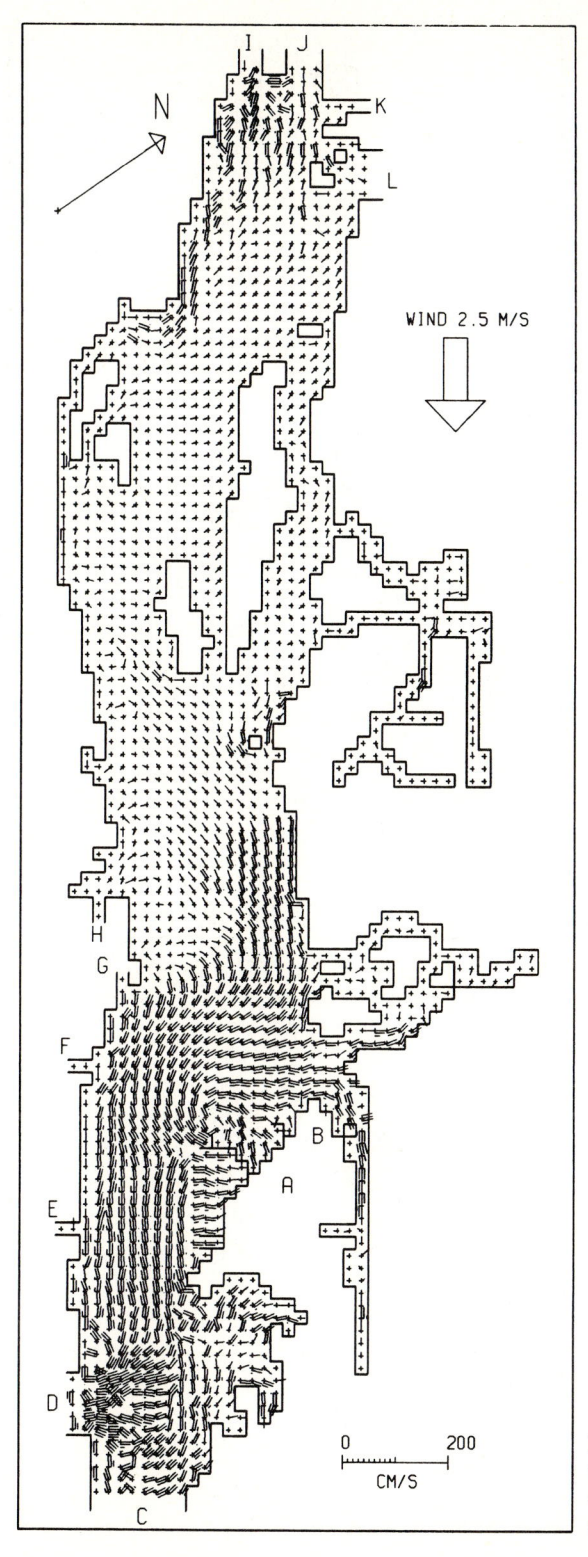

Fig. 11.8a Distributions of velocity (cm/s) for flood tide and wind of 2.5 m/s from the northwest.

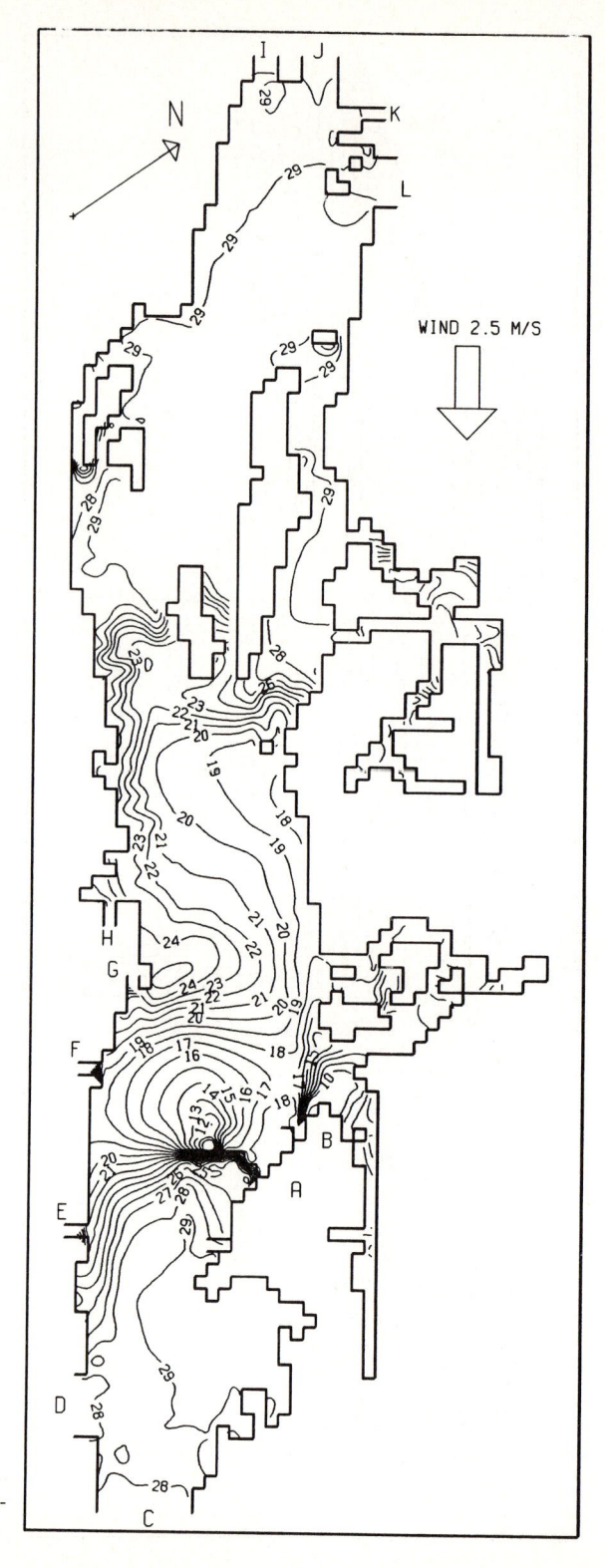

Fig. 11.8b Distributions of salinity (⁰/₀₀) for flood tide and wind of 2.5 m/s from the northwest.

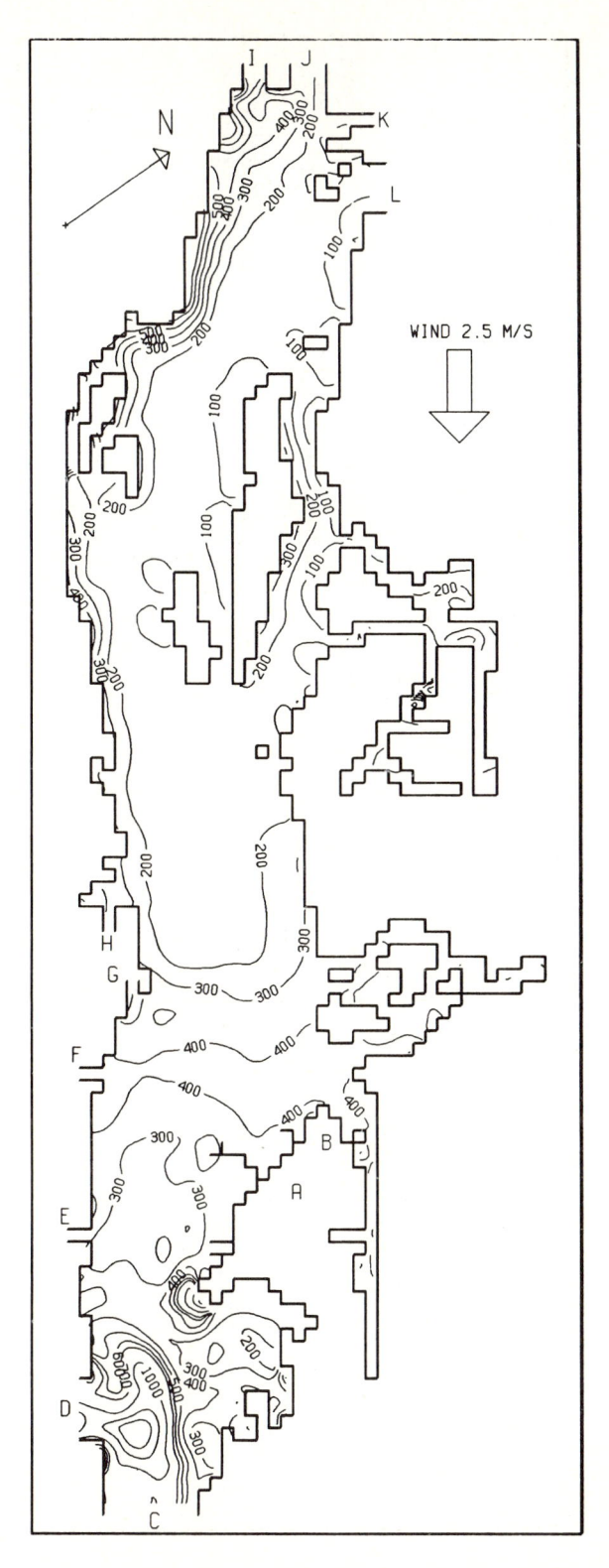

Fig. 11.8c Distributions of layer depth (cm) for flood tide and wind of 2.5 m/s from the northwest.

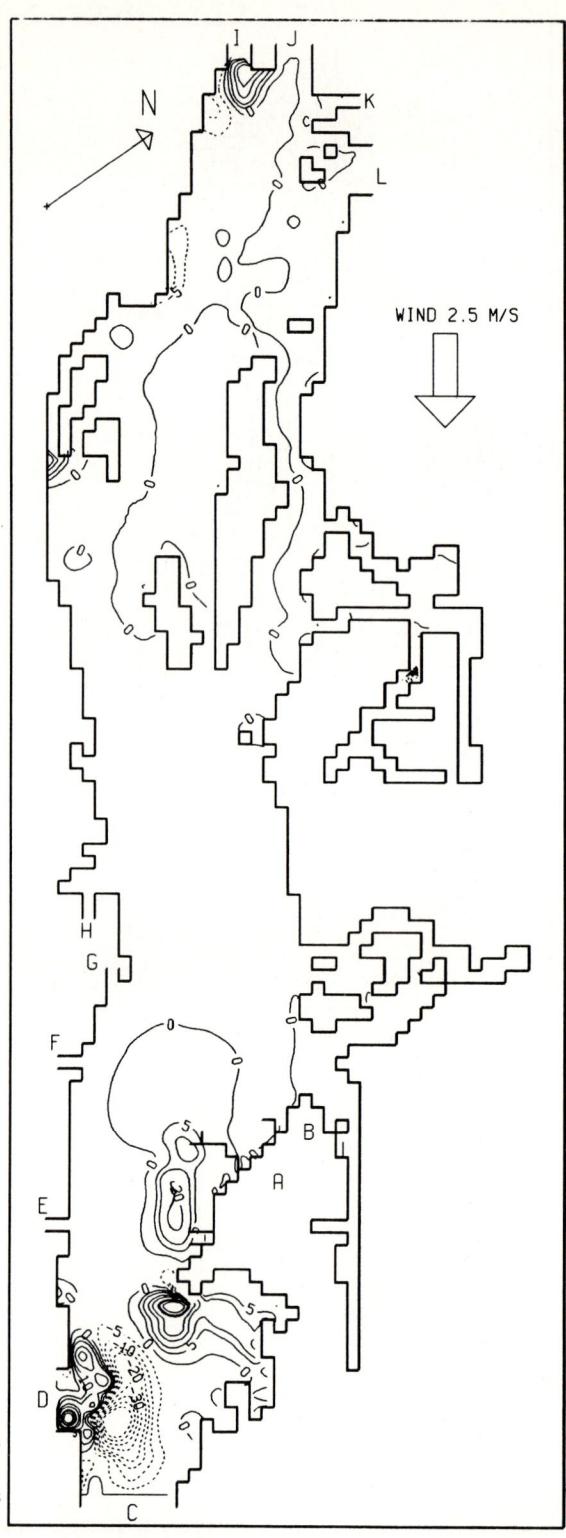

Fig. 11.8d Distributions of relative vertical velocity ($\times 10^3$ cm/s) for flood tide and wind of 2.5 m/s from the northwest.

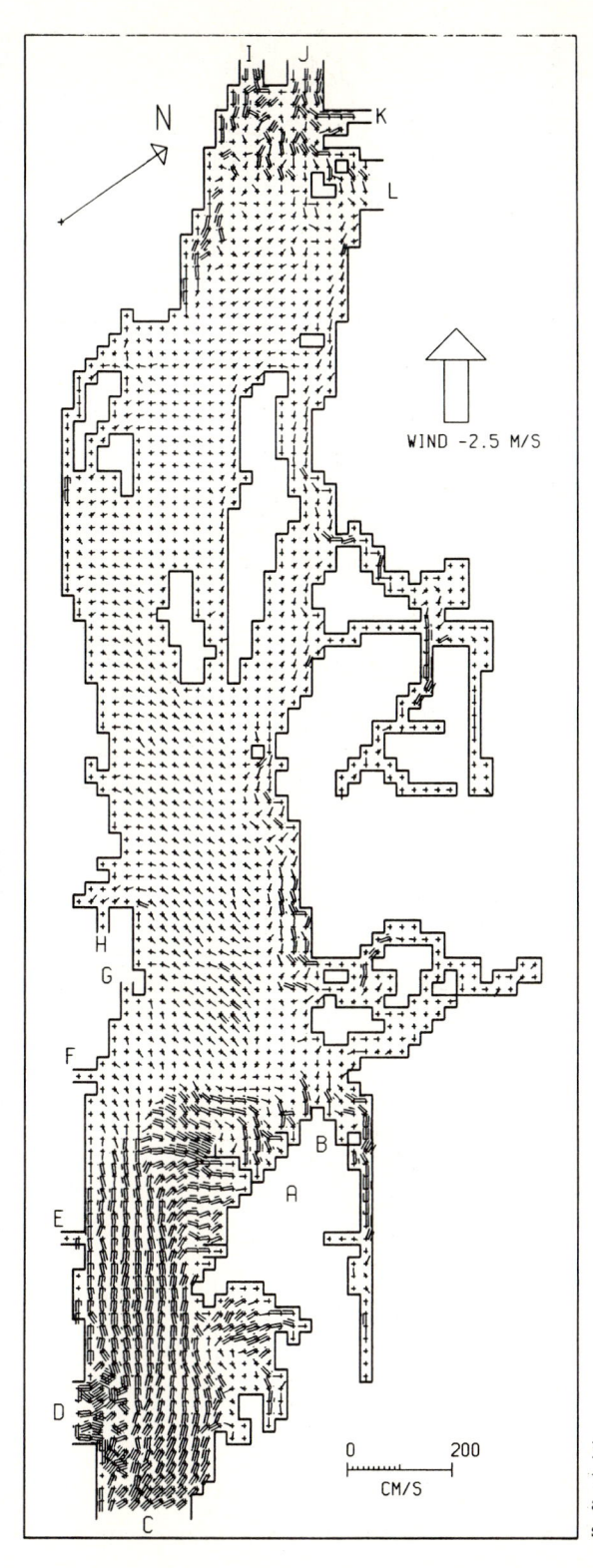

Fig. 11.9a Distributions of velocity (cm/s) for ebb tide and wind of 2.5 m/s from the southeast.

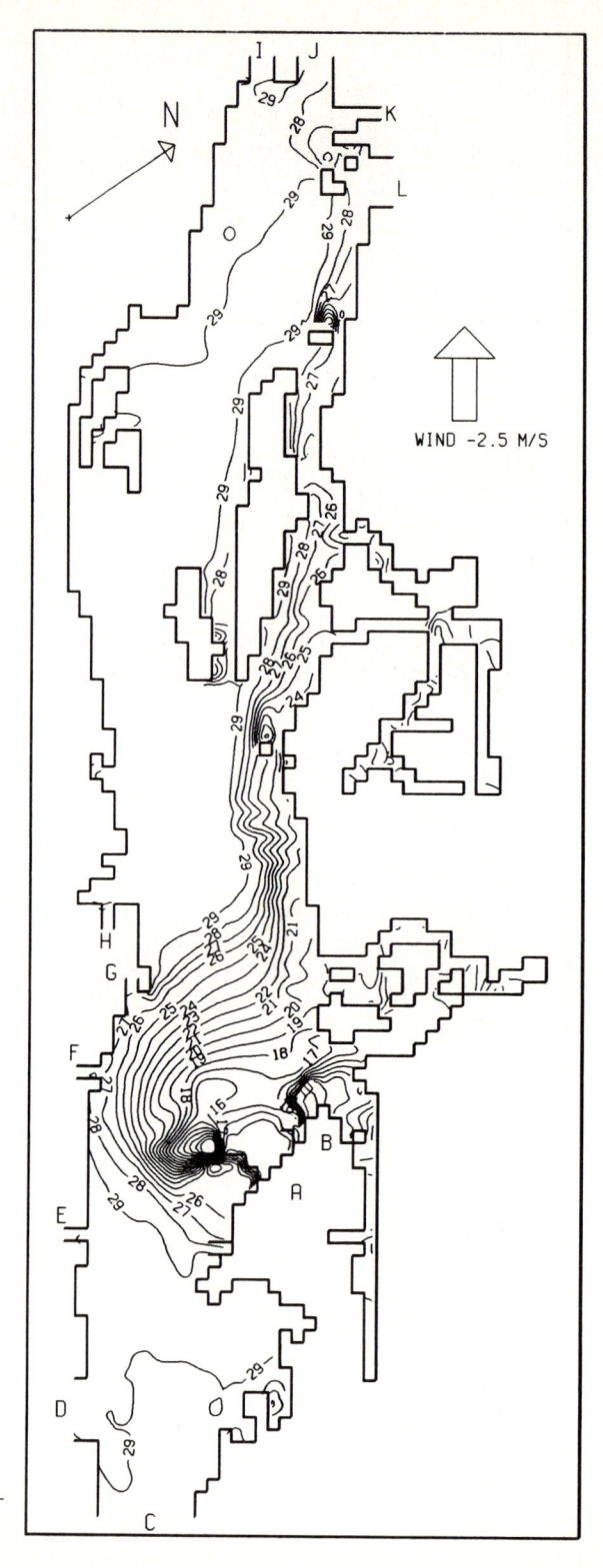

Fig. 11.9b Distributions of salinity ($^0/_{00}$) for ebb tide and wind of 2.5 m/s from the southeast.

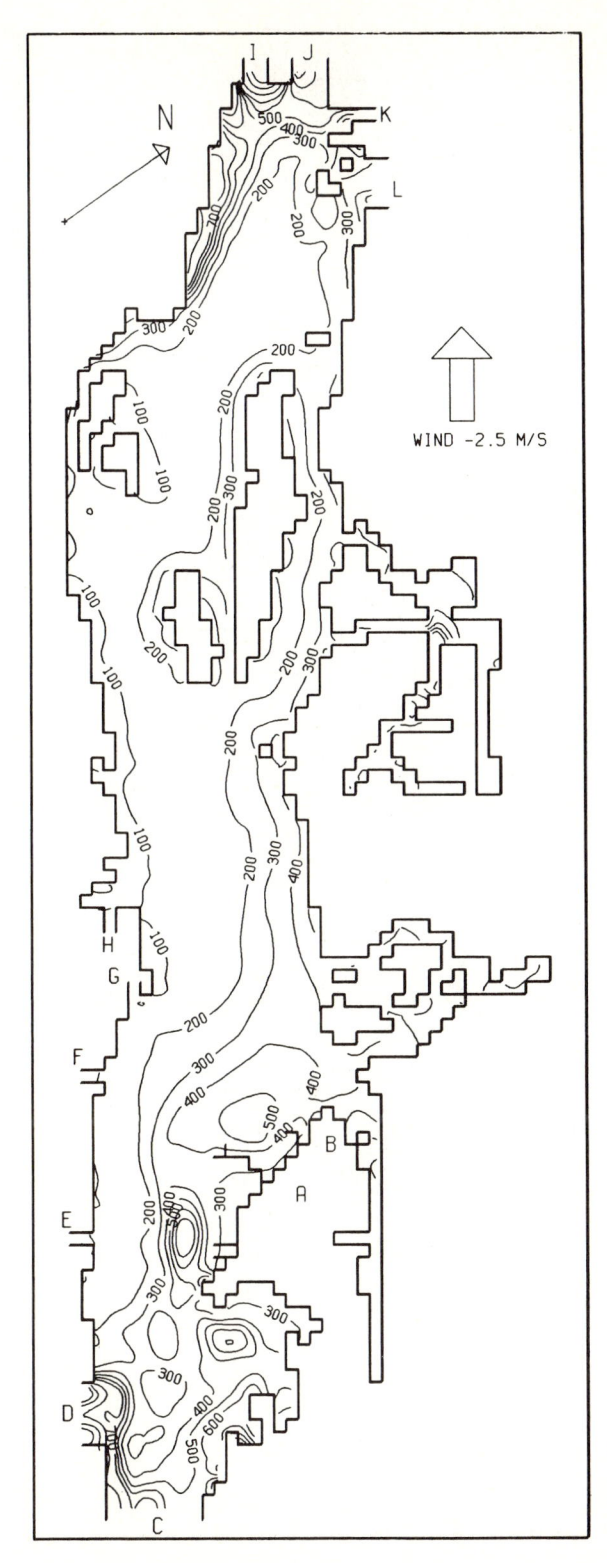

Fig. 11.9c Distributions of layer depth (cm) for ebb tide and wind of 2.5 m/s from the southeast.

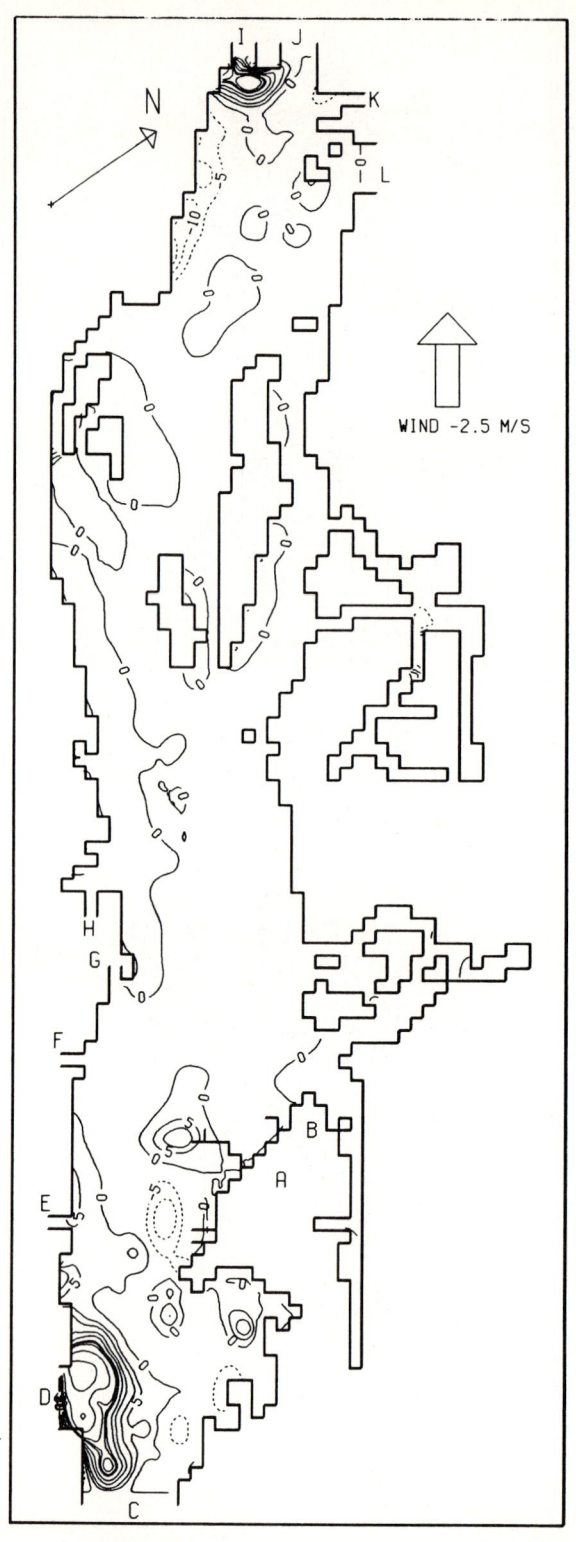

Fig. 11.9d Distributions of relative vertical velocity ($\times 10^3$ cm/s) for ebb tide and wind of 2.5 m/s from the southeast.

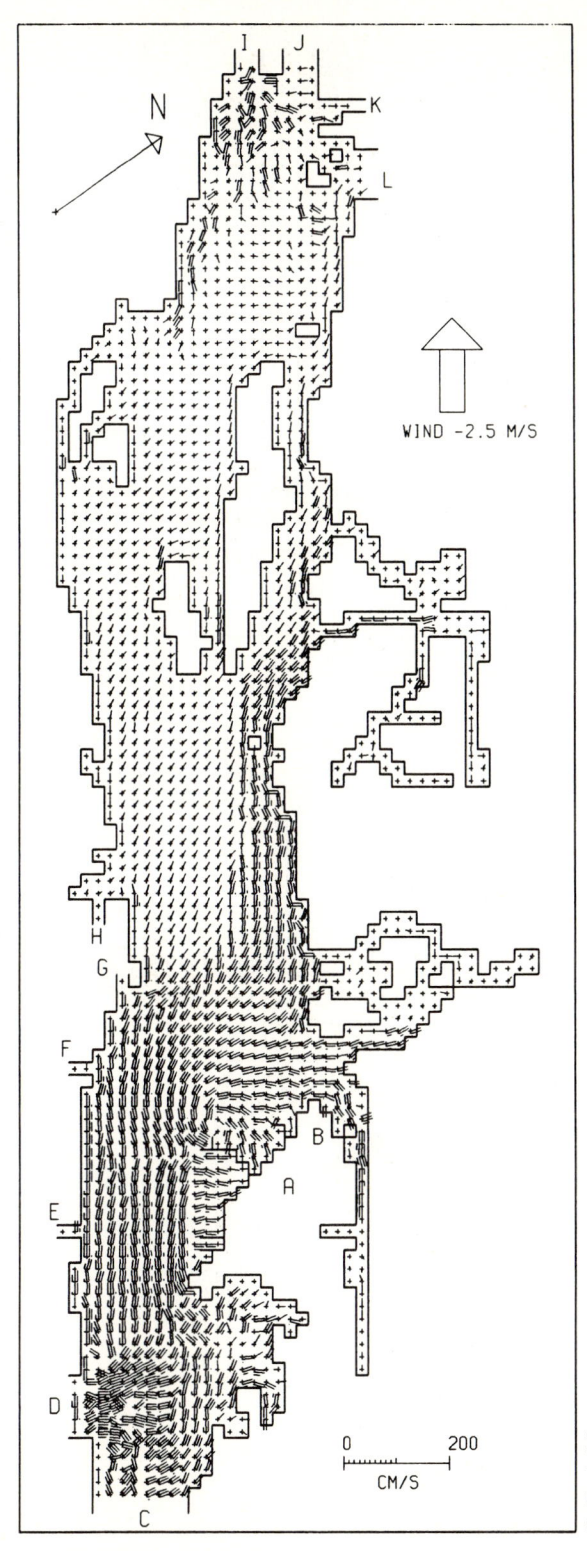

Fig. 11.10a Distributions of velocity (cm/s) for flood tide and wind of 2.5 m/s from the southeast.

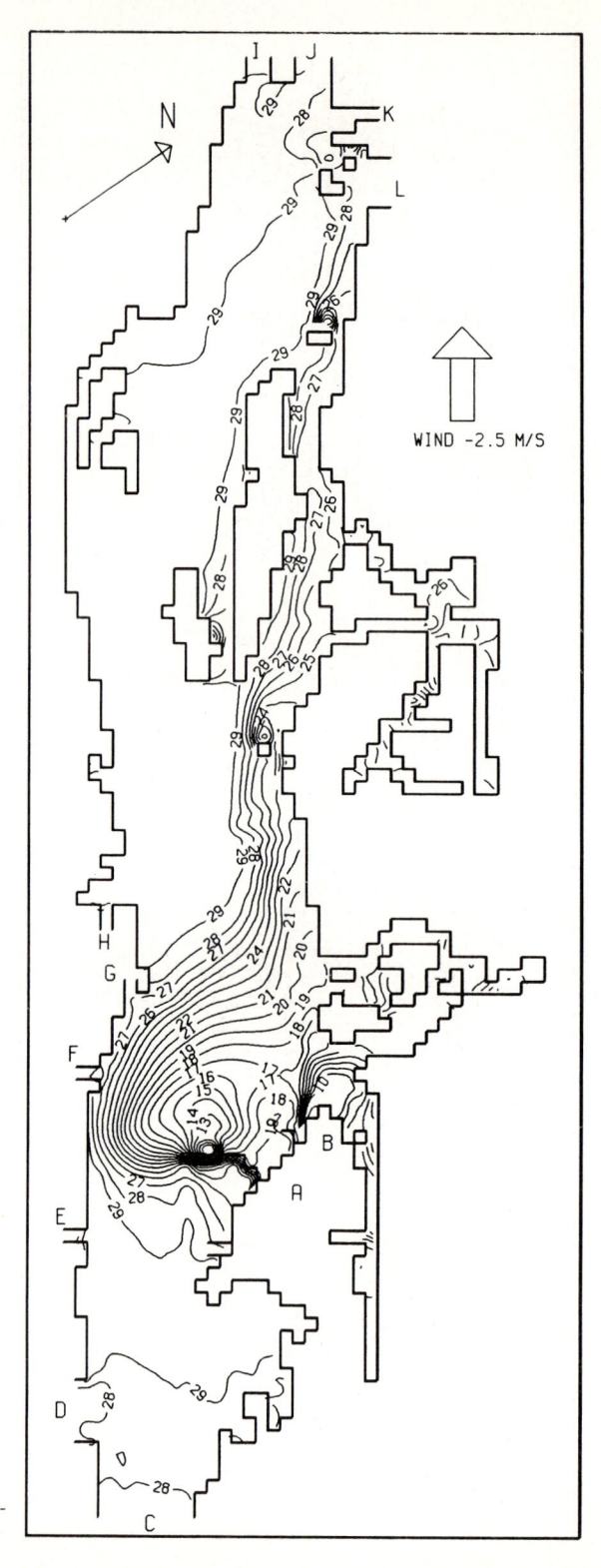

Fig. 11.10b Distributions of salinity (⁰/₀₀) for flood tide and wind of 2.5 m/s from the southeast.

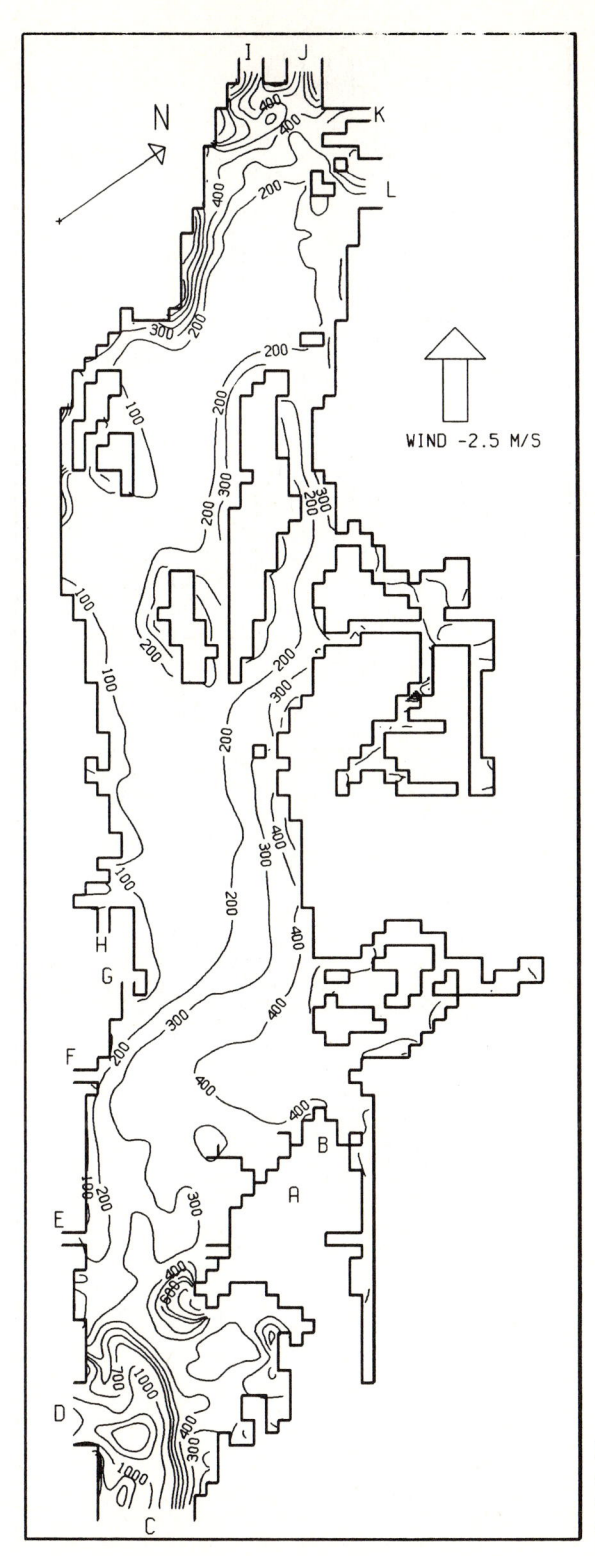

Fig. 11.10c Distributions of layer depth (cm) for flood tide and wind of 2.5 m/s from the southeast.

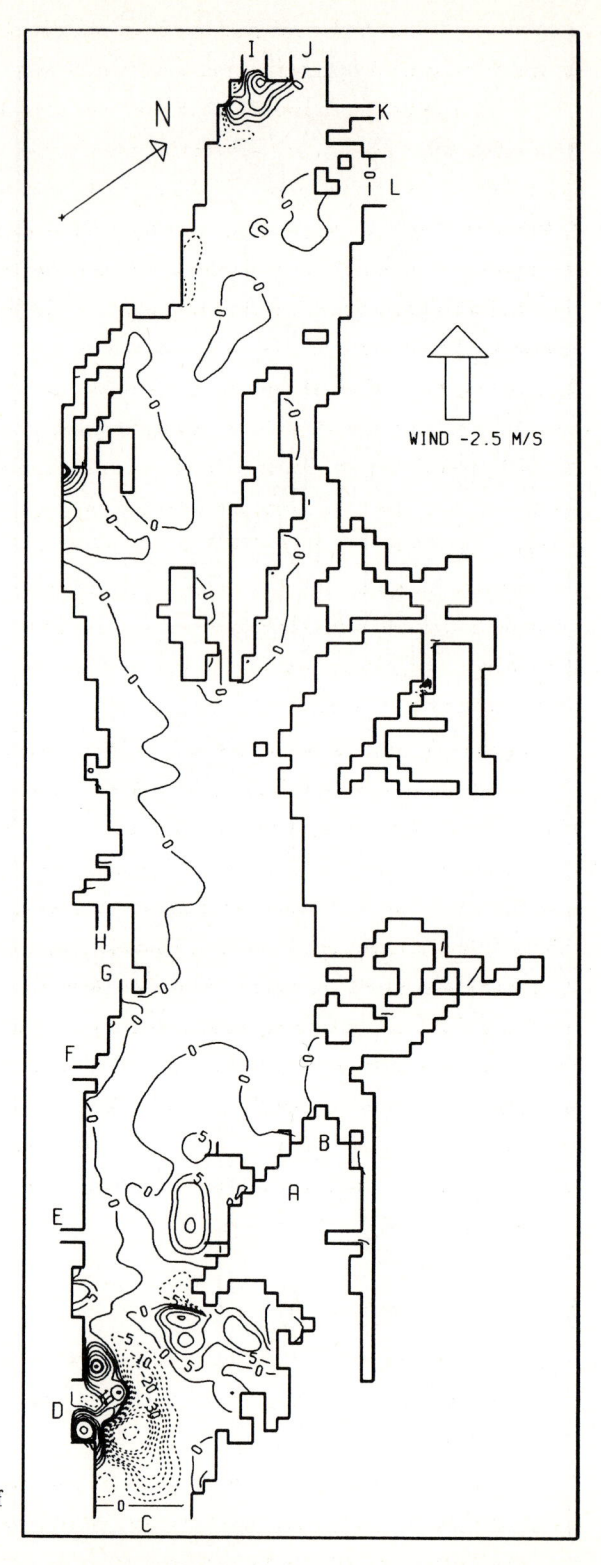

Fig. 11.10d Distributions of relative vertical velocity ($\times 10^3$ cm/s) for flood tide and wind of 2.5 m/s from the southeast.

part of the Strait off the entrance to Howe Sound coincides with the local boundary formed when wind-dominated surface flow from the north opposes the northward movement of river discharge driven by the previous flood tide, shown later (Fig. 11.12a). This boundary is evident in the salinity distributions (Fig. 11.11b) which also show the main area of river dilution extending down to Boundary Pass (D). The thickness distribution (Fig. 11.11c) shows a well defined thickening of the layer closely confined to the coastline to the right of the wind direction in the northern and central portions of the Strait. Of particular interest is the tongue of deeper water extending out from the Vancouver Island shore into the Strait towards the northern entrance to Howe Sound. This is located at the convergence of river flow moving northward on a flooding tide and the wind-driven surface flow moving down from the northern Strait of Georgia. The net vertical velocity distribution (Fig. 11.11d) shows the development of a net negative vertical component along the shoreline to the left and a net positive component along the shoreline to the right of the wind direction. The depletion velocity is perhaps over-estimated by the present *ad hoc* numerical formulation of the w_- term. The formulation of w_- was primarily governed by the requirement to approximately balance the intense entrainment near the strongly tidal passes. Under conditions of moderate wind forcing, w_- should probably not exceed w_+.

The simulated formation of the jets and layer deepening due to coastal convergence of Coriolis-deflected wind-driven surface flow is also a feature of theoretical solutions of the simplified governing equations applied to an infinite two-layer channel (Simons, 1980). These solutions indicate that the longshore current is in geostrophic balance with the pycnocline excursion and that both increase linearly with time and decrease rapidly with distance from the shore. Furthermore, the transverse velocity components in the surface layer are compensated for by a flow in the opposite direction in the lower layer. Thus, the net vertical velocities relative to the lower boundary surface, evident in Fig. 11.11d, are consistent with such a circulation. For the greater part of the central and northern portions of the Strait the surface speeds conform well to the empirical value of 2–3% of the wind speed. Over the most exposed part of the central Strait of Georgia, the flow is directed about 40° to the right of the wind direction which is consistent with simple Ekman theory.

Flood Tide, Northwest Wind of 5 m/s

The flood velocity vectors (Fig. 11.12a) show a dramatic change from those of the ebb since over the greater part of the Strait the wind is opposing the tidal streams in the upper layer.

There is a general area of convergence in the central Strait of Georgia between the wind-induced southerly flows from the north and the sharply curtailed field of river velocities apparent across the Strait near the entrance to Howe Sound. This

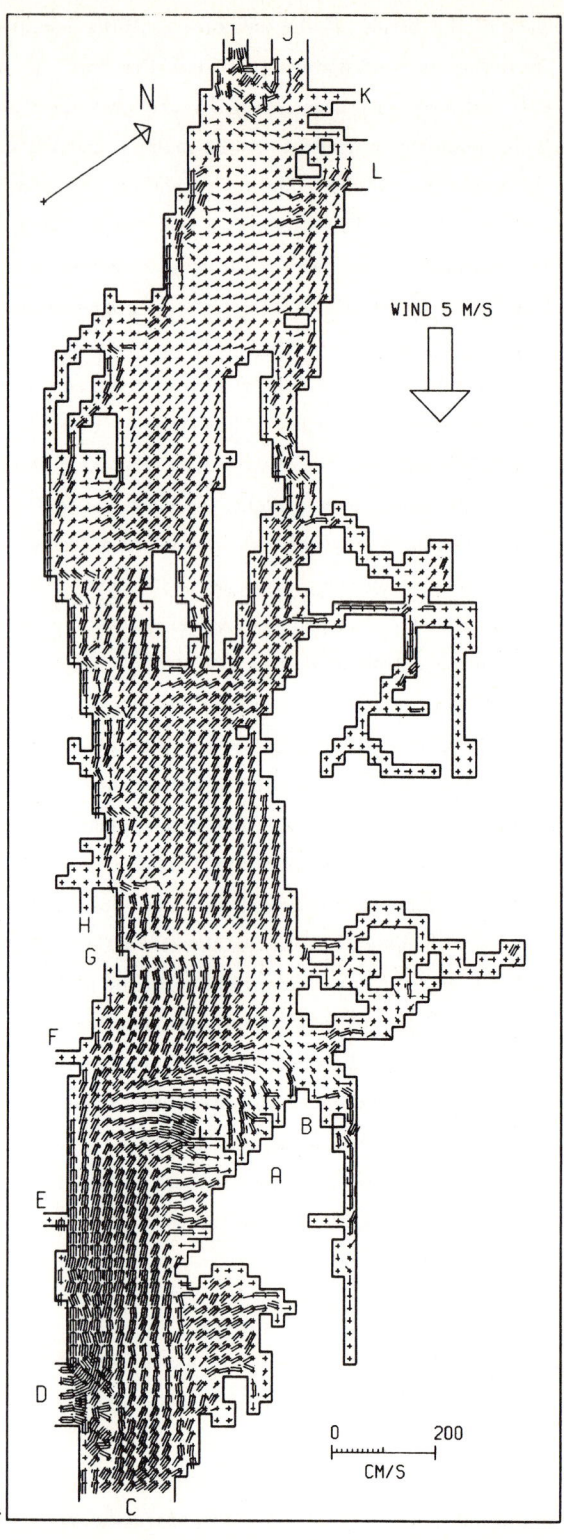

Fig. 11.11a Distributions of velocity (cm/s) for ebb tide and wind of 5 m/s from the northwest.

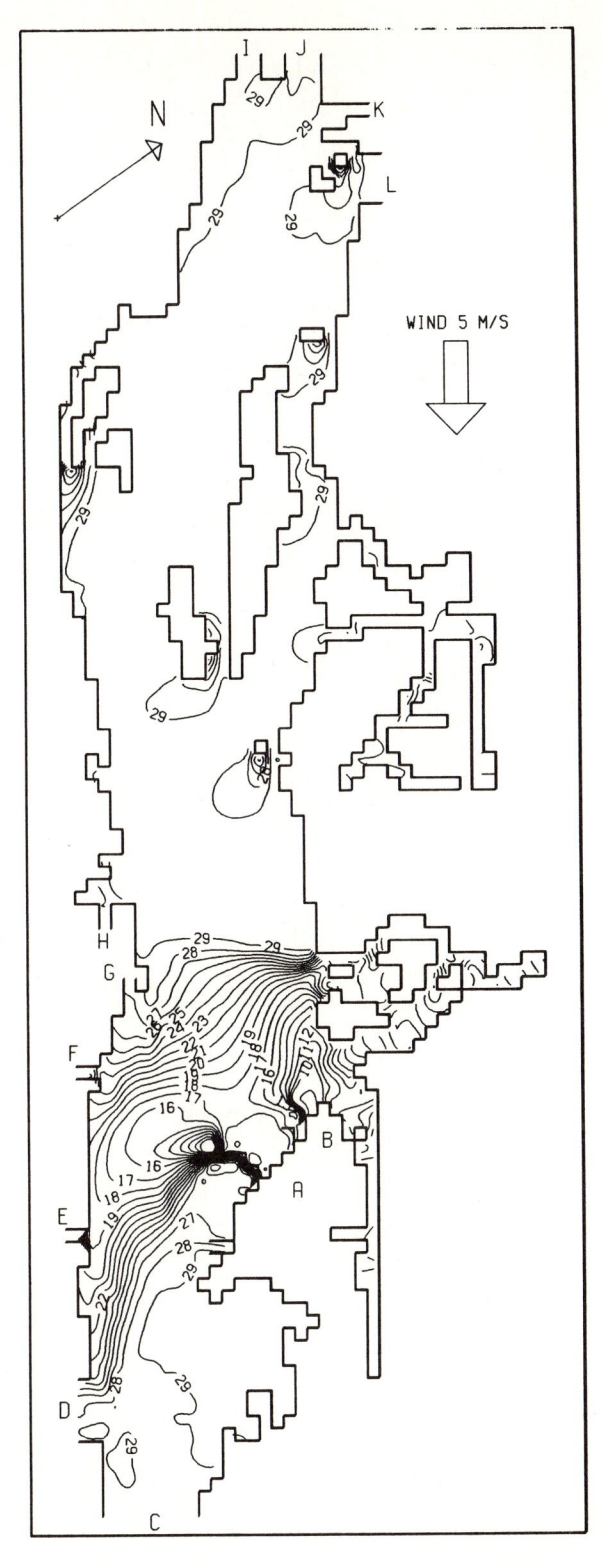

Fig. 11.11b Distributions of salinity ($^0/_{00}$) for ebb tide and wind of 5 m/s from the northwest.

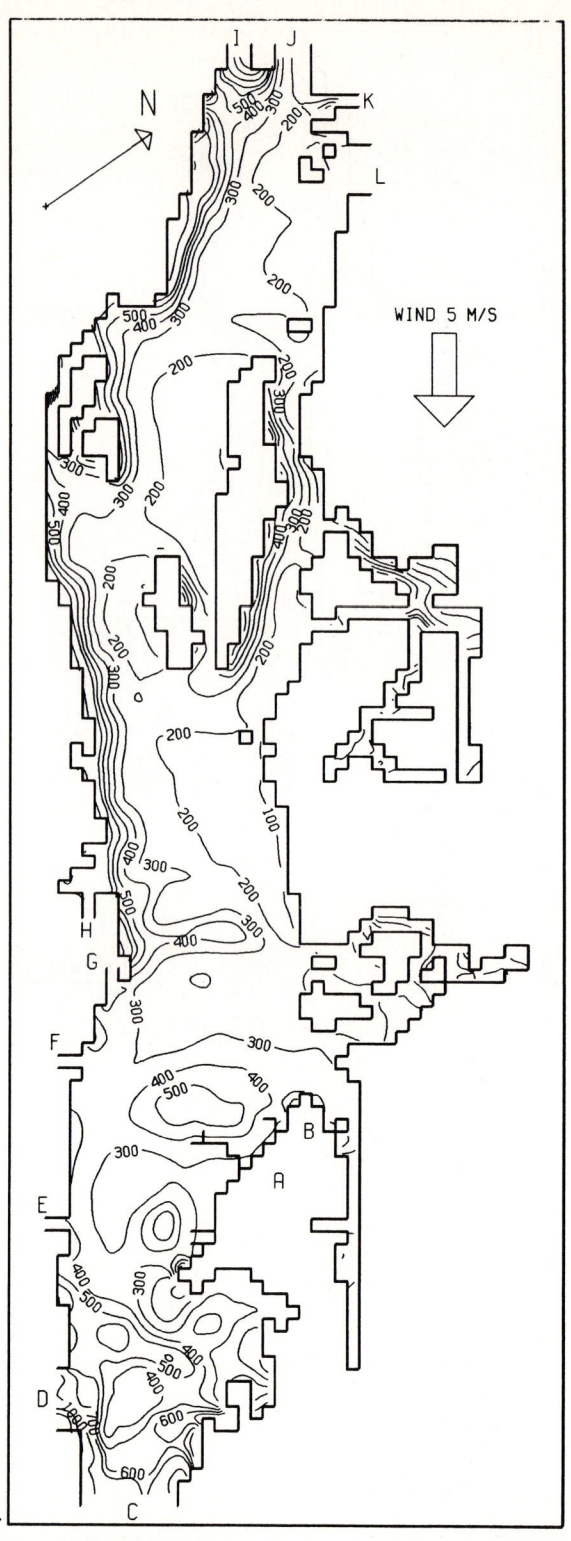

Fig. 11.11c Distributions of layer depth (cm) for ebb tide and wind of 5 m/s from the northwest.

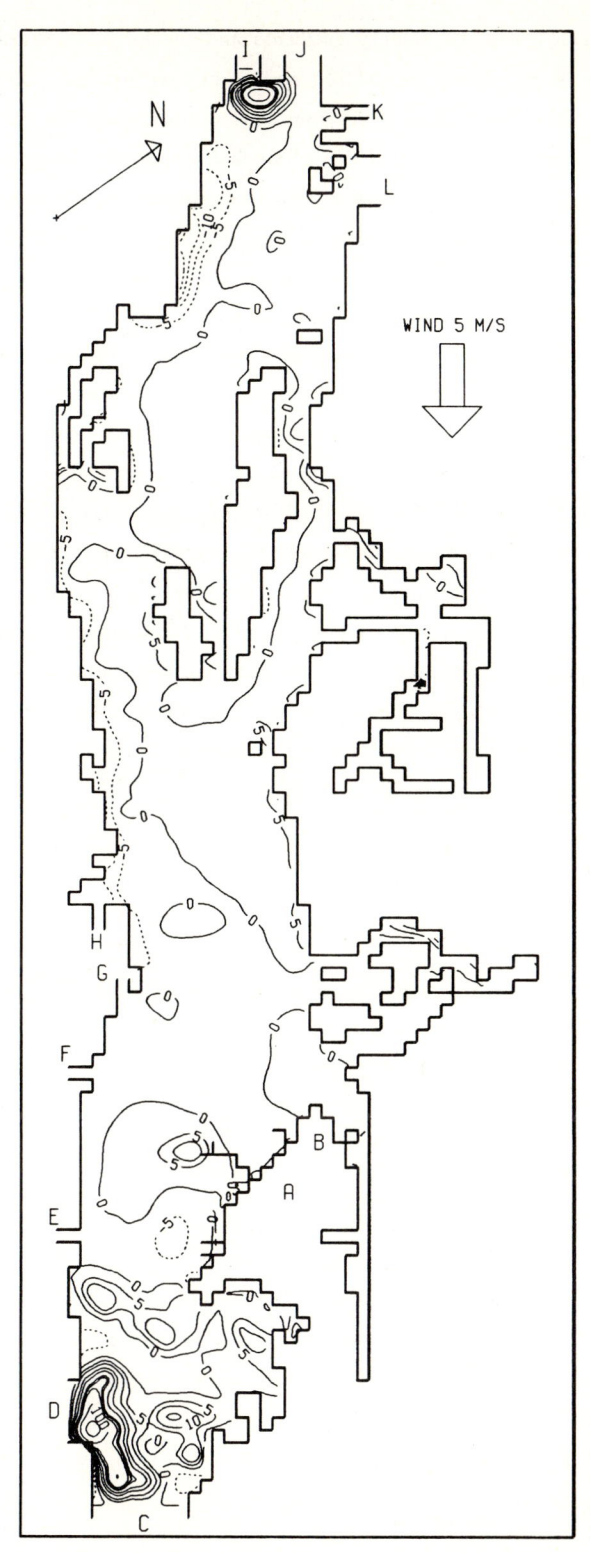

Fig. 11.11d Distributions of relative vertical velocity ($\times 10^3$ cm/s) for ebb tide and wind of 5 m/s from the northwest.

accords with the sharp transverse horizontal salinity gradients and the increased layer thickness along the Vancouver Island shore (Fig. 11.12b and c) in this vicinity. This layer thickening along western shorelines differs little from that occurring on the ebb tide which is consistent with the slow time scales of these baroclinic processes. The net relative vertical velocity (Fig. 11.12d) differs negligibly from that obtained in the case of the ebb tide in the vicinity of these coastal features.

Ebb Tide, Southeast Wind of 5 m/s

The ebbing tide velocity vectors (Fig. 11.13a) are now confined to the part of the Strait south of the major source of freshwater discharge and to the extreme local northern region of the model where the winds are now assisting the ebbing tidal streams. Over much of the remainder of the northern and central portions of the Strait there is a marked component of flow towards the mainland coast. Along this coast a local northward stream results from the wind-induced coastal jet which acts in concert with the northward-moving river discharge. The salinity distribution (Fig. 11.13b) is also confined along these shores. The thickness distribution (Fig. 11.13c) complements these features in the fields of velocity and salinity, showing a marked local deepening of the layer along the coastline to the right of the wind direction. The distribution of net vertical velocity (Fig. 11.13d) again displays a net negative component along the coast to the right of the wind direction and a net positive component to the left along the coast.

Flood Tide, Southeast Wind of 5 m/s

The strong northerly flow (Fig. 11.14a) throughout all but the northernmost part of the model and its intensification along the mainland coast in the central Strait of Georgia, reflects an augmentation of the flooding streams and the Coriolis-deflected river discharge by the imposed winds. The distributions of thickness, salinity, and net vertical velocity (Fig. 11.14b–d) vary little from those described above for the ebbing tide.

Ebb Tide, Northwest Wind of 10 m/s

Except for the ebb tide at the Discovery Passage boundary opening, a strong southerly flow with a lateral component directed toward the Vancouver Island shore prevails throughout the Strait (Fig. 11.15a). Significant dilution is restricted to a relatively local area around the river mouths (Fig. 11.15b). Flow intensification, a marked local layer thickening (Fig. 11.15c), and components of net velocity (Fig. 11.15d) attest to an enhanced behaviour similar to that described for the case of 5 m/s winds and the ebbing tide.

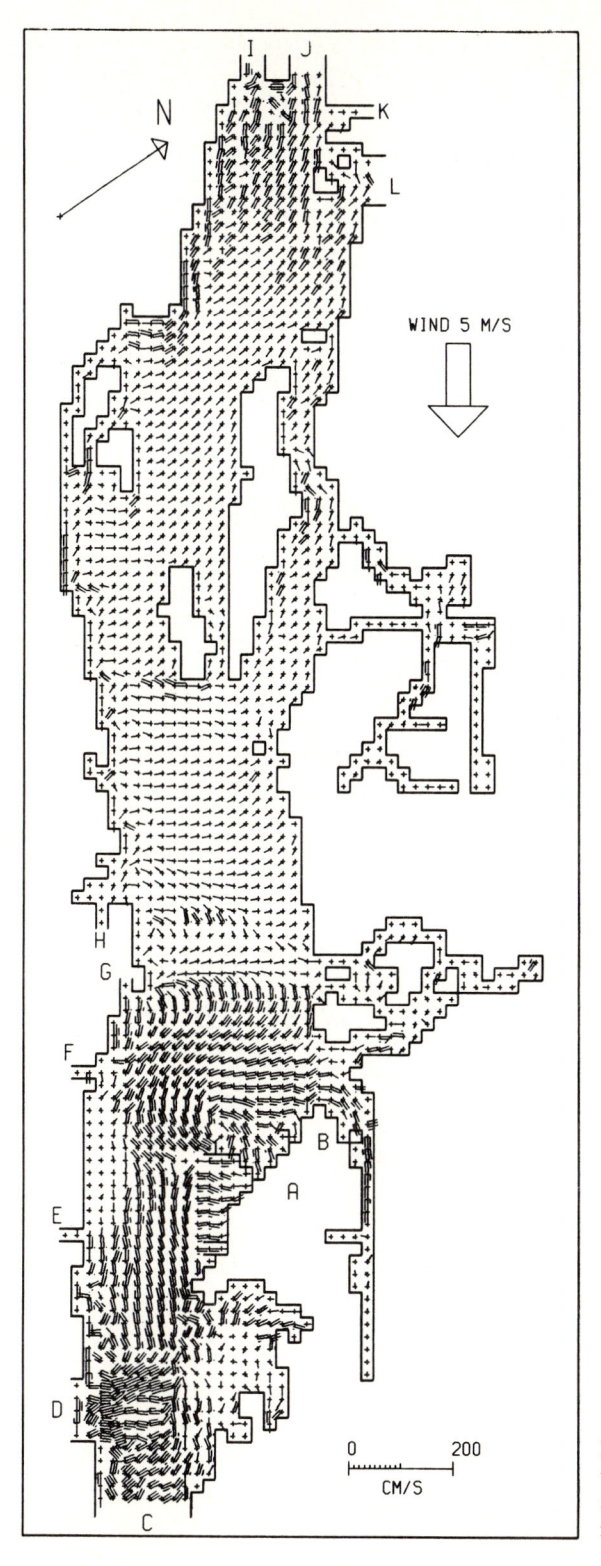

Fig. 11.12a Distributions of velocity (cm/s) for flood tide and wind of 5 m/s from the northwest.

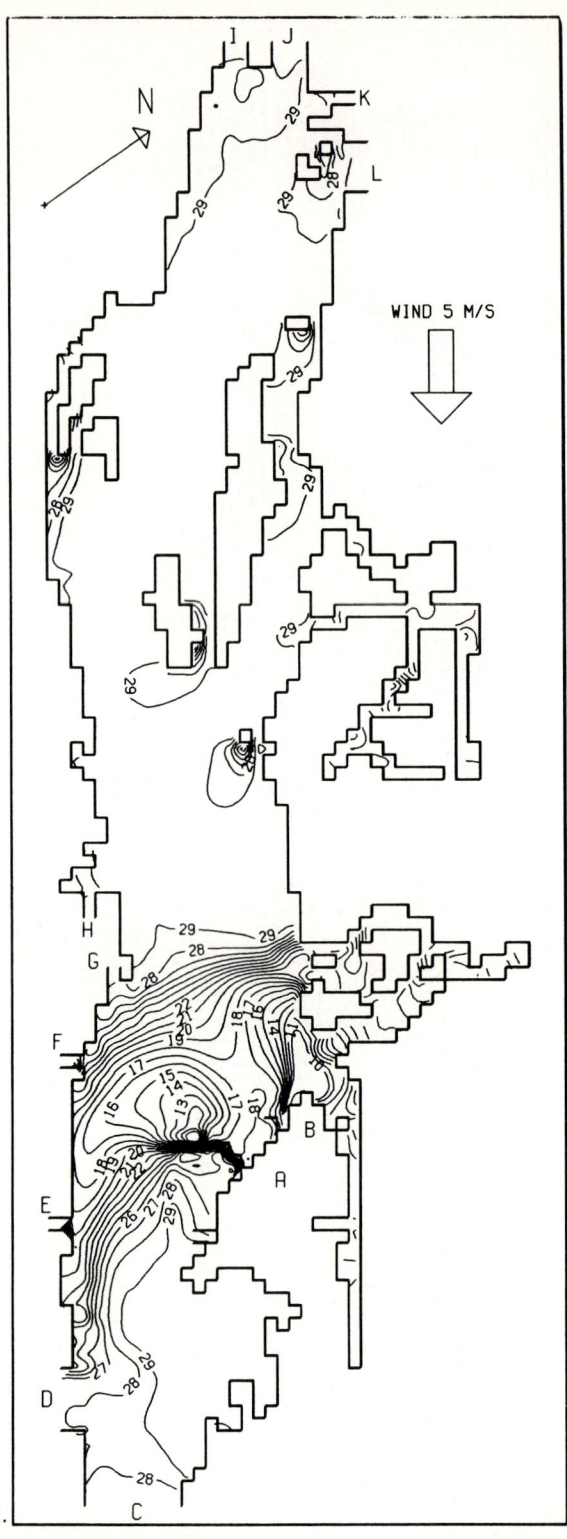

Fig. 11.12b Distributions of salinity ($^0/_{00}$) for flood tide and wind of 5 m/s from the northwest.

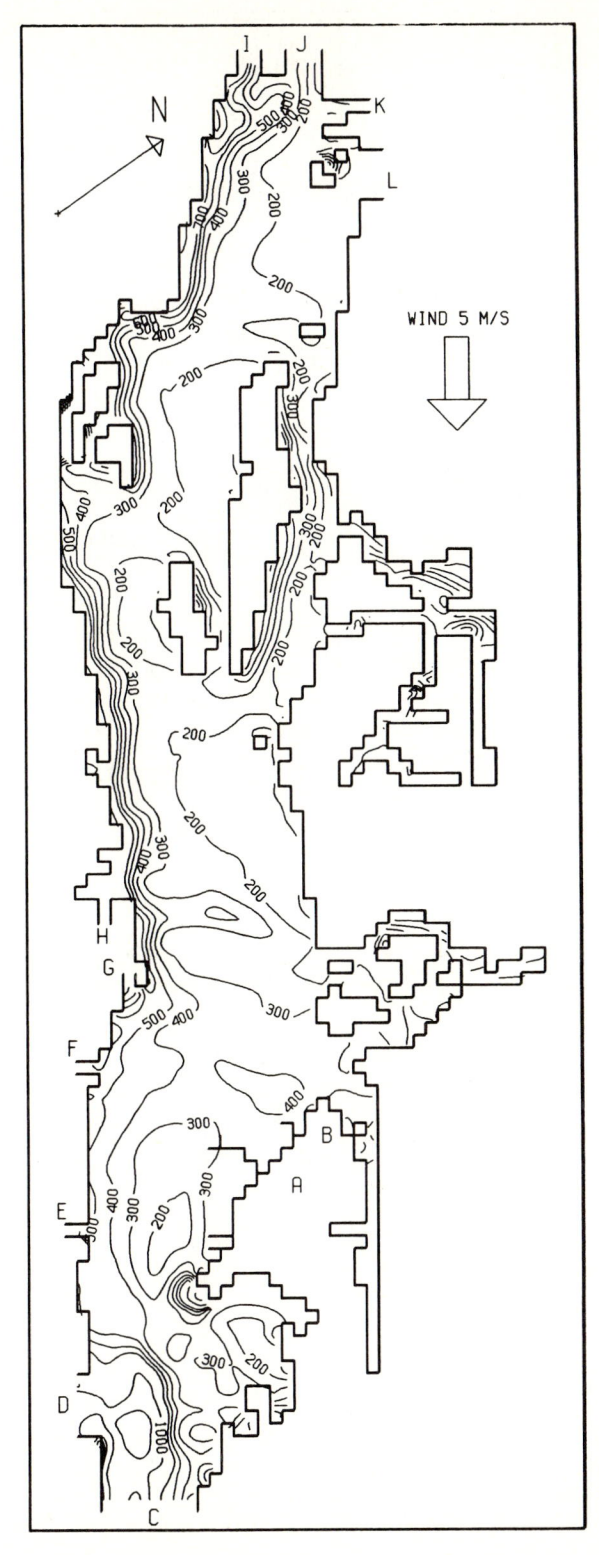

Fig. 11.12c Distributions of layer depth (cm) for flood tide and wind of 5 m/s from the northwest.

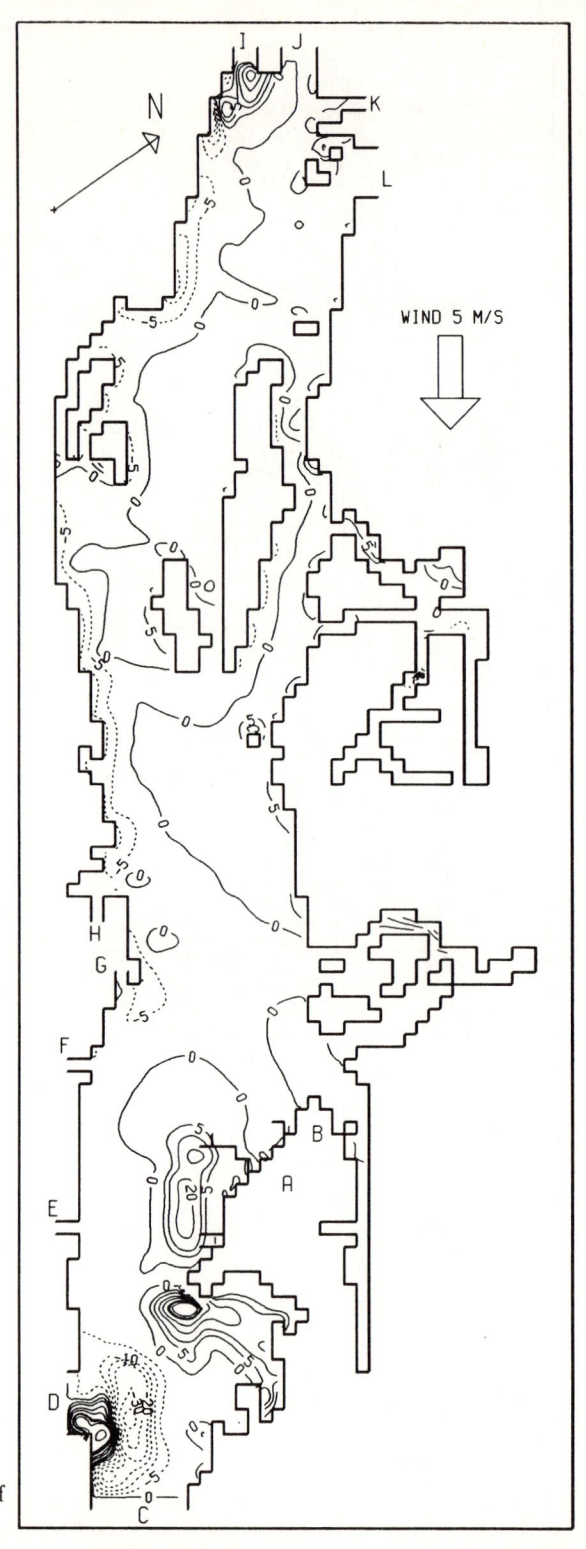

Fig. 11.12d Distributions of relative vertical velocity ($\times 10^3$ cm/s) for flood tide and wind of 5 m/s from the northwest.

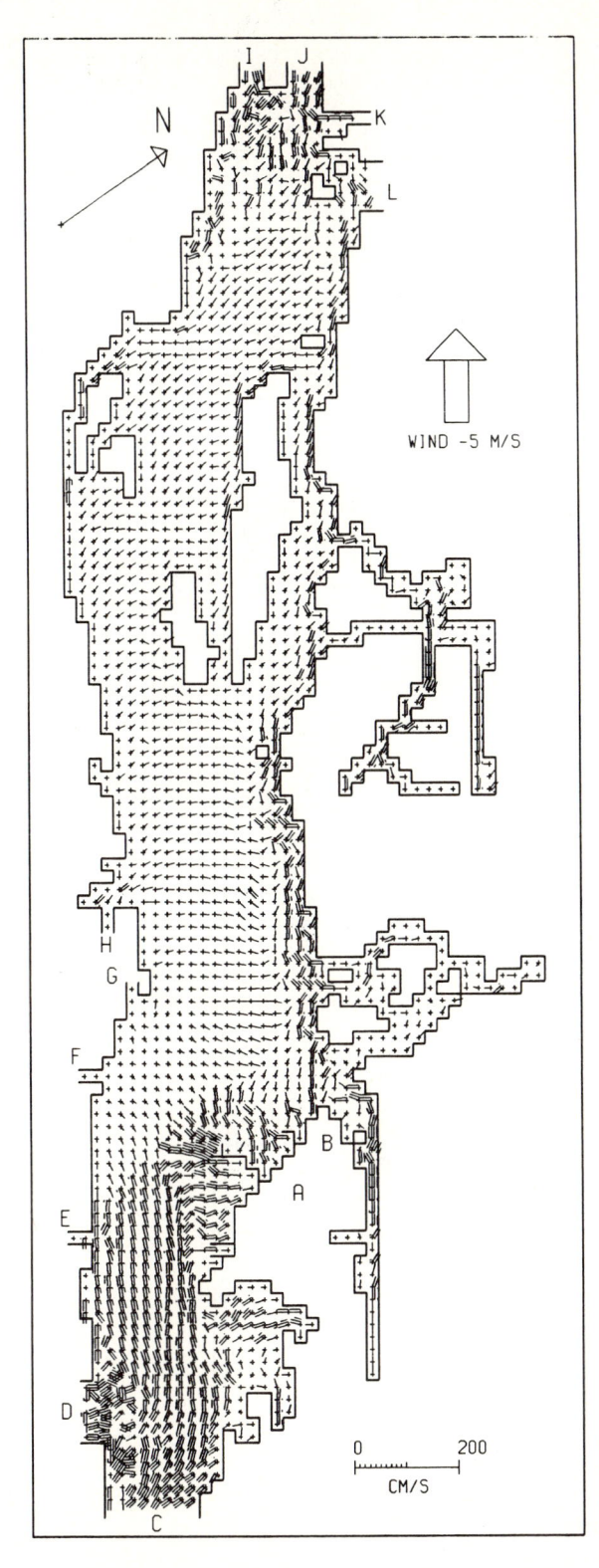

Fig. 11.13a Distributions of velocity (cm/s) for ebb tide and wind of 5 m/s from the southeast.

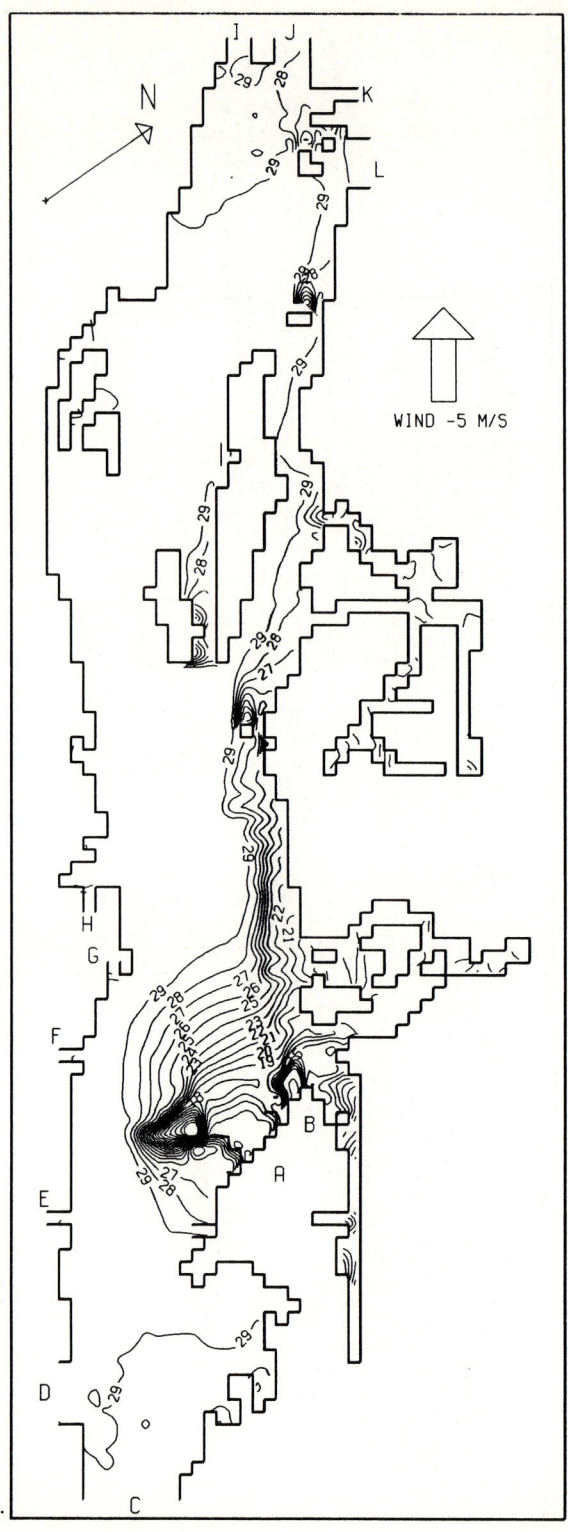

Fig. 11.13b Distributions of salinity ($^0/_{00}$) for ebb tide and wind of 5 m/s from the southeast.

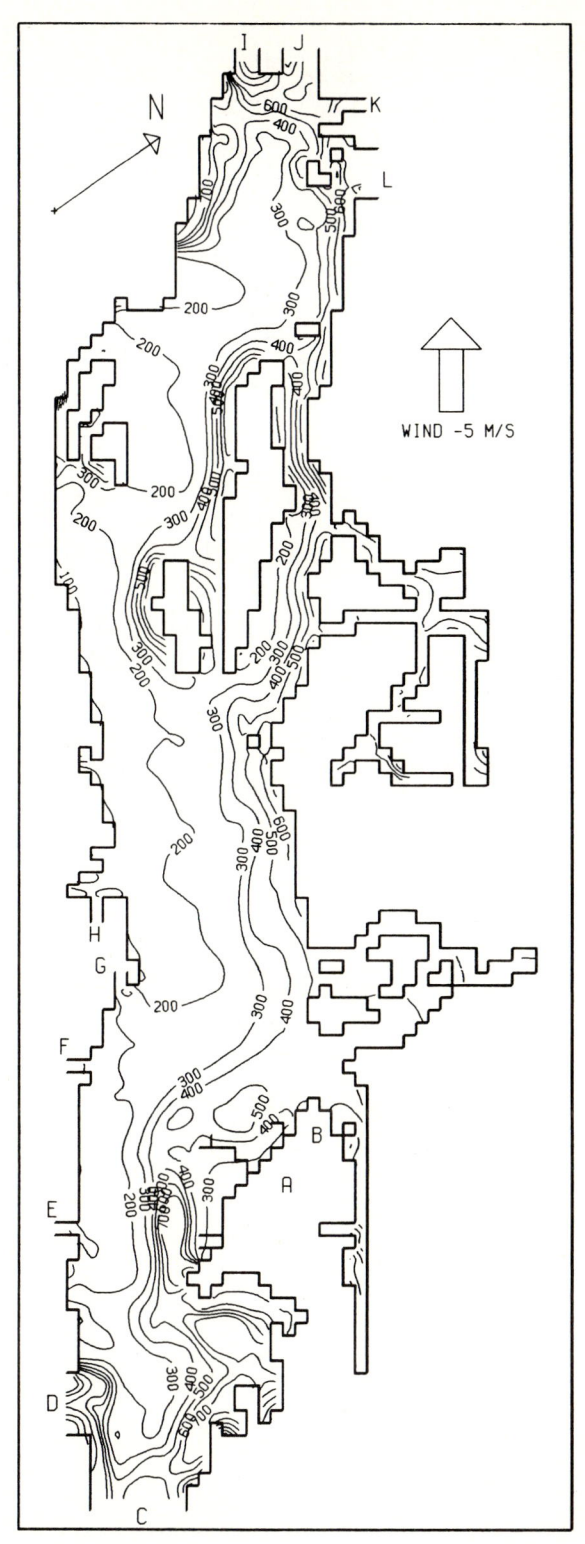

Fig. 11.13c Distributions of layer depth (cm) for ebb tide and wind of 5 m/s from the southeast.

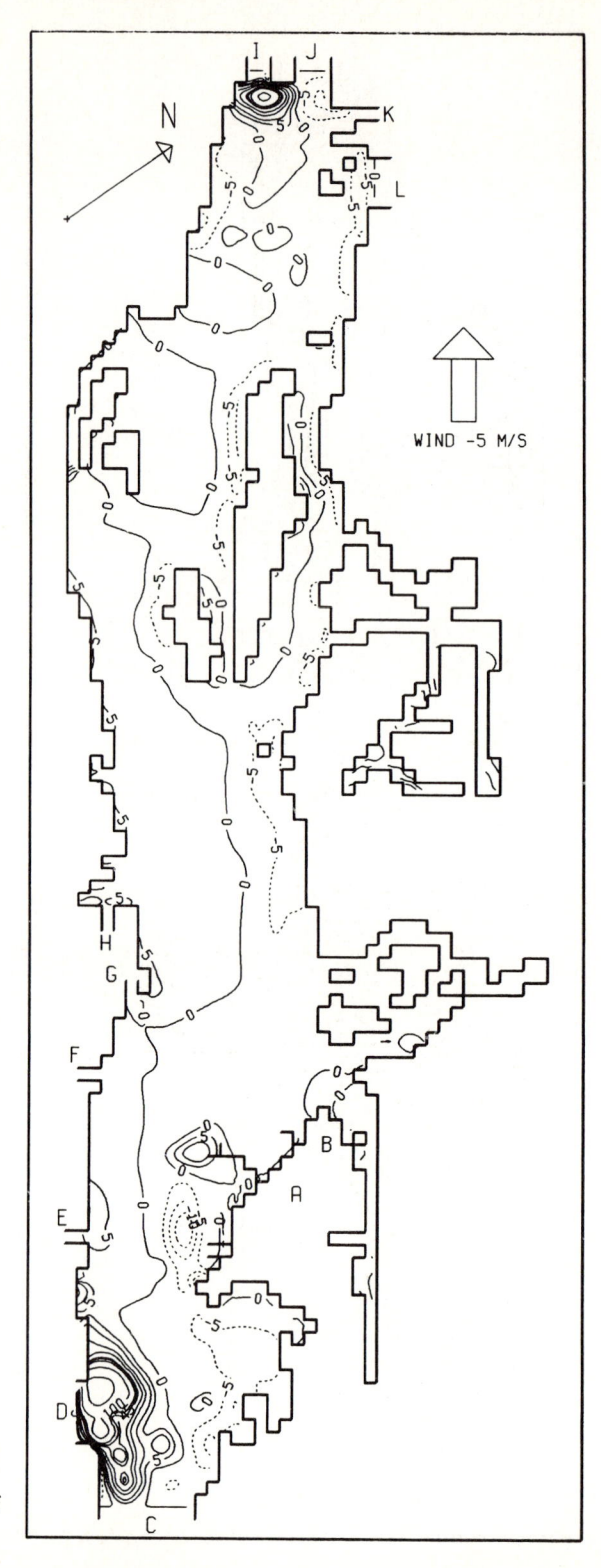

Fig. 11.13d Distributions of relative vertical velocity ($\times 10^3$ cm/s) for ebb tide and wind of 5 m/s from the southeast.

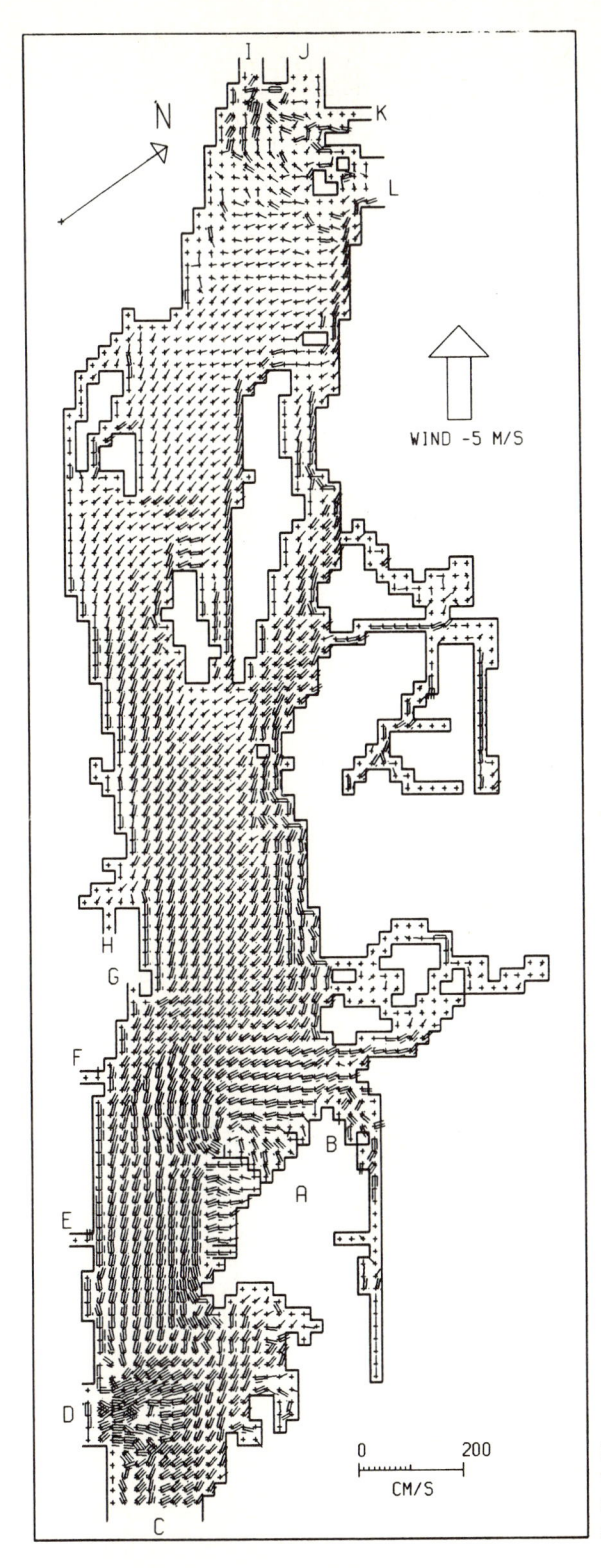

Fig. 11.14a Distributions of velocity (cm/s) for flood tide and wind of 5 m/s from the southeast.

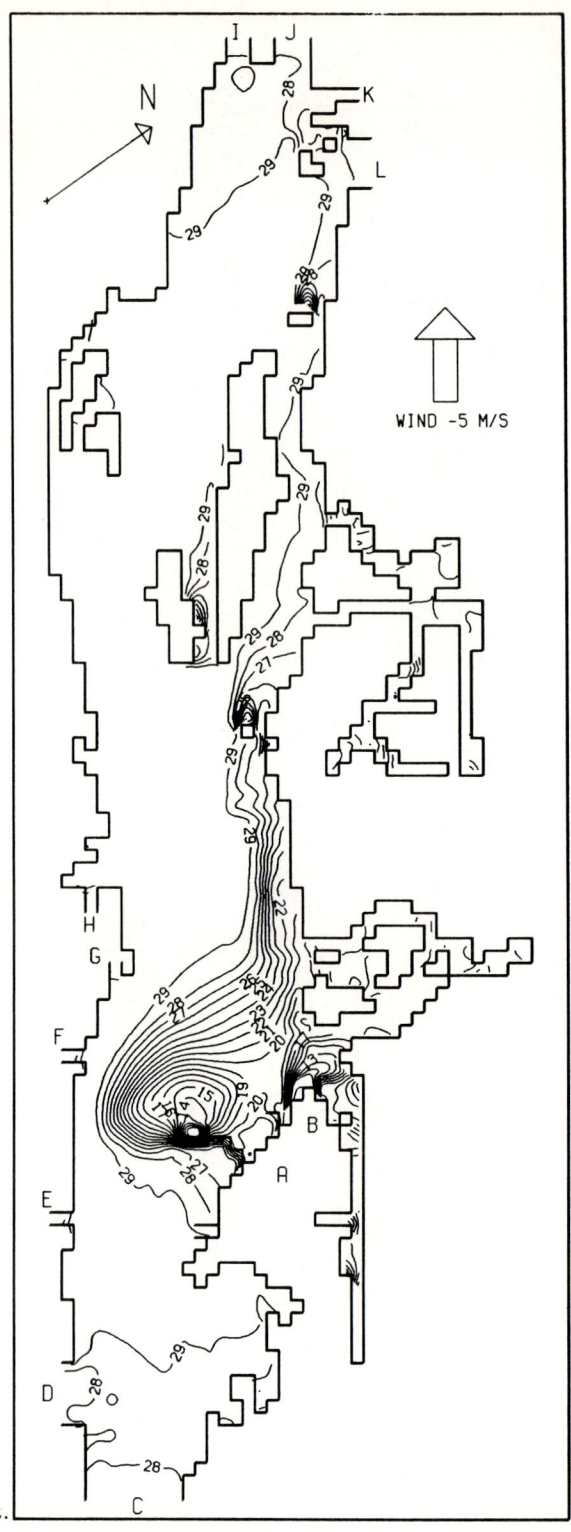

Fig. 11.14b Distributions of salinity (⁰/₀₀) for flood tide and wind of 5 m/s from the southeast.

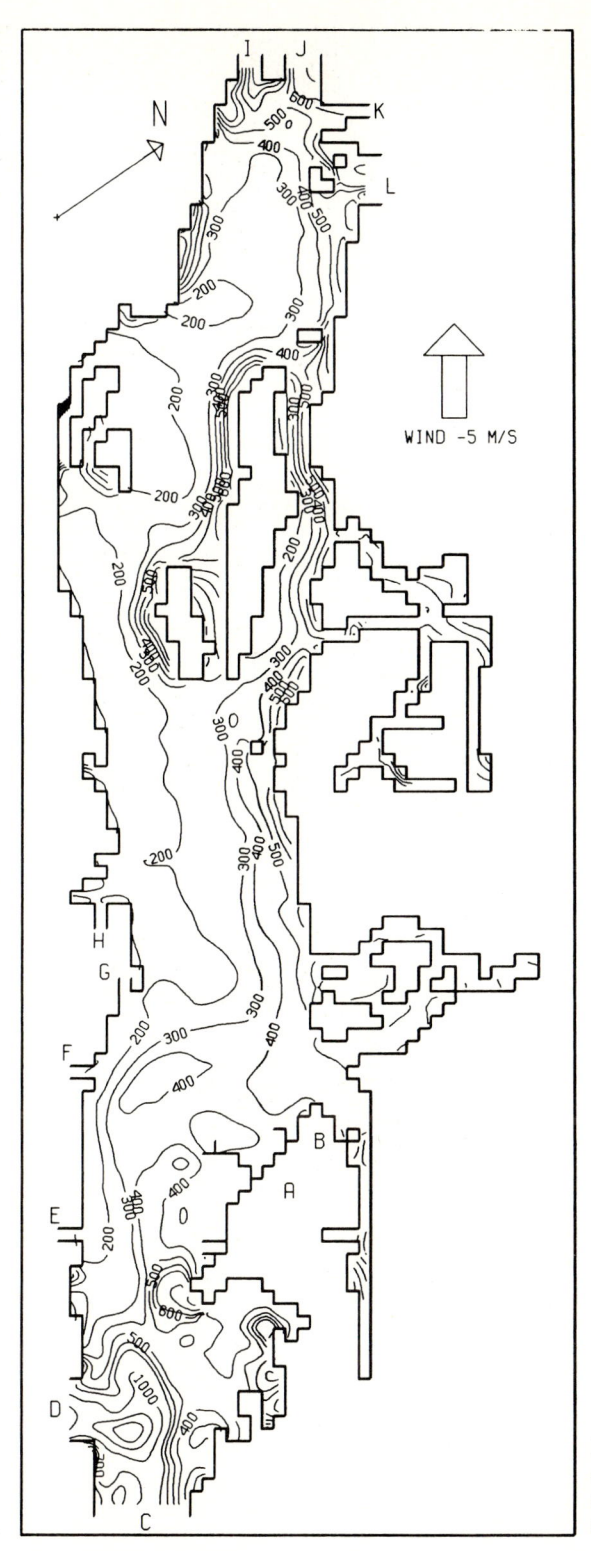

Fig. 11.14c Distributions of layer depth (cm) for flood tide and wind of 5 m/s from the southeast.

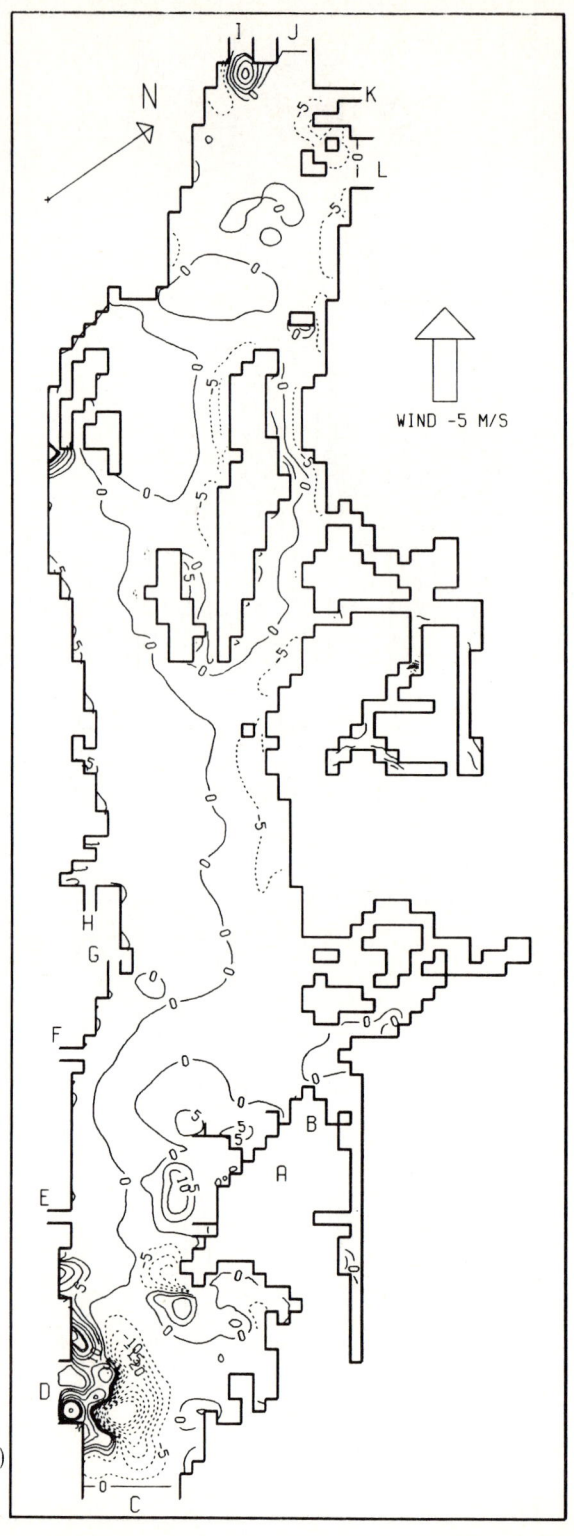

Fig. 11.14d Distributions of relative vertical velocity ($\times 10^3$ cm/s) for flood tide and wind of 5 m/s from the southeast.

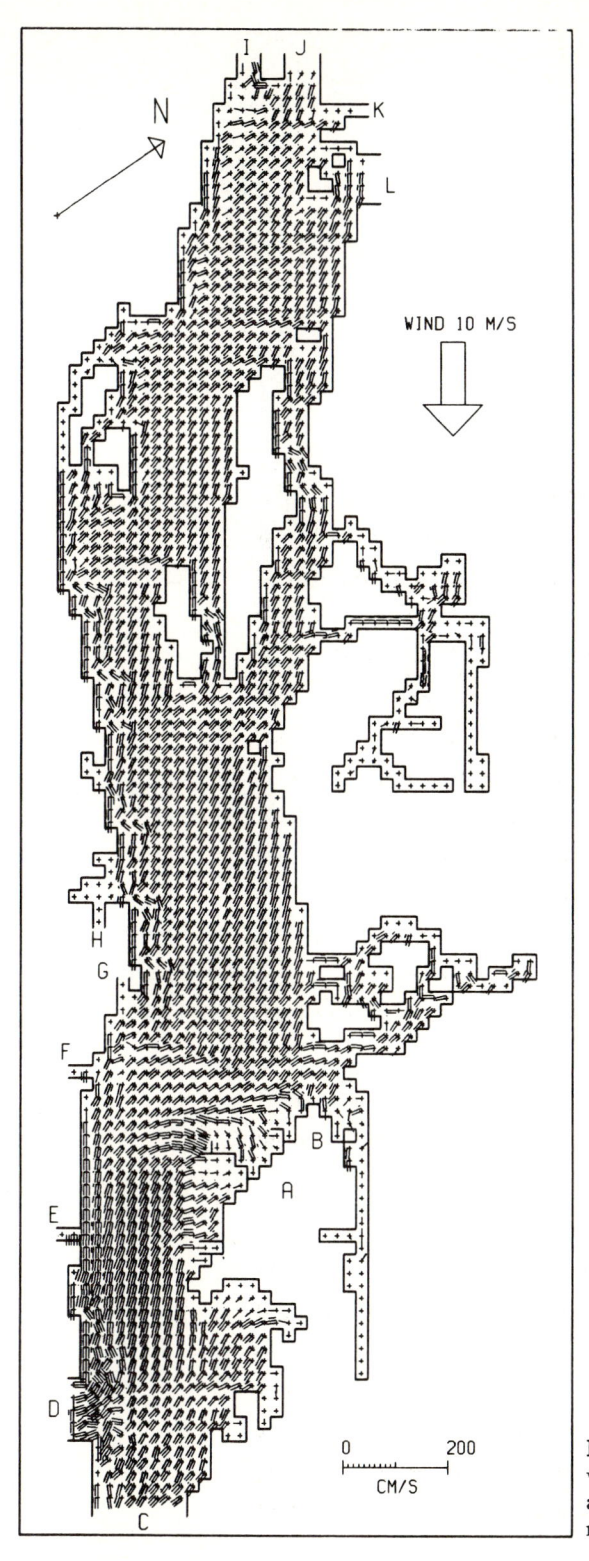

Fig. 11.15a Distributions of velocity (cm/s) for ebb tide and wind of 10 m/s from the northwest.

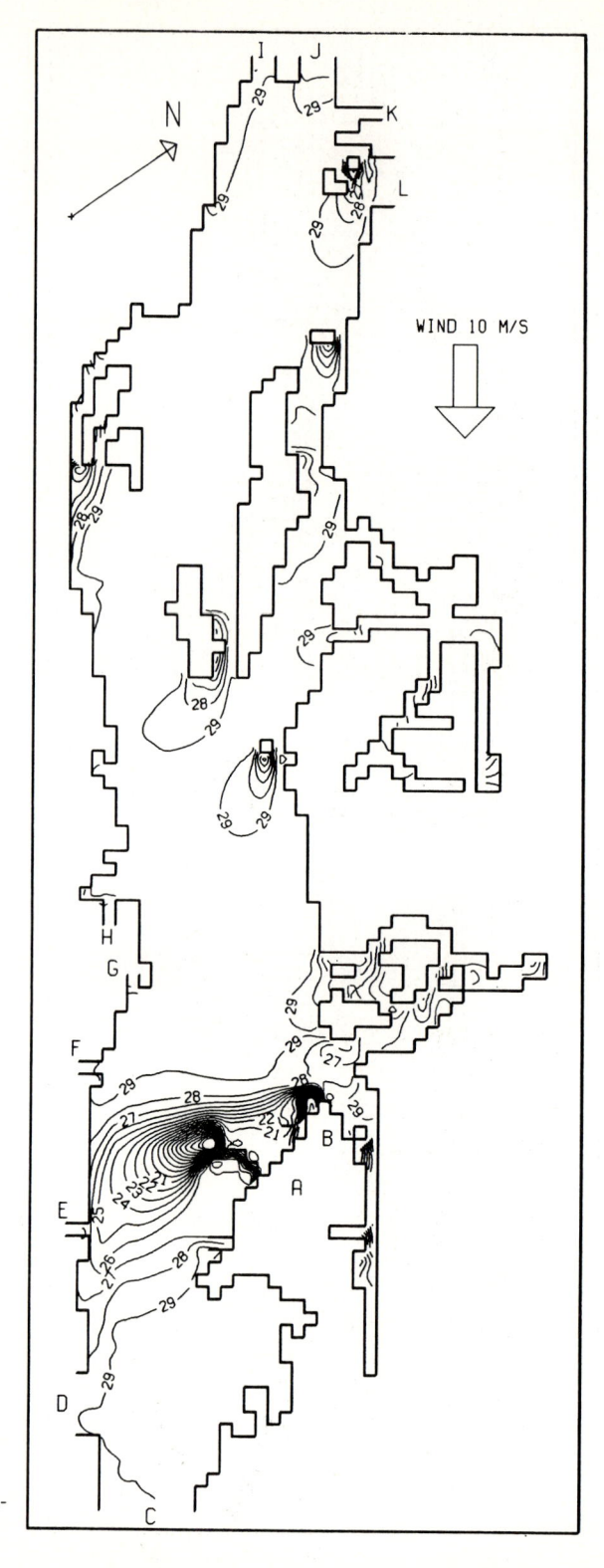

Fig. 11.15b Distributions of salinity (‰) for ebb tide and wind of 10 m/s from the northwest.

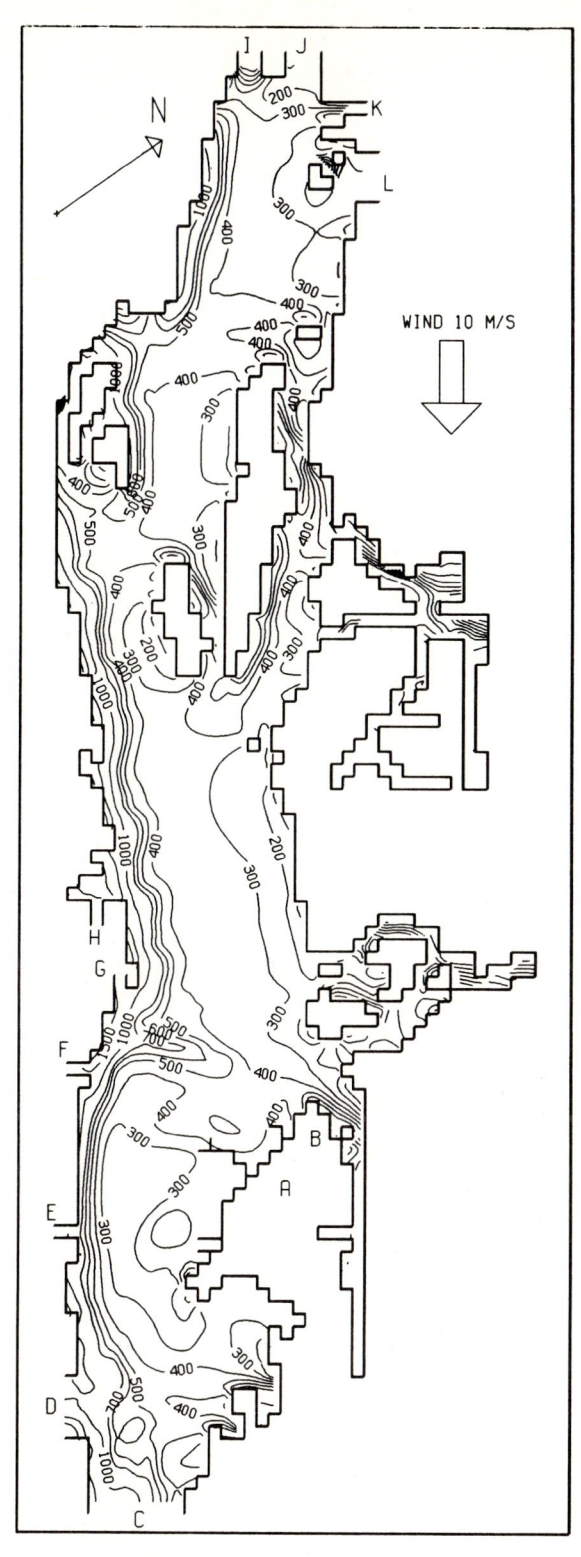

Fig. 11.15c Distributions of layer depth (cm) for ebb tide and wind of 10 m/s from the northwest.

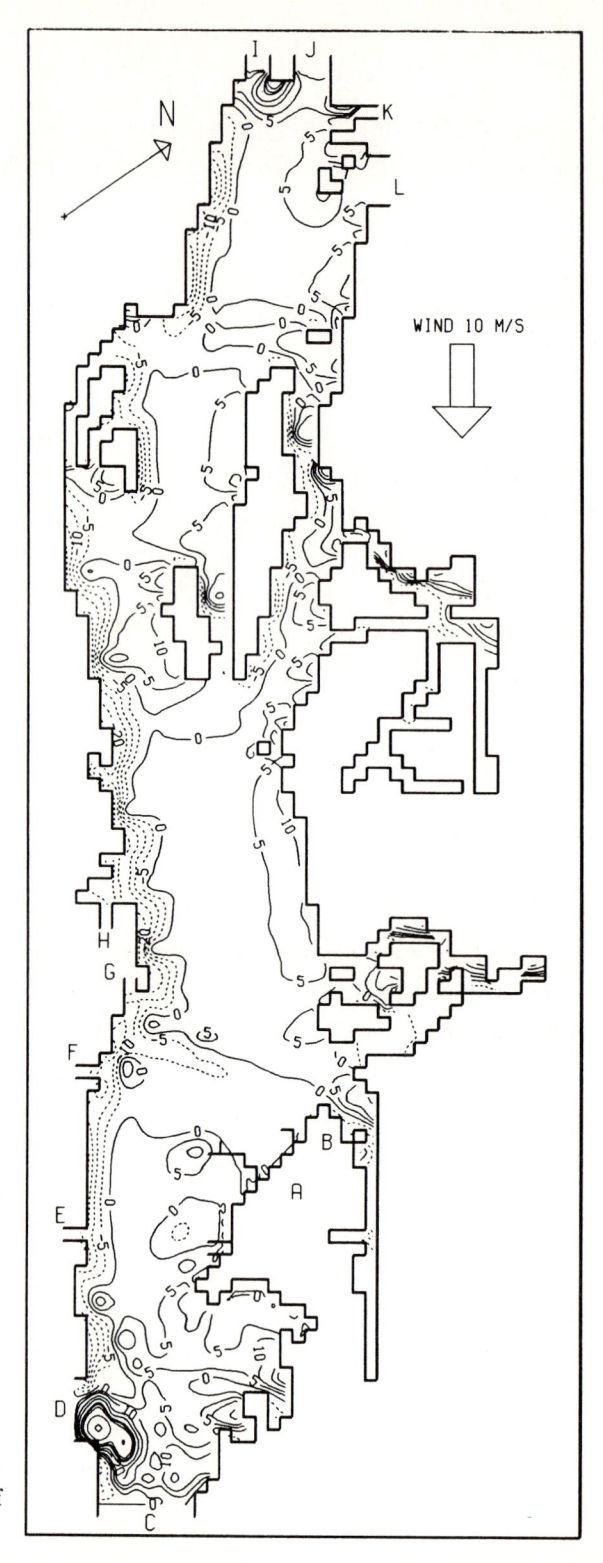

Fig. 11.15d Distributions of relative vertical velocity ($\times 10^3$ cm/s) for ebb tide and wind of 10 m/s from the northwest.

Flood Tide, Northwest Wind of 10 m/s

In this case the strong flood tidal streams from the south and the northward Coriolis-deflected river flow encounter strong southerly-directed wind-induced surface flow in the southern part of the central Strait of Georgia near the entrance to Howe Sound (Fig. 11.16a). A strong horizontal salinity gradient results across the channel close to the freshwater source (Fig. 11.16b) approximately coincident with the marked thickening of the layer (Fig. 11.16c). The vertical velocity distribution (Fig. 11.16d) differs little from that previously described for an ebbing tide.

Ebb Tide, Southeast Wind of 10 m/s

Over the central and northern portions of the Strait of Georgia the predominantly northerly flow is strongly deflected towards the mainland coast (Fig. 11.17a). Ebb streams in the southern part of the model are confined to the narrowed part of the Strait south of the main freshwater source of discharge where the barotropic tidal streams are strongest. Significant surface dilution is confined close to a relatively small area in the vicinity of the freshwater source (Fig. 11.17b). Marked layer deepening close to coasts to the right of the wind direction is clearly in evidence (Fig. 11.17c). A deep pool of mixed water has appeared in the upper northern corner of the model. The pattern of net vertical velocity (Fig. 11.17d) again shows the development of a strong negative component to the right of the wind direction near the coast while alongshore to the left of the wind direction there are weaker positive net components.

Flood Tide, Southeast Wind of 10 m/s

Except for the flood stream entering the Discovery Passage boundary opening, surface flow in the model (Fig. 11.18a and b) is strongly dominated by a northeasterly flow in which the flooding streams, the prescribed winds, and Coriolis-deflected river discharge act in concert. Again, the formation of locally thickened layers (Fig. 11.18c) along coastlines to the right of the wind direction, confinement of significant dilution to an area close to freshwater sources, and the associated distribution of net relative vertical velocity (Fig. 11.18d) are similar to those obtained on the ebbing tide under the same conditions of wind.

These results show that when the model plume is subjected to a selection of winds encompassing the range of most likely occurrence and greatest oceanographic significance, the model reproduces those features of the surface flow field that might reasonably be anticipated on available grounds of both observation and theory. As such, it constitutes a unique and relatively inexpensive heuristic tool for estimating actual behaviour for a large number of possible combinations of tide, wind, and river discharge.

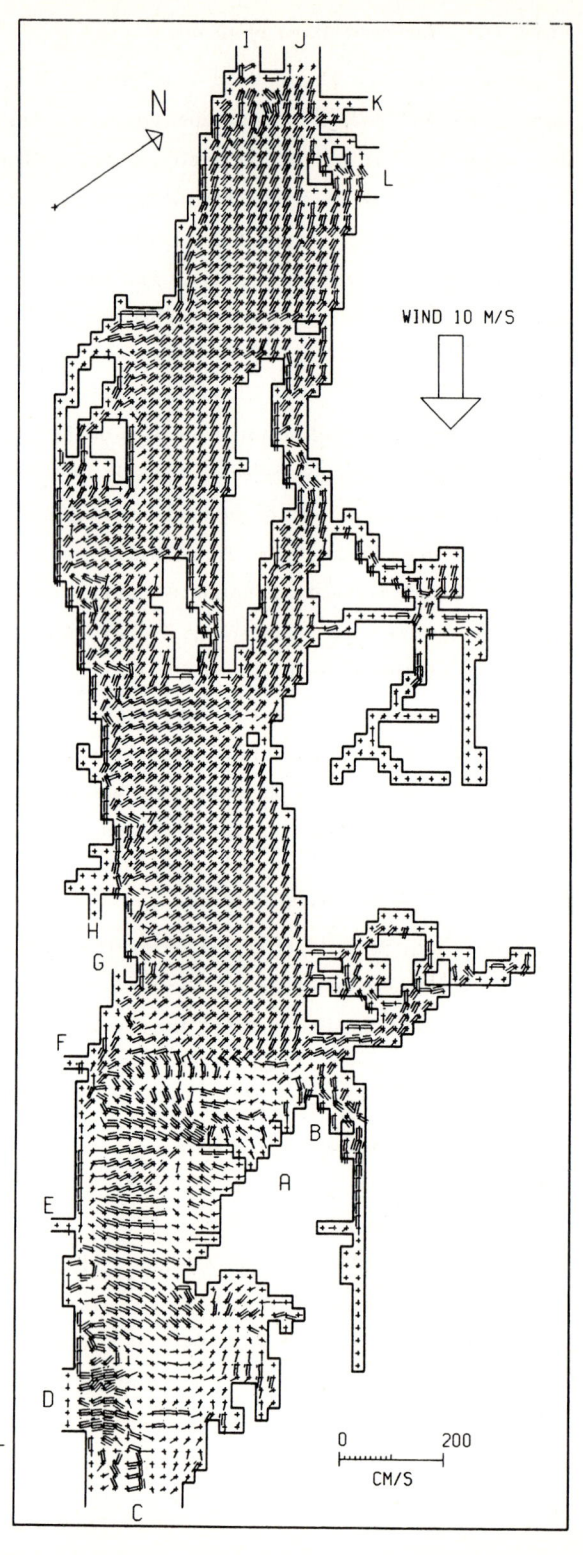

Fig. 11.16a Distributions of velocity (cm/s) for flood tide and wind of 10 m/s from the northwest.

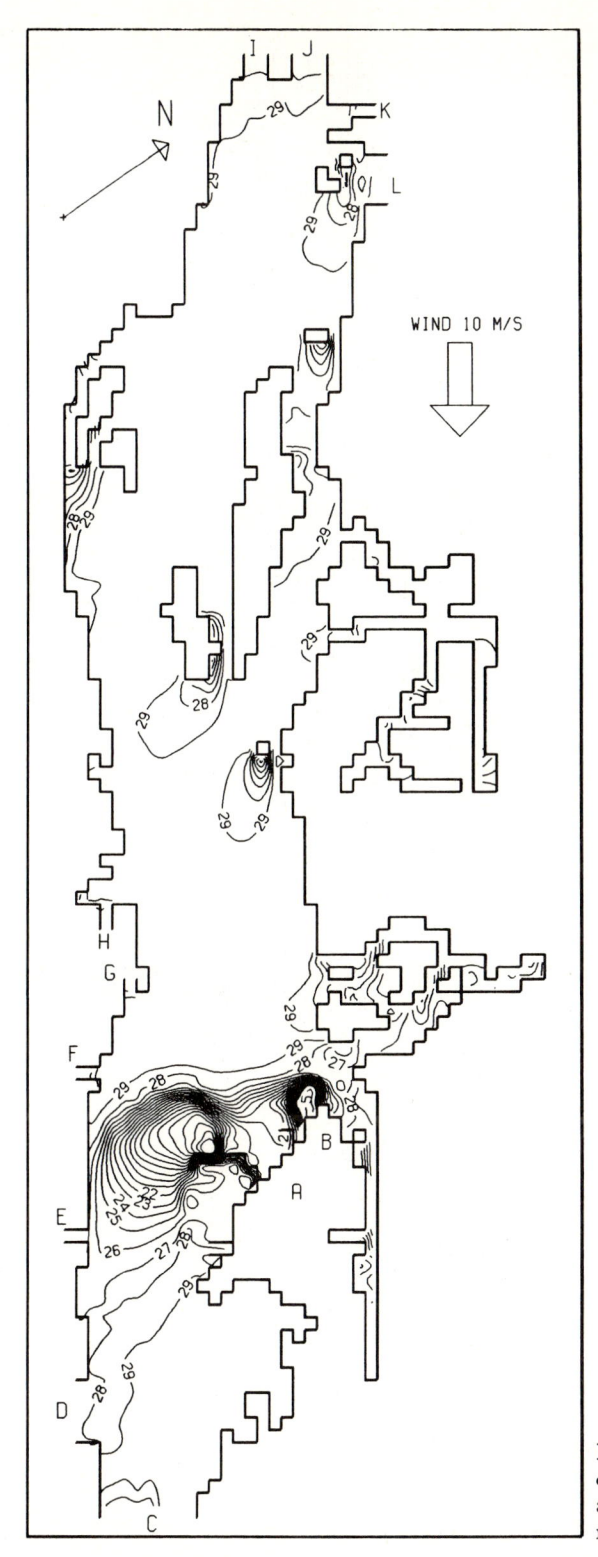

Fig. 11.16b Distributions of salinity ($^0/_{00}$) for flood tide and wind of 10 m/s from the northwest.

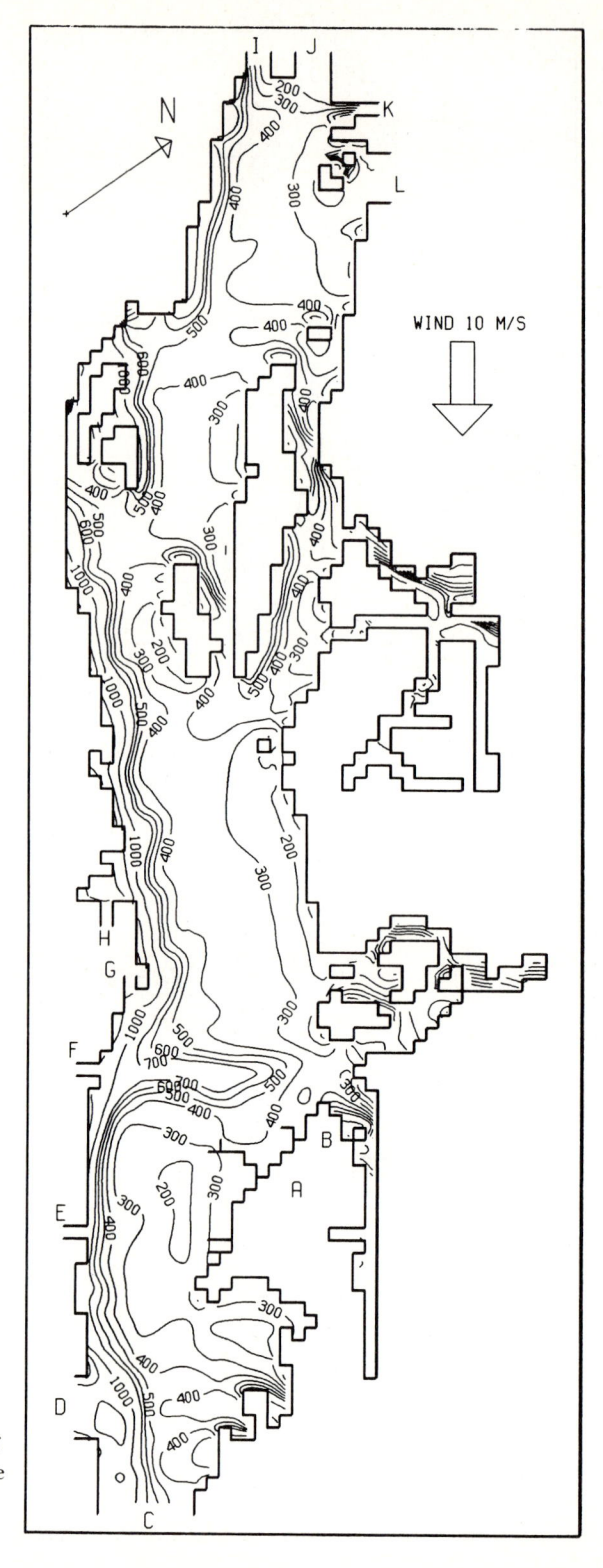

Fig. 11.16c Distributions of layer depth (cm) for flood tide and wind of 10 m/s from the northwest.

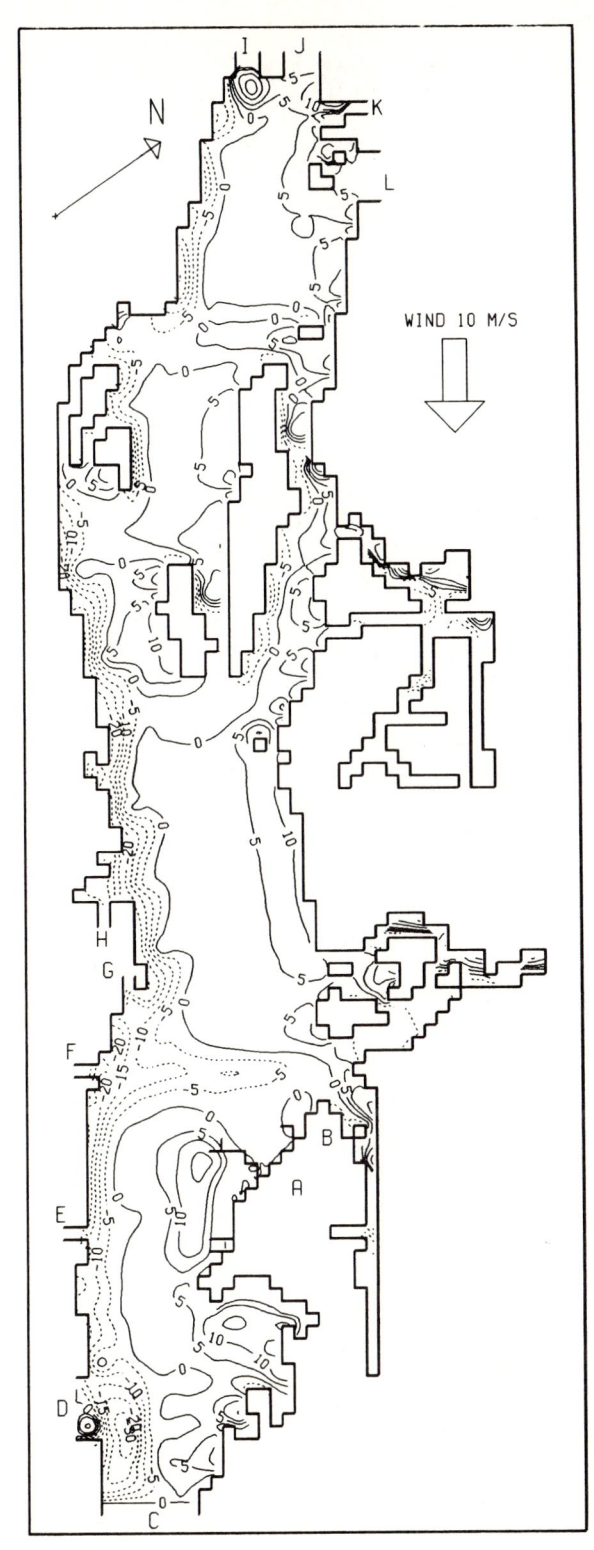

Fig. 11.16d Distributions of relative vertical velocity ($\times 10^3$ cm/s) for flood tide and wind of 10 m/s from the northwest.

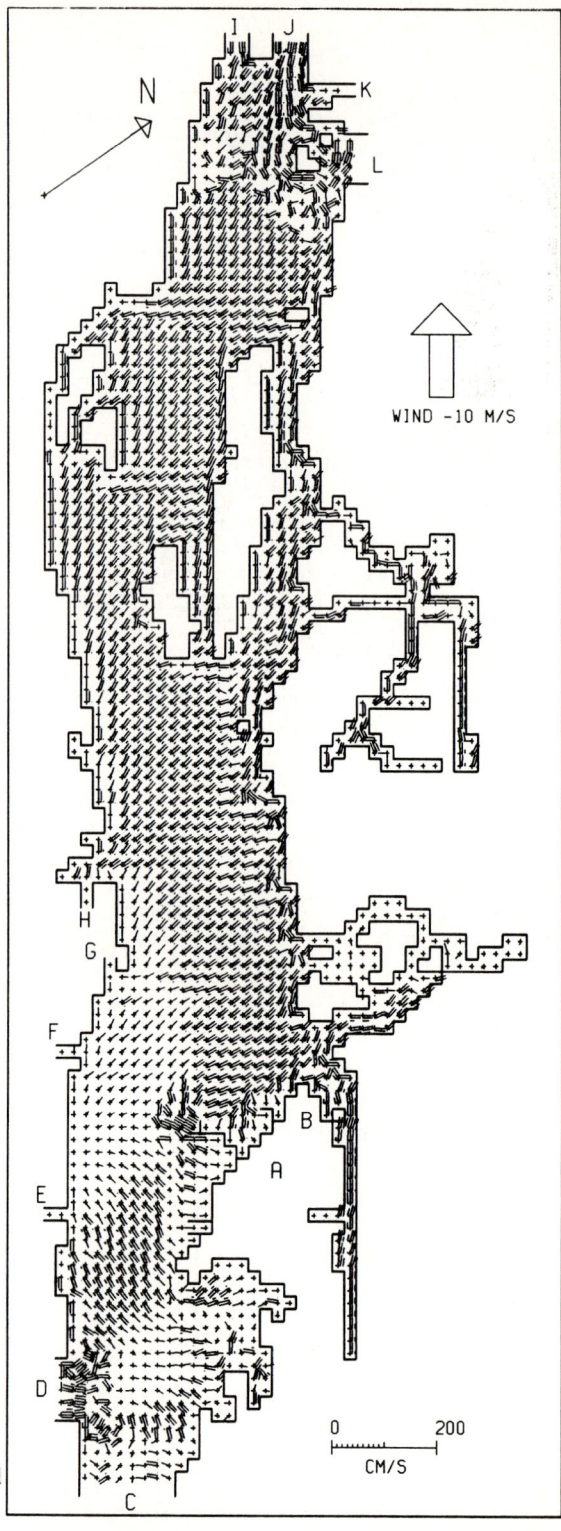

Fig. 11.17a Distributions of velocity (cm/s) for ebb tide and wind of 10 m/s from the southeast.

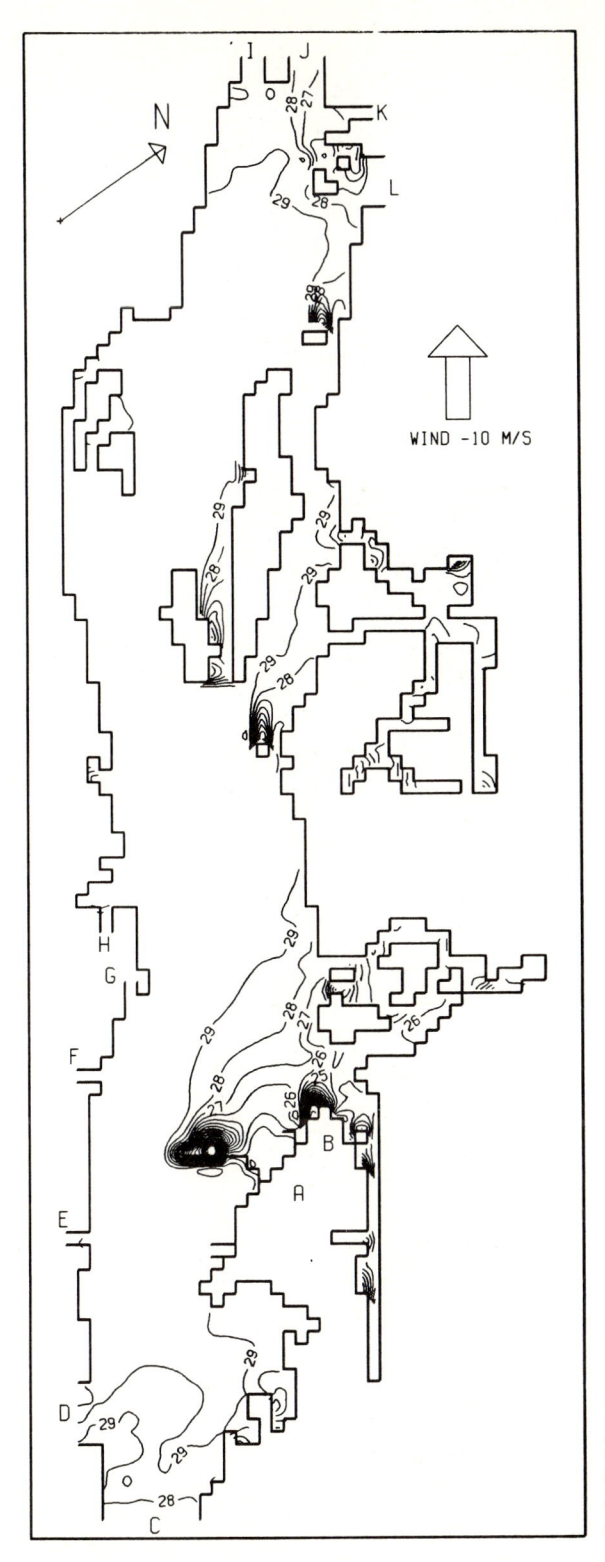

Fig. 11.17b Distributions of salinity (⁰/₀₀) for ebb tide and wind of 10 m/s from the southeast.

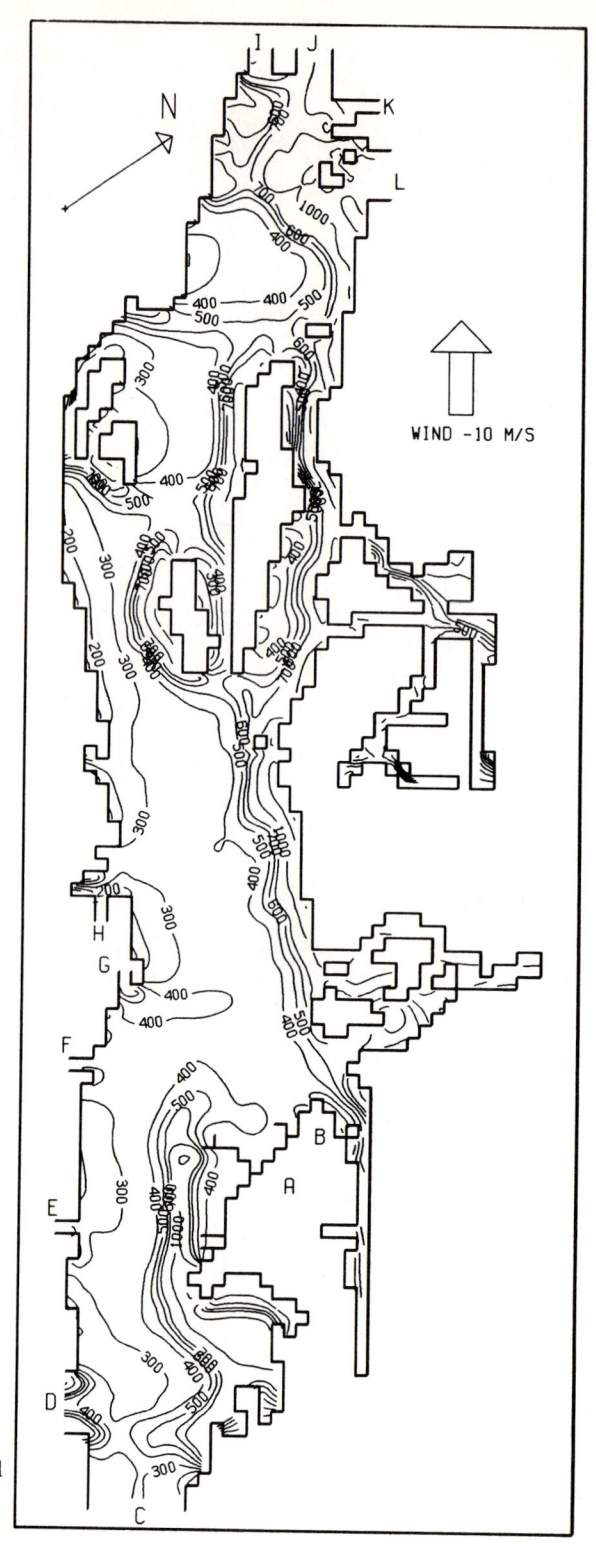

Fig. 11.17c Distributions of layer depth (cm) for ebb tide and wind of 10 m/s from the southeast.

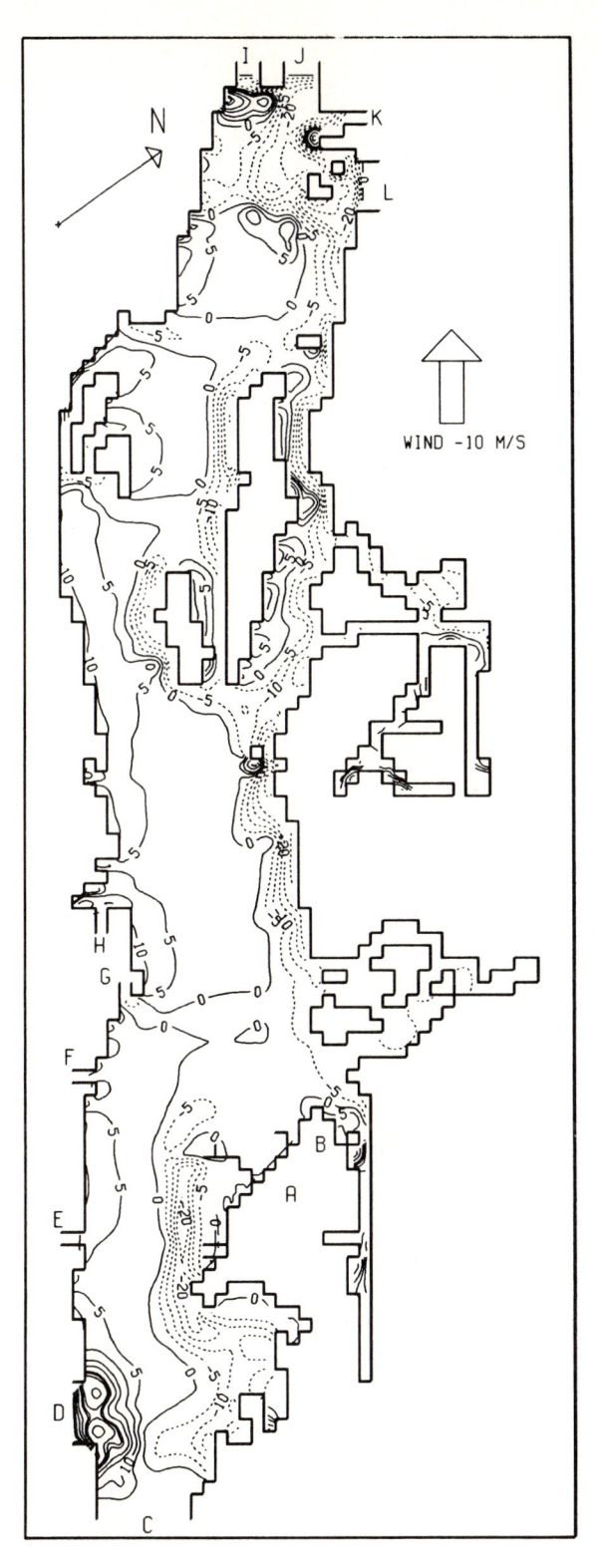

Fig. 11.17d Distributions of relative vertical velocity ($\times 10^3$ cm/s) for ebb tide and wind of 10 m/s from the southeast.

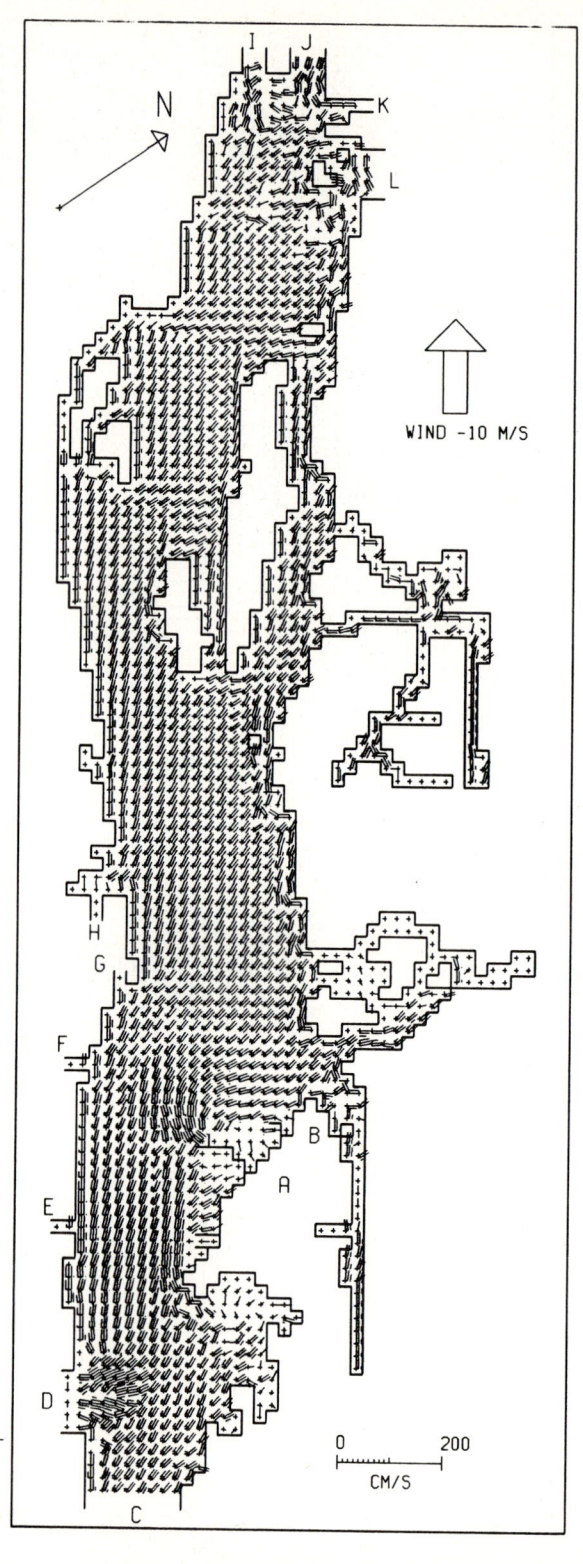

Fig. 11.18a Distributions of velocity (cm/s) for flood tide and wind of 10 m/s from the southeast.

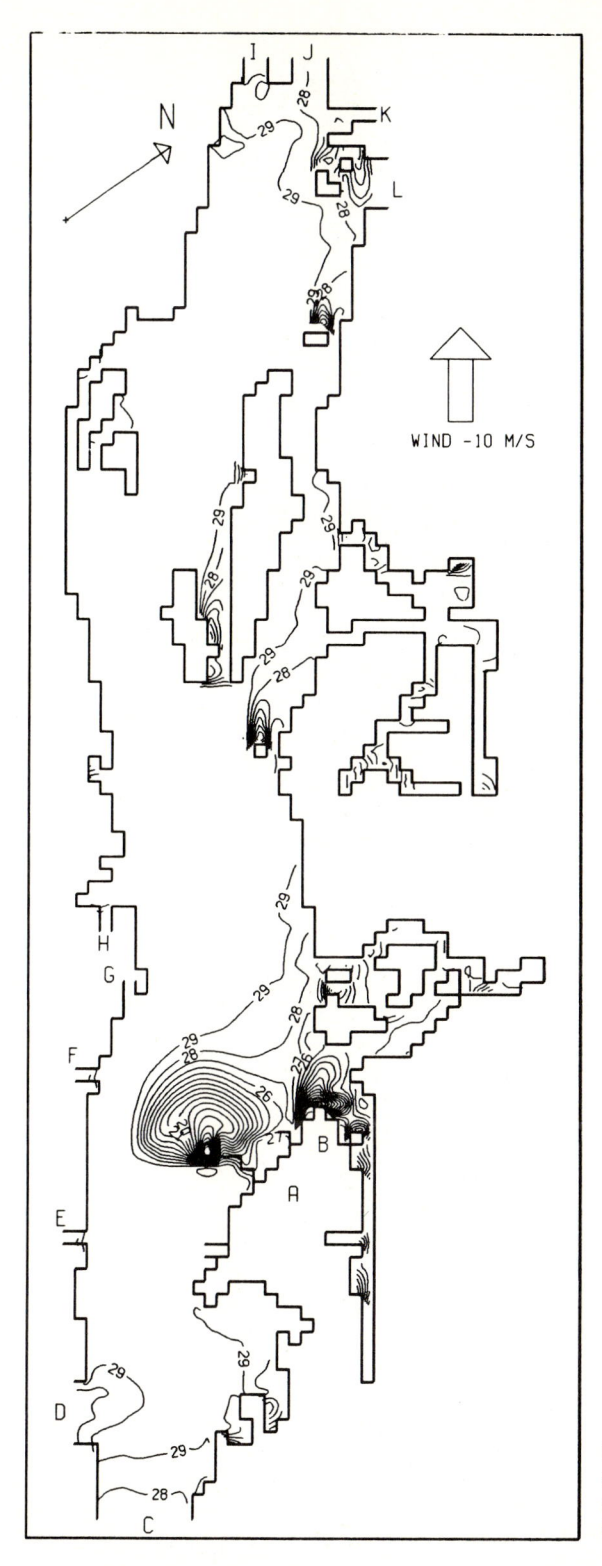

Fig. 11.18b Distributions of salinity (⁰/₀₀) for flood tide and wind of 10 m/s from the southeast.

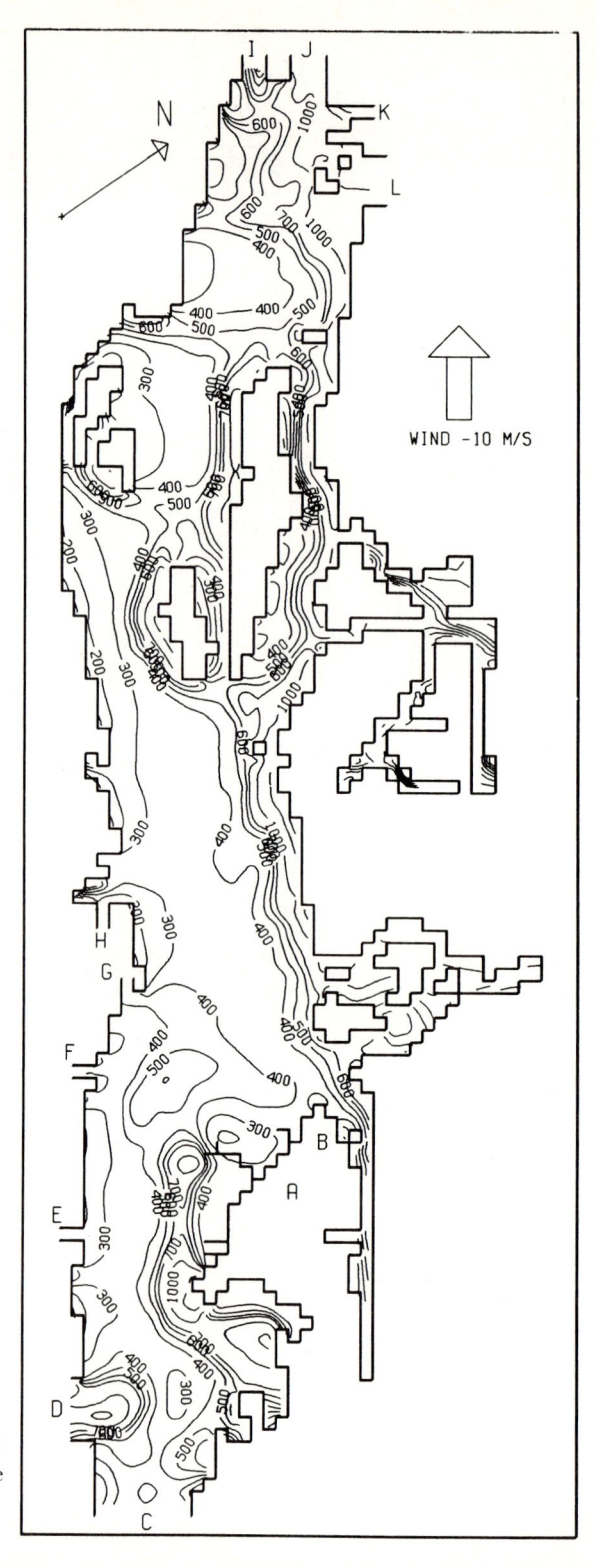

Fig. 11.18c Distributions of layer depth (cm) for flood tide and wind of 10 m/s from the southeast.

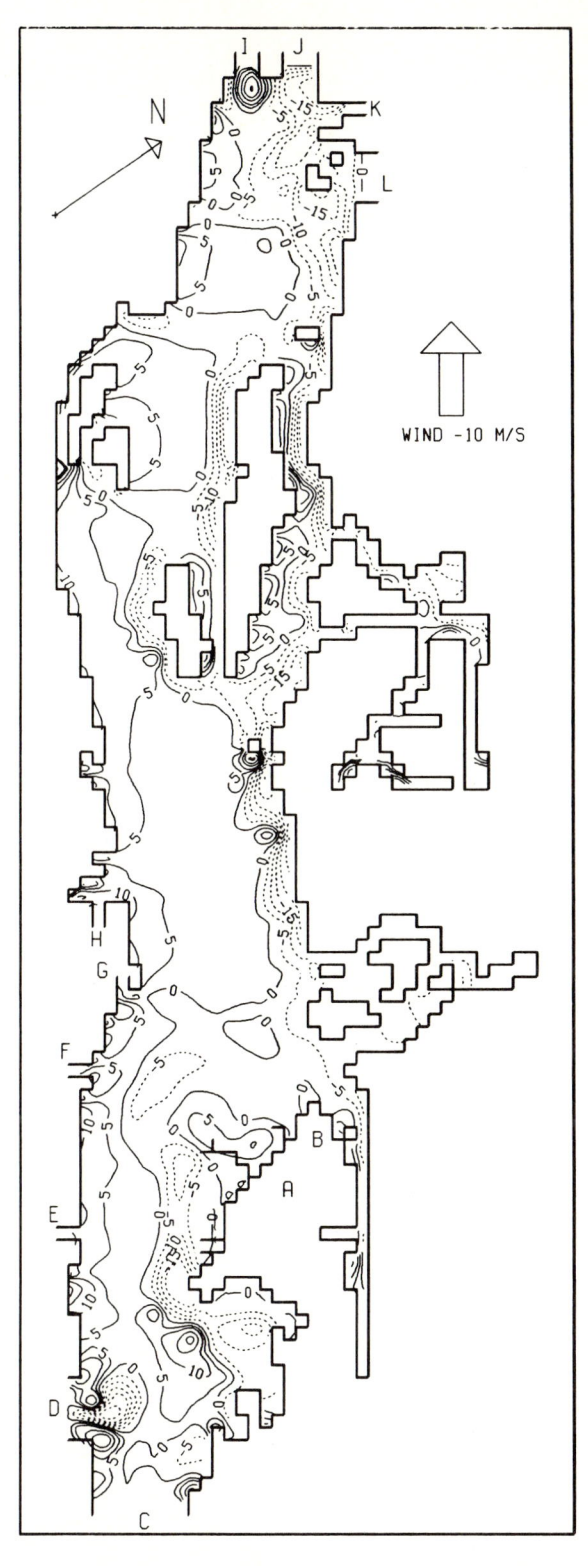

Fig. 11.18d Distributions of relative vertical velocity ($\times 10^3$ cm/s) for flood tide and wind of 10 m/s from the southeast.

11.5.3 Residual Circulation

Figure 11.19a presents the residual velocities produced by the upper layer model in the absence of wind forcing but with tidal forcing for that part of the Strait of Georgia most strongly affected by the river discharge. The vectors represent the low-pass filtered average velocities calculated over the last 15 days of a 30-day run. The most striking feature is the well defined stream of water moving out from the Fraser River and turning northward. A smaller part of the jet moves south along the Gulf Islands. Another major feature is the stagnant zone between mouths of the two Arms.

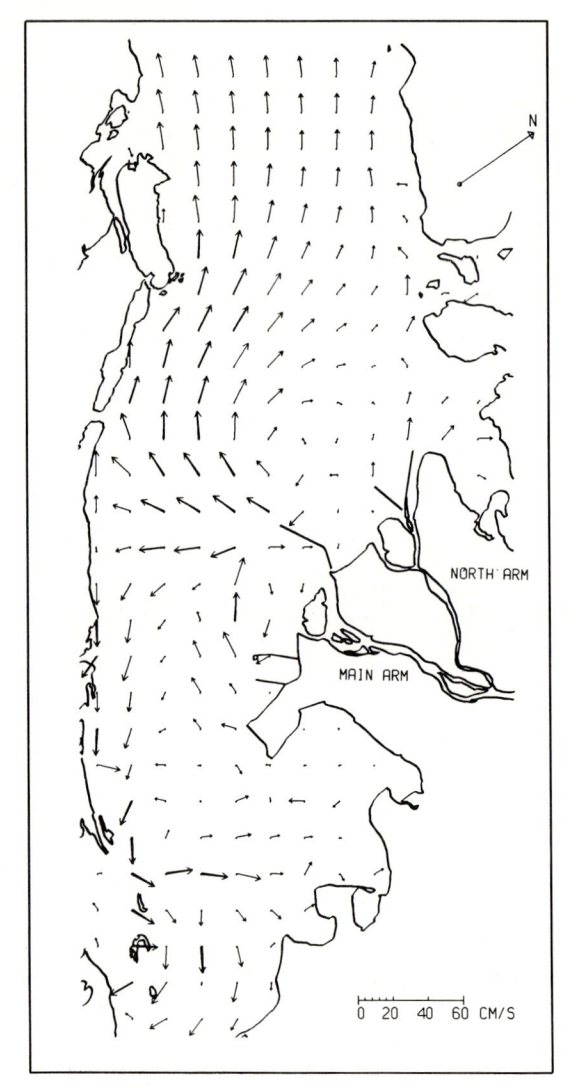

Fig. 11.19a Residual velocity averaged over the last 15 days of a 30-day run for no wind but with the inclusion of tide and Fraser River discharge.

The prime driving force for the residual current is the residual surface slope. Figure 11.19b shows the low-pass filtered surface elevation $(gh(\Delta\rho))/\rho$. When the Bernoulli function $(u^2 + v^2)/2g$ is added to the surface elevation there are no significant differences. It is evident that over much of the region the residual flow is in geostrophic balance as shown by the alignment of the velocities with the iso-elevation contours. Comparison of these residual surface elevations with those obtained later from the fully three-dimensional model GF6 (Chapter 13) under similar conditions of tide and river discharge display a highly pertinent and interesting distribution. In general, the elevations throughout the Strait of Georgia in the three-dimensional model GF6 are considerably higher than those obtained from the layer model GF4. This increase is clearly due to a major process of deep water renewal, fully operative in the case of the GF6 results but absent from GF4. Thus, the baroclinic pressure gradients present at depth in GF6, due to the prescribed observed initial density field, bring about a marked deep inflow. The resulting increase in surface elevations drives the surface outflow as required by continuity considerations. There remains, however, the actual distribution of surface elevations in the Strait of Georgia due to the observed density distribution. This distribution of residual surface elevation, illustrated later in Fig. 13.17, is determined by the distribution of observed density and is negligibly affected by river discharge. This distribution, in turn, will be affected by the net gradients moving water seaward from the river mouth as well as the buoyant-spreading pressure gradients of the type operative in the layer model GF4. Thus, the drop between a representative elevation near the river mouth and the southern boundary opening for the GF6 model is 11 cm whereas in the case of the GF4 model the corresponding drop is 7 cm. For the northern boundary opening, the drop is 14 cm for the model GF6 and 5 cm for the model GF4. It would appear that GF4 will tend to underestimate the net flow velocities away from the river mouth.

11.5.4 Comparison with Observations

An initial consideration concerns the distribution of salinity along the major axis of the emergent jet from the river mouth as it curves around towards, and then along, the mainland coast. Figure 11.20 shows comparisons of observed (Crean and Ages, 1971) and computed salinities at the sea surface under conditions of high (July 1968) and low (December 1967) river flow. It will be recalled that the model starts from rest assuming an arbitrary field of uniform salinity. The agreement would appear entirely satisfactory.

11.5.4.1 Comparison of Observed and Computed Velocity Vectors

Reference has been made in Chapter 1 to the extensive programs of field observations in which free-floating surface current drifters were used to track flow patterns characteristic of the Fraser River plume. The most extensive geographical coverage was

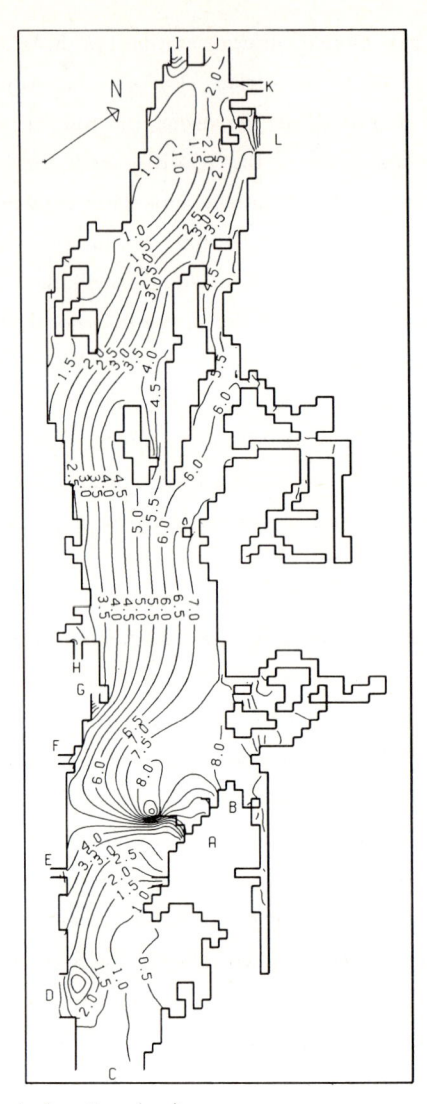

Fig. 11.19b Residual elevation (cm).

obtained in a monthly series of 2-day operations over a 12-month period during 1981–82. It will be recalled that these operations were scheduled when lower low water occurred in the middle of the day. The flood-ebb sequence of tracks was obtained for a region south of the river mouth on the first day while on the second day tracks for the region north of the mouth were determined. Thus, if it could be assumed that the influence of wind was either similar for each day or negligible then it is possible, given relatively small changes in day-to-day discharge and in the sequence of tidal streams, to determine a composite set of tracks over almost the entire region of the active plume. From the tracks observed, velocity vectors representative of a large ebb-large flood sequence may be obtained. However, sustained calm conditions are relatively infrequent.

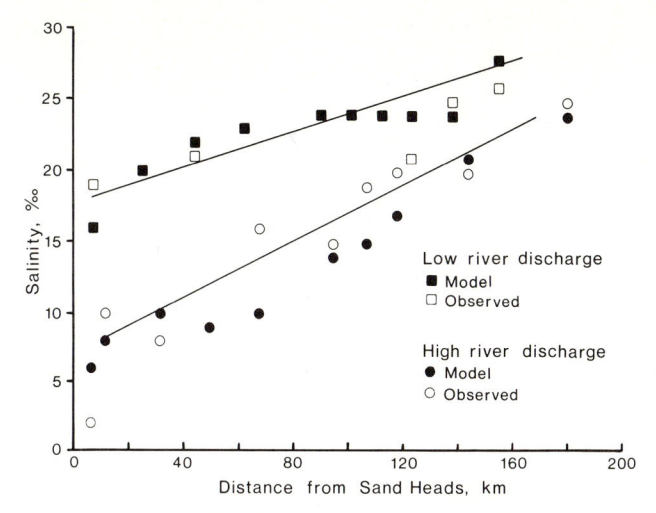

Fig. 11.20 Comparison of modelled and observed salinities.

To include effects from the wind in the model, the varying wind field must be approximated from a sparse distribution of coastal wind observations. This has been done for the comparisons for the relatively light winds that occurred during the observations. It is reasonable to suppose that an appreciable measure of uncertainty will exist in that part of the model where the tidal streams are weak north of the main freshwater source and where wind interactions will be relatively more important.

Examples of such comparisons have been drawn from drogue tracking experiments carried out from 22–23 June 1982. For clarity, the modelled vectors have been plotted only for alternate meshes. Using wind velocities obtained at shore stations over the period of the drogue tracking operations, an interpolated wind field was prescribed at each time step over a 5-day computation (19–23 June). The initial fields of velocity, salinity, and thickness were taken from the final values of an earlier 30-day simulation in which a similar river discharge rate (11,000 m^3/s) had been employed. The vector comparisons in Fig. 11.21a–f use a sequence of modelled velocity fields at hourly intervals (−2, −1, 0, 1, 2, 3) with respect to lower low water on 23 June. The observed velocities were obtained from drogue tracks at corresponding times with respect to lower low water from both 22 and 23 June.

On the days in question winds over the southern part of the modelled region were variable in direction and intensity though generally from the south and of order 2–3 m/s. Over the upper left hand part of the figure winds were generally from the west and north with speeds of 5–10 m/s. In the upper right hand half of the figure winds were weak (about 2 m/s) and variable both in speed and direction. In summary, little effect from winds other than modest and rather variable flows in a general southerly or southeasterly direction, in the upper left hand half of the diagram, would be expected.

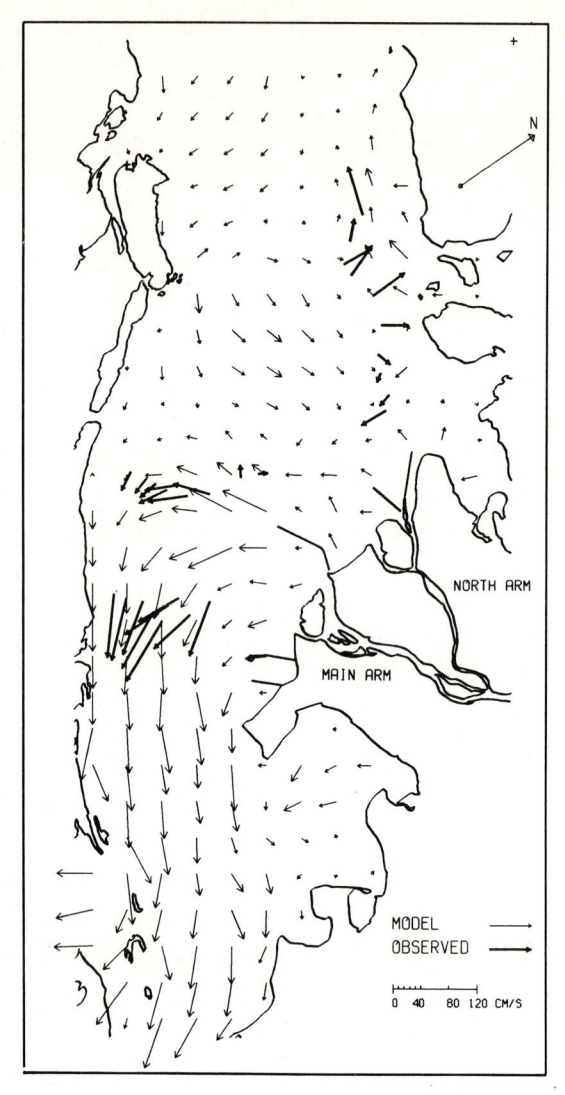

Fig. 11.21a Comparison of modelled horizontal velocities (thin vectors) and observed velocities (thick vectors) for LLW−2 inferred from surface drift measurements taken on 23 June 1982.

In general, comparisons of velocities inferred from the movements of the surface drifters carrying drogues of 2 m depth, with the averaged velocities over the layer thickness obtained from the model, indicate the model velocities to be substantially less than the observed. As discussed earlier, this model takes no account of surface slopes deriving from deep water renewal. Corrections for vertical current shear, wave-induced effects, and wind drag on radar reflectors could be applied but such refinements will be undertaken in connection with a more detailed program of numerical adjustment when the effects of deep water renewal will be included and better wind data will become

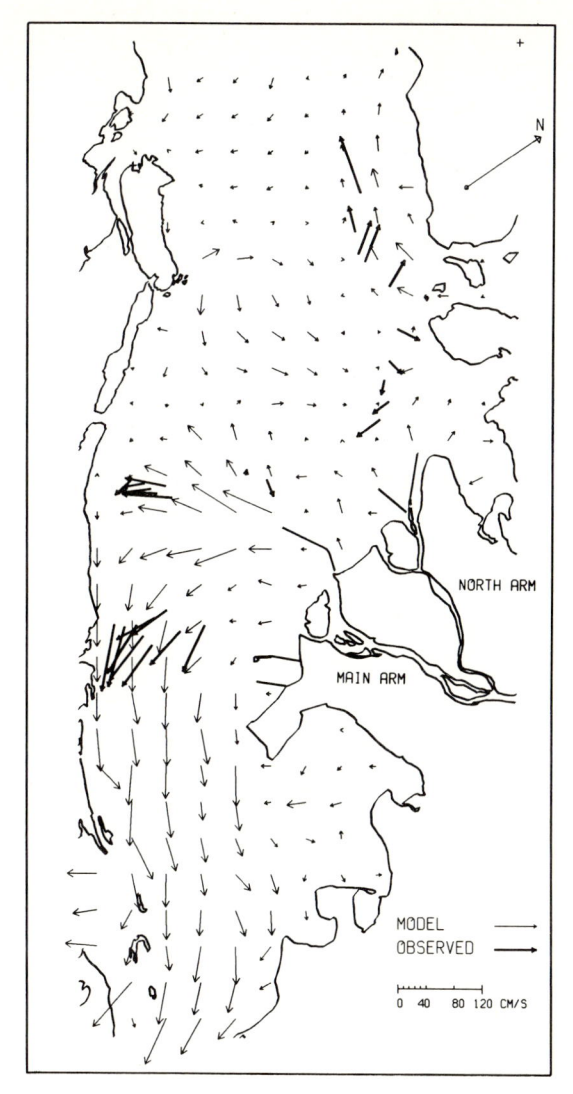

Fig. 11.21b Comparison of modelled horizontal velocities (thin vectors) and observed velocities (thick vectors) for LLW−1 inferred from surface drift measurements taken on 23 June 1982.

available. Of primary concern in the present context is variation of the surface flow pattern as the tide ebbs and floods in the Strait of Georgia. References to the river mouth imply the Main Arm in view of its dominant influence.

At LLW−2 h and LLW−1 h, (Fig. 11.21a and b, respectively) closer to the Vancouver Island side of the Strait of Georgia the southerly curvature of the modelled velocity field of Fraser River discharge agrees well with that displayed by the observed velocities. South of the river mouth the modelled and observed streams ebb strongly. In the region just north the currents are relatively weak and confused while farther

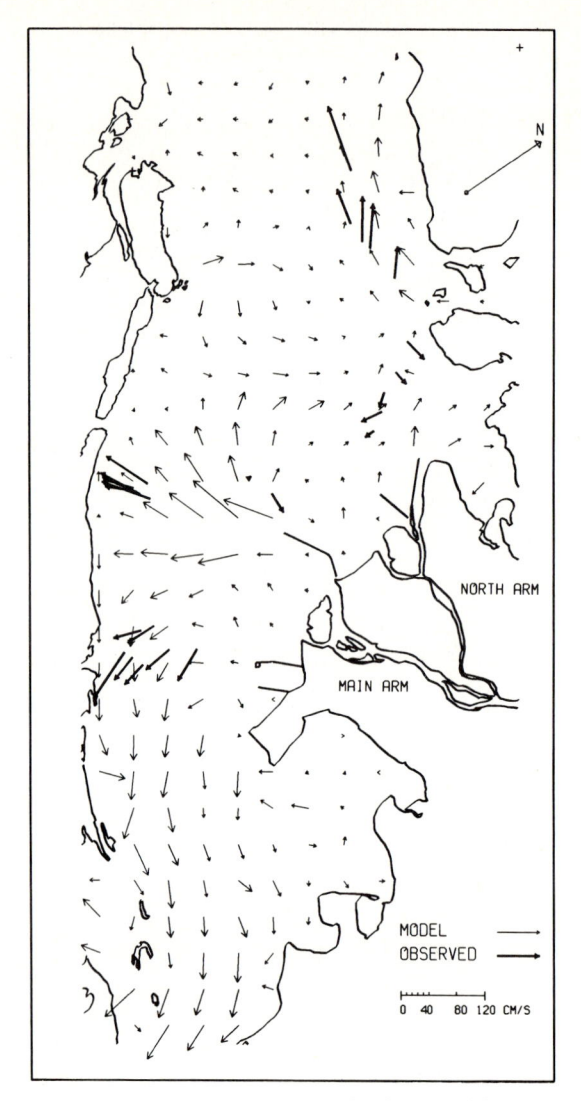

Fig. 11.21c Comparison of modelled horizontal velocities (thin vectors) and observed velocities (thick vectors) for LLW inferred from surface drift measurements taken on 23 June 1982.

north along the mainland coast there is evidence of a northerly flow of river discharge in both modelled and observed velocities.

At LLW (Fig. 11.21c) the river discharge velocities are maximal and maintain entry direction almost to the opposite shore in both model and observations. To the south the ebbing streams are weak though the observed velocities display a stronger component of velocity towards the Gulf Islands shore than those of the model.

At LLW+1 (Fig. 11.21d), the entrant stream velocities close the opposite shore have assumed a more northerly component in both model and observations while to the

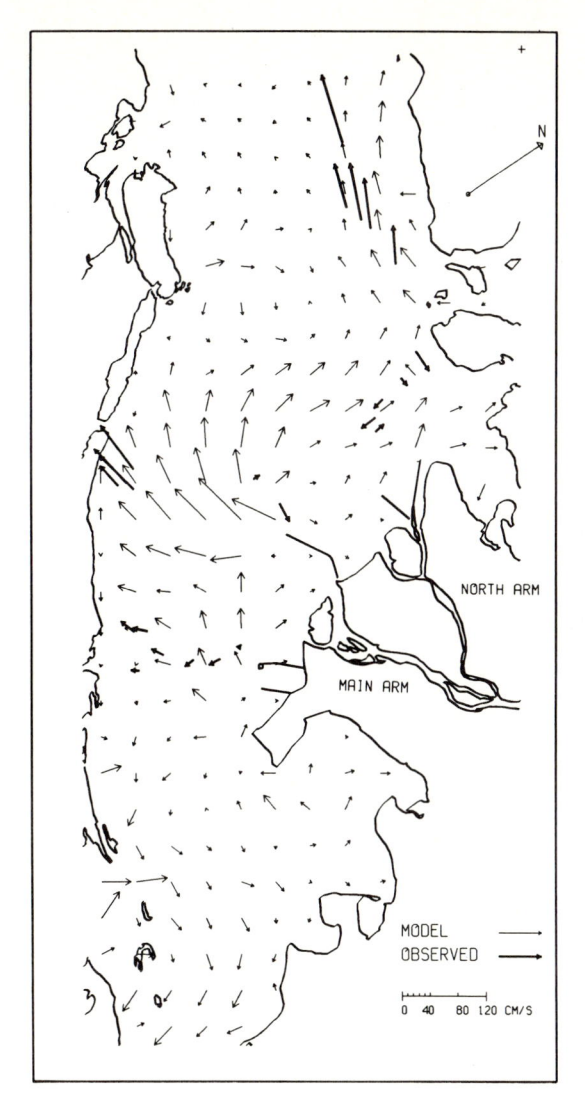

Fig. 11.21d Comparison of modelled horizontal velocities (thin vectors) and observed velocities (thick vectors) for LLW+1 inferred from surface drift measurements taken on 23 June 1982.

south the streams are turning. The observed velocities off Howe Sound are rather weak and confused but along the mainland coast there is a northerly flow, more strongly developed in the observations than in the model.

At LLW+2 h, (Fig. 11.21e) the flood has commenced south of the River and there is good accord between modelled and observed velocities. The entrant stream curves more strongly around towards the north and a weak return flow back towards the banks, north of the Steveston jetty training wall, at the mouth of the Main Arm, results. The three observed velocity vectors agree with the features in the model. Off

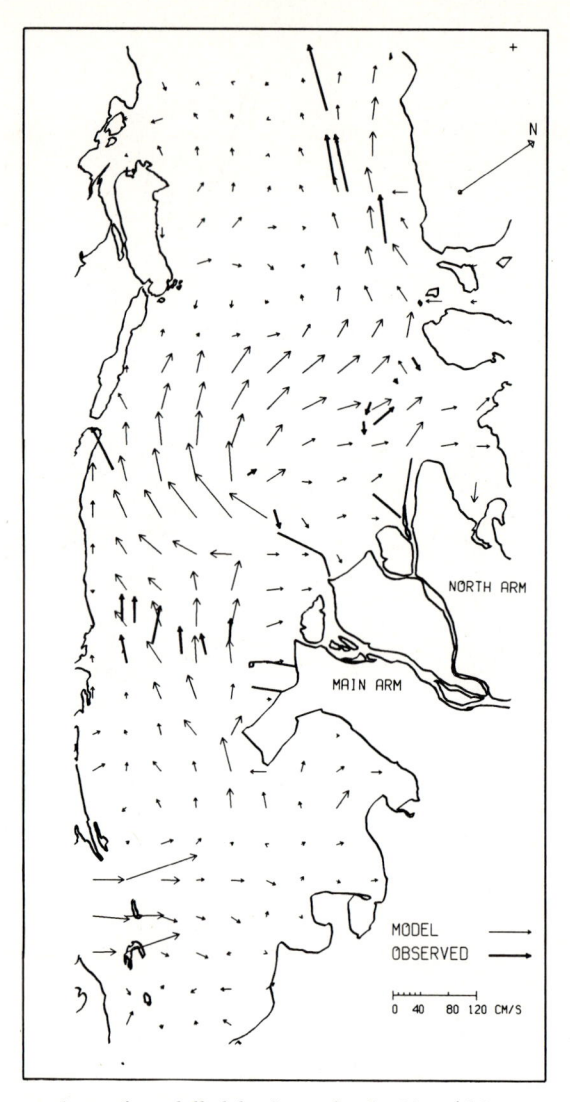

Fig. 11.21e Comparison of modelled horizontal velocities (thin vectors) and observed velocities (thick vectors) for LLW+2 inferred from surface drift measurements taken on 23 June 1982.

the mouth of Howe Sound the observed velocities suggest a weak eddying motion, not evident in the modelled field. The northwesterly flow along the mainland coast is well developed but faster in the observations than in the model.

At LLW+3 h (Fig.11.21f), the basic pattern described above is maintained though velocities generally increase as the flooding tide augments the Coriolis-redirected river flow. The particularly large observed velocity vectors (about 125 cm/s) directed away from the river mouth in the central part of the diagram clearly conform to surface flows following the passage of a front described in Chapter 1. The thin rapid surface flow

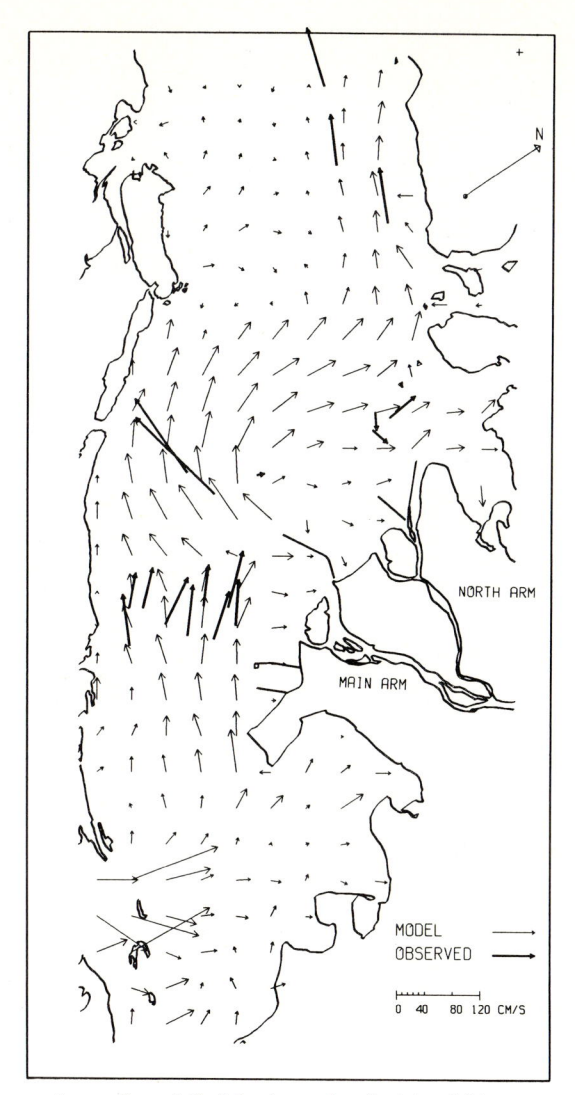

Fig. 11.21f Comparison of modelled horizontal velocities (thin vectors) and observed velocities (thick vectors) for LLW+3 inferred from surface drift measurements taken on 23 June 1982.

that so strongly affects the drogues is not reproducible in the model.

The distribution of thickness (Fig. 11.21g) generally obtained throughout the preceding vectors plots is consistent with significant winds from the northwest over the northern part of the Strait and which lead to a layer thickening along the Vancouver Island side of the Strait. Along the mainland coast the increased layer thickness is due to the rotationally-deflected high river discharge.

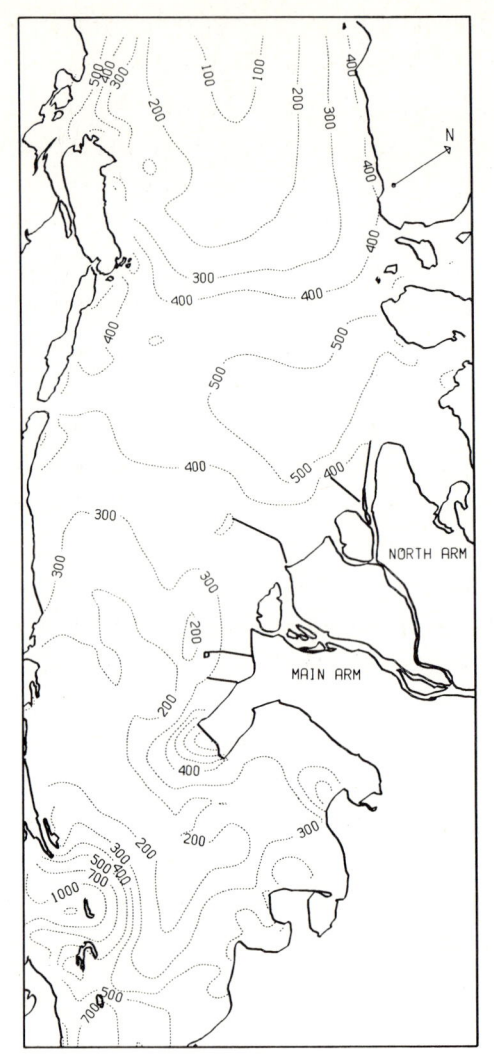

Fig. 11.21g Layer thickness (cm) at LLW+3 on 23 June 1982.

11.6 CONCLUDING DISCUSSION

A vertically-integrated numerical model of the brackish upper layer in the Strait of Georgia has been described. The model satisfactorily simulates the basic flow features under a variety of representative prevailing wind conditions that is associated with the tidally and seasonally modulated, Coriolis-deflected large buoyant jet entering the deep Strait with divergent tidal cross-flows.

Processes at the lower interface of this layer model vary with distance from the river mouth and with river discharge. Near the river mouth the dominant process is entrainment. Where the isohaline surface chosen for this interface is located in a strong vertical pycnocline and relatively weak vertical shear, the latter plays a significant role in the dynamics of the upper layer by inhibiting vertical transfers of momentum, volume, and salt. This is the case for a large part of the Strait. In regions of the model remote from the river mouth where the pycnocline is weak, advective and diffusive processes come into play. These processes are parameterized in terms of the major upper layer flow variables which permit the model to yield patterns of horizontal flow, salinity, and layer thickness in accord with the available data base. A further requirement at the present stage of the work is that the chosen isohaline interface shall not intersect the free surface. Through extensive numerical experimentation, it was possible to achieve such an adequate parameterization. It is of interest to note that the fields of velocity, salinity, and thickness are not highly sensitive to these parameterizations. This is consistent with the initial hypothesis that the observed surface motions evince relatively few major flow features determined by combinations of tide, river discharge, and wind.

Considerable economy results from the fact that the time step limitation required to maintain numerical stability is based on the much slower internal gravitational wave speed rather than on the fast free-surface wave speed appropriate to such a deep system. The elimination of the surface wave criterion is accomplished by prescribing barotropic tidal pressure gradients taken from a prior barotropic tidal model (GF7) rather than recalculating their gradients and associated streams. Typical results have been presented for conditions of high river discharge though equally satisfactory results can be obtained for much lower rates of discharge.

Although the phasing and directions of the computed flow features agree well with observations, the modelled speeds tend to be less than those observed. This is considered due, at least in part, to the absence of gradients of the free surface associated with deep water renewal.

12 THE LATERALLY-INTEGRATED MODEL: GF5

12.1 INTRODUCTION

The simplest conceptual model (Redfield, 1950a; Waldichuk, 1957) of circulation in the Georgia/Fuca system consists of inner and outer straits with a connecting region of strong tidal mixing. In this region surface water from the Strait of Georgia and bottom water from Juan de Fuca Strait are mixed with part of the mixed water being returned to form bottom water in the Strait of Georgia and the remainder moving seaward at the surface in Juan de Fuca Strait. This classical model of a generally persistent estuarine exchange of water with the adjacent ocean must be modified somewhat in light of more recent observations. Frequent major flow reversals in the outer strait are observed to occur under the influence of fall and winter southerly winds along the open coast (Holbrook *et al.*, 1983). Such wind events tend to move Columbia River-diluted surface waters to the mouth of Juan de Fuca Strait. This can result in an intrusion of surface water into the Strait. Such an intrusion is probably superimposed on an internal Kelvin wave (Proehl and Rattray, 1984) which has its origin in the response of the stratified shelf waters to the wind. The net consequence is a reversal of the estuarine circulation in Juan de Fuca Strait. Although the data are insufficient to prescribe these boundary influences, and rotation is omitted from the GF5 model, the observed salinity field used to illustrate the winter condition readily demonstrates such a reversal. The most marked change in the water of the outer strait, however, is associated with the seasonal shift in prevailing offshore winds from the winter southeasterlies to the northwesterlies in summer. It is thus of prime interest to obtain some estimate of the seasonal variability in residual circulation and sea surface elevations associated with the characteristic longitudinal vertical distributions of salinity under winter and summer conditions (see Fig. 1.5a and b).

Since the total length-to-width ratio of the inner and outer straits is of the order 15:1, GF5 took the form of a single-channel, laterally-integrated model having variable widths and depths. The northern end of the Strait of Georgia was assumed closed and M_2 and K_1 tidal elevations were prescribed at the open entrance. Initial salinity distributions were assigned corresponding to the winter and summer distributions, respectively, referred to above. These were employed in both diagnostic and prognostic contexts, the latter instance providing some insight into the numerical problems likely to be encountered in salinity computations when applied to such a complex non-linear system.

12.2 GOVERNING EQUATIONS

The basic laterally-integrated equations governing the conservation of volume, mass (salt), and momentum in a channel of varying width and depth (Blumberg, 1975; Elliot, 1976) are

$$\frac{\partial}{\partial x}(Bu) + \frac{\partial}{\partial z}(Bw) = 0, \tag{12.1}$$

$$\frac{\partial}{\partial t}(Bs) + \frac{\partial}{\partial x}(Bus) + \frac{\partial}{\partial z}(Bws) - \frac{\partial}{\partial x}\left(BK_x \frac{\partial s}{\partial x}\right) - \frac{\partial}{\partial z}\left(BK_z \frac{\partial s}{\partial z}\right) = 0, \tag{12.2}$$

$$\frac{\partial}{\partial t}(Bu) + \frac{\partial}{\partial x}(Buu) + \frac{\partial}{\partial z}(Buw) - \frac{\partial}{\partial x}\left(BN_x \frac{\partial u}{\partial x}\right) - \frac{\partial}{\partial z}\left(BN_z \frac{\partial u}{\partial z}\right) = -\frac{B}{\rho}\frac{\partial p}{\partial x} + BX, \tag{12.3}$$

$$\frac{\partial p}{\partial z} = \rho g \tag{12.4}$$

where

x, z = co-ordinates in the undisturbed surface along the channel axis and vertically down, respectively,

t = time,

$u(x, z, t), w(x, z, t)$ = corresponding width-averaged components of velocity,

$B(x, z)$ = channel width,

$s(x, z, t)$ = salinity,

$p(x, z, t)$ = pressure,

$\rho(x, z, t)$ = density of water,

K_x, K_z = respective coefficients of horizontal and vertical diffusion,

N_x, N_z = coefficients of horizontal and vertical eddy viscosity, respectively,

X = body force, e.g., tide-generating force = $X(x, t)$,

g = gravitational acceleration.

Vertically integrating the continuity equation (12.1),

$$\int_\zeta^h \frac{\partial}{\partial x}(Bu)\, dz + \int_\zeta^h \frac{\partial}{\partial z}(Bw)\, dz = 0,$$

and using the surface and bottom conditions while neglecting cross-channel variations,

$$\frac{D}{Dt}(z + \zeta(x,t)) = 0 \quad \Rightarrow \quad w(-\zeta) + \frac{\partial \zeta}{\partial t} + u(-\zeta)\frac{\partial \zeta}{\partial x} = 0,$$

$$\frac{D}{Dt}(z - h(x)) = 0 \quad \Rightarrow \quad w(h) - u(h)\frac{\partial h}{\partial x} = 0$$

where $h(x)$ is the depth below the undisturbed water surface. Thus, $\zeta(x, t)$ is the elevation of the free surface,

$$\frac{\partial}{\partial x} \int_{-\zeta}^{h} Bu\, dz + B(-\zeta)\frac{\partial \zeta}{\partial t} = 0. \tag{12.5}$$

Vertically integrating the pressure term (Hamilton, 1975) down to some depth z,

$$\frac{\partial p}{\partial x} = g\bar{\rho}\frac{\partial \zeta}{\partial x} + g(z+\zeta)\frac{\partial \bar{\rho}}{\partial x} \tag{12.6}$$

where

$$\bar{\rho} = \frac{1}{(z+\zeta)} \int_{\zeta}^{z} \rho\, dz'.$$

12.3 MODEL DESCRIPTION

The schematization for this simple model is an adaptation of that used for Juan de Fuca, Haro. and Rosario Straits and the Strait of Georgia in the one-dimensional tidal model GF1. A single channel having mean widths varying with depth in each of the sections is thus obtained. The layout of these sections is shown in Fig. 12.1. In view of the relatively small net fluxes through the northern openings from the Strait of Georgia, as discussed in Chapters 1 and 4, this end of the Strait is assumed closed. The mouth of the Puget Sound system is also assumed closed. Haro (sections 11 and 12) and Rosario (sections 19 and 20) Straits are combined into a single channel. The resulting arrangement of channel widths and depths depicted as a cutaway section along a median line (Fig. 12.2) shows a deep outer strait (Juan de Fuca) shoaling towards a broad deep sill joined by a constriction (Haro and Rosario Straits) containing a deep narrow trench (Haro Strait) connected to a deep inner basin (Georgia Strait). This scheme is considered minimally adequate for the estimation of residual circulation derived from observed longitudinal density gradients. The computational grid (Fig. 12.3) employs a section length $(2\Delta x)$ of 22.2 km. vertical spacing $(2\Delta z)$ of 20 m, and a time step of 150 s. The vertical dashed line denotes the location of the narrow, deep passage (Boundary Pass) 0.5 km wide and 200 m deep at the eastern end of Haro Strait.

12.4 FINITE-DIFFERENCE EQUATIONS

The notation employed essentially follows that of Elliot (1976). The scheme is of the leap-frog type involving three time levels $t = (n-1)\Delta t$, $n\Delta t$, and $(n+1)\Delta t$, respectively, where n is the number of time steps. Since channel widths B (Fig. 12.3) are defined at the w points in each channel section, it is necessary to define a mean width appropriate to a particular horizontal velocity point. Referring to the numerical grid scheme (Fig. 12.3), the width $B'_{i,j}$ at $u_{i,j}$ is defined by

$$B'_{i,j} = \tfrac{1}{4}(B_{i,j-1} + B_{i+1,j-1} + B_{i,j} + B_{i+1,j})$$

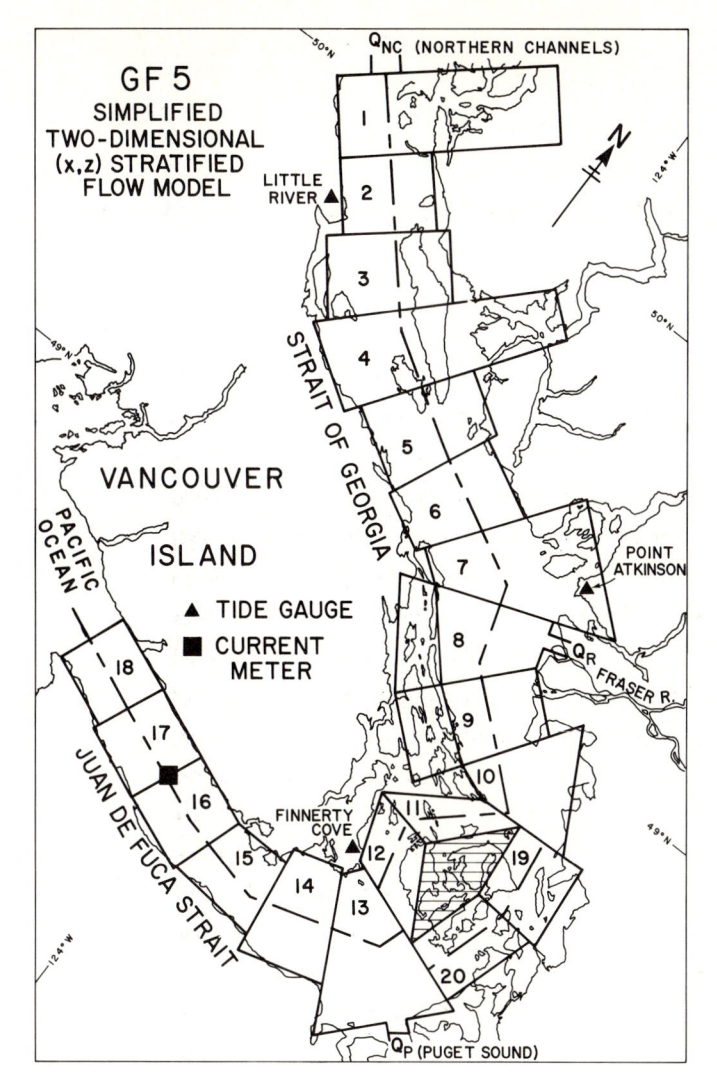

Fig. 12.1 System of channel sections used in the simplified two-dimensional (x, z) stratified flow model GF5. In the version described, sections 19 and 20 (Rosario Strait) were combined with sections 11 and 12 (Haro Strait), respectively, and flows Q_p entering the Puget Sound system were set equal to zero.

where $j = 1$ must be evaluated separately.

The finite-difference form of Eq. (12.1) is

$$[w_{i,j}]_{n+1} = \left[\frac{(\Delta z/\Delta x)\left(uB'_{i,j+1} - uB'_{i,j}\right) + wB_{i+1,j}}{B_{i,j}} \right]_{n+1} . \quad (12.7)$$

Using the boundary condition at the sea bottom,

$$w = 0 \quad \text{at} \quad z = h,$$

the above equation is solved from the bottom upward to yield values at the various levels. Where the product of variables is followed by an indexed bracket, this index is

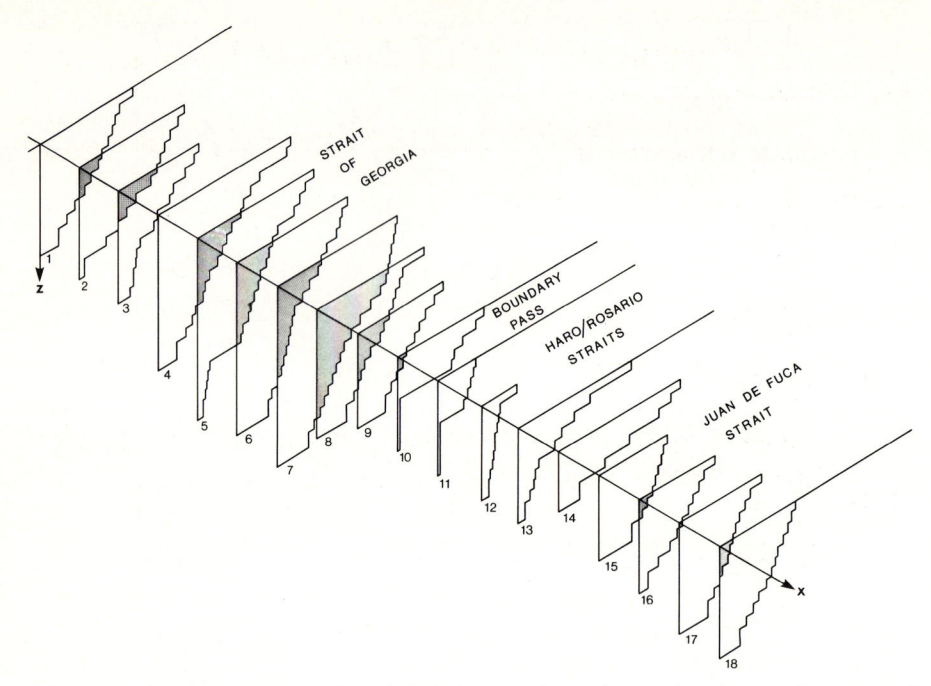

Fig. 12.2 Cutaway section of modelled depths along the median line and associated channel half widths.

taken to apply to both the preceding variables. The finite-difference form of the salt conservation equation is formulated as follows:

$$\text{SB1} = \frac{\partial}{\partial t}(Bs)$$

$$= \left[(B_{i,j} + B_{i+1,j})([s_{i,j}]_{n+1} - [s_{i,j}]_{n-1})\right]\frac{1}{4\Delta t},$$

$$\text{SB2} = \left[B'_{i,j+1}[(u_{i,j+1} + |u_{i,j+1}|)s_{i,j} + (u_{i,j+1} - |u_{i,j+1}|)s_{i,j+1}]\right.$$

$$\left. - B'_{i,j}[(u_{i,j} + |u_{i,j}|)s_{i,j-1} + (u_{i,j} - |_{i,j}|)s_{i,j}]\right]_n \frac{1}{4\Delta x},$$

$$\text{SB3} = \left[B_{i+1,j}[(w_{i+1,j} + |w_{i+1,j}|)s_{i,j} + (w_{i+1,j} - |w_{i+1,j}|)s_{i+1,j}]\right.$$

$$\left. - B_{i,j}[(w_{i,j} + |w_{i,j}|)s_{i-1,j} + (w_{i,j} - |w_{i,j}|)s_{i,j}]\right]_n \frac{1}{4\Delta z},$$

$$\text{SB4} = -\frac{\partial}{\partial x}\left(BK_x \frac{\partial s}{\partial x}\right)$$

$$= -\left[[B'_{i,j+1}K_{x_{i,j+1}}(s_{i,j+1} - s_{i,j}) - B'_{i,j}K_{x_{i,j}}(s_{i,j} - s_{i,j-1})]_{n-1}\right]\frac{1}{4\Delta x^2},$$

$$\text{SB5} = -\frac{\partial}{\partial z}\left(BK_z \frac{\partial s}{\partial z}\right)$$

$$= -\left[[B_{i+1,j}K_{z_{i+1,j}}(s_{i+1,j} - s_{i,j}) - B_{i,j}K_{z_{i,j}}(s_{i,j} - s_{i-1,j})]_{n-1}\right]\frac{1}{4\Delta z^2}.$$

Hence, salinity values are updated using

$$[s_{i,j}]_{n+1} = [s_{i,j}]_{n-1} + \frac{4\Delta t(-[\text{SB2}]_n - [\text{SB3}]_n - [\text{SB4}]_{n-1} - [\text{SB5}]_{n-1})}{B_{i,j} + B_{i+1,j}}. \tag{12.8}$$

The finite-difference form of the momentum equation (12.3) is

$$\text{MB1} = \frac{\partial}{\partial t}(Bu)$$

$$= \frac{B'_{i,j}}{2\Delta t}([u_{i,j}]_{n+1} - [u_{i,j}]_{n-1}),$$

$$\text{MB2} = \frac{\partial}{\partial x}(Buu)$$

$$= \left[[(B_{i+1,j} + B_{i,j})(u_{i,j} + u_{i,j+1})^2 - (B_{i+1,j-1} + B_{i,j-1})(u_{i,j-1} + u_{i,j})^2]_n\right]\frac{1}{16\Delta x},$$

$$\text{MB3} = \frac{\partial}{\partial z}(Buw)$$

$$= \left[[(B_{i+1,j-1}w_{i+1,j-1} + B_{i+1,j}w_{i+1,j})(u_{i+1,j} + u_{i,j})\right.$$

$$\left. - (B_{i,j-1}w_{i,j-1} + B_{i,j}w_{i,j})(u_{i,j} + u_{i-1,j})]_n\right]\frac{1}{8\Delta z},$$

$$\text{MB4} = -\frac{\partial}{\partial x}\left(BN_x\frac{\partial u}{\partial x}\right)$$

$$= \left[[N_{x_{i,j}}(B_{i+1,j} + B_{i,j})(u_{i,j+1} - u_{i,j})\right.$$

$$\left. - N_{x_{i,j-1}}(B_{i+1,j-1} + B_{i,j-1})(u_{i,j} - u_{i,j-1})]_{n-1}\right]\frac{1}{8\Delta x^2},$$

$$\text{MB5} = -\frac{\partial}{\partial z}\left(BN_z\frac{\partial u}{\partial z}\right)$$

$$= -\left[[(B_{i+1,j-1} + B_{i+1,j})N_{z_{i,j-1}}(u_{i+1,j} - u_{i,j})\right.$$

$$\left. - (B_{i,j-1} + B_{i,j})N_{z_{i-1,j-1}}(u_{i,j} - u_{i-1,j})]_{n-1}\right]\frac{1}{8\Delta z^2},$$

$$\text{MB6} = ku|u|\left(1 + \left(\frac{\partial B}{\partial z}\right)^2\right)^{1/2}$$

$$= k_j[u_{i,j}|u_{i,j}|]_{n-1}\left[1 + \left(\frac{B_{i,j-1} + B_{i,j} - B_{i+1,j-1} - B_{i+1,j}}{4\Delta z}\right)^2\right]^{1/2},$$

$$\text{MB7} = gB\frac{\bar{\rho}}{\rho}\frac{\partial \zeta}{\partial x}$$

$$= gB'_{i,j}\left[\left(\frac{\bar{\rho}_{i,j-1} + \bar{\rho}_{i,j}}{\rho_{i,j-1} + \rho_{i,j}}\right)(\zeta_j - \zeta_{j-1})\right]_n\frac{1}{2\Delta x},$$

$$\text{MB8} = gB(z + \zeta)\frac{1}{\rho}\frac{\partial \rho}{\partial x}$$

$$= gB'_{i,j}\left[\left(2i\Delta z + \frac{(\zeta_{j-1} + \zeta_j)}{2}\right)\left(\frac{\bar{\rho}_{i,j} - \bar{\rho}_{i,j-1}}{\Delta x(\rho_{i,j} + \rho_{i,j-1})}\right)\right]_n.$$

Hence,

$$[u_{i,j}]_{n+1} = [u_{i,j}]_{n-1} + \frac{2\Delta t}{B'_{i,j}}(-[\text{MB2}]_n - [\text{MB3}]_n$$

$$- [\text{MB4}]_{n-1} - [\text{MB5}]_{n-1} - [\text{MB6}]_{n-1} - [\text{MB7}]_n - [\text{MB8}]_n). \tag{12.9}$$

In view of the broad steep-sided character of the conveying channels the lateral stress terms MB6 are considered negligible (except in the San Juan Island passages) with respect to the stresses acting over the upper and lower horizontal faces of the schematized channel sections.

Surface elevations are computed using the following finite-difference equation,

$$\zeta_{j,n+1} = \zeta_{j,n-1} - \frac{\Delta t}{\Delta x}\left[\left(2\Delta z + \frac{\zeta_j + \zeta_{j+1}}{2}\right)B'u_{1,j+1} + 2\Delta z \sum_{\text{bottom}}^{i=2} B'_{i,j+1}u_{i,j+1}\right.$$
$$\left. - \left(2\Delta z + \frac{\zeta_j + \zeta_{j-1}}{2}\right)B'_{i,j}u_{1,j} - 2\Delta z \sum_{\text{bottom}}^{i=2} B'_{i,j}u_{i,j}\right]_n \frac{1}{B_{1,j}}. \qquad (12.10)$$

12.5 BOUNDARY CONDITIONS

At the sea surface the advective and diffusive fluxes of salt and momentum are set equal to zero. It is assumed that the stress at the sea bottom entering into the term MB5 in Eq. (12.9) is given by

$$N_{z_{i,j-1}}(u_{i+1,j} - u_{i,j}) = ku_{i,j}|u_{i,j}|. \qquad (12.11)$$

A problem arises concerning the salinity and sea level to be prescribed at the open boundary of the model. Thus, the proportion of water dispelled on an ebb and then returned on the following flood is unkown. This proportioning will affect the static head associated with the water column on the open boundary. In these preliminary trials it is assumed that an observed ocean salinity is advected into the system on an inflow while outflow is of a salinity known from computations within the model itself. Tidal elevations M_2 and K_1 are prescribed about a plane of zero mean sea level at the open boundary. A steady state mean monthly river flow, corresponding to the month of the salinity observations used as an initial condition, is prescribed at grid column 8 which corresponds to the mouth of the Fraser River (Fig. 12.3).

12.6 REVIEW OF LONGITUDINAL CIRCULATION

Prior to discussing the results obtained from this model, it is desirable to review elements of the basic longitudinal circulation that might be anticipated. The basic circulation is governed by two interacting pressure distributions, one due to surface slopes, seaward-directed, and depth-invariant; the other landward-directed and increasing with depth, dependent on the density distribution. The latter can be strongly modified by mixing processes within the system.

For the simple case of a two-layered closed rectangular basin in which strong vertical mixing is introduced at one end, the resulting circulation will, in general, form a three-layered system. The static head associated with the lighter fluid will move surface

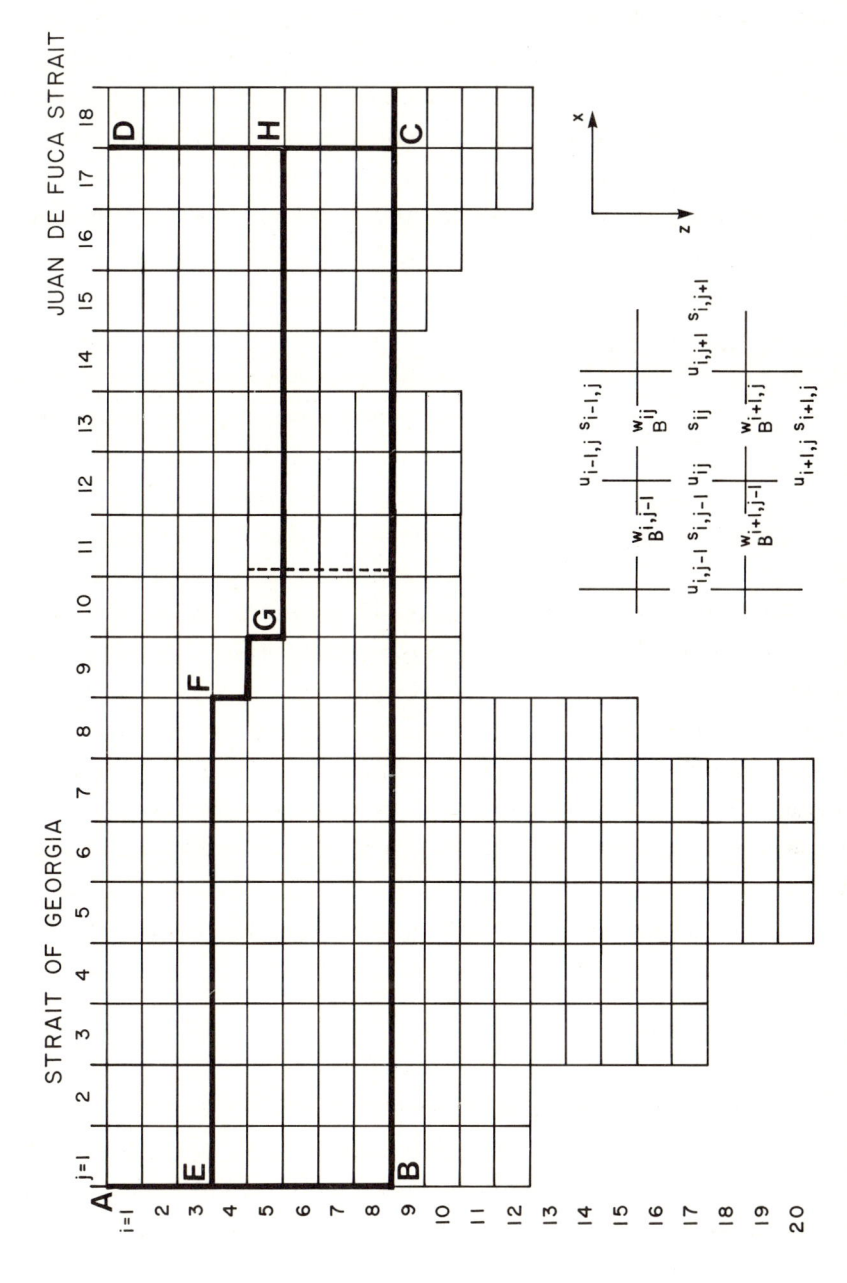

Fig. 12.3 Side view of the numerical scheme. The inset shows the arrangement of subscripted variables. The significance of the letters is explained in the text.

water towards the mixing site. A deeper movement away from the mixing site will occur at some depth below which the baroclinic pressure gradient exceeds the barotropic gradient. In order to supply this mixed water, there must be a compensating, deep flow of dense water towards the mixing site along the floor of the basin.

For the case of a rectangular gulf with a freshwater source at one end and open to the sea at the other, there will be a seaward movement of brackish surface water entraining salt water from an underlying compensating inflow of saline water. If strong local mixing occurs at some location between the freshwater source and the open sea, the estuarine exchange with the open ocean will be impeded. The observed longitudinal distributions of salinity (see Fig. 1.5a and b) are presumably a superposition and a seasonal modification of these dynamic processes. This seasonal modification is due in part to a large peak in the annual discharge rate from the Fraser River in late May and early June, and to the changing water masses and circulation at the entrance to Juan de Fuca Strait. In the latter case, winter surface water associated with a wind-driven coastal jet, incorporating Coriolis-deflected Columbia River water, can have a surface salinity less than that in Juan de Fuca Strait. This coastal water is not subject to the strong tidal mixing processes in the inner part of Juan de Fuca Strait. If the static head associated with this coastal water is of sufficient magnitude, a surface slope downward into the Strait will oppose the normal estuarine circulation and bring about an inflow at the surface. As noted by Tabata *et al.* (1986), halosteric effects dominate the annual cycle in total steric height on the open coast. This cycle correlates strongly with the annual cycle in monthly mean sea levels which, after correction for the inverse barometer effect, are primarily associated with annual variations in the coastal currents. Thus, opposition to the estuarine exchange near the mouth of Juan de Fuca can derive both from Columbia River-induced static sea level increases at the entrance and sea level increases due to onshore Ekman transport. The latter should provide a barotropic pressure gradient tending to raise overall levels in the Georgia/Fuca system.

12.7 MODEL RESULTS

12.7.1 Tidal Residual Circulation for a Homogeneous Density Field

Since the primary object of this elementary model concerns delineation of basic processes and estimates of the longitudinal circulation associated with observed winter and summer salinity fields, little attempt has been made to optimize the agreement between the observed and computed M_2 and K_1 harmonic constants other than to employ large coefficients of bottom and side friction (0.1) in the turbulent region between the inner and outer straits.

Harmonic constants were determined by analysis (Foreman, 1977, 1978) of the last 25 h of computed elevations and velocities from a 10-day simulation. Comparisons of the observed and computed tidal amplitudes and phases suggest that frictional dissipation is too small and modelled velocities are somewhat overestimated. This apparent model shortcoming is offset by the fact that only two tidal constituents are being prescribed rather than the larger number used to simulate actual tides and streams. Thus, the energy in the semi-diurnal tidal band is simulated reasonably well.

Since the feature of primary interest is the residual field of baroclinic circulation, it is desirable to determine initially the field of residual circulation that results when the tidal motion is computed for the case of a homogeneous field of density. River input is set equal to zero. The small residual surface elevations and velocities obtained from the harmonic analysis are shown in Fig. 12.4. (Note the different horizontal and vertical velocity scales.) There is a small downward slope of the water surface from the Strait of Georgia to the ocean entrance and a weak net seaward flow. This is consistent with, and offsets, the landward drift deriving from the phase relation between tidal surface elevations and velocities. Considering only the tidal velocities, slightly more water enters the system on a flood tide than is carried out on an ebb. Thus, the residual current is required to satisfy conservation of volume in the system. For the frictionless case, stream maxima occur as the surface elevations pass through zero midway between high and low water and accordingly, the outflow balances the inflow. With increasing friction, the stream maxima occur later, the flood tide being accompanied by somewhat higher elevations than those accompanying the ebb tide.

Relatively large vertical motions occur near the narrow deep passage simulating Boundary Pass denoted by the vertical dashed line. This location undoubtedly plays a key role in terms of salt water mixing and exchange. Representative vertical profiles of horizontal tidal velocities, shown in this case for the vertical section at the inner end of the sill (column 14), are presented in Fig. 12.5. These correspond to a flood maximum (negative values) between higher high water and lower low water, and the following ebb maximum between lower low water and lower high water. The profiles indicate a plausible distribution of the effects of bottom stress over the water column through the vertical eddy viscosity term in the momentum equation.

12.7.2 Baroclinic Residual Circulation

In these trials, initial salinity fields representative of summer and winter conditions, respectively, were prescribed in the model. Appropriate discharges from the Fraser River ($8,000$ m^3/s in summer and $1,500$ m^3/s in winter) for the month in question were also entered but were found to have negligible effect over the time scale (10 days) of these runs. In the observed data, the overall longitudinal salinity fields appear to change fairly slowly (Crean and Ages, 1971). This is certainly true of the data obtained in early July 1968 when compared to those obtained at the end of the same month even though conditions were maximally conducive to the intrusion of bottom water at depth

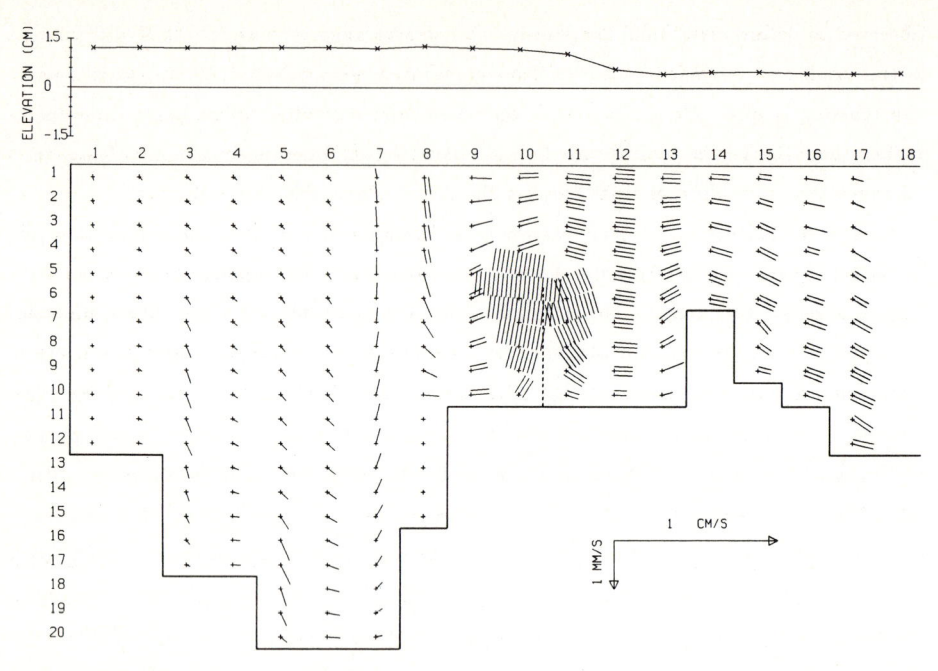

Fig. 12.4 Distribution of residual velocity vectors over the median vertical section of the channel when the summed M_2 and K_1 tides are prescribed at the ocean entrance. Note difference in scales. Shown at the top of the diagram is the residual surface elevation.

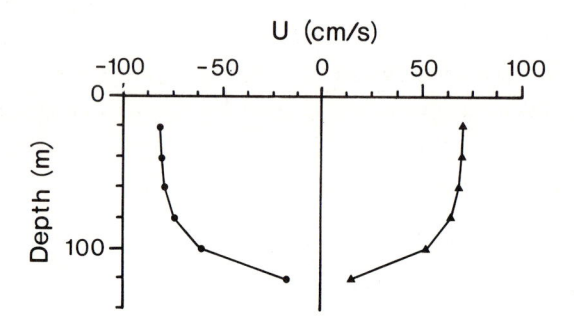

Fig. 12.5 Representative vertical profiles of ebb and flood horizontal tidal velocities for section 14 located at the inner end of the sill in Juan de Fuca Strait.

depth from Juan de Fuca Strait into the Strait of Georgia. It is thus instructive to use GF5 to determine the distribution of static head and its associated residual circulation along the Strait using a fixed salinity field (a diagnostic approach). For completeness, the M_2 and K_1 barotropic tides were included in the calculations although the tidal residual circulation, as shown on p. 311, is small. These calculations could then be repeated but with computation of the changing salinity field (a prognostic approach). Some trials were undertaken to determine a satisfactory numerical formulation of the salt conservation equation. The use of centred-difference type advective terms required large values of horizontal eddy diffusivity to maintain stability. Even so, unrealistic gradual net gains or losses of salt from the model occurred, depending on the value of the parameters employed. However, use of upwind differencing in the advective terms insured that a satisfactory salt balance to within 5% could be maintained over runs of at least 30 days. In spite of the upwind differencing, large values of horizontal eddy diffusivity (10^8 cm^2/s) were required to maintain stable salt calculations. A subsequent application of this scheme to a British Columbia inlet indicated that these problems can be overcome by occasional recurrent time smoothing of the computed fields, use of an Euler-backward scheme to restart, and use of double precision computation (Dunbar, 1985).

12.7.3 Diagnostic Trials

In these trials, the tidal computation described above was first applied to a fixed salinity field based on the winter observations (December 1967) and then to a salinity field for summer observations (July 1968). These dates correspond to the longitudinal vertical salinity sections shown on pages 10 and 11 although some averaging has been employed to take into account secondary cross-channel salinity variations. A 10-day simulation was carried out. The residual surface elevations and velocities were then determined by tidal harmonic analyses of the last 25 h of computed values. Since vertical advection is a topic of primary interest, the velocity vectors which employ averaged u and w components at the centre of a mesh include a large vertical exaggeration of the w component. The general order of magnitude of the horizontal residual circulation is illustrated by selected vertical velocity profiles through the mesh side of particular water columns.

12.7.3.1 Winter circulation

The prescribed initial salinity field is shown in Fig. 12.6 (Crean and Ages, 1971). In general, there is extensive moderate vertical stratification except where near uniformity exists in the deep basin of the Strait of Georgia and in the strong mixing region between the inner and outer straits. At the time of the observations the near surface waters at the entrance to Juan de Fuca Strait were less saline than those in its strongly mixed eastern end, a common condition in the winter months associated with the northerly movement of water from the Columbia River along the coast. In the 3 days prior to the taking of these observations, a deep low pressure system gave rise to strong southerly winds over the exposed coastal waters.

Considering initially the residual circulation that is set up within the model when the salinity field is held constant, the associated field of density determines a field of pressure within the interior of the modelled fluid. Through the equations of motion and

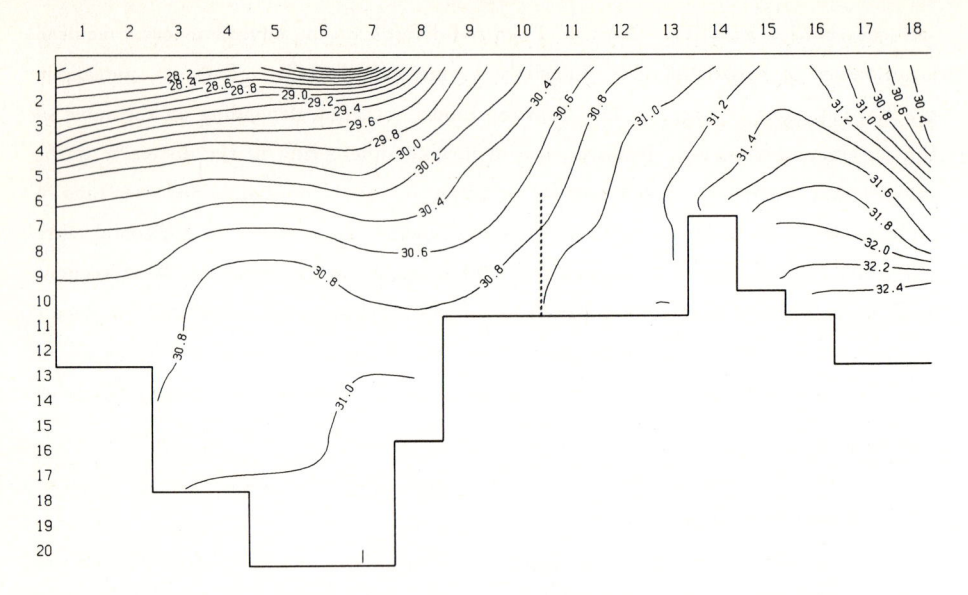

Fig. 12.6 Contoured distribution of the prescribed initial salinity field ($^0/_{00}$) over a vertical section through the channel median of the model using observations obtained in December 1967.

continuity, an adjustment of fluid volumes occurs such that a compensatory distribution of surface elevations is established. It was shown by earlier trials that in virtue of the geographical extent of the system and the relatively short time scale of these model runs, the effect of river discharge has little direct influence on the estuarine circulation. It was also shown that the tidal residual circulation is very small. Thus, the primary factors in determining the residual circulation at a point within the model are the balance between the baroclinic pressure gradients deriving from the fixed density field which tend to increase with depth and the barotropic depth-invariant pressure gradients due to the residual surface slopes. A net circulation results in which the frictional stresses opposing the motions and any contribution from the advective accelerations, balance the net pressure gradient at each point within the fluid. The primary source of frictional dissipation in the overall model is in the passages connecting the inner and outer straits where tidal velocities are high. Since the residual and tidal motions are coupled through a square law formulation of bottom stress, it may be supposed that the effect of the tides will be to retard the overall strength of residual circulation. The distribution of static surface elevations and residual velocity vectors that result is shown in Fig. 12.7. Three basic cells of vertical circulation are evident. In each of the cells the barotropic pressure gradients due to the surface slopes dominate the near-surface motions while the baroclinic gradients dominate at depth.

At the entrance to Juan de Fuca Strait there is a net inflow at the surface and outflow at depth even though any elevation of the water surface due to the presence of Columbia River water is not included in the prescribed boundary elevations. The density field contains sufficient information for the model to produce a net downward sea-surface slope into Juan de Fuca Strait. Thus, there is a reversal of the normal estuarine surface outflow and deeper inflow.

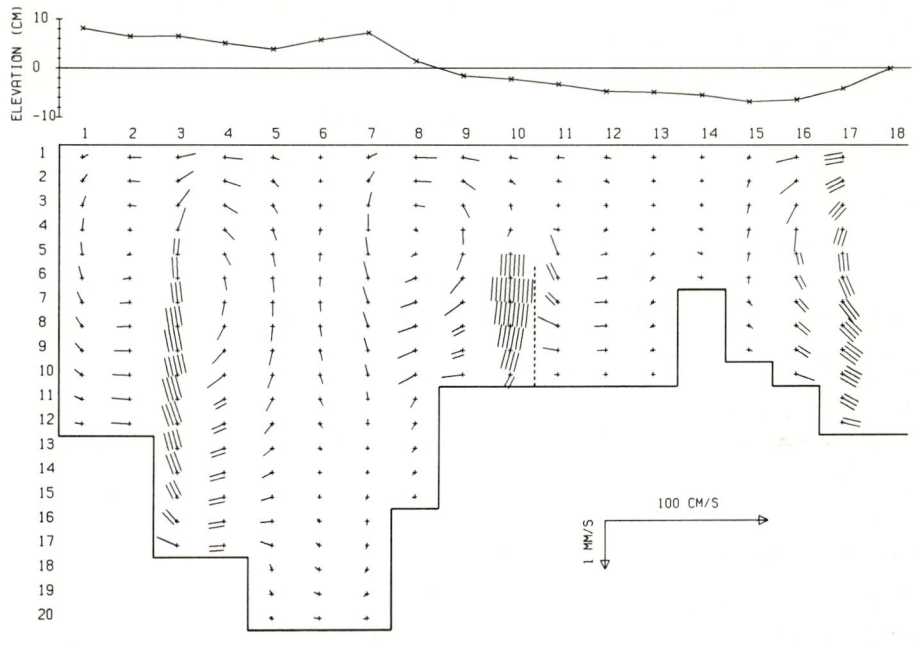

Fig. 12.7 Residual surface elevation (top) and distribution of residual velocity vectors (December 1967) over a vertical section through the channel median.

In the Strait of Georgia there are two basic regions of vertical recirculation which clearly show a strong topographic influence on the vertical velocities. In the southern region there is a marked downward slope of the free surface in the seaward direction. At depth there are inward-directed horizontal pressure gradients deriving from the salinity field (Fig. 12.6). The summed barotropic (surface slope) and baroclinic gradients give rise to this vertical recirculation. The large downward velocity components do not imply large volume transports since the channel widths where they occur are small. The upward vertical velocities at middepth in this recirculation, which are relevant to considerations of deep water replacement, are of order 15-25 m/day.

In the northern region there is a second vertical recirculation having the same sense of rotation as that to the south. Again, there is a surface slope downward towards the southern end of the Strait of Georgia. This is presumably due to the earlier movement of lower salinity surface waters into the northern Strait by the southeast winds which occurred prior to the time of the observations. The southerly surface flow is accompanied by a deep return flow deriving from the horizontal pressure gradients associated with the salinity field. The strong vertical upwelling in column 3 (Fig. 12.7) is presumably a consequence of a marked narrowing of the channel (Fig. 12.2). The vertical velocities in the lower part of the water column are of order 100–200 m/day. It is of interest to speculate as to whether or not an extensive patch of cool high salinity surface water observed in this general vicinity along the Vancouver Island shore in June 1950 (Waldichuk, 1957) might have resulted from such upwelling.

Of particular interest, in a qualitative context, are the distributions of horizontal velocity (Fig. 12.8a and b) over the vertical cross-sections 7 and 8, respectively, (flanking the elevation maximum near the Fraser River mouth, Fig. 12.7). In the first case (Fig. 12.8a) the pressure gradients resulting from the summed barotropic and baroclinic contributions give rise to northerly flows near the surface and bottom and southerly flows at middepth. In the second (Fig. 12.8b), the southerly flow extends from the surface to middepth. The upward vertical velocities (of order 25 m/day) at middepth in Fig. 12.7 are a consequence of this divergence near the surface and convergence at middepth. No account has been taken of the northerly Coriolis deflection of the river flow, a major factor in the overall circulation shown by observations and model simulations. The result is of direct relevance in considering the dynamics governing the net southerly flow shown by observations to occur underneath much of the Fraser River plume (see Fig. 1.27).

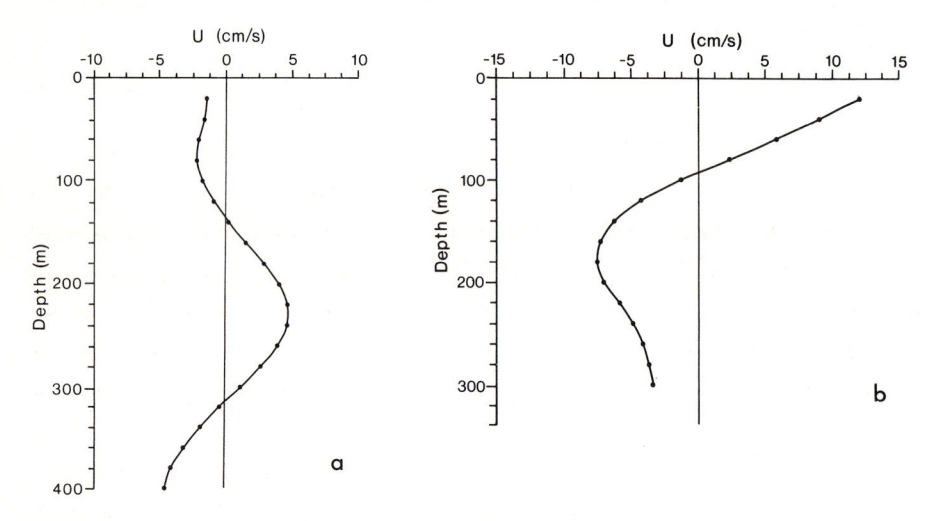

Fig. 12.8 Vertical profiles of horizontal velocity (December 1967) at (a) section 7 and (b) section 8, the sections flanking the elevation maximum in section 7.

The field of pressure derived from the time-invariant salinity distribution establishes a compensatory distribution of sea surface elevations. In effect, these define the small differences in local mean sea levels about which the vertical tidal oscillations are occurring and might have some small influence on the propagation of barotropic tides. No significant difference was found between the harmonic constants obtained from this trial and those for the homogeneous salinity case.

The vertical profiles of flood and ebb velocities at the inner end of the sill (column 14) for the homogeneous case (Fig. 12.5) are almost identical with those obtained for the winter constant salinity trial. The vertical profiles of residual currents differ considerably, however. For the homogeneous case (Fig. 12.4) the residual velocities of magnitude, approximately 0.3 cm/s, are seen to be uniformly directed seaward. For the winter baroclinic case (Fig. 12.9) a two-layered flow with speeds about 1–2 cm/s is evident. There is a small surface outflow and small inflow at depth, presumably due to the reversal of the normal estuarine type flow in Juan de Fuca Strait.

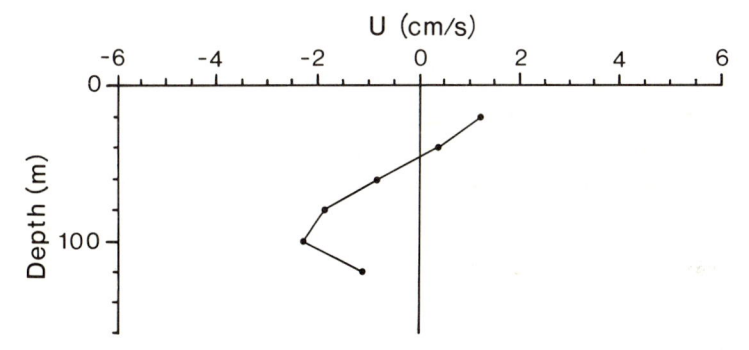

Fig. 12.9 Vertical profile of residual flow velocities (December 1967) at the inner end (section 14) of the sill in Juan de Fuca Strait (winter).

12.7.3.2 Summer circulation

The preceding simulation was repeated but with salinities typical of summer conditions. The increased surface dilution in the Strait of Georgia (Fig. 12.10) and higher salinities at depth in Juan de Fuca Strait lead to a considerable increase in the net water surface levels in the system (Fig. 12.11). The highest values (about 30 cm) occur off the mouth of the Fraser River and slope downward towards the open sea and also toward the closed northern end of the Strait of Georgia. As in the winter case, three basic vertical residual circulations are in evidence. A well developed outflow at the surface and inflow at depth is evident in Juan de Fuca Strait, consistent with the general sense of estuarine exchange with the open ocean. The Strait of Georgia shows two large vertical circulation patterns which can be considered as a general divergence near the river mouth in the near surface layers, and a general convergence at depth. The slopes of the sea surface are clearly consonant with such a circulation. Although the northern end of the Strait of Georgia is closed in the model, it is important to note that the prescribed salinity field is determined by the actual dynamical processes within the system including the strong tidal mixing that occurs in the northern passages. This topic will be discussed at greater length in connection with the three-dimensional model GF6.

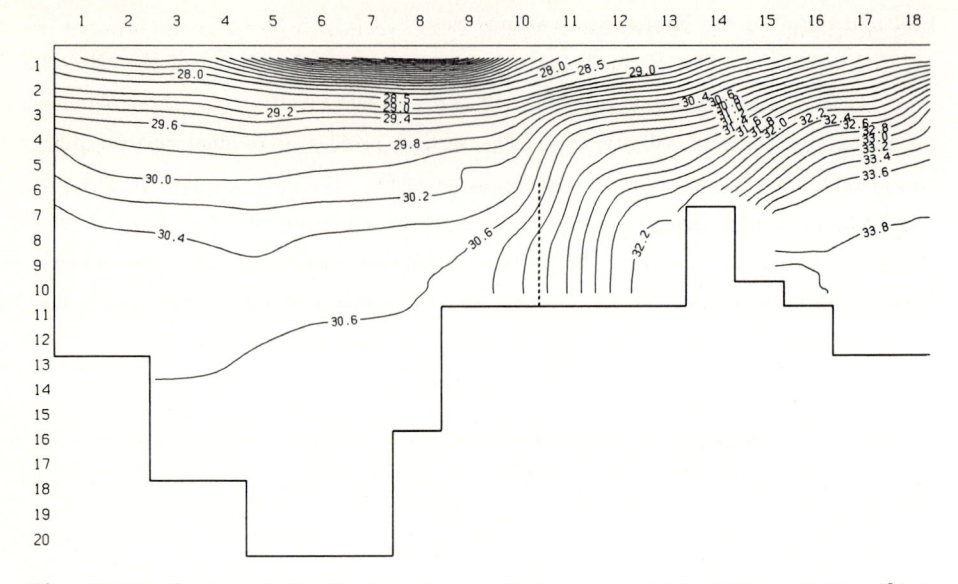

Fig. 12.10 Contoured distribution of prescribed summer (July 1968) salinities (⁰/₀₀) over a vertical section through the channel median.

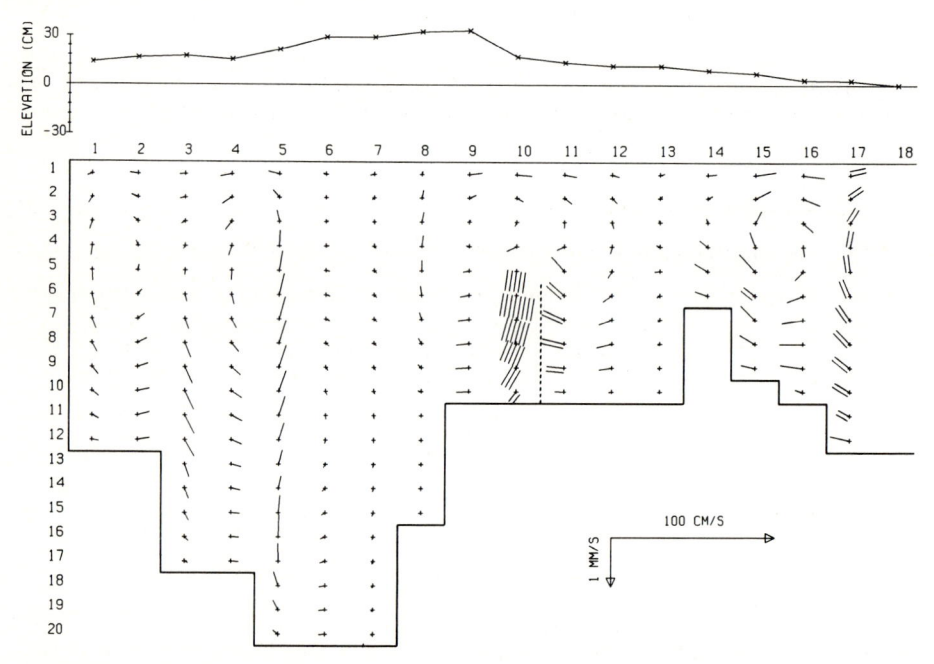

Fig. 12.11 Distribution of summer residual velocity vectors (July 1968) over a vertical section through the channel median. Also shown is the residual surface elevation.

The maximum surface elevations near the Fraser River mouth are much more pronounced than in the earlier winter calculation. Representative vertical distributions of horizontal residual velocities on either side of this maximum (sections 7 and 10) are shown in Fig. 12.12a and b, respectively. Again, there is evidence of a horizontal divergence near the surface and a convergence at middepth. A representative value of the vertical velocity for the deep basin in the Strait of Georgia is about 5 m/day. The resulting process of slow vertical advection associated with replenishment of the deep basin indicates a time scale of some 2–3 months. This accords well with that inferred from the time series shown in Fig. 1.28 (p. 39).

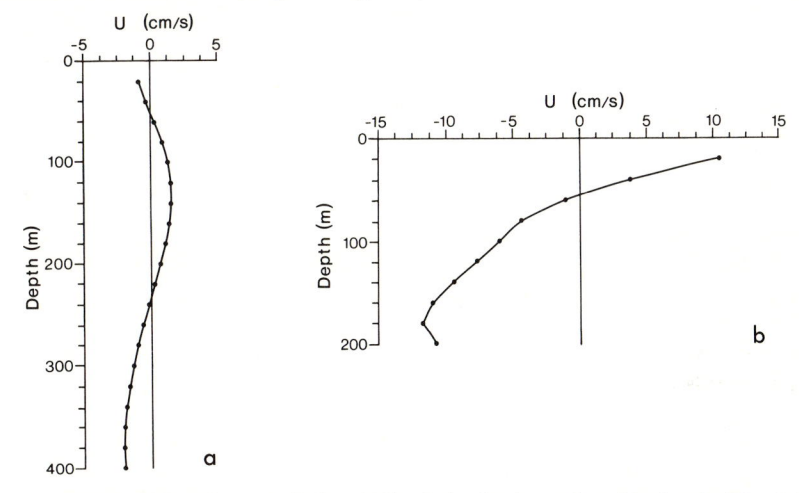

Fig. 12.12 Vertical distribution (July 1968) of the horizontal residual velocities in (a) section 7 (north of the Fraser River) and (b) section 10 (Boundary Pass).

The vertical distribution of residual horizontal velocities over the sill (section 14) shows surface outflow and deep inflow characteristic of estuarine circulation (Fig. 12.13).

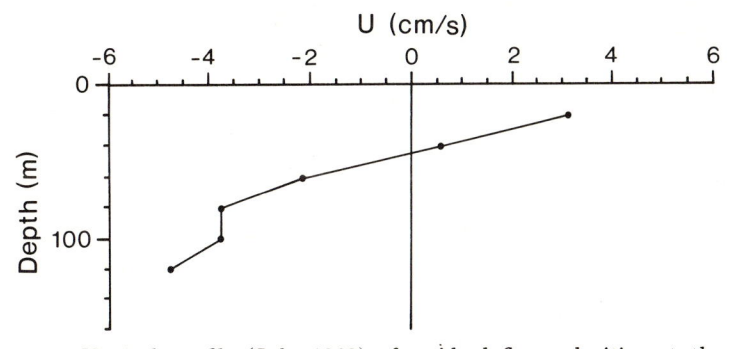

Fig. 12.13 Vertical profile (July 1968) of residual flow velocities at the inner end (section 14) of the sill in Juan de Fuca Strait.

12.7.3.3 Volume fluxes and recirculation

GF5 is a simple model, laterally integrated with large horizontal grid spacing and taking no account of the earth's rotation. As such, it can only provide a rough approximation to the actual process occurring in the Georgia/Fuca system. However, the general patterns of residual velocities and surface elevations would appear to be in reasonable accord with expectation based on available observations. It would thus be useful to estimate the net volume exchanges occurring between the open ocean and the modelled region and also the net ratio of mixed water moving seaward through Juan de Fuca Strait to that returned at depth to the deep basin of the Strait of Georgia.

Estimates of these net volume fluxes employ vertical profiles of residual horizontal velocities (Fig. 12.14a and b) through section 17 which is closest to the position of line 11 (Fig. 2.1) where extensive currrent meter records have been obtained. In the case of the winter profile (Fig. 12.14a) there is essentially a balance between the net inflow (21.6 km^3/day) occurring this time at the surface and outflow at depth. This is presumably characteristic of a winter flow reversal event rather than any net winter estuarine circulation. Although this value may seem large compared to the observed values referred to above, it is worth noting that the maximum inflow velocities recorded by the current meters associated with two reversal events, each of about 5 days duration, were 50 and 40 cm/s, respectively. The laterally-averaged inflow velocities shown in Fig. 12.14a appear to be quite reasonable.

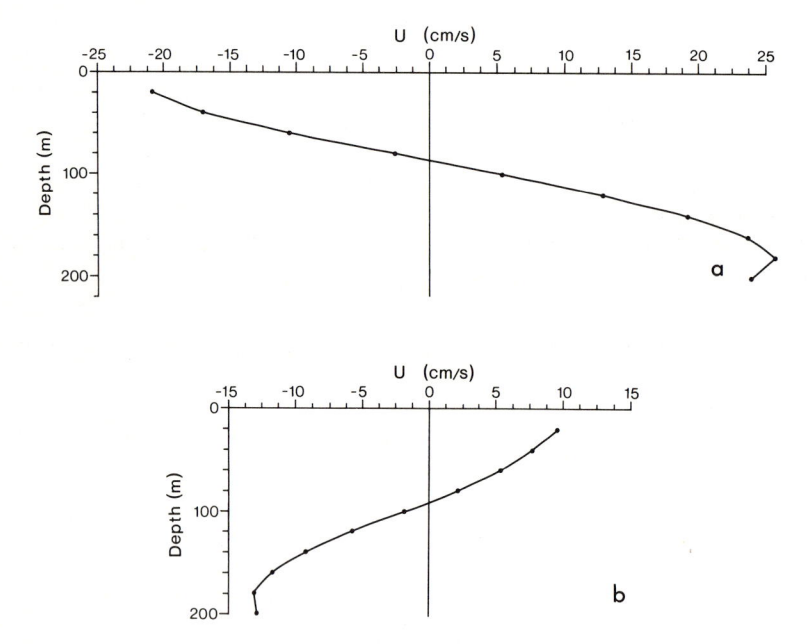

Fig. 12.14 Vertical profiles of residual velocity at section 17 close to the current meter moorings (line 11, Fig. 2.1) for (a) December 1967 and (b) July 1968.

In summer (Fig. 12.14b) there is again a balance of outflow (9.3 km^3/day) against inflow. If it is assumed that the outflow consists of mixed freshwater and sea-water of salinity 33.8‰, the freshwater proportion (0.7 km^3/day) moving seaward is readily determined. Estimates of observed net volume transports based on the March–April

1973 current meter measurements (line 11) range from 7.78–13.8 km^3/day (Godin et al., 1981). These residual volume transports were essentially in counterbalance, consistent with the absence of any significant net sea level changes in the system for the months in question.

The relative proportion of freshwater moving seaward to that being returned at depth to the Strait of Georgia, as a consequence of the strong mixing processes in the Gulf and San Juan Island passages, is readily obtained from the model. Assuming winter and summer steady state conditions, this proportioning may be checked against the relevant monthly freshwater volume distributions at different depths in the Strait of Georgia with the aid of a simple volume and freshwater fraction balance over the appropriate cross-sections flanking Haro Strait (Fig. 12.15).

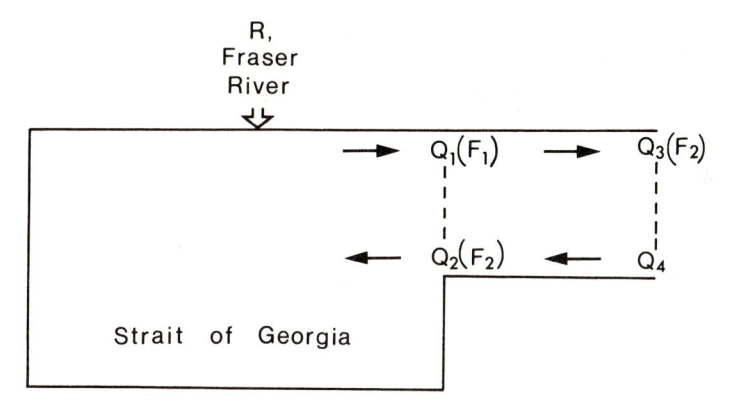

Fig. 12.15 Schematic arrangement of volume fluxes and freshwater fractions entering into the calculation of the ratio of surface seaward outflow to returned deep inflow.

Thus, the volume and freshwater fraction balances applied to the Strait of Georgia are

$$R + Q_2 = Q_1,$$
$$Q_2 F_2 + R = Q_1 F_1.$$

Hence,
$$Q_1 = R\left(\frac{1-F_2}{F_1-F_2}\right), \qquad Q_2 = R\left(\frac{1-F_1}{F_1-F_2}\right).$$

Similarly, in Haro Strait,

$$Q_1 + Q_4 = Q_3 + Q_2,$$
$$Q_1 F_1 = Q_2 F_2 + Q_3 F_2,$$
$$Q_3 = Q_4 + R.$$

Hence,
$$\frac{Q_2}{Q_3} = \frac{F_2(1-F_1)}{F_1 - F_2}$$

where

R = rate of Fraser River discharge,

Q_1, F_1 = volume outflow and its freshwater fraction from the Strait of Georgia,

Q_2, F_2 = volume inflow rate and its freshwater fraction of mixed water entering the Strait of Georgia,

Q_3 = outflow volume of mixed water from Haro Strait,

Q_4 = inflow rate of sea water to Haro Strait (having a zero freshwater fraction).

The depths of the horizontal interfaces between the various levels are determined from the zero crossings of the residual current profiles calculated by GF5. Using Waldichuk's (1957) values for F_1 and F_2 for winter and summer, respectively, to evaluate the right hand side of the preceding equation and using the appropriate inflow and outflow rates from GF5 to evaluate the left hand side, the ratios $Q_2:Q_3$ obtained from data and from the model are:

	Observation	Model
Summer	2.3	2.4
Winter	5.2	4.9

Accordingly, about 83% of the mixed water in the Gulf and San Juan Island passages is returned at depth to the Strait of Georgia in winter while 70% is returned in summer.

12.7.3.4 Prognostic salinity calculations

As noted in connection with the finite-difference formulations, the salt conservation equation introduces a high degree of numerical diffusion. Furthermore, it is necessary to introduce additional horizontal diffusive mixing in the system in order to ensure numerical stability. A prognostic calculation requires that both salinity and velocity fields be allowed to evolve freely, as they do in nature. If the model scheme and grid are adequate, the modelled fields of salinity and velocity will evince a realistic evolution. It is necessary then to clarify which aspects of this model must be revised to allow a full prognostic calculation. Thus, it is of interest, as a first trial, to minimize mixing in the model by omitting the tidal oscillations from the calculation and determine the net circulation derived from the summer density (i.e., salinity) field. The resulting vertically-contoured salinity field (Fig. 12.16) compared with Fig. 12.10 shows the relatively small changes in salinity that occur over a 10-day simulation. The principle change is associated with the near surface waters in the system, with an attenuation of the intense vertical salinity gradient in the southern Strait of Georgia, and an intensification of that gradient in Juan de Fuca Strait. At depth there is a small increase in salinity in the Strait of Georgia near Boundary Pass and an increased extent over the lower part of the water column of a vertically-uniform horizontal salinity gradient.

Considering now the associated field for the residual velocity vectors (Fig. 12.17), there is a single large vertical circulation of water in the Strait of Georgia with a surface

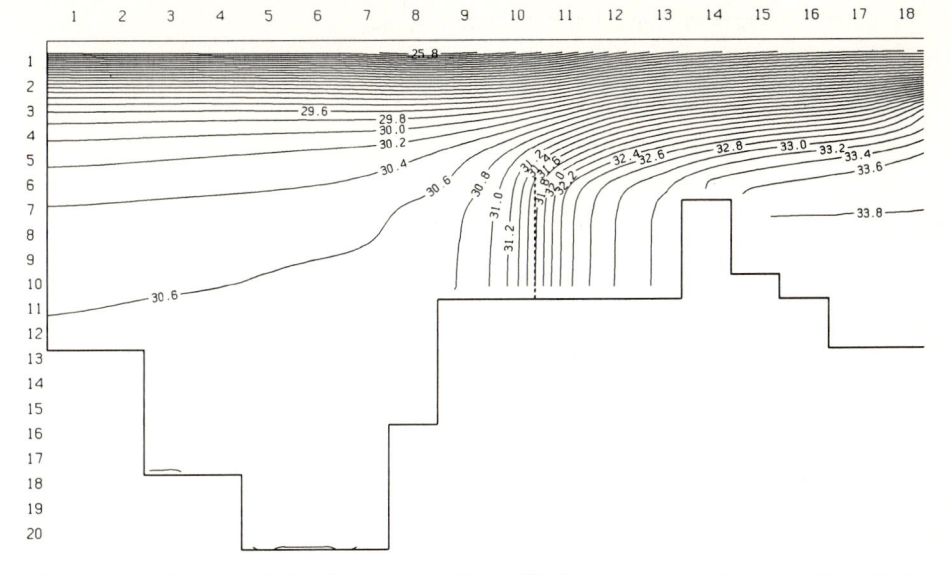

Fig. 12.16 Contoured distribution of salinity ($^0/_{00}$) over a vertical section through the channel median averaged over the last 25 h of a 10-day prognostic computation in the absence of tides and using the initial salinity field of July 1968.

flow directed seaward and a net flow directed inward at depth. This contrasts with the circulation shown in Fig. 12.11 where the field of constant salinity is maintained and two distinctive vertical circulations are in evidence. This simplified vertical recirculation is clearly associated with levelling off of residual sea levels in the Strait of Georgia (Fig. 12.17) and sharp seaward gradient at its southern end. Furthermore, it will be noted that a substantial reduction in these surface elevations has occurred. The general simplification in the residual velocity field for the prognostic case indicates that there are features in the prescribed diagnostic salinity field which are eliminated by increasing the freedom of response of the system through salt mixing and advection.

In comparing the vertical profiles of horizontal velocity (Fig. 12.18a and b) at section 7 near the Fraser River mouth and section 14 on the sill, respectively, with those determined earlier for the diagnostic case (Figs. 12.12a and 12.13), the velocities possess essentially the same order of magnitude. The most significant difference is the elimination of the three-layer flow regime near the river mouth and a substantial increase in the surface outflow and deep water inflow in column 7. At the sill (column 14) the velocity profile has undergone little change. It would appear from these results that the estuarine circulation may be estimated in this manner over short periods, provided the effects of tides are omitted, thereby compensating to a reasonable degree for the excessive diffusion inherent in the finite-difference formulation of the salt equation.

These results are now compared when the prognostic computation is repeated, but with the inclusion of the M_2 and K_1 tidal co-oscillation. If the contoured vertical salin-

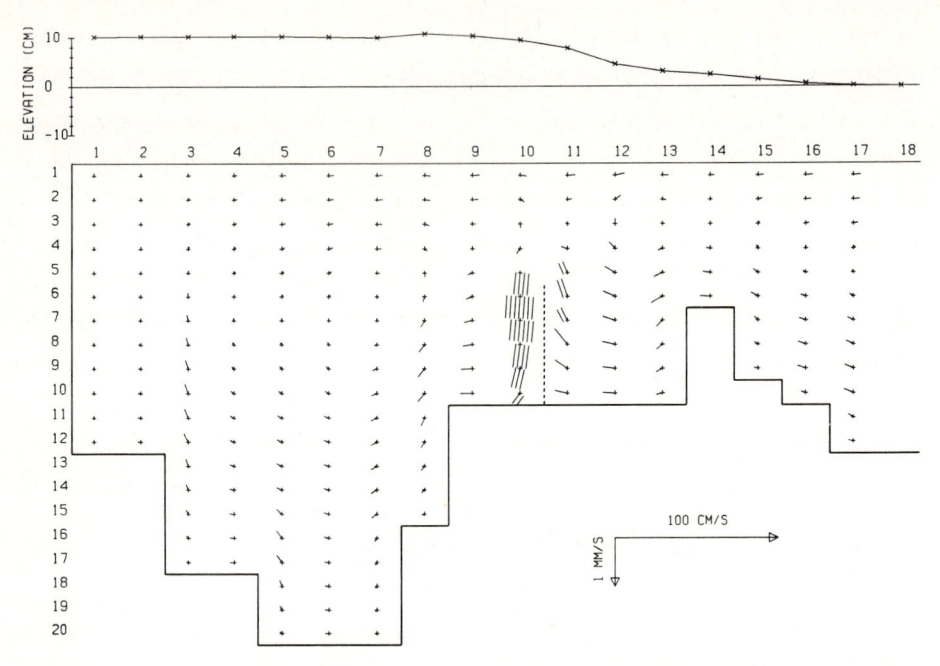

Fig. 12.17 Distribution of summer residual velocity vectors (July 1968) over a vertical section through the channel median. Also shown is the residual surface elevation.

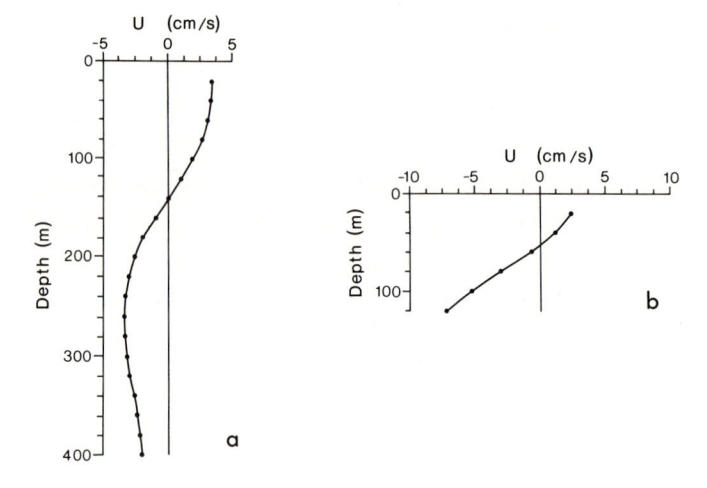

Fig. 12.18 Vertical distribution of the horizontal residual velocities in the absence of tides in sections (a) 7 and (b) 14 for the prognostic salinity trial.

ity distribution (Fig. 12.19) is compared with the earlier non-tidal case (Fig. 12.16), it is clear that large changes have occurred in that part of the model where the tidal currents are strong. Thus, extensive vertical and horizontal mixing has occurred in the vicinity of the sill and salinity has greatly increased at depth throughout the Strait of Georgia. At its southern end, near Boundary Pass, there is a greatly increased extent, over the lower part of the water column, of a vertically-uniform horizontal salinity gradient. Comparison of the fields of residual circulation vectors (Figs. 12.20 and 12.17) show that inclusion of the tides greatly increases the vertical recirculation in the Strait of Georgia while reducing the vertical circulation in Juan de Fuca Strait. This is associated with a general increase in mean sea levels in the Strait of Georgia and decrease to negative values of sea levels in Juan de Fuca Strait. This is further illustrated by comparisons of the corresponding vertical profiles of horizontal velocity near the Fraser River mouth (Figs. 12.21 and 12.18). It is evident that in the Strait of Georgia, the inclusion of tides has led to an almost tenfold increase in the velocities (Fig. 12.21a) of the vertical recirculation in the Strait of Georgia whereas on the sill (Fig. 12.21b) both the surface outflow and deep inflow have been substantially decreased.

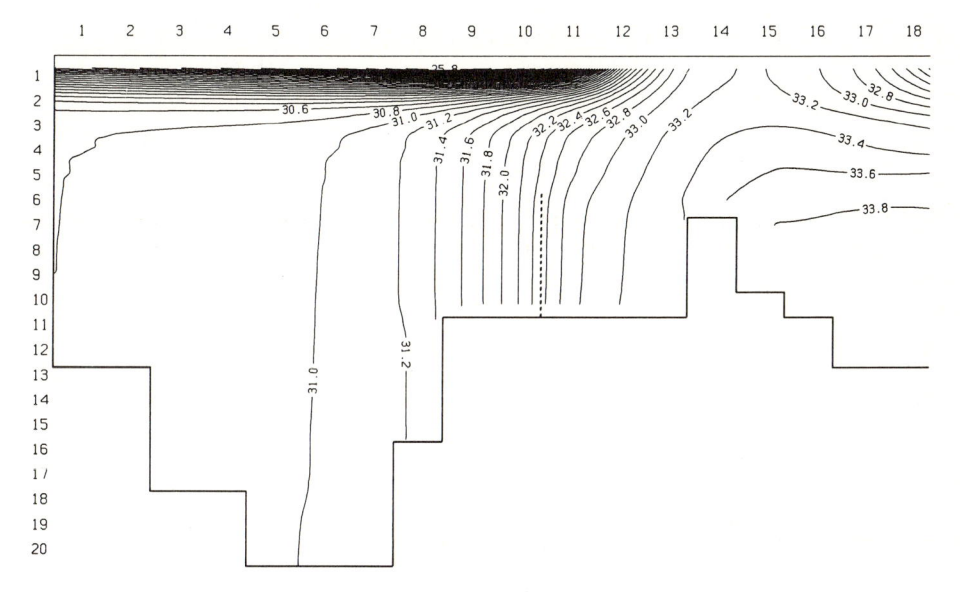

Fig. 12.19 Contoured distribution of salinity ($^0/_{00}$) over a vertical section through the channel median, averaged over the last 25 h of a 10-day prognostic computation when M_2 and K_1 tides are included.

A question arises as to the much more rapid modelled vertical recirculation occurring in the Strait of Georgia when salinities are computed in the presence of tides in contrast to their absence. A comparison of the salinity distributions for the diagnostic case (Fig. 12.10) and the non-tidal prognostic case (Fig. 12.16) show relatively small differences in the overall salinity distribution. On the other hand, in the prognostic case in which tides are included, there is a much more marked attenuation of both vertical and horizontal salinity gradients in that part of the model where the tidal currents are strong (from approximately column 10 to the mouth of Juan de Fuca Strait).

Two considerations arise. The first of these concerns the mesh Reynolds number $u\Delta x/\mu$ where μ is the horizontal eddy diffusivity. The second is a provision in the nu-

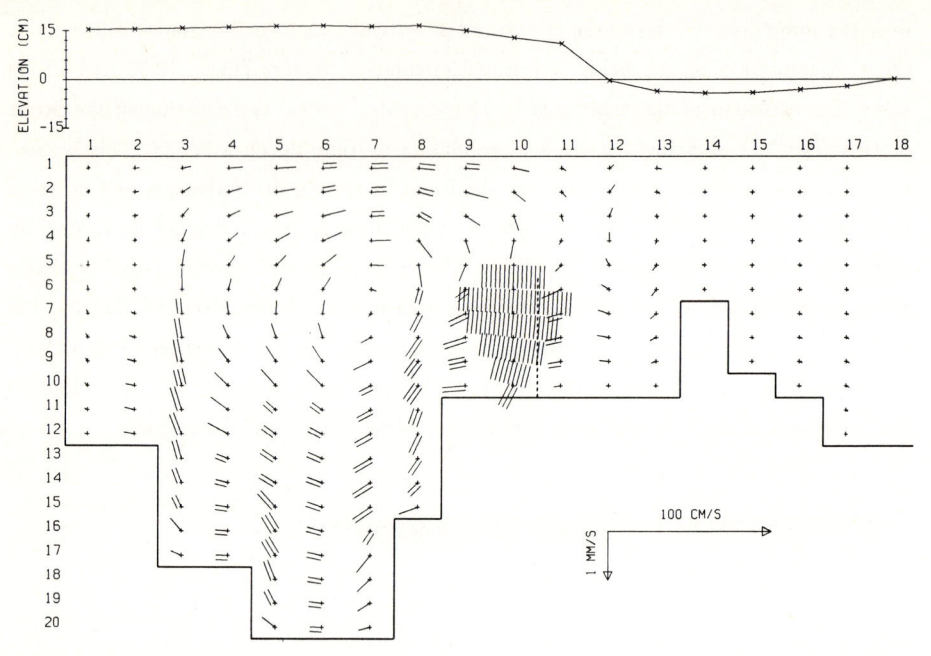

Fig. 12.20 Distribution of residual velocity vectors over a vertical section through the channel median, averaged over the last 25 h of a 10-day prognostic trial when the M_2 and K_1 tides are included.

merical scheme for vertical mixing whenever tidal advection brings about the physically unrealistic superposition of water in a given mesh with a density or salinity greater than that of the density in the mesh immediately below.

For low order schemes, when the Courant number

$$\frac{u\Delta x}{\Delta t} \neq 1 \quad \text{or} \quad 0,$$

then numerical solutions are subject to a spurious numerical diffusion (Fromm, 1969), the numerical diffusion coefficient being proportional to $u\Delta x$. The ratio of this factor to the actual physical diffusivity ν is the mesh Reynolds number. This quantity must be much less than some critical value (of order 1) for acceptable solutions to be obtained. For the type of scheme employed in GF5, the critical mesh Reynolds number is 4. In the non-tidal trials, the mesh Reynolds number was less than 1, resulting in a plausible solution. In the tidal model this critical value is exceeded and the temporal evolution of the salinity field is dominated by numerical diffusivity with a wholly unrealistic rapid approach to uniformity and associated unrealistic residual circulations.

When the physically unrealistic condition of water density in a mesh exceeding that of the mesh below occurs, the densities in each mesh are replaced by their volume-weighted average. Such a procedure should also include mixing of the associated momentum. However, this is considerably more complicated, particularly in virtue of the

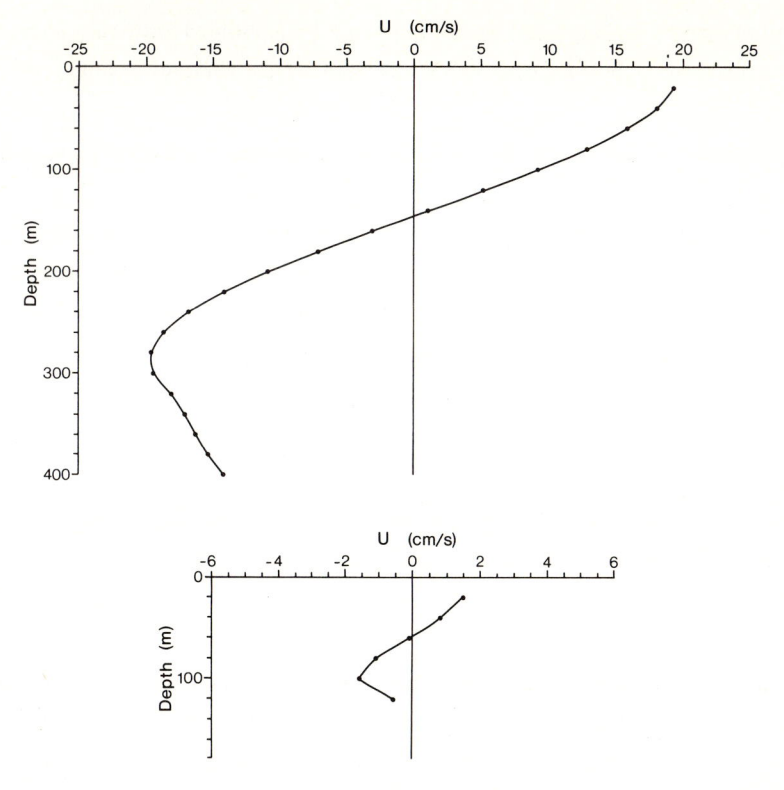

Fig. 12.21 Vertical distribution of the horizontal residual velocities in sections (a) 7 and (b) 14 for the prognostic trial run when the M_2 and K_1 tides are included.

bottom topography, and in view of the simple exploratory nature of this model, was not included. A point of consideration arising in models of this type concerns the effect of such density mixing where strong parabolic vertical profiles of tidal velocity occur in the presence of strong vertically-uniform horizontal density gradients along the flow direction resulting in a highly vigorous mixing process of questionable derivation.

It would thus appear that the net effect of these enhanced mixing procedures in the region of swift tidal streams has given rise to a strong horizontal pressure gradient acting over a greater subsurface cross-sectional area of the water column. The absence of part of the appropriate vertical momentum transfer has led to a much more rapid inflow of more saline water at depth into the Strait of Georgia and accompanying marked increase in sea levels. In the tidal case, this results in a more rapid surface flow that occurring in the non-tidal case (Figs. 12.17 and 12.20). It is also noted that the effect of this intensified mixing has led to negative residual sea levels in Juan de Fuca Strait. It was noted earlier that the introduction of strong vertical mixing at one end of an initial two-layer closed rectangular basin would generally tend to give rise to a three-layer system of vertical recirculation. The strong vertical residual velocity distribution shown in Fig. 12.20 would appear to be associated with this process, but with the need for a bottom counterflow obviated by an alternative supply of denser water.

It is clearly evident that the mesh Reynolds number is an important parameter in the selection of mesh size. In the three-dimensional model GF6, the mesh length is less than a quarter of that employed in GF5. Though problems of numerical diffusion are not eliminated, the three-dimensional scheme permits quite realistic trials, though of limited duration, which include computation of salinity, thus permitting the salinity field to achieve an approximate and plausible dynamic equilibrium.

12.7.3.5 Sea levels

The residual sea levels (Figs. 12.7 and 12.11) obtained in the course of the diagnostic trials derive primarily from the density-driven circulation when a fixed mean sea level of zero elevation at the entrance to Juan de Fuca Strait is assumed. These modelled levels are now discussed with reference to observed sea levels in the modelled region.

A particular problem in connection with interpretation of observed sea levels in British Columbia coastal waters, attributable to the difficult nature of the terrain, concerns the dearth of suitably-located tide gauges levelled against a geodetic datum. For Vancouver Island, the geodetic datum is taken as long term mean sea level at the Victoria tide gauge. At the present time this datum has not been levelled against the national levelling network on the mainland. Three gauges levelled into this Vancouver Island net enter into this discussion: Victoria in the inner part of Juan de Fuca Strait, Tofino which is taken as representative of conditions on the open coast of Vancouver Island, and Little River which may be used to characterize conditions in the northern Strait of Georgia.

Sea levels are, in general, much more readily accessible to measurement than the circulation and can afford a useful index of major changes taking place. The major problem is the separation of the many factors that can affect the record of a tide gauge operating correctly at a particular location. The dominant factors are the gravitational tides with periods of about a day. Longer periods, fortnightly, monthly, biannual, and annual also occur. The latter two can be obscured by meteorological effects (large-scale changes in sea level atmospheric pressure field and winds) deriving essentially from solar radiation. More specifically in the present context, elevations can be altered by changes in the speed and/or direction of the flow, geostrophic deflection near a closed boundary, atmospheric pressure changes and wind setup, density changes in the water column, and discharge from a freshwater source.

Monthly mean sea levels in the coastal waters of British Columbia undergo a marked annual cycle. When sea levels are corrected for the effect of local atmospheric pressure changes, this cycle is primarily determined by the variation in the coastal alongshore current, associated with seasonal oceanic wind changes, and concomitant changes in density over the water column (Tabata et al., 1986). The annual range is of order 20 cm of which some 11 cm is due to the change in density occurring between winter and summer. Propagation of changes on the open coast into the Georgia/Fuca system occur on two time scales, a barotropic scale characterized by speeds of order 40–50 m/s and a baroclinic scale associated with low density surface water having speeds of order 20–30 cm/s (Holbrook et al., 1983). It is this latter scale which is of interest in connection with effects on the estuarine and density-driven circulation. A rough estimate of the surface slopes associated with the baroclinic scale may be made using monthly mean gauge data from Tofino and Little River. Taking into account the error in levelling, long term mean sea level at Little River is 4 ± 4 cm higher than at Tofino. Changes in annual mean sea level from long term mean sea level are generally small (1–2 cm) and the departures from one gauge to another over the modelled region are similar for a particular year. Furthermore, monthly mean atmospheric pressures are essentially the same over the region. Thus, if the monthly mean sea levels at Tofino are expressed as departures from the long term mean sea level and monthly means at Little River are similarly adjusted, the slow changing net difference in elevation of the water surface between the northern end of the Strait of Georgia and the open sea may be approximated. A plot of this difference in adjusted monthly means over a 3-yr period (Fig. 12.22) shows a distinctive annual cycle characterized by maximal values in summer. Unfortunately the Little River gauge was not in operation at the time when the salinity data used in the winter diagnostic trial were obtained. The winter reversal of the vertical circulation in Juan de Fuca Strait compared to the corresponding summer circulation, is consistent with the possibility of somewhat higher winter sea levels on the open coast and consequent negative values.

In the case of the July 1969 data, the observed monthly mean elevation difference between Little River and Tofino is 13 ± 4 cm. The modelled value, assuming Tofino is representative of conditions at the entrance to Juan de Fuca Strait, is 16 cm. It would thus appear that this elementary baroclinic model gives a reasonable approximation to the water surface slope driving the circulation.

12.8 PRESCRIBED LID NUMERICAL TRIALS

In view of the strong influence of barotropic tides and possible dominant role of internal gravitational convection in flushing events, it is desirable to determine if a significant saving in computing costs might result from adapting the well known notion of a 'rigid lid' used in ocean models in the form of a 'prescribed lid' for a modelled, strongly tidal coastal sea where the changing shape of the lid is known for all time from a single previous homogeneous-fluid barotropic tidal calculation. The basic

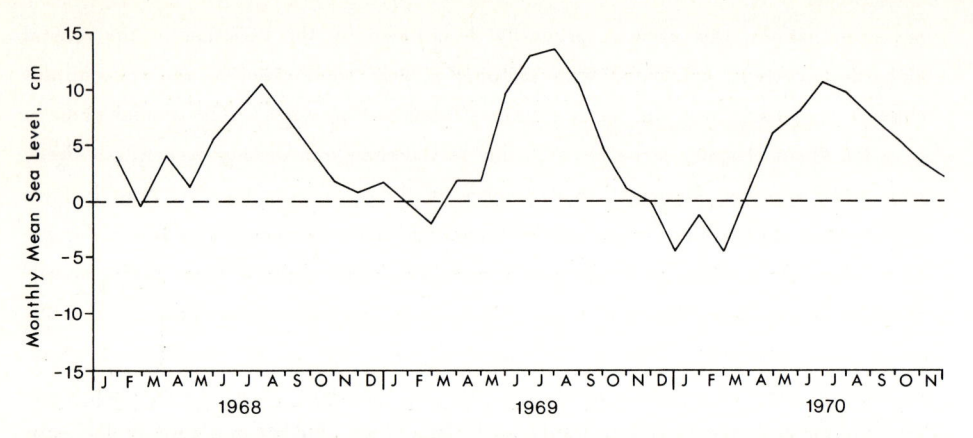

Fig. 12.22 Time series plot of the difference between the departure of monthly mean from long term mean sea levels at Little River and the departure of monthly mean from long term sea levels at Tofino when it is assumed (from gauge levelling) that long term mean sea level is 4 cm higher in the Strait of Georgia (Little River) than on the open coast (Tofino).

assumption which seems reasonable from an inspection of the extensive tidal current data available for the system, is that the barotropic tides are not significantly affected by the baroclinic flows but that the latter are much affected by the barotropic tides. Rigid lid approximations are discussed by Simons (1980) who notes that for application of the method in a three-dimensional model of such a coastal sea, the square of the ratio of the fundamental free surface seiche period to the inertial period should be substantially less than unity. The first normal mode for the overall system is, in fact, close to that of the appropriate inertial period as shown in Chapter 8. There appears, however, to be a weak response of the system at that period, presumably due to weak coupling between its various parts.

For the development of an algorithm suitable for a prescribed lid calculation, the dynamics of a damped diffusive uninodal internal seiche in a rectangular basin of a length and depth corresponding roughly to the Georgia/Fuca system was simulated. For simplification, the basin is assumed of unit width. The closed ends and bottom of the test basin are denoted by the thick lines A, B, C, and D in Fig. 12.3 from which the sill has been removed. The lines E, F, G, and H denote the initial position of the interface between the upper ($20^0/_{00}$) and lower ($33^0/_{00}$) salinity layers. As suggested by earlier studies of the Strait of Georgia and to enhance the non-linear effect of bottom stress, an unusually high coefficient of friction (0.01) was employed. The horizontal and vertical coefficients of viscosity and diffusivity were, respectively, $N_x = 10^6$ cm^2/s, $N_z = 10$ cm^2/s, $K_x = 10^6$ cm^2/s, and $K_z = 10$ cm^2/s. Illustrative of the seiche motion, a representative velocity field is shown in Fig. 12.23.

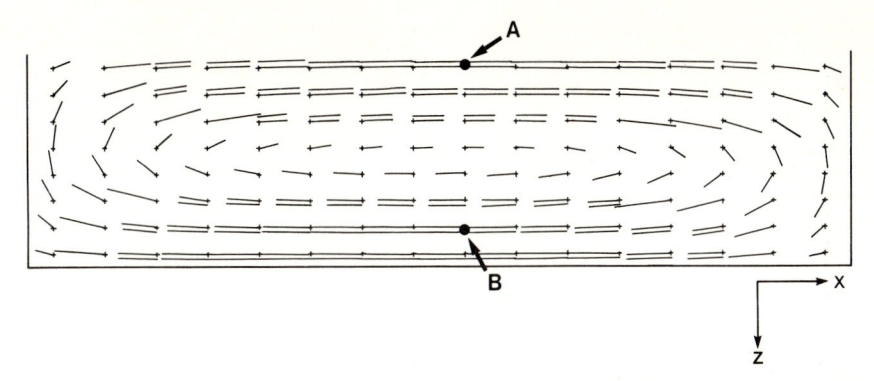

Fig. 12.23 Representative velocity vector field associated with the internal seiche. A and B denote locations of velocity points used in the time series comparison of free surface and prescribed lid computations.

12.8.1 The Numerical Application of the Prescribed Lid

Though capable of considerable refinement the procedure presently in use, which has proved quite satisfactory for these initial experiments, may by outlined as follows. Suppose that the barotropic pressure gradients within the fluid can be approximated by a part, $g\bar{\rho}(\partial Z/\partial x)$ due to the surface tide and a part, $\partial p_s/\partial x$ due to the gradient of the pressure p_s exerted by the prescribed lid. Equation (12.6) then becomes

$$\frac{\partial p}{\partial x} = g\bar{\rho}\left(\frac{\partial Z}{\partial x}\right) + \frac{\partial p_s}{\partial x} + gz\frac{\partial \bar{\rho}}{\partial x}. \tag{12.12}$$

Assuming unit width and substituting from (12.12), Eq. (12.3) becomes

$$\frac{\partial u}{\partial t} = -\left[\frac{\partial u^2}{\partial x} + \frac{\partial}{\partial z}(uw) - \frac{\partial}{\partial x}\left(N_x\frac{\partial u}{\partial x}\right) - \frac{\partial}{\partial z}\left(N_z\frac{\partial u}{\partial z}\right)\right] - \frac{1}{\rho}\left[g\bar{\rho}\frac{\partial Z}{\partial x} + \frac{\partial p_s}{\partial x} + gz\frac{\partial \bar{\rho}}{\partial x}\right]. \tag{12.13}$$

Equation (12.13) may be solved numerically using a time step procedure in which the time step is not limited by the Courant-Friedrichs-Lewy criterion, provided the slopes of the water surface $\partial Z/\partial x$ are always known for all u-point calculations from an earlier barotropic tidal calculation.

This direct introduction of the tidal slopes of the water surface is, however, incompatible with the procedure, referred to above, used in the case of the rigid lid to prevent the accumulation of small numerical errors over the course of the time integration. Hence, the following procedure is employed.

For the homogeneous fluid case, assuming unit width, Eq. (12.13) may be arranged as

$$g\frac{\partial Z}{\partial x} = -\frac{\partial U}{\partial t} - T(U,W) \tag{12.14}$$

where $T(U,W)$ includes the advective and diffusive terms and U and W are the barotropic tidal velocities, distinct from the total velocity fields given by u and w. The numerical values of both terms on the right hand side of Eq. (12.14) are known from the earlier calculation.

Substituting from (12.14), Eq. (12.13) may be written as

$$\frac{\partial u}{\partial t} = A + \frac{\partial U}{\partial t} \qquad (12.15)$$

where

$$A = -\left[\frac{\partial u^2}{\partial x} + \frac{\partial uw}{\partial x} - \frac{\partial}{\partial x}\left(N_x \frac{\partial u}{\partial x}\right) - \frac{\partial}{\partial z}\left(N_z \frac{\partial u}{\partial z}\right)\right] + \frac{\bar{\rho}}{\rho} T(U,W) - \frac{1}{\rho}\left[\frac{\partial p_s}{\partial x} + gz\frac{\partial \bar{\rho}}{\partial x}\right]. \qquad (12.16)$$

By expanding the overall velocities in Eq. (12.16) into a sum of barotropic tidal and baroclinic parts, the addition of the term $T(U,W)$ to the right hand side of Eq. (12.15) can readily be seen to cancel out, in effect, a redundant contributory increment of barotropic velocity in the increment of overall velocity computed over a time step.

The time-centred finite-difference formulation of Eq. (12.15) is given by

$$u_{i,j}(t+\Delta t) = u'_{i,j} + U_{i,j}(t+\Delta t) \qquad (12.17)$$

where

$$u'_{i,j} = u_{i,j}(t-\Delta t) + 2\Delta t\, A - U_{i,j}(t-\Delta t)$$

and is the purely baroclinic velocity (including the effect of interactions with the barotropic tide) computed over the time step. Small numerical errors are now removed by subtracting the vertically-averaged velocity from the jth column from each velocity in that column by replacing

$$u'_{i,j} \quad \text{with} \quad u'_{i,j} - \overline{u'_{i,j}}$$

where

$$\overline{u'_{i,j}} = \frac{1}{N}\sum_{i=1}^{N} u'_{i,j}$$

where N is the number of velocity points in a column. The overall small $u_{i,j}$ are then computed from Eq. (12.16) using the adjusted values of $u'_{i,j}$.

12.8.2 Numerical Experiments

The following numerical experiments were carried out where, in each, the changing velocity and salinity fields were computed for a 15-day period. Throughout these numerical trials, the total salt content of the basin was shown to be conserved.

In experiment I an initial distribution of salinity was employed such that all values of salinity were $20^0/_{00}$ above the sequence of thick lines E, F, G, H in Fig. 12.3 and $33^0/_{00}$ below the lines. The subsequent motion and changing salinity distribution over 15 days were computed using the finite-difference forms of Eqs. (12.1–12.4).

Using the same initial conditions, the calculation was repeated but assuming the presence of a rigid lid (experiment II). Thus, surface elevations were not calculated and the essentially equivalent pressure gradients exerted by the lid were computed from the divergence of the vertically-integrated terms in Eq. (12.3) (Paul and Lick, 1974).

Discrepancies between the results of experiments I and II, obtained in the initial trials, showed distortion of the velocity field apparently due to the accumulation of small numerical errors. These could be removed by subtracting small spurious vertically-averaged horizontal velocities from the individual horizontal velocities in the same column (in which the net flow through a vertical cross-section under the rigid lid must be zero). Excellent agreement was then obtained between the computed velocities and salinities obtained from experiments I and II, other than for a very small damped free surface seiche in the case of experiment I which died out in the course of the computation.

Experiment I (free surface) was now repeated but with the inclusion of an arbitrary tide-generating force, applied uniformly throughout the basin, varying with the M_2 tidal period (12.4 h), resulting in experiment III. The computed velocity fields now showed the combined M_2 barotropic and internal seiche oscillations. These were obtained for comparison with the prescribed lid trials, experiments V and VI.

Experiment III (free surface) was repeated but assuming a constant density field, the result (experiment IV) being the pure forced barotropic tidal oscillation. All computed elevations Z and velocities U were recorded on tape for inclusion in experiment V.

It was now assumed that the surface elevations obtained in experiment III consisted entirely of a barotropic tidal component Z and a baroclinic part ζ' given as

$$\zeta = Z + \zeta'.$$

The quantities Z are known from experiment IV while those ζ' are replaced in Eq. (12.6) for the pressure gradient by the pressure gradient exerted by the lid as in experiment II. The normal conditions of $w = 0$ at a rigid lid ($z = 0$) can be replaced by setting w equal to the known vertical velocity of the water surface which is also given by experiment IV. The changing volumes of the upper meshes required in Eq. (12.6) are determined by the known barotropic surface elevations from experiment IV. A simple test of the scheme was to carry out the complete sequence of calculations using a constant density fluid when a simple reconstitution of the purely barotropic tidal solution was obtained. The computed velocity and density fields agreed well with those obtained from experiment III though the maximal values of the internal seiche velocity computed by the prescribed lid were slightly smaller than those obtained from the free surface calculation. The final calculation showed insignificant differences.

Since the free surface wave speed now no longer is a major limitation on the length of the time step, experiment V was repeated using double this time step. The results conformed to those obtained in experiment V and are illustrated in Fig. 12.24 by time sequences of computed velocities for $u_{1,9}$ and $u_{7,9}$ (A and B, Fig. 12.23). The superposition of the barotropic tidal oscillations on the much longer period damped internal seiche is clearly evident. The velocities of the latter, which were derived by the prescribed lid method, as in the previous experiment V, are slightly smaller than those obtained for the free surface. This is possibly due to the omission of some smaller terms in computing the pressure gradient exerted by the lid. A more rigorous derivation of the governing equations by R.A. Flather is contained in Appendix I.

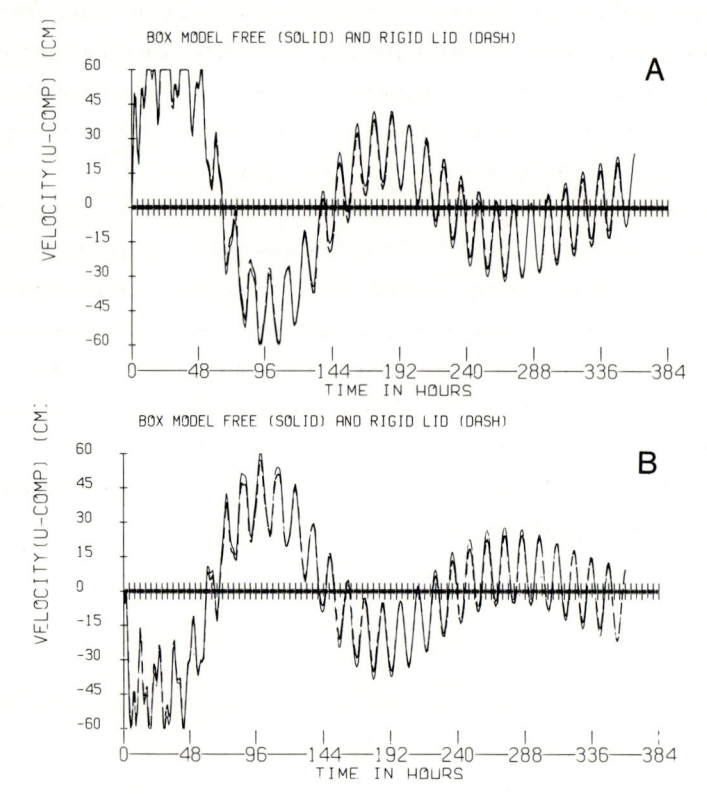

Fig. 12.24 Time series of horizontal velocities computed by the free surface method (solid line) and by the prescribed lid method (dashed line) using a double time step at locations A and B. Damped internal seiche under strong semi-diurnal tidal forcing in a two-dimensional (x, z) rectangular basin approximating the Georgia/Fuca system.

These results suggest the feasibility of significant reductions in computing costs by the extension of this method to fully three-dimensional computations though extensive further development work will need to be undertaken.

12.9 CONCLUDING SUMMARY

It has been shown that a simple two-dimensional model (x, z) of the main conveying channels can be used in a diagnostic context to bring out some interesting features of the system. Residual velocities associated with the overall estuarine circulation tend to be relatively small (generally < 10 cm/s in the Strait of Georgia). The static head plays a dominant role in moving surface water into the mixing region between the inner and outer straits. A return flow moves into the deep basin of the southern Strait of Georgia. A general three-level flow regime off the Fraser River mouth section results, northward at the surface, due to the downward slope north of the static head, a southerly flow at middepth due to the baroclinic pressure field, and further flow northward at depth, consisting of more saline water. A rough estimate of the removal time (3–4 months) of deep water through vertical advection from the main central basin of the Strait of Georgia accords with observation. The reversal of the estuarine circulation in Juan de Fuca Strait, presumably due to the effect of winter winds and the presence of Columbia River-diluted surface waters can readily be shown to occur. A large proportion (70–80%) of the water mixed by the strong tidal streams in the vicinity of the Gulf and San Juan Islands is returned to form deep water in the Strait of Georgia. These results show the feasibility of obtaining useful results when employing the observed salinity fields in a largely diagnostic context. It would appear that the coarse section length employed leads to a gross exaggeration of mixing in the modelled region.

The next stage of development in these studies thus called for a three-dimensional scheme employing a smaller mesh size. Such a scheme can include the important effects associated with the earth's rotation and topography on the basic longitudinal residual circulation indicated by this simple model.

13 THE THREE-DIMENSIONAL MODEL: GF6

13.1 INTRODUCTION

It is clearly evident that a fully three-dimensional model is required to simulate barotropic and baroclinic processes within these waters. The system is demanding of any scheme for a variety of reasons including the topographical complexity of both the sea bottom and coastal boundaries, the strong tidal streams which result in many eddies and associated interactions, and the variously stratified estuarine circulation. Work had been initiated on such a model when interest was expressed by Dr. J. Backhaus of the Institut für Meereskunde, University of Hamburg, in assessing the performance of a three-dimensional numerical scheme, successfully employed in his own North Sea studies, under markedly different dynamical requirements. The two-dimensional grid of the 4-km mesh model GF2 was thus adapted to a seven-level three-dimensional operation, the GF6 model. To obviate problems that were encountered at open boundaries in preliminary trials, the grid was extended to include crude approximations of adjoining channels, thus increasing the distance between and reducing the number of boundary openings. The number of levels was subsequently increased to eight with the provision of a relatively thin upper layer. This facilitated comparisons with the net circulation obtained from the buoyant-spreading upper layer model GF4. This model (GF6) is able to include the effects of both river discharge and deep water renewal.

The work is considered in two parts. In the first, a variety of topics is considered in connection with the performance and adjustment of the model based on simulation of the summed M_2 and K_1 tidal co-oscillations. These topics include location of open boundaries, adjustment of vertical and horizontal eddy viscosities and bottom friction, the role of advective accelerations and associated horizontal and vertical residual circulations, partial- versus full-slip conditions at lateral boundaries, effect of grid-size reduction, use of strain rate-related eddy viscosities, and the effects of numerical diffusion when a simple upwind differencing formulation of the density-conservation equation is included in the simulation. It is not proposed to provide a detailed presentation of all the results obtained but rather to illustrate salient points in comparisons with observations and results from the earlier two-dimensional vertically-integrated 4-km mesh model GF2. Some of these results have been presented in Backhaus et al. (1987).

In the second part, it is shown that useful insights into the basic barotropic and baroclinic residual circulation within the system were obtained from relatively short model runs. In the baroclinic case, an initial density field was employed which corresponds to the salinity field observed in early July 1968 (see Fig. 1.5b). The results are thus considered in a semi-diagnostic context. In effect, the computed density field assumes a configuration in general geostrophic equilibrium. This gives rise to reasonably stable distributions of both surface elevations, residual circulation, and balanced net inflows and outflows at the open boundaries. Comparisons of this modelled residual circulation with that obtained from interpretation of hydrocast data and moored current meter and cyclesonde records show good agreement. The model is then used to provide estimates of the changes brought about in this circulation by two representative, oppositely-directed characteristic wind fields (southeast and northwest) parallel to the major axis (y) of the model.

13.2. FUNDAMENTAL HYDRODYNAMICAL EQUATIONS

The equations of continuity and momentum, integrated over a discrete vertical interval h are

$$\frac{\partial U}{\partial x} + \frac{\partial V}{\partial y} + \Delta w = 0. \tag{13.1}$$

$$U = \int_h u\, dz, \qquad V = \int_h v\, dz.$$

The elevation of the water surface is given by the vertical velocity at the sea surface,

$$\frac{\partial \zeta}{\partial t} = w_{\text{surface}}, \tag{13.2}$$

$$\frac{\partial U}{\partial t} + \frac{\partial}{\partial x}\left(\frac{U^2}{h}\right) + \frac{\partial}{\partial y}\left(\frac{VU}{h}\right) + \Delta\left(\frac{wU}{h}\right) - fV + \frac{h}{\rho}\frac{\partial p}{\partial x}$$
$$= \frac{\partial}{\partial x}\left(A_h \frac{\partial U}{\partial x}\right) + \frac{\partial}{\partial y}\left(A_h \frac{\partial U}{\partial y}\right) + \Delta \tau^z \tag{13.3}$$

where $\tau^z = \dfrac{\partial}{\partial z}\left(\dfrac{A_v U}{h}\right)$,

$$\frac{\partial V}{\partial t} + \frac{\partial}{\partial x}\left(\frac{UV}{h}\right) + \frac{\partial}{\partial y}\left(\frac{V^2}{h}\right) + \Delta\left(\frac{wV}{h}\right) + fU + \frac{h}{\rho}\frac{\partial p}{\partial y}$$
$$= \frac{\partial}{\partial x}\left(A_h \frac{\partial V}{\partial x}\right) + \frac{\partial}{\partial y}\left(A_h \frac{\partial V}{\partial y}\right) + \Delta \tau^y \tag{13.4}$$

where $\tau^y = \dfrac{\partial}{\partial z}\left(\dfrac{A_v V}{h}\right)$.

In the above equations, velocity fluctuations with depth over the vertical interval are ignored and the hydrostatic approximation is assumed.

The equation for the conservation of density employs velocities derived from the transports and is formulated as

$$\frac{\partial \rho}{\partial t} + \frac{\partial}{\partial x}(u\rho) + \frac{\partial}{\partial y}(v\rho) + \frac{\Delta}{h}(w\rho) = 0. \tag{13.5}$$

The pressure at any depth z in the water column is given by the hydrostatic relation,

$$\frac{\partial p}{\partial z} + \rho g = 0. \tag{13.6}$$

In the above set of equations, the notation is

- x, y, z = Cartesian co-ordinates, z being taken vertically upward,
- U, V = components of vertically-integrated velocity over the length h of the vertical side of a mesh,
- ζ = elevation of the water surface,
- p = pressure,
- u, v, w = components of velocity,
- A_h, A_v = the horizontal and vertical eddy viscosities, respectively,
- f = 0.113×10^{-3}/s and is the Coriolis parameter, assumed uniform over the modelled region,
- g = gravitational constant,
- Δ = difference taken between values at the lower and upper surfaces, respectively, of a mesh,
- ρ = density.

Considering now the lateral boundaries of the model, the velocity component normal to a coastline is set equal to zero. At open boundaries, elevations determined from the tide gauge data are prescribed. In the case of inflow, densities at appropriate levels of the water column are prescribed on the basis of field observations. For outflow, an upstream condition is applied since the density of the outgoing water is already known.

At the sea surface, the following approximation to the wind stress is applied,

$$(\tau_x, \tau_y) = \lambda (W_x, W_y) \sqrt{W_x^2 + W_y^2}, \qquad \lambda = 3.2 \times 10^{-6}. \tag{13.7}$$

At the sea bottom, the stress is given by

$$(\tau_x, \tau_y) = r(U, V) \frac{\sqrt{U^2 + V^2}}{h^2}, \qquad r = 0.003 \tag{13.8}$$

where

- W_x, W_y = components of the wind vector,
- λ = tangential wind stress coefficient,
- r = coefficient of bottom friction, spatially adjustable.

13.2.1 Explicit Finite-Difference Equations

The finite-difference solutions of Eqs. (13.1–13.4) follow essentially the three-dimensional extension by Sündermann (1971) of an earlier two-dimensional scheme introduced by Hansen (1956). Using the notation of Backhaus (1980), finite-difference formulation of Eqs. (13.1–13.4) is now presented using the grid notation shown in Fig.

13.1. Indices in the orthogonal co-ordinate directions x, y, z are taken as i, k, j, respectively. (Note that x increases with increasing k but y decreases with increasing i.) Time is expressed as $t = n\Delta t$. For economy of presentation, subscripts employed to denote a particular variable are omitted unless differing from i, k, j, n.

The finite-difference approximation of the continuity equation (13.1), assuming that the vertical velocity at the sea bottom is zero, is

$$w = w_{j+1} - h\left(\frac{U_{k+1} - U}{\Delta x} + \frac{V_{i-1} - V}{\Delta y}\right). \tag{13.9}$$

This equation is solved sequentially upward starting from the lowest layer. Using Eq. (13.2), the finite-difference approximation for the elevation of the sea surface is given by Eq. (13.10),

$$\zeta_{n+1/2} = \zeta_{n-1/2} + \Delta t\, w_{i,k,1}. \tag{13.10}$$

$$U_{n+1} = U + \Delta t[-T_{1x} + fV^* - T_{2x} + T_{3x} + \tilde{\tau}^x - \tilde{\tau}^x_{j+1}], \tag{13.11}$$

$$V_{n+1} = V + \Delta t[-T_{1y} - fU^* - T_{2y} + T_{3y} + \tilde{\tau}^y - \tilde{\tau}^y_{j+1}] \tag{13.12}$$

where

$$T_{1x} = \left[\frac{U\left(\frac{1}{2}h^*(U_{k+1} - U_{k-1}) - U(h_{k+1} - h)\right)}{\Delta x}\right.$$
$$\left. + \frac{V^*\left(\frac{1}{2}h^*(U_{i-1} - U_{i+1}) - \frac{1}{4}U(h_{i-1} + h_{i-1,k+1} - h_{i+1} - h_{i+1,k+1})\right)}{\Delta y}\right]\frac{1}{h^{*2}}$$
$$+ \frac{w^*}{2}\left(\frac{U_{j-1}}{h_{j-1}} - \frac{U_{j+1}}{h_{j+1}}\right)$$

and

$$T_{1y} = \left[\frac{U^*\left(\frac{1}{2}h^*(V_{k+1} - V_{k-1}) - \frac{1}{4}V(h_{k+1} + h_{i+1,k+1} - h_{k-1} - h_{i+1,k-1})\right)}{\Delta x}\right.$$
$$\left. + \frac{V\left(\frac{1}{2}h^*(V_{i-1} - V_{i+1}) - V(h - h_{i+1})\right)}{\Delta y}\right]\frac{1}{h^{*2}}$$
$$+ \frac{w^*}{2}\left(\frac{V_{j-1}}{h_{j-1}} - \frac{V_{j+1}}{h_{j+1}}\right)$$

The pressure gradient terms defined at time $t = (n + \frac{1}{2})\Delta t$ are

$$T_{2x} = \frac{h^*(p_{k+1} - p)}{\frac{1}{2}\Delta x(\rho + \rho_{k+1})},$$

$$T_{2y} = \frac{h^*(p - p_{i+1})}{\frac{1}{2}\Delta y(\rho + \rho_{i+1})}.$$

Ignoring atmospheric pressure, the pressure in layer j at the position i, k is

$$p_j = g\rho_1\left(\zeta + \frac{h_1}{2}\right) + g\sum_{l=2}^{j}\left(\frac{\rho_l + \rho_{l-1}}{2}\right)\left(\frac{h + h_{l-1}}{2}\right).$$

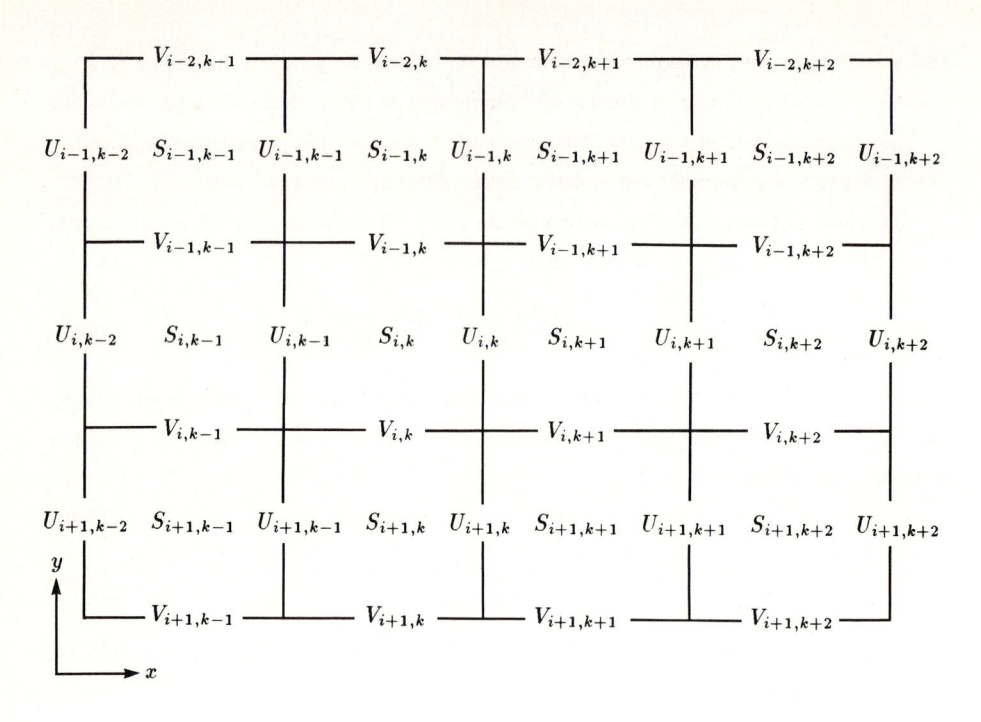

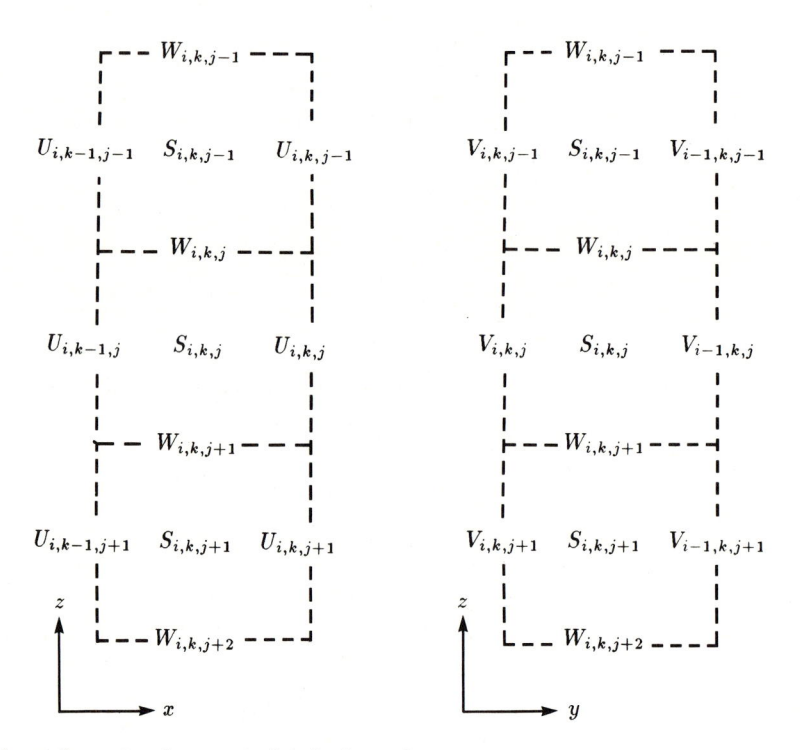

Fig. 13.1 Three-dimensional numerical indexing scheme.

The terms for horizontal eddy stress are

$$T_{3x} = \frac{h^* A_h (U_{k-1} + U_{k+1} + U_{i-1} + U_{i+1} - 4U)}{(\Delta \ell)^2},$$

$$T_{3y} = \frac{h^* A_h (V_{k-1} + V_{k+1} + V_{i-1} + V_{i+1} - 4V)}{(\Delta \ell)^2}.$$

At the ocean surface, the wind stress terms are

$$\tilde{\tau}_1^x = \lambda W^x \sqrt{W^{x2} + W^{y2}},$$

$$\tilde{\tau}_1^y = \lambda W^y \sqrt{W^{x2} + W^{y2}}.$$

Within the water column,

$$\tilde{\tau}_j^x = 2A_v \left(\frac{U_{j-1}/h_{j-1} - U/h}{h + h_{j-1}} \right),$$

$$\tilde{\tau}_j^y = 2A_v \left(\frac{V_{j-1}/h_{j-1} - V/h}{h + h_{j-1}} \right).$$

At the sea bottom,

$$\tilde{\tau}_{J+1}^x = \frac{rU\sqrt{U^2 + V^{*2}}}{h^{*2}},$$

$$\tilde{\tau}_{J+1}^y = \frac{rV\sqrt{U^{*2} + V^2}}{h^{*2}}.$$

Averaged values at U point are

$$V^* = \frac{V_{i-1} + V_{i-1,k+1} + V + V_{k+1}}{4},$$

$$w^* = \frac{w + w_{k+1} + w_{j+1} + w_{k+1,j+1}}{4},$$

$$h^* = \frac{h + h_{k+1}}{2},$$

Averaged values at V point are

$$U^* = \frac{U_{k-1} + U + U_{i+1} + U_{i+1,k-1}}{4},$$

$$w^* = \frac{w + w_{i+1} + w_{j+1} + w_{i+1,j+1}}{4},$$

$$h^* = \frac{h + h_{i+1}}{2}.$$

A simple upwind differencing scheme was used to compute density transports. The finite-difference form of the density equation (13.5) is

$$\begin{aligned}\rho_{n+1} = \rho_n &- \frac{\Delta t}{2} \left(\frac{(U + |U|)\rho}{\Delta \ell} + \frac{(U - |U|)\rho_{k+1}}{\Delta \ell} \right) \left(\frac{1}{h + h_{k+1}} \right) \\
&- \left(\frac{(U_{k-1} + |U|_{k-1})\rho_{k-1}}{\Delta \ell} + \frac{(U_{k-1} - |U|_{k-1})\rho}{\Delta \ell} \right) \left(\frac{1}{h + h_{k-1}} \right) \\
&+ \left(\frac{(V_{i-1} + |V|_{i-1})\rho}{\Delta \ell} + \frac{(V_{i-1} - |V|_{i-1})\rho_{i-1}}{\Delta \ell} \right) \left(\frac{1}{h_i + h_{i-1}} \right) \\
&- \left(\frac{(V + |V|)\rho_{i+1}}{\Delta \ell} + \frac{(V - |V|)\rho}{\Delta \ell} \right) \left(\frac{1}{h_i + h_{i+1}} \right) \\
&+ \left([(w + |w|)\rho + (w - |w|)\rho_{j-1}] \right. \\
&\left. - [(w_{j+1} + |w_{j+1}|)\rho_{j+1} + (w_{j+1} - |w_{j+1}|)\rho] \right) \frac{1}{2h} \end{aligned} \quad (13.13)$$

where the additional notation is

Δt = time step,
$\Delta x = \Delta y = \Delta \ell$ = grid size in the horizontal direction,
n = number of time steps,
h = layer thickness,
d = undisturbed upper layer thickness,
J = maximum number of layers,
$J+1$ = sea bottom.

It will be noted that terms approximating horizontal and vertical eddy diffusion have been omitted from this equation. Two considerations arise. First, strong mixing processes exist in the system. Secondly, a substantial degree of mixing results from the upwind formulation of the advective terms. Thus, for present purposes which are considered exploratory in nature, the density equation formulated above is considered adequate. Numerical diffusion will take the place of turbulent diffusion.

The thickness of the upper layer is given by

$$h_{n+1/2} = d + \zeta_{n+1/2}. \tag{13.14}$$

The explicit finite-difference procedure described above imposes two stability criteria on the time step. Thus, in accordance with the Courant-Friedrichs-Lewy criterion,

$$\Delta t < \frac{\Delta \ell}{\sqrt{2g\,H_{\max}}} \tag{13.15}$$

where $H_{\max}$ is the maximum total depth occurring in the modelled region. Furthermore, with respect to the vertical diffusion process,

$$\Delta t < \frac{h^2}{2A_v}. \tag{13.16}$$

These limitations on the time step can be mitigated through the use of implicit finite-difference procedures. More recent developments of the model employ implicit procedures in both horizontal and vertical contexts (Backhaus, 1980). In the present version of the model, while an explicit procedure in the horizontal context is retained, an implicit procedure in the vertical context is employed. This is illustrated for the x-directed equation of motion which includes formulation of the vertical stress terms at the future time step. Thus, terms in the momentum equation involving velocities at the future time step $n+1$ may be separated from the remaining terms as follows:

$$U_{n+1} - \tfrac{1}{2}\Delta t(\tilde{\tau}^x_{n+1} - \tilde{\tau}^x_{j+1,n+1}) = b_j \tag{13.17}$$

where

$$b_j = U + \Delta t\bigl(-T_{1x} + fV^* - T_{2x} + T_{3x} + \tfrac{1}{2}(\tilde{\tau}^x - \tilde{\tau}^x_{j+1})\bigr).$$

Introducing the following terms,

$$a_j = \frac{\Delta t \, A_{v_j}}{h_{j-1} + h_j}$$

and

$$c_j = 1 + \frac{a_j + a_{j+1}}{h_j},$$

Eq. (13.17) may be written as

$$-\frac{a_j U_{j-1}}{h_{j-1}} + c_j U_j - \frac{a_{j+1} U_{j+1}}{h_{j+1}} = b_j \qquad (13.18)$$

in which transport components at time $(n+1)\Delta t$ are confined to the left hand side of the equation and those at time $n\Delta t$ to the right. Assuming known stresses at the sea surface and bottom, the system of equations in the vertical can be readily solved by Gaussian elimination provided the determinant of the system of equations (13.18) from $j = 1$ to J is not zero.

At the ocean surface let

$$a_1 = 0 \quad \text{and} \quad b'_1 = b_1 + \bar{\tau}_1^x \frac{\Delta t}{2}.$$

At the sea bottom let

$$a_{J+1} = 0 \quad \text{and} \quad c'_J = c_J + R$$

where R is the friction term evaluated at time $t = n\Delta t$ to avoid introduction of a quadratic term and where

$$R = \tau_{J+1}^x \frac{\Delta t}{2}.$$

The Gaussian elimination procedure is simplified by using the following recurrence relations which evaluate each term in a sequence from its predecessors,

$$\begin{aligned} \alpha_j &= b'_j + \frac{a_{j+1} \alpha_{j+1}}{\beta_{j+1} h_{j+1}}, \\ \beta_j &= c'_j - \frac{(a_{j+1}/h_{j+1})^2}{\beta_{j+1}}. \end{aligned} \qquad (13.19)$$

The terms α and β then permit calculation of the x-directed transport components in the following equation,

$$U_j = \frac{\alpha_j + (a_j U_{j-1})/h_{j-1}}{\beta_j} \quad \text{for} \quad j = 2, \ldots, J. \qquad (13.20)$$

where $a_1 = 0$ and $U_1 = \alpha_1/\beta_1$.

The basic configuration of the model (Fig. 13.2) is essentially a machine adaptation of the 4-km vertically-integrated two-dimensional model GF2. Subsidiary channels, represented by one-dimensional schemes in GF2, were replaced by simple extensions of the three-dimensional grid selected to give approximately correct dynamic responses. An initial configuration which had boundary openings at A, C, and D was altered to that having openings at A and B. A vertical section through line OO′, shown in the inset, illustrates the grid scheme in the vertical. The intervals in the vertical were selected to provide increased resolution of the more intense stratification nearer to the sea surface. The number of levels was subsequently increased from seven to eight. Depths were prescribed at the centre of each column, the depth of the lowest level being taken to the sea bottom. The numbers and layer thicknesses (m) were, respectively, 1(15), 2(15), 3(30), 4(30), 5(60), 6(100), and 7(150). In later runs the uppermost layer was divided into two layers, one of 5 m and the other 10 m.

13.3 GENERAL TRIALS OF THE MODEL

Two types of trial were undertaken. In the first, elevations determined from the summed M_2 and K_1 tides, were prescribed at openings A, C, and D. Unrealistic flows at two of the boundary openings, C and D, resulted. In the second type, it was found that these could be eliminated, consistent with earlier experience obtained from GF2, by reducing the dynamical dependence of the computed elevations on prescribed elevations at boundary openings. This was done by extending the model to include the Puget Sound system to the south and a single channel to the north (denoted as the northern channel in this chapter) which would simulate the complex of heavily dissipative passages leading northward from the Strait of Georgia. Elevations in this latter case were prescribed at openings A and B.

The basic procedure in these trials was to run the model until constant values were obtained for amplitudes and phases of the M_2 and K_1 tidal elevations and velocity fields and, in the case of the latter, for the field of residual circulation. A preliminary run of four cycles for a particular tidal constituent was generally required before data suitable for analysis were obtained. This could be reduced somewhat by the introduction of initial elevation and velocity fields obtained at the conclusion of an earlier run rather than starting from rest.

Energy is lost from the barotropic tidal co-oscillations to frictional dissipation, generation of internal modes, and by increasing the potential energy of stably stratified fluid through vertical mixing within the system. Estimates of the summed global M_2 barotropic to baroclinic tidal energy conversion through the generation of internal tidal energy fluxes on continental shelves and slopes is about 0.3% of the estimated tidal energy dissipation required from astronomical observations of the secular acceleration of the moon (Baines, 1982). For a local West Coast fjord, it has been estimated that of the energy dissipated from the M_2 barotropic tide, 24% is attributable to the excitation of baroclinic motions (Gade and Edwards, 1980).

The basic assumption at this stage of the present work is that an empirical adjustment of frictional dissipation in a correctly formulated model giving a satisfactory

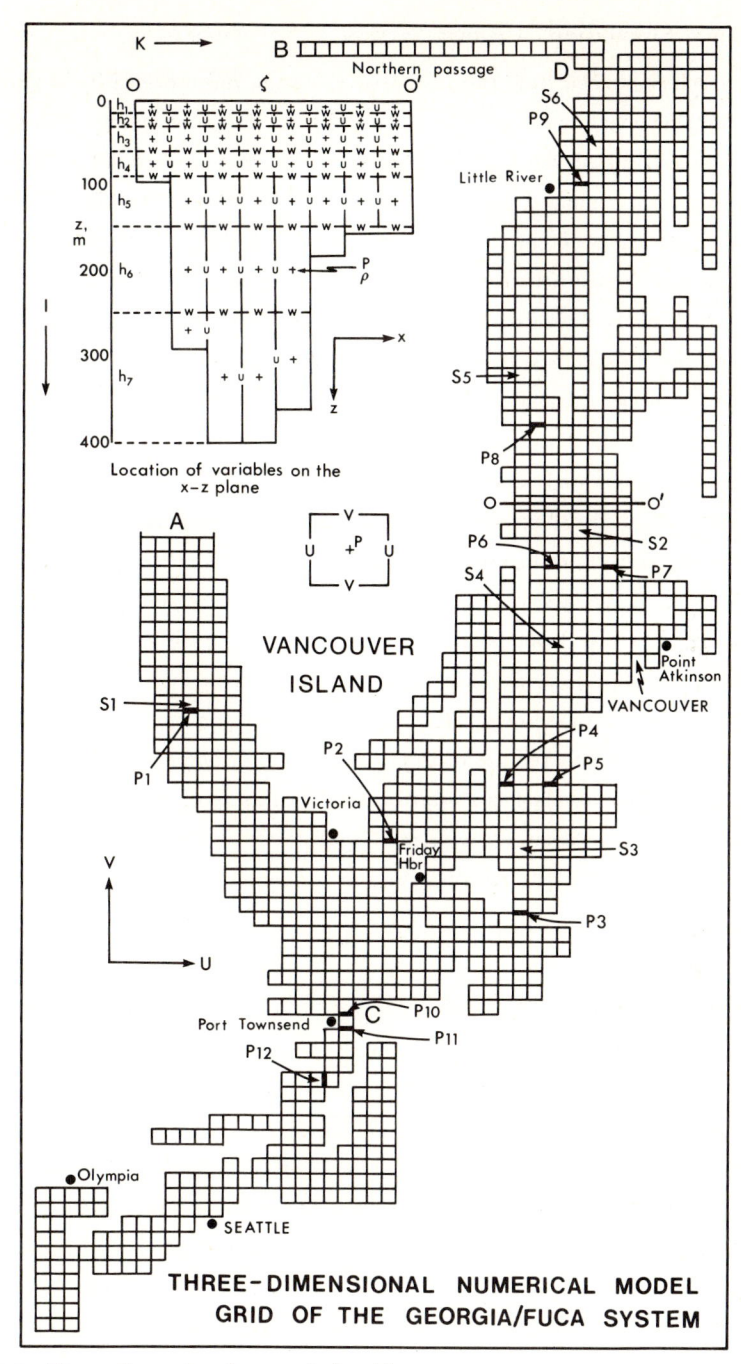

Fig. 13.2 Three-dimensional numerical grid.

reproduction of the tides and streams will also provide a good estimate of the tidal energy dissipation in the system.

13.3.1 Tidal Simulation — Short Version

The results obtained from the tidal trial using the constant-density shortened version of GF6 (boundary openings at A, C, and D) are discussed briefly. This shortened-version trial constituted the first preliminary test of the computational scheme. The model satisfactorily simulated the main features of the M_2 and K_1 tides. The primary shortcoming was apparent in the flows at certain boundary openings. The values of model parameters employed were $A_H = 8 \times 10^6$ cm^2/s, $A_V = 200$ cm^2/s, and $r = 0.003$.

Illustrative of these results, Table 13.1 shows the computed flood volume transports (total transport through a section on a flood tide from low water slack to high water slack) and Greenwich phases of the associated flow maxima normal to the boundary openings A, C, and D. For comparison, the results obtained for these locations in the earlier vertically-integrated model GF2 which was well calibrated, are also included. At the entrance to the Puget Sound system, the flood volumes are to be taken as leaving the model in order to provide the tidal complement in that system. Significant discrepancies exist. the largest being that of the flood volume and phase of the K_1 tide at the entrance to the Puget Sound system. Large errors are also evident in the K_1 flood volume entering Juan de Fuca Strait and in the phase of the M_2 as well as in the flood volume and phase of the K_1 tides at the northern opening. A further major discrepancy that arises in the use of this configuration of boundary openings is a large and highly unrealistic residual flow which enters the northern opening and which leaves through Juan de Fuca Strait. It is present at all levels of the model. The presence of such a net flow is particularly unsatisfactory in the context of density-driven circulation studies.

13.3.2 Tidal Simulation — Long Version

To reduce these shortcomings, the numerical scheme was extended to include the full configuration shown in Fig. 13.2. The shape and depths of these additions were individually adjusted to approximate actual topography and thus ensure reasonably correct tidal volume transports and energy fluxes at their openings. Several trial runs were made to determine the sensitivity of the model to selected values of the eddy viscosities and coefficients of bottom friction. The values finally selected were $A_H = 2 \times 10^6$ cm^2/s (northern channel, $A_H = 4 \times 10^6$ cm^2/s), $A_V = 800$ cm^2/s, $r = 0.003$ (northern channel, $r = 0.03$). The flood volume transports and Greenwich phases of the flow maxima for this extended version of the model indicate that the worst deficiencies apparent in values at the boundary openings of the short version have been much reduced (Table 13.1). This is substantiated by the M_2 co-amplitude and co-phase lines shown in Fig. 13.3. The degenerate amphidromic distribution of these lines in the inner part of Juan de Fuca Strait, the standing wave character of the tides in the Strait of Georgia and Puget Sound, the large changes in amplitude and phase in the northern channel are all reproduced in the model. Harmonic constants derived from tide gauge observations for representative locations are included within the square brackets in the figure. Distributions of the K_1 amplitude and phase (Fig. 13.4) show that the main features of this tide within the system are also reproduced.

Table 13.1 Flood volume transports (km^3) and Greenwich phases g (°) of the flow maxima for open boundary cross-sections, A, C, and D (Fig. 13.2) of the numerical models GF2, GF6 (short), and GF6 (long).

Location	Model	M_2		K_1	
		Vol.	g	Vol.	g
Juan de Fuca Strait, A	GF2	18.90	62	25.20	76
	GF6 (short)	16.30	55	18.90	71
	GF6 (long)	15.40	65	29.20	76
Puget Sound, C	GF2	4.93	53	4.06	69
	GF6 (short)	5.15	29	0.50	20
	GF6 (long)	4.47	45	4.39	70
Discovery Passage, D	GF2	3.78	53	1.94	43
	GF6 (short)	4.19	108	3.16	109
	GF6 (long)	2.04	48	1.22	24

A more detailed discussion of the calibration of this model is considered using the harmonic constants, derived from observations, GF2 and GF6, respectively (Table 13.2). Constants for Victoria, Port Townsend, and Friday Harbor (Fig. 13.2) are representative of the tidal regime in the inner part of Juan de Fuca Strait. Constants for Point Atkinson and Little River describe the tides in the Strait of Georgia; Seattle and Olympia, the tides in the Puget Sound system. In all cases the observed constants are derived from at least 1 yr of tide gauge records.

It is desirable to review briefly theoretical notions pertaining to the effects of frictional dissipation and partial transmission on Kelvin wave reflection in a narrow co-oscillating bay. Prescinding from the smaller-area Puget Sound system and in the absence of damping and topographic detail, the superposition of the M_2 incoming and perfectly reflected Kelvin waves should give an amphidromic point centrally located in the inner part of Juan de Fuca Strait. The effect of frictional dissipation and imperfect reflection is to shift the amphidromic point towards the Vancouver Island shore thereby increasing the amplitude in the inner part of Juan de Fuca Strait (Hendershott and Speranza, 1971). Further effects of increasing friction are to reduce the tidal amplitudes and decrease the Greenwich phase lags in the Strait of Georgia. For the frictionless

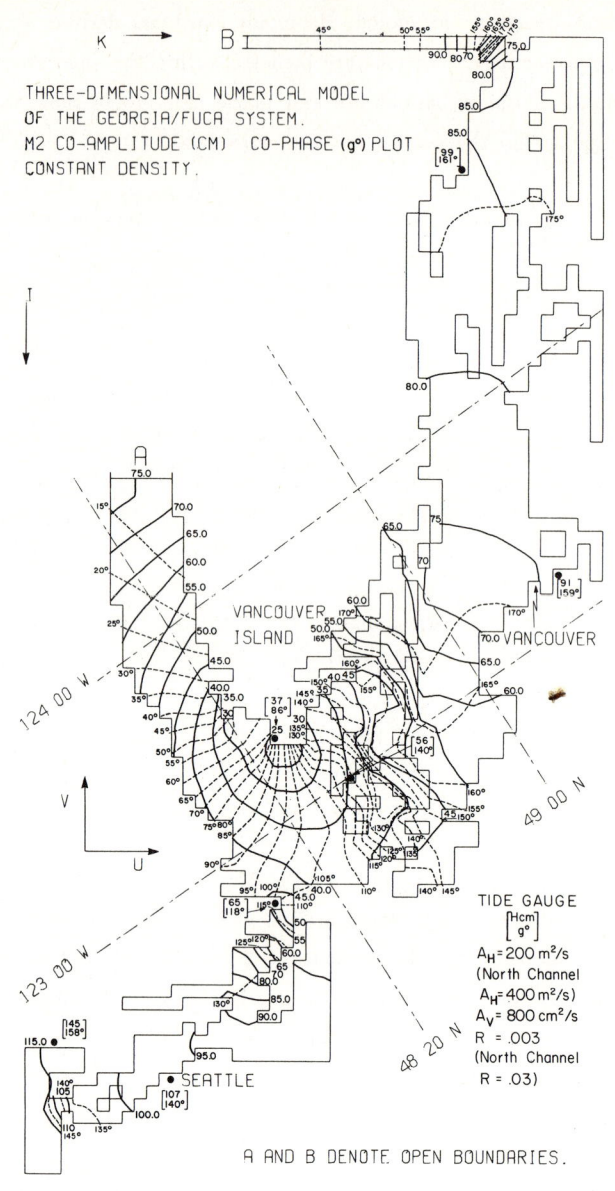

Fig. 13.3 M_2 co-amplitude (solid lines) and co-phase (dashed lines) determined from modelled elevations.

non-rotating case, there should be a 180° phase lag between the outer part of Juan de Fuca Strait and the northern part of the Strait of Georgia. For the case of the K_1 tide, though increased dissipation will lead to reduced amplitudes, the phase lags should increase. In the absence of dissipation, no difference in phase should exist between Juan de Fuca Strait and the Strait of Georgia.

Considering now the harmonic constants obtained from GF6 (Table 13.2) it is

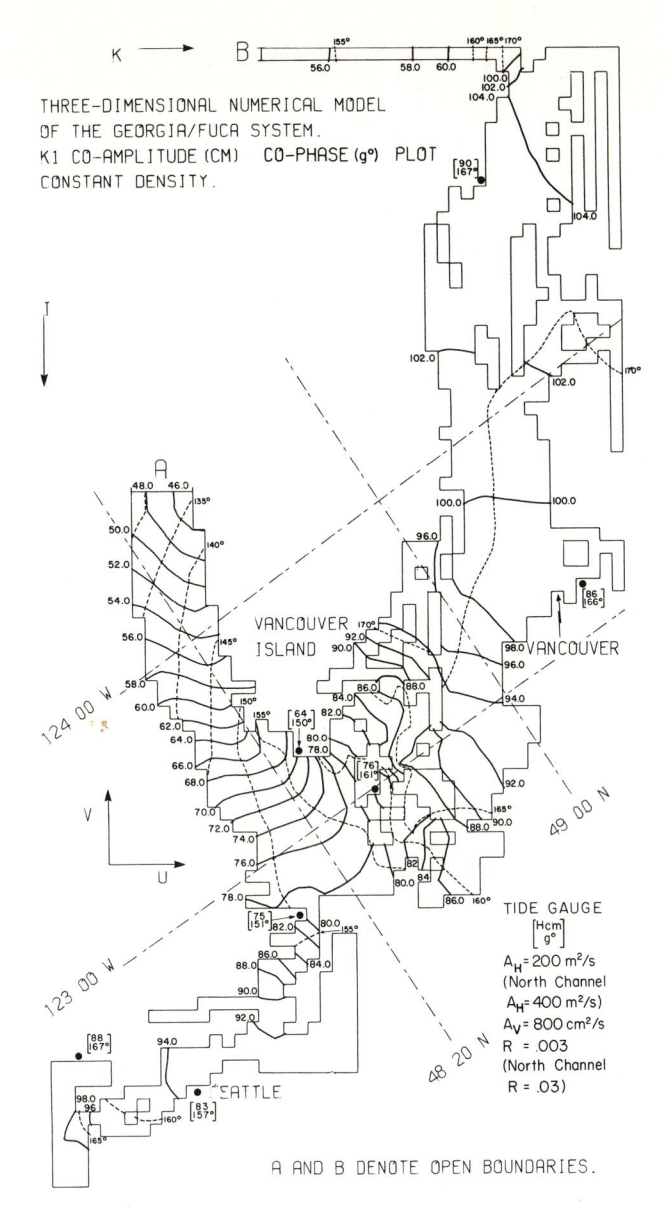

Fig. 13.4 K_1 co-amplitude (solid lines) and co-phase (dashed lines) determined from modelled elevations.

evident that no adjustment of frictional dissipation can simultaneously improve the agreement between the computed and observed M_2 and K_1 harmonic constants. Thus, if the coefficient of bottom friction is reduced to increase the amplitude of the M_2 tide at Point Atkinson, the discrepancies between the respective observed and computed M_2 phase lags and K_1 amplitudes will be increased.

The small M_2 amplitudes in the inner part of Juan de Fuca Strait and large M_2

Table 13.2 The M_2 and K_1 harmonic constants, amplitude H (cm) and phase g (°) derived from observations and from the vertically-integrated (GF2) and three-dimensional (GF6) numerical models (4-km mesh).

Location	Tide gauge number	M_2		K_1	
		H	g	H	g
Victoria	7120				
observed		37	86	64	150
GF2		32	83	65	153
GF6		22	97	72	159
Port Townsend	7160				
observed		65	118	75	151
GF2		62	114	74	152
GF6		58	115	81	155
Friday Harbor	7240				
observed		56	140	76	161
GF2		52	143	75	162
GF6		39	130	82	159
Point Atkinson	7795				
observed		93	158	86	166
GF2		93	164	88	170
GF6		75	171	99	169
Little River	7793				
observed		99	161	90	167
GF2		101	167	93	170
GF6		85	175	104	171
Seattle	7180				
observed		107	140	83	157
GF2		102	140	82	160
GF6		99	134	94	160
Olympia	—				
observed		145	158	88	167
GF2		137	161	89	171
GF6		121	147	99	167

tidal phases in the Strait of Georgia indicate a degree of energy dissipation which is too low while the M_2 amplitudes in the Strait of Georgia suggest it to be excessive. The K_1 tidal amplitudes are too high, indicative of insufficient dissipation. Unfortunately the good agreement between the computed and observed K_1 tidal phases is misleading since a proper non-linear bottom frictional coupling with the M_2 tide, referred to in

Chapter 4 in connection with the model GF2, should lead to a substantial increase in phase of the K_1 tide.

From these results it is concluded that, though the numerical scheme satisfactorily reproduces the main features of the M_2 and K_1 tides within the system, there is need for considerable improvement in the formulation of dissipative processes within the model.

To illustrate the general character of the tidal streams, Fig. 13.5 shows the velocity vectors obtained for an average tidal ebb. The strong flows in the southern Strait of Georgia, Juan de Fuca Strait, Admiralty Inlet, and the northern channel are evident. Throughout most of the system the maximum ebb (and flood) currents occur within about 1 h.

To illustrate the vertical distributions of tidal velocity amplitudes and phases computed by the model GF6 for representative locations, Fig. 13.6a shows the major M_2 velocity component directed along the median line of the channel as well as the associated Greenwich phases for the six levels of the model at location S1 in Juan de Fuca Strait (Fig. 13.2). Also shown are the observed values obtained from moored current meter observations, each record being in excess of 1 month (Huggett et al., 1976b). The velocities are somewhat smaller than those observed and also smaller than the corresponding vertically-averaged value (47 cm/s) obtained from GF2. The large scatter in the M_2 phase values based on observations makes comparison difficult with those computed. From general inspection of the observed distributions of tidal phase over a number of cross-sections, the phase differential over the water column would appear reasonable. The phases of the two uppermost layers agree well with that obtained from GF2 (67°).

Similar plots for the K_1 tidal streams over the greater part of the water column (Fig. 13.6b) show larger values compared to the vertically-averaged velocity obtained from GF2 (29 cm/s) in addition to those observed. This is consistent with the rather large K_1 tidal amplitudes in the Strait of Georgia (Table 13.2). The vertically-averaged phase leads the computed K_1 phase of the surface tide in the Strait of Georgia by some 90° and agrees well with that obtained for GF2 (82°). Both models, however, yield phase lags which are larger than the observed values. A further interesting feature is the occurrence of larger phase lags in the lower part of the water column.

13.3.3 Energy Fluxes

The energy fluxes averaged over a tidal cycle through a particular cross-section can be obtained by integrating the following expression over the mesh sides in the cross-section (Taylor, 1919),

$$W \simeq \tfrac{1}{2} \rho g \, H U \, h \, \cos(g_H - g_U)$$

where

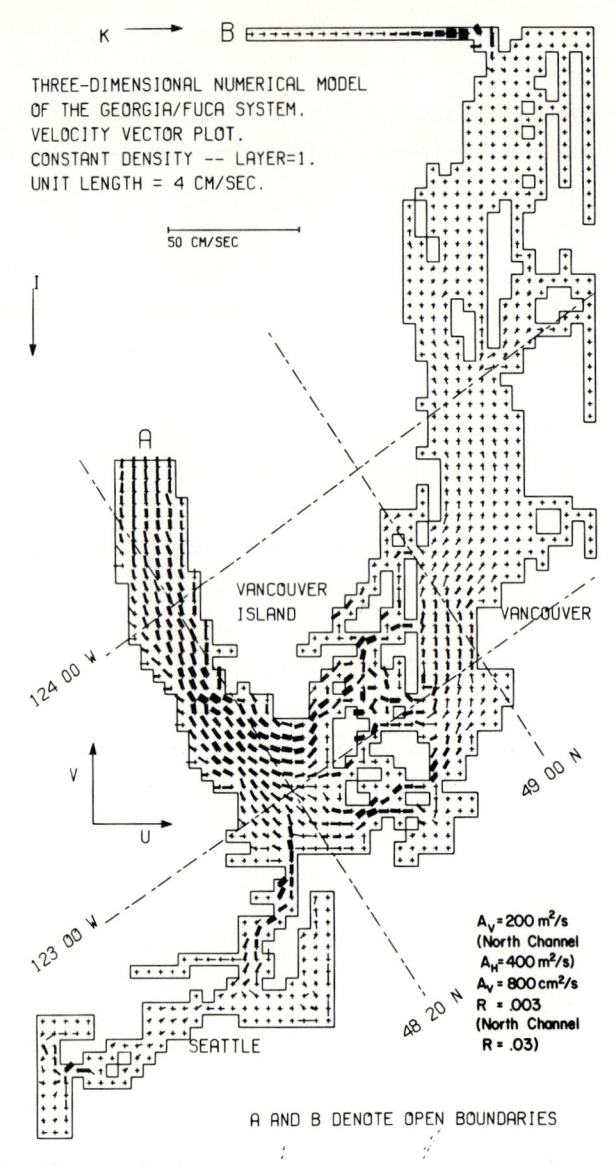

Fig. 13.5 Distribution of velocity vectors illustrating the typical distribution of ebb tidal streams in the top layer.

W = flux of barotropic tidal energy (ergs/s) normal to the direction of wave propagation,

H, U = surface elevation (cm) and velocity (cm/s) amplitudes, respectively, of a particular constituent,

g_H, g_U = surface and velocity Greenwich phases, respectively, of that one constituent.

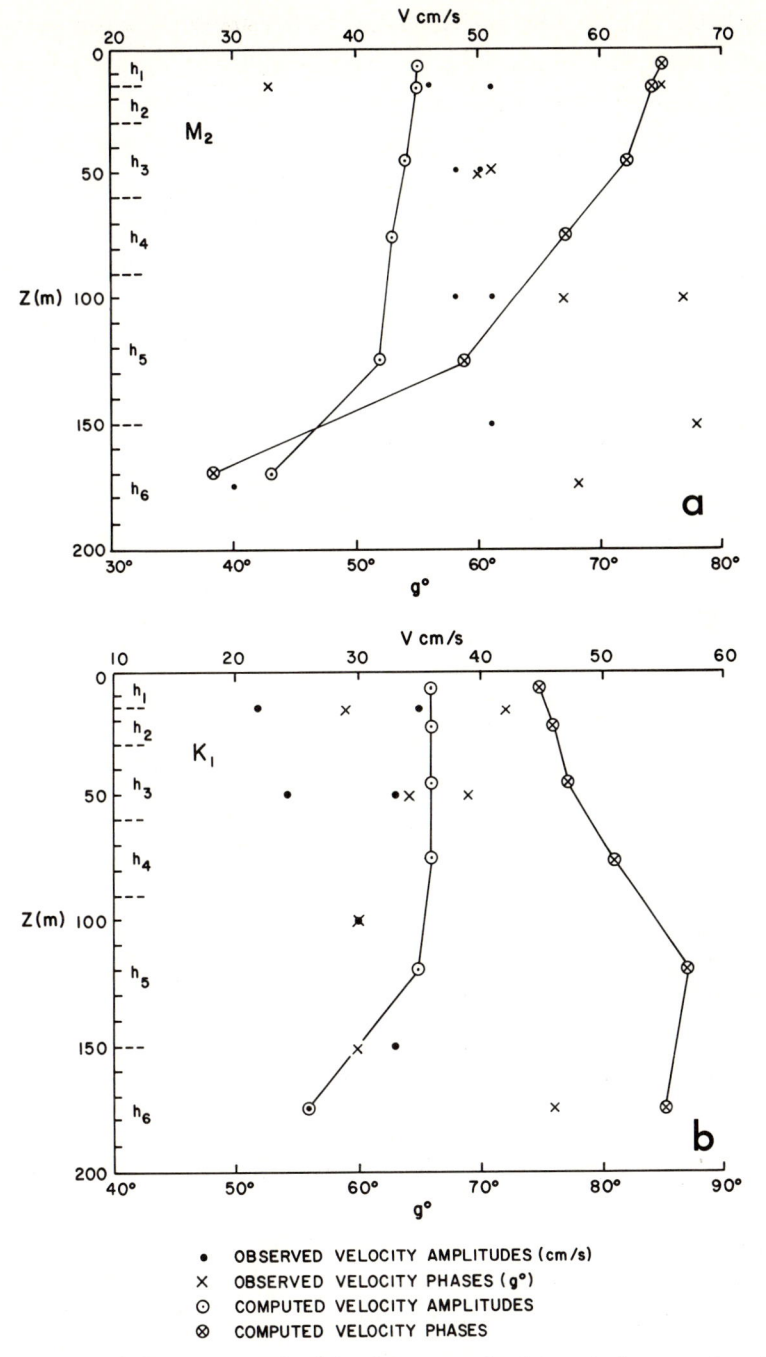

Fig. 13.6 Vertical distribution of tidal velocity amplitudes and phases at location S1 (Fig. 13.2) for (a) M_2 and (b) K_1 tidal constituents.

Further illustrative of the model's performance, the mean energy fluxes for the M_2 and K_1 constituents at the various boundary openings are shown for both GF2 and

GF6 in Table 13.3. In GF6, the M_2 energy fluxes entering the system through Juan de Fuca Strait and leaving to enter the Puget Sound region and the passages leading northward from the Strait of Georgia, respectively, are too small, consistent in part with the low M_2 amplitudes and associated streams. For the diurnal tide, the energy flux entering Juan de Fuca Strait is too large, consistent in part with the large K_1 amplitudes and associated streams. The proportioning of the original energy entering into the Puget Sound system and the northern channel from Juan de Fuca Strait would, however, appear to be satisfactory, justifying the rather crude numerical schematization used for these regions.

Table 13.3 M_2 and K_1 mean energy fluxes* (10^{15} erg/s) through boundary openings A, C, and D (Fig. 13.2) of the vertically-integrated (GF2) and three-dimensional (GF6) numerical models (4-km mesh) of the Georgia/Fuca system. Negative signs denote fluxes out of Juan de Fuca Strait and the Strait of Georgia.

Location	M_2		K_1	
	GF2	GF6	GF2	GF6
Juan de Fuca Strait, A	32.10	24.60	11.90	13.90
Puget Sound, C	−4.72	−2.78	−0.72	−0.61
Discovery Passage, D	−4.71	−3.29	−1.94	−1.93

* Computed from amplitudes and Greenwich phases of elevations and inward-directed normal stream components at selected flow cross-sections.

13.3.4 Puget Sound Model Trials

To investigate the role of horizontal eddy viscosity and related matters and to obtain guidance in optimizing the performance of the model, additional tidal trials were carried out. These calculations were limited for economy and simplicity to the Puget Sound part of the overall model. Mixed tidal (M_2 and K_1) elevations were prescribed at the single open boundary C (Fig. 13.2) for this smaller model. The general character of the tides in this part of the system is shown in Figs. 13.3 and 13.4. The amplitudes and phases of both constituents increase progressively from the mouth to the closed end of the system (Olympia). For present purposes it is sufficient to present the harmonic constants obtained from the model for this latter location to illustrate the effects of advective accelerations, changing horizontal and vertical eddy viscosities, and partial- and full-slip conditions at the coastal boundaries on the computed tidal regime within the system (Table 13.4).

It is apparent from Table 13.2 that the M_2 tidal amplitude computed by the overall model GF6 for the entrance to the Puget Sound system is 10% lower than the observed

Table 13.4 Effects of advective accelerations, horizontal eddy viscosity, and slip conditions on the M_2 and K_1 amplitudes H (cm) and phase g (°) for Olympia (Fig. 13.2) computed by a three-dimensional model of the Puget Sound system. The differences between observed and modelled values are tabulated. Unless otherwise specified, values of $A_V = 200$ cm^2/s and $r = 0.003$ are used.

Trial	Tide	M_2				K_1			
		H	g	ΔH	Δg	H	g	ΔH	Δg
	Prescribed tides at Port Townsend	65	118	—	—	75	151	—	—
	Observed tides at Olympia	145	158	—	—	88	167	—	—
	Modelled tides at Olympia								
1	with advection, partial-slip. $A_H = 8 \times 10^6$ cm^2/s	94	202	−51	44	84	188	−4	21
2	with advection, full-slip. $A_H = 8 \times 10^6$ cm^2/s	142	176	−3	18	89	175	1	8
3	no advection, partial-slip. $A_H = 8 \times 10^6$ cm^2/s	95	202	−50	44	85	188	−3	21
4	no advection, $A_H = 0$	161	171	16	13	94	169	6	2
5	no advection, $A_H = 0$ $A_V = 10$ cm^2/s	165	172	20	14	94	169	6	2

value. If the observed M_2 amplitude is prescribed at the entrance to this smaller model, the horizontal eddy viscosity must be increased by about 30% to supress the non-linear instability resulting from the increased velocities in the system. Following some minor local modifications to coastal boundaries, the revision of the Puget Sound model used in these tests required a horizontal eddy viscosity of 8×10^6 cm^2/s to maintain stability. The high value of horizontal eddy viscosity found necessary to ensure stability, as in the earlier trials of the overall model, GF6, would appear to extend an unrealistic controlling influence on the tides within the system. This is notably at variance with experience obtained in the use of models GF2 and GF3 where large changes in the value of horizontal eddy viscosity brought about no appreciable change in the character of the computed tides within the system (Crean, 1978) and where horizontal eddy viscosity terms are not required to maintain stability.

The formulation of horizontal stresses near coastal boundaries assumes that velocities located on, or landward of, mesh sides defining the coast, are zero. Thus, energy is lost from the modelled fluid through work done on these boundaries. An additional source of energy loss is through the smoothing of spatial changes in velocities in the interior of the fluid by the horizontal stress terms. The first two trials illustrate the effect of replacing a partial-slip with a full-slip condition. The increased amplitudes and decreased phase lags that result show an appreciable reduction in energy dissipation, particularly in the case of the M_2 tide. A further significant result, not illustrated here, is a considerable change in the distribution of residual velocities.

The effect of omitting the advective accelerations for the partial-slip case is demonstrated by comparing the first and the third trials. The net result is a relatively small increase in the amplitude of the M_2 tide. A concomitant effect is the virtual supression of the residual circulation. Comparing the results of trials 3 and 4 shows the effect of eliminating horizontal stresses for the partial-slip case used in the overall model. A large increase in amplitude and decrease in phase lag is evident for the M_2 tide while the K_1 tide is relatively unaffected. It would appear that some improvement could result from using a full-slip condition in GF6.

The small effect of a large reduction in the vertical eddy viscosity from $A_V = 200$ cm^2/s to 10 cm^2/s on the computed barotropic M_2 and K_1 tides within the system is apparent when comparing the results from trials 4 and 5. In fact, very much larger values of vertical eddy viscosity must be employed before any significant effect on the computed velocities results. This is discussed further in Section 13.3.7.

If these runs are repeated following an adjustment of the model coastline to eliminate locations where the bulk of the tidal volume transport is forced to change direction through a right angle over a single vertical column of meshes, then a much smaller horizontal eddy viscosity may be employed. The result is that the sensitivity of the computed tides to varied factors in the above trials is then greatly reduced. It would thus be desirable to employ a smaller grid size to provide a better resolution of the complex coastal topography characteristic of these waters. In addition to the above, further trials were carried out using the Puget Sound test model. Since these tests did not yield significant improvements, the results are summarized briefly.

A mixed tidal experiment was carried out in which the existing central difference advective acceleration in the equation of motion was reformulated in terms of upwind differencing. It was found, however, that the numerical dissipation of this latter scheme, while obviating the need for a large horizontal eddy viscosity in the context of maintaining stability, itself provided a comparable degree of dissipation.

Another experiment concerned use of a horizontal eddy viscosity which is dependent on the strain rate in the modelled fluid, following early practice in atmospheric models (Smagorinsky, 1963). Alternatively, one could relate such horizontal eddy viscosity to vorticity. However, since the expression tends to be dominated by one of the horizontal velocity gradients, the methods are essentially equivalent. Adjustment of the empirical parameters included in the formulation resulted in a degree of energy dissipation fully equivalent to that required when a constant horizontal eddy viscosity sufficient to maintain stability, was imposed.

It has been noted earlier that adequate resolution of the major features of the tidally-induced residual flow field would require a 2-km mesh resolution. A number of trials were carried out in which the 4-km mesh test model of the Puget Sound system was modified to yield a 2-km mesh model while maintaining the same configuration of coastal boundaries and depth. Under the same tidal forcing where the summed M_2 and K_1 tidal constituents were prescribed at the single open boundary, it was found that the improved resolution of the tidally-driven residual circulation that resulted could be obtained only through the use of much larger horizontal eddy viscosities compared to those used in the 4-km mesh trials. This latter is presumably due to more intense horizontal gradients of velocity that occur in the higher resolution model. This is consistent with experience obtained in comparing particular eddies in the 4-km mesh (GF2) and the 2-km mesh (GF3) models. Trials involving a potential enstrophy and energy conserving scheme (Arakawa and Lamb, 1981), reported to improve the simulation of the non-linear aspects associated with sharp changes in topography, were not pursued in virtue of the uncertain effect of the predominance of complex coastlines and the large number of horizontal grid points entering into calculation of the variable close to such boundaries. While employing the simple numerical formulation of the advective terms, comparison of observed and modelled residual velocities displayed a quite remarkable accord in the case of the vertically-integrated models GF3 and GF7.

13.3.5 Numerical Diffusion

A fundamental problem in the numerical simulation of time-dependent density-driven circulation concerns the introduction of a spurious process of numerical diffusion. Other than under the unrealistic requirement that the Courant number equal unity, *i.e.*, that an advecting velocity results in a displacement of exactly one grid length over one time step, the stable formulation of the advective terms generally entails some measure of numerical diffusion. If this purely numerical diffusion exceeds the actual mixing in the original system, it becomes impossible to regulate these density-driven flows through adjustment of coefficients in the eddy diffusion terms of the density conservation equation and hence, to simulate events involving significant changes in the density field such as deep water renewal, the effects of major wind events, or the persistent estuarine exchange.

An examination of the computed density fields, including density computations at the end of an 8-day run, shows the presence of extensive vertical mixing throughout the model considerably in excess of that observed after a period of approximately a month in the actual system. This is illustrated (Fig. 13.7a and b) by representative vertical profiles of computed and observed density at locations S1 (Juan de Fuca Strait) and S2 (Strait of Georgia) (Fig. 13.2). At S1 the observed profiles show that no significant change occurred over the month of July. In contrast, the computed profile indicates a much increased density in the upper part of the column and, despite its proximity to the major source of inflowing dense water, a considerable dilution at depth. Similar profiles for S2 show a dilution throughout the greater part of the column which does not accord with the negligible change evident in the observed values. From these results it is concluded that a significant level of numerical diffusion in the model exists that is not present in the real system. This then limits its present usefulness to relatively short semi-diagnostic trials.

13.3.6 Tidally-Induced Residual Circulation in a Vertical Plane

Over and above the pattern of tidally-induced residual circulation in the horizontal described for the vertically-integrated models GF2, GF3, and GF7, discussion of such circulation in the three-dimensional model GF6 must take into account the possibility of tidally-induced residual circulation in the vertical. The existence of such circulation for a constriction in a tidal channel has been theoretically demonstrated by Ianiello (1979). In the present numerical scheme the formation of a residual vertical eddy is illustrated for a test channel one mesh (4 km) in width, 40 meshes (160 km) in length, possessing the eight-level vertical spacings and containing a single thin shallow sill. Tidal elevations (M_2 and K_1) are prescribed at the elevation points on the open boundaries at either end of the channel. The model parameters for vertical and horizontal eddy viscosities, and coefficient of bottom friction are, respectively, $A_H = 10^7$ cm^2/s, $A_V = 800$ cm^2/s, and $r = 0.003$. The reversing tidal streams with their associated eddies (a large eddy in the lee of the sill for each flood and ebb tidal stream, respectively) give rise to a net tidal residual circulation shown in Fig. 13.8. Thus, a net divergence occurs in the surface layers over the sill and a net flow in the deeper layers towards the sill. The presence of small direction changes in alternate vectors in regions of high velocities indicate diversion of energy to grid-scale wavelengths by the field accelerations.

A further feature of these trials concerns the vertical distributions of amplitudes and phases of the M_2 and K_1 tidal velocities near the constriction. Of particular interest are the large differences in phase occurring over the length of the water column for the K_1 tide. The tidal stream turns when the time-acceleration term in the momentum equation is zero, i.e., when the pressure gradient term is equal and opposite to the sum

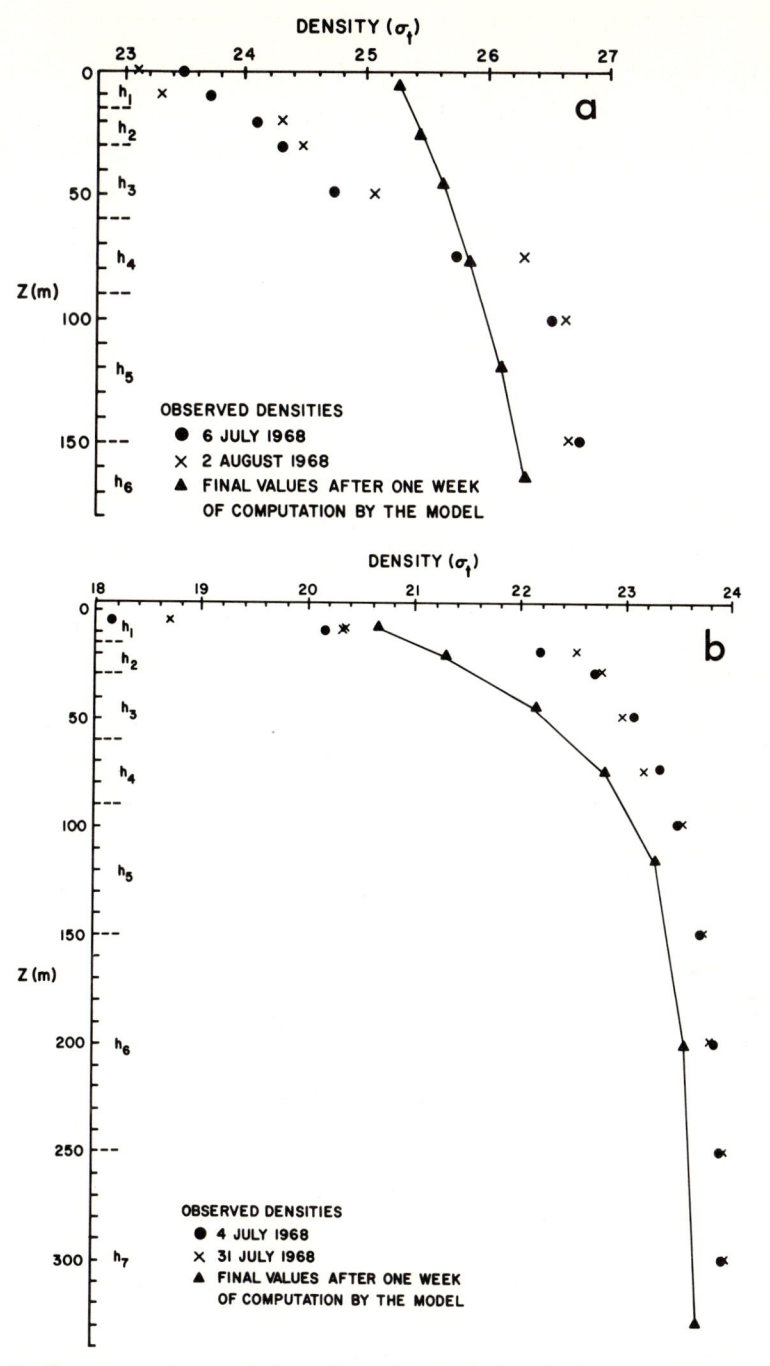

Fig. 13.7 Representative vertical profiles of computed and observed densities at (a) location S1 (Juan de Fuca Strait) and (b) S2 (Strait of Georgia).

of the remaining terms on the right hand side of the momentum equation. Qualitative

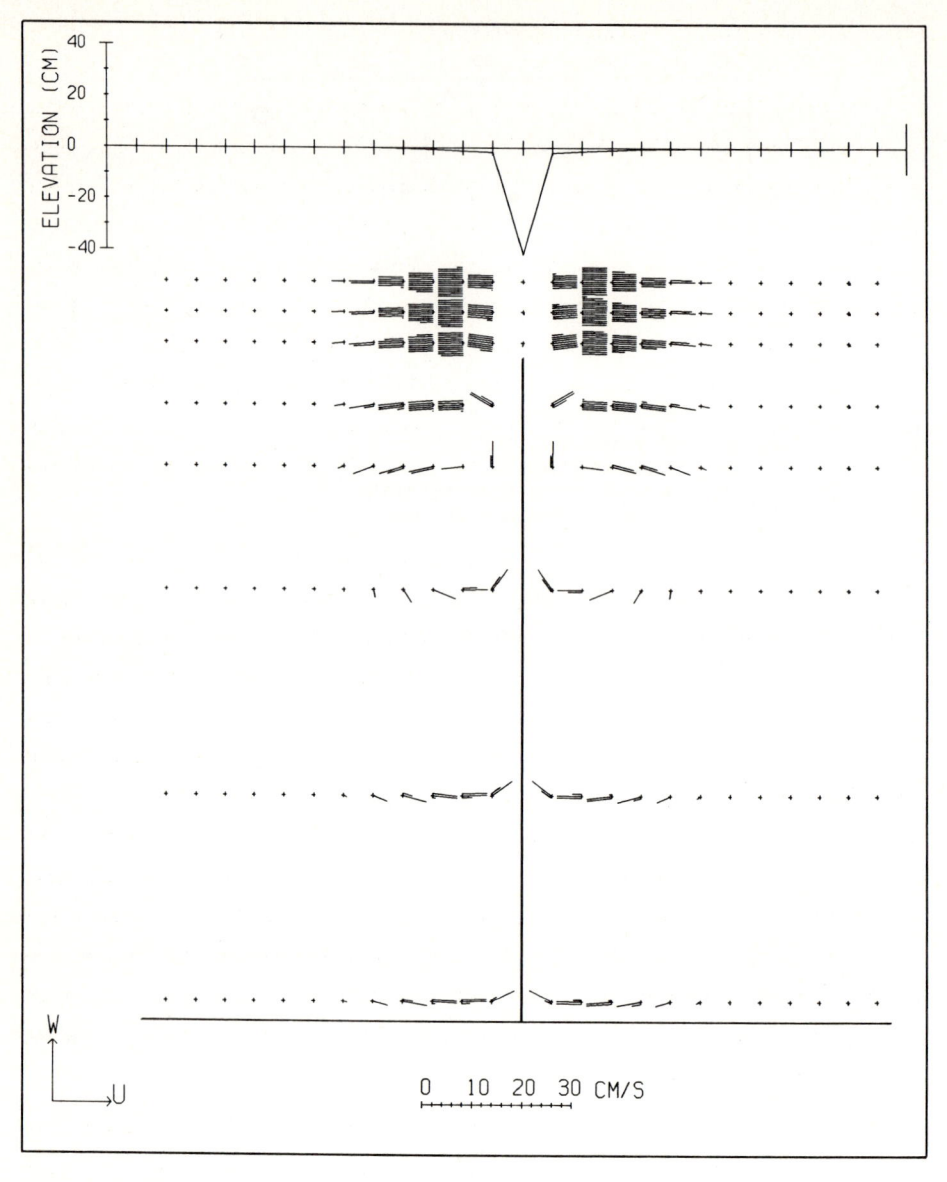

Fig. 13.8 Distribution of tidal residual velocity vectors on a vertical plane through a narrow (one mesh wide) test channel containing a narrow sill.

arguments with respect to adjacent pairs of velocity points in the neighbourhood of the sill can readily demonstrate how the advective terms $\partial u^2/\partial x + \partial u\, w/\partial z$, for a given stream direction, will tend to augment the pressure gradient in one layer while opposing it in another layer. The elevations of the water surface included in this figure shows the net depression of the water surface over the sill.

13.3.7 Vertical Eddy Viscosity

Further trials of the full model were carried out in order to determine the effect of vertical eddy viscosity on the simulation of tidal co-oscillations within the system. First, as noted in Section 13.3.4, it is impossible to achieve a satisfactory tidal calibration due to the excessive dissipation occasioned by the high value of horizontal eddy viscosity. Secondly, in view of the large width-to-depth ratios of stacked mesh boxes in a water column, it appears physically implausible that, at least over the greater part of the modelled region, lateral stresses and horizontal eddy viscosity should predominate over the bottom stress and vertical eddy viscosity. Results obtained from the barotropic tidal model GF_2 show that it is impossible to reproduce the tidal constituent non-linear interactions shown in earlier work without the employment of high square-law bottom friction in certain locations.

A number of trials were carried out in which the coefficient of bottom friction was increased to a high value (0.1) in passages among the Gulf and San Juan Islands, a region of strong tidal energy dissipation. Estimates of the order of magnitude of vertical eddy viscosity required to introduce any significant retardation of flow in layers above the bottom layer, due to high friction acting on the bottom layer, would require unusually large values. A series of trials in which such large values were employed resulted in vertical velocity profiles which did not agree with those observed at a central location in Haro Strait. The highest velocities actually occurred in the layer above the bottom layer. This suggested that account should be taken of the vertical mesh spacing which increases with increasing thickness of the layers. Results obtained from runs in which the vertical eddy viscosity employed in determining the stress at the horizontal interface between two adjacent layers was made proportional to the square of the average thickness of the two layers, yielded vertical velocity profiles in rather better agreement with those observed.

Although the complexity of bottom topography makes it difficult to compare vertical velocity profiles in this region, Figs. 13.9 and 13.10 show some results of these exploratory trials. Thus, profile A shows the vertical distribution of M_2 and K_1 velocity amplitudes, respectively, when a constant vertical eddy viscosity of $2,000 \text{ cm}^2/\text{s}$ is employed. Profile B shows the result of using values proportional to the squared average thickness of adjacent layers downwards from the interface of layers 1 and 2, of magnitudes 56, 156, 506, 900, 2025, 6400, and $10,000 \text{ cm}^2/\text{s}$, respectively. Also shown for these depths are velocity amplitudes obtained from moored current meter records. It was further found that an increased lag of about 1 h in the K_1 tide occurred in the Strait of Georgia when K_1 was prescribed alone on the boundaries compared to the value obtained when the run included both M_2 and K_1 tides. This is consistent with the non-linear interactions of tidal constituents referred to above.

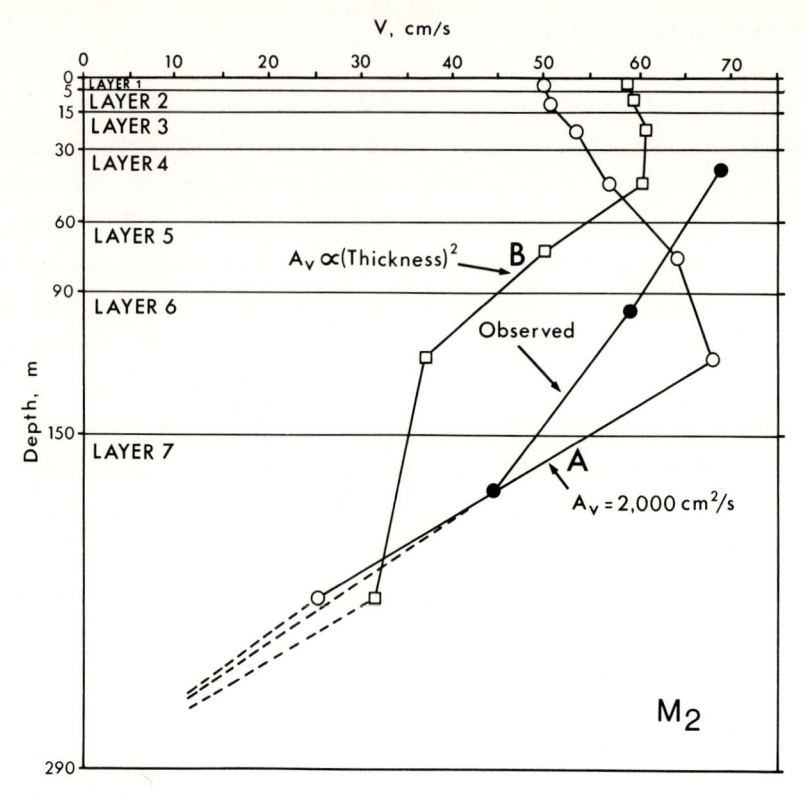

Fig. 13.9 Comparison of observed M_2 vertical velocity profile in Haro Strait and modelled profiles using vertical eddy viscosities of (a) $A_V = 2,000$ cm^2/s and (b) A_V proportional to (thickness)^2.

13.3.8 Prescription of Pre-Computed Advective Accelerations

A simple and economical approach to solving the problem of non-linear instability is suggested by the result noted earlier in which the advective terms have little effect on the overall tidal propagation. Thus, a tidal computation in which advective terms, formulated as centred differences, are computed, stored, but not included in the dynamical calculations, can be repeated with the stored advective terms now being prescribed for each solution of a velocity component equation over the model grid. Horizontal eddy viscosity may be introduced on grounds of physical adjustment rather than numerical necessity. Preliminary results indicated that satisfactory simulation of the tidally-driven residual circulation could be obtained in this manner.

When the averaged advective acceleration (or tidal stress) at each particular velocity point is obtained and then prescribed as a time-invariant stress field in the subsequent tidal simulation, the residual circulation also agrees well with that from the full calculation. Although further testing is required, the approach may significantly expedite future baroclinic trials of the model.

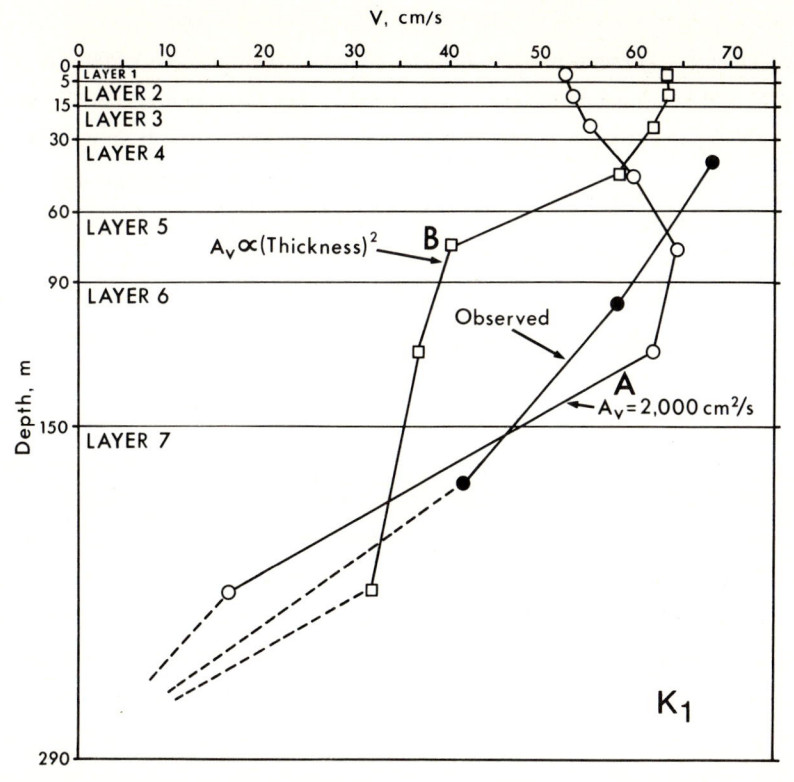

Fig. 13.10 Comparison of observed K_1 vertical velocity profile in Haro Strait and modelled profiles using vertical eddy viscosities of (a) $A_V = 2,000$ cm^2/s and (b) A_V proportional to (thickness)^2.

13.4 RESIDUAL CIRCULATION

If the emphasis above has been on the shortcomings of the model, it will now be shown that the model can provide information on the basic net circulation in reasonable accord with observation. Furthermore, it can provide estimates of the way in which this circulation is altered by the major winds that prevail, respectively, in summer and winter.

Two trials were carried out in which 6 days of the summed M_2 and K_1 tidal co-oscillations were simulated. The first trial assumed a constant and uniform field of density. In the second, an initial field of density was prescribed corresponding to that observed in early July 1968 and was allowed to evolve according to Eq. (13.13). (Associated salinities contoured over a longitudinal vertical section through the system, are shown in Fig. 1.5b, p. 11.) The last 2 days of the second trial were then repeated, in one case with a uniform wind of -10 m/s, parallel to the y axis (approximately from the northwest), in a second case with the wind field reversed (approximately from the southeast). The residual circulation was determined by removing the tidal components from the computed fields of velocity, density, and elevation over the last 25 h of each run. The values of the model parameters employed were $A_H = 8 \times 10^6$ cm^2/s, $A_V = 450$ cm^2/s (except in the northern channel where $A_V = 1{,}000$ cm^2/s), and $r = 0.003$. A full-slip condition was applied at the lateral boundaries. These values gave a reasonable approximation to the tides (though differing appreciably from those shown in Figs. 13.3 and 13.4) within the system and are considered adequate for the simulation of residual circulation associated with baroclinicity and wind effects. The density field was used in an essentially diagnostic context, long enough for geostrophic equilibration to occur but not so long as to excessively prejudice the outcome by numerical diffusion. It will be shown later that the results accorded reasonably well with observations.

In presenting these results it is necessary to take into account scales of motion differing roughly by a factor of 10. The first of these is associated with the regions of swift tidal streams in Juan de Fuca Strait, the Gulf and San Juan Island passages, the southern Strait of Georgia, Puget Sound, and the northern passages. (In some instances, the deepest layers are omitted from the illustrations due to the lack of significant features in an area of limited extent.) The second scale of motion is primarily associated with the deep circulation in the Strait of Georgia. Thus, the first set of vector plots presented employs scales suitable for the higher residual velocities and is given for the overall system. Another set of vector plots is presented later for the slower residual circulation in the Strait of Georgia. This latter presentation further facilitates comparison of the vector fields at the various levels. It is important to note that the vectors consist of the averaged u and v components at the centre of a mesh. This can be misleading in the vicinity of sills where a strong horizontal divergence or convergence in a single column of meshes can result in an apparent zero net horizontal velocity through a mesh. Where such a case is discussed, vertical profiles of velocity through each mesh side are employed.

13.4.1 Barotropic Tidal Residual Circulation

Consistent with adequacy of presentation, the modelled residual circulation associated with the strong tidal streams in Juan de Fuca Strait and southern Strait of Georgia is considered initially, deferring until later that occurring elsewhere in the model.

Features of the tidally-driven three-dimensional residual circulation shown in Fig. 13.11a-h are discussed with reference to the higher-resolution vertically-integrated model GF7 results shown in Fig. 7.9 (p. 151). Thus, the eddies numbered 1, 2, 4, 5, 7, 8, and 10 are found to be present and extend down to the level where an eddy is either obviated or rendered vestigial by the bottom topography. Of particular interest is the residual flow associated with the Admiralty Inlet ebb jet (layers 1–4). This is a local external circulation initiated by the jet and is essentially independent of any net exchange with the Puget Sound system itself.

Remaining features of the residual circulation conform essentially to the tidally-

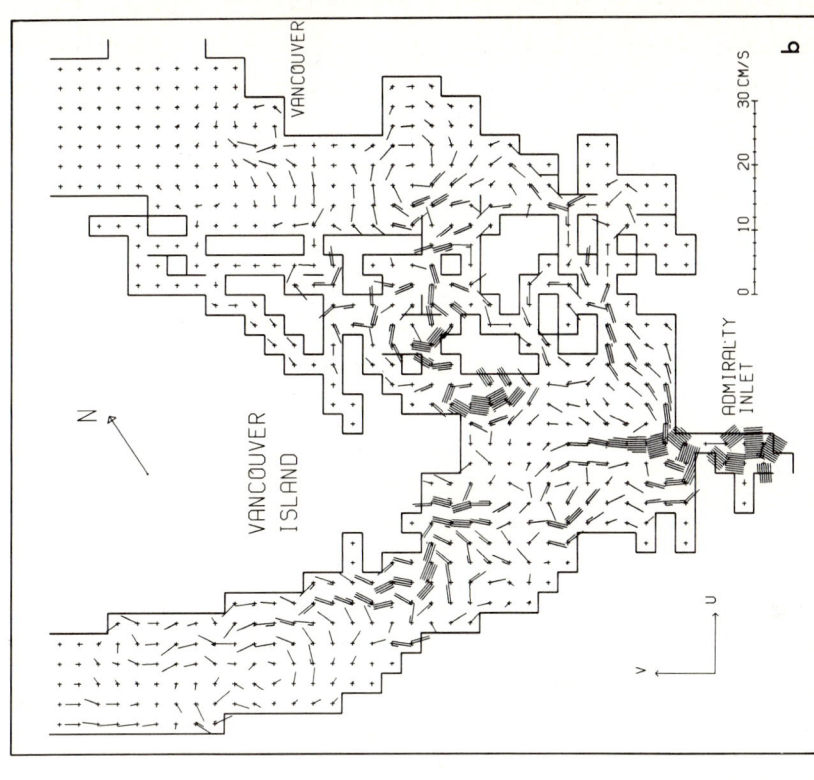

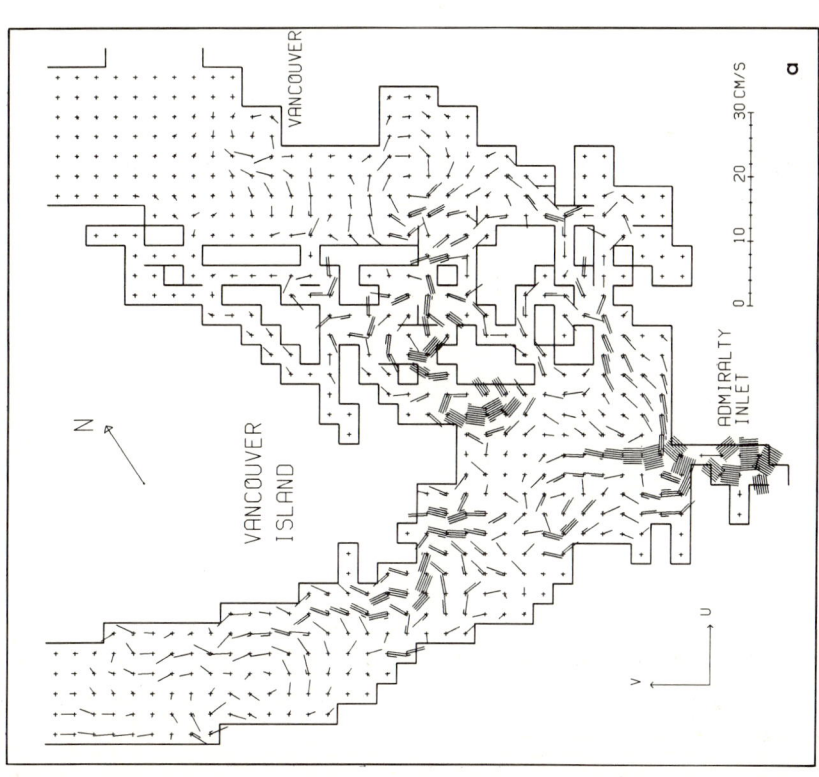

Fig. 13.11 Field of mixed (M_2 and K_1) barotropic tidal residual current vectors for (a) layer 1 and (b) layer 2, obtained from a harmonic analysis of 25 h of computed velocites displayed for the Juan de Fuca Strait and the southern Strait of Georgia.

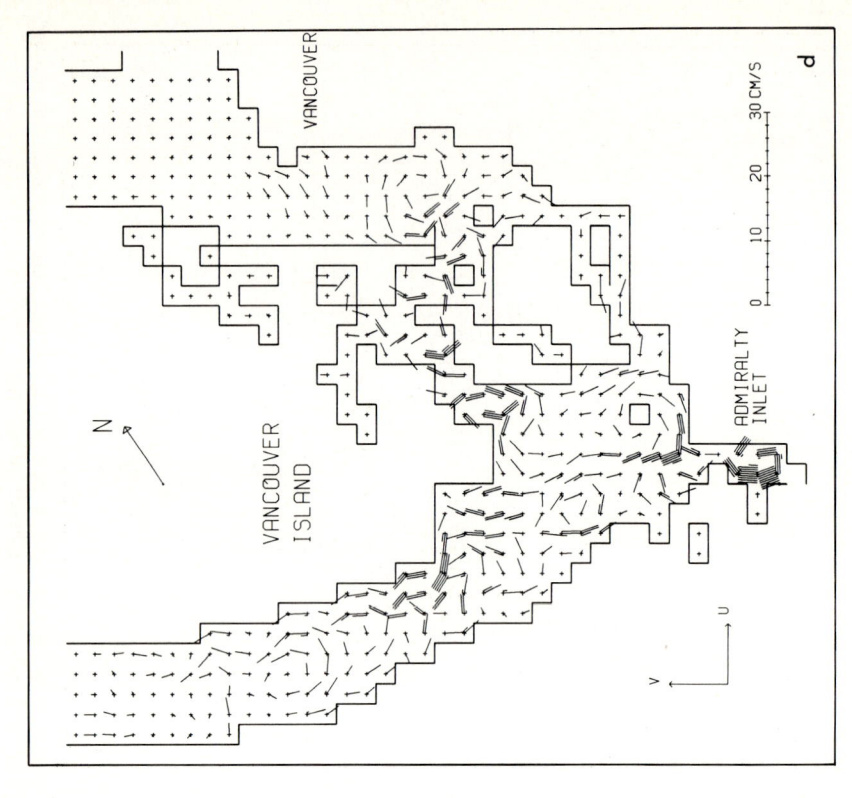

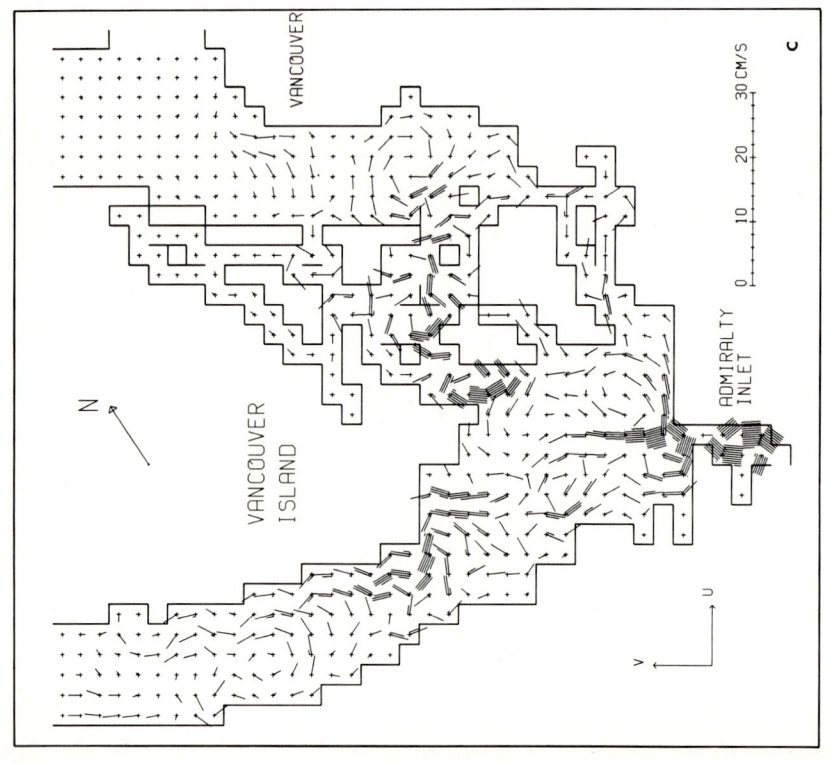

Fig. 13.11 (c) Layer 3 and (d) layer 4.

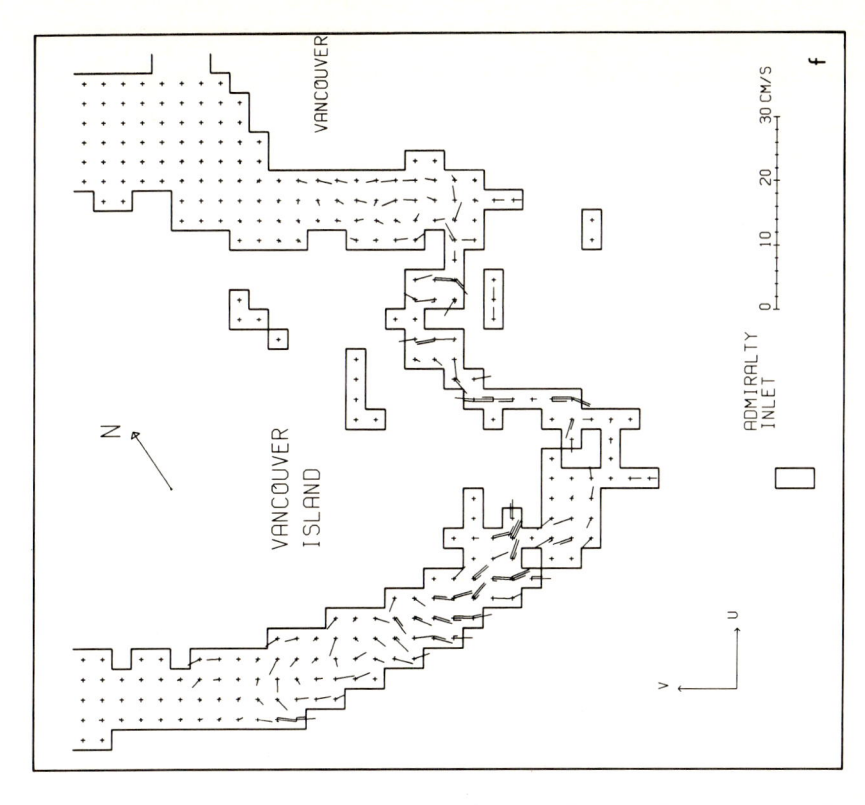

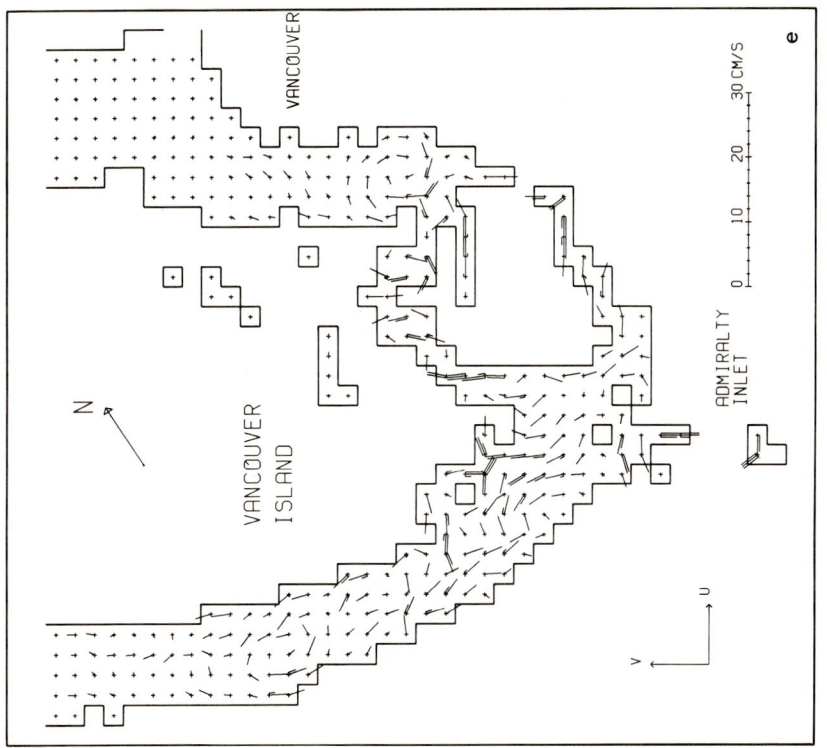

Fig. 13.11 (e) Layer 5 and (f) layer 6.

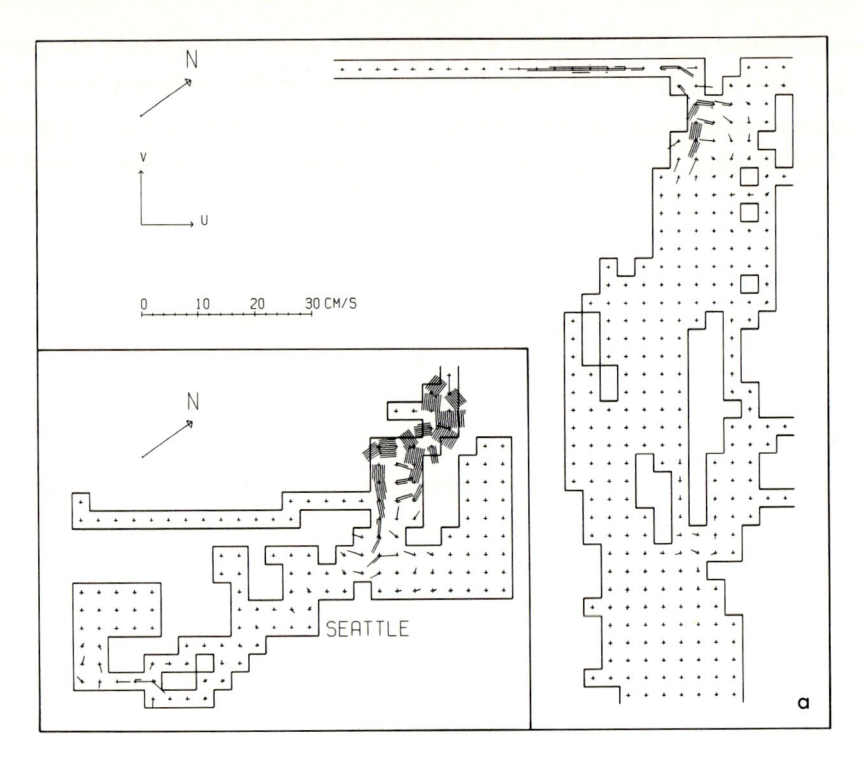

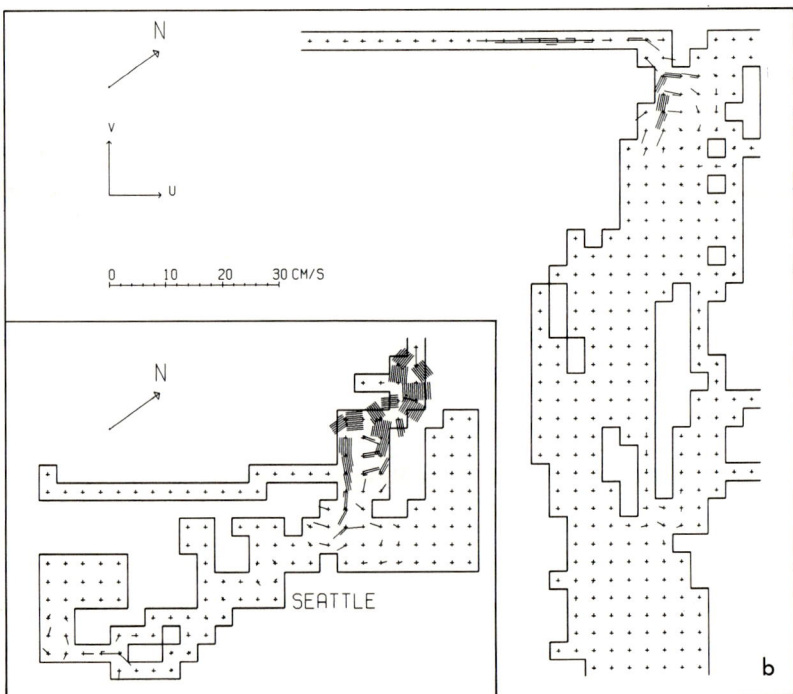

Fig. 13.12 Field of mixed (M_2 and K_1) barotropic tidal residual current vectors for (a) layer 1 and (b) layer 2 obtained from a harmonic analysis of 25 h of computed velocities displayed for Puget Sound and the northern Strait of Georgia.

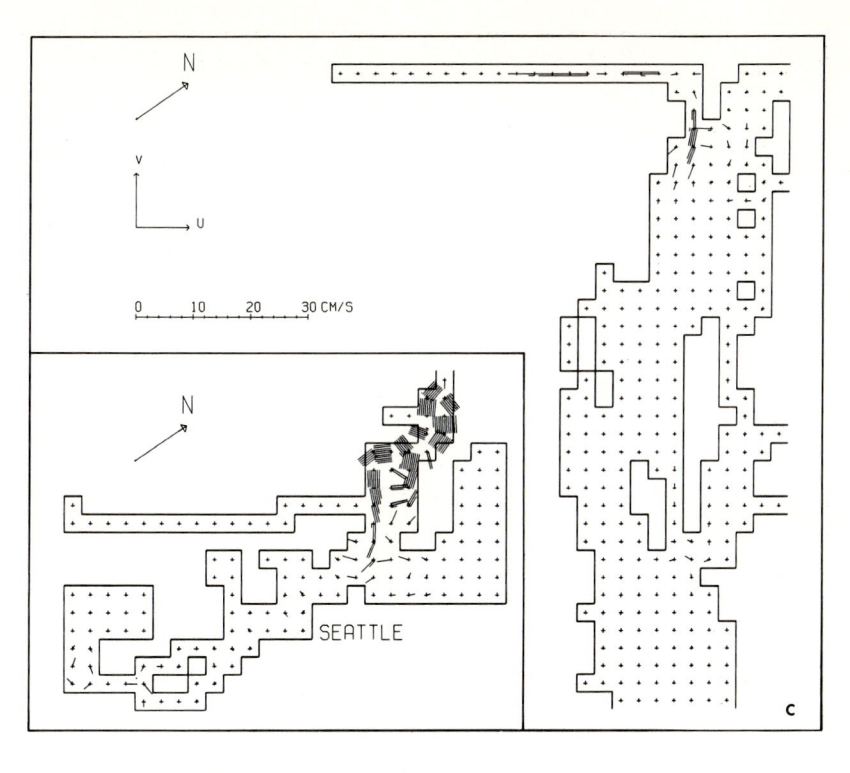

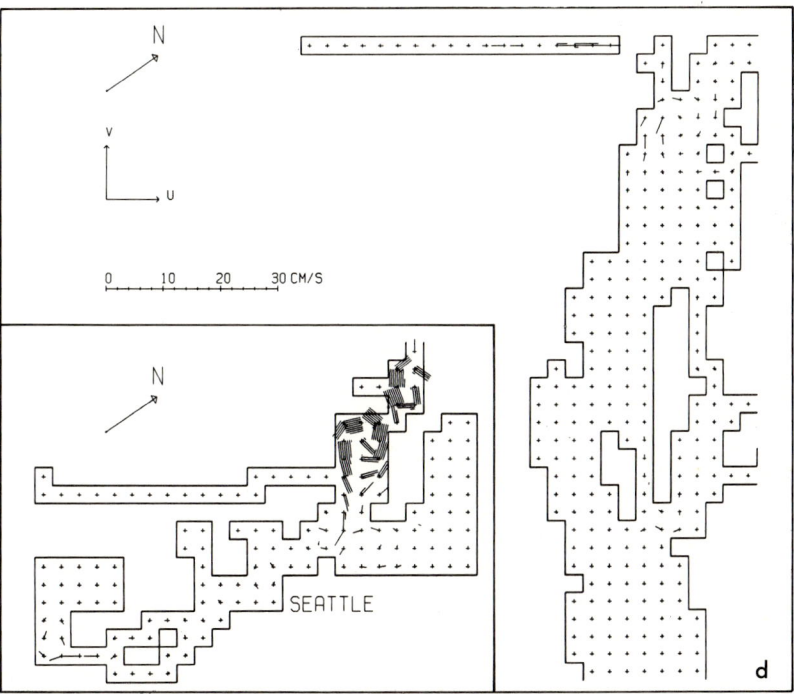

Fig. 13.12 (c) Layer 3 and (d) layer 4.

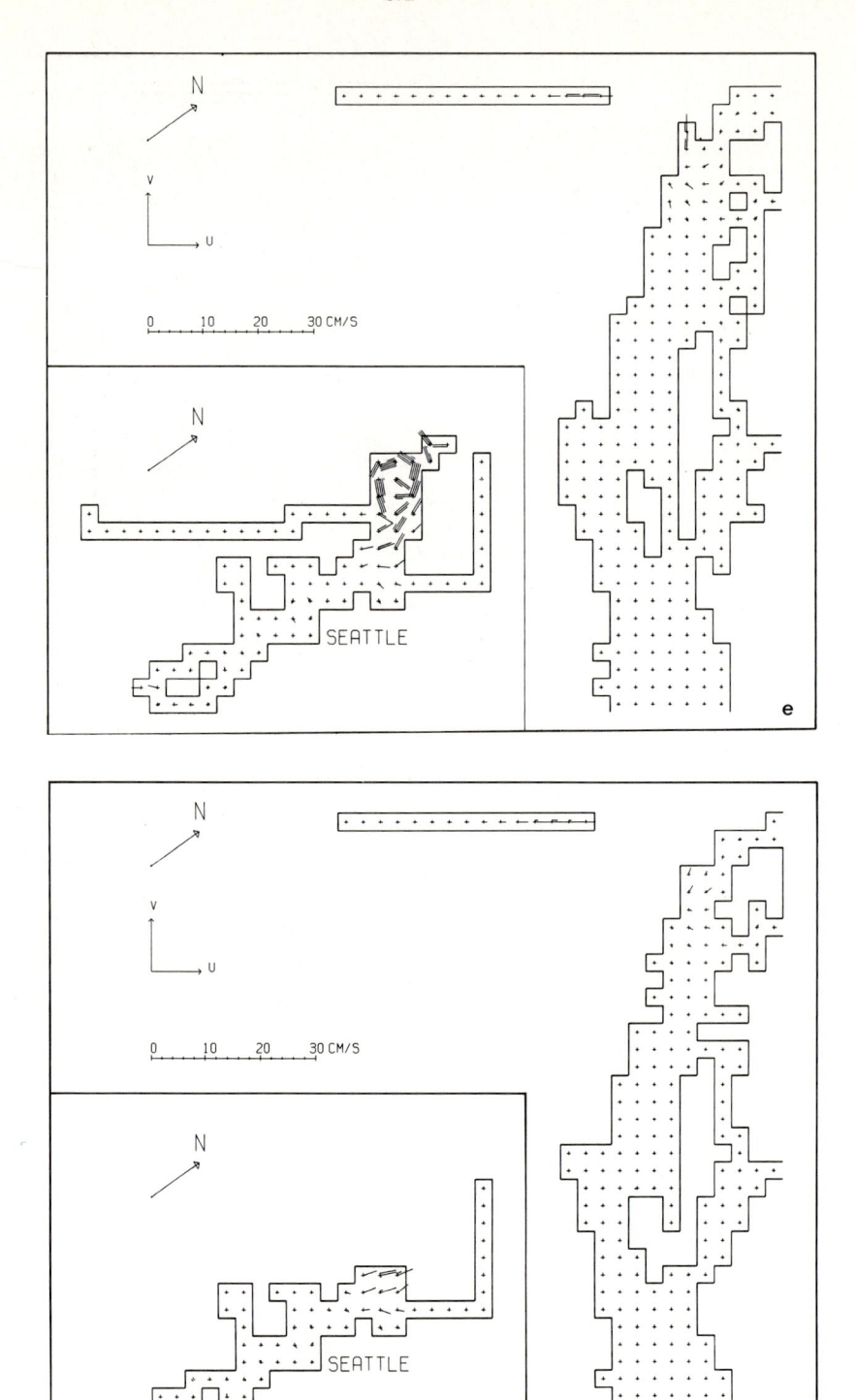

Fig. 13.12 (e) Layer 5 and (f) layer 6.

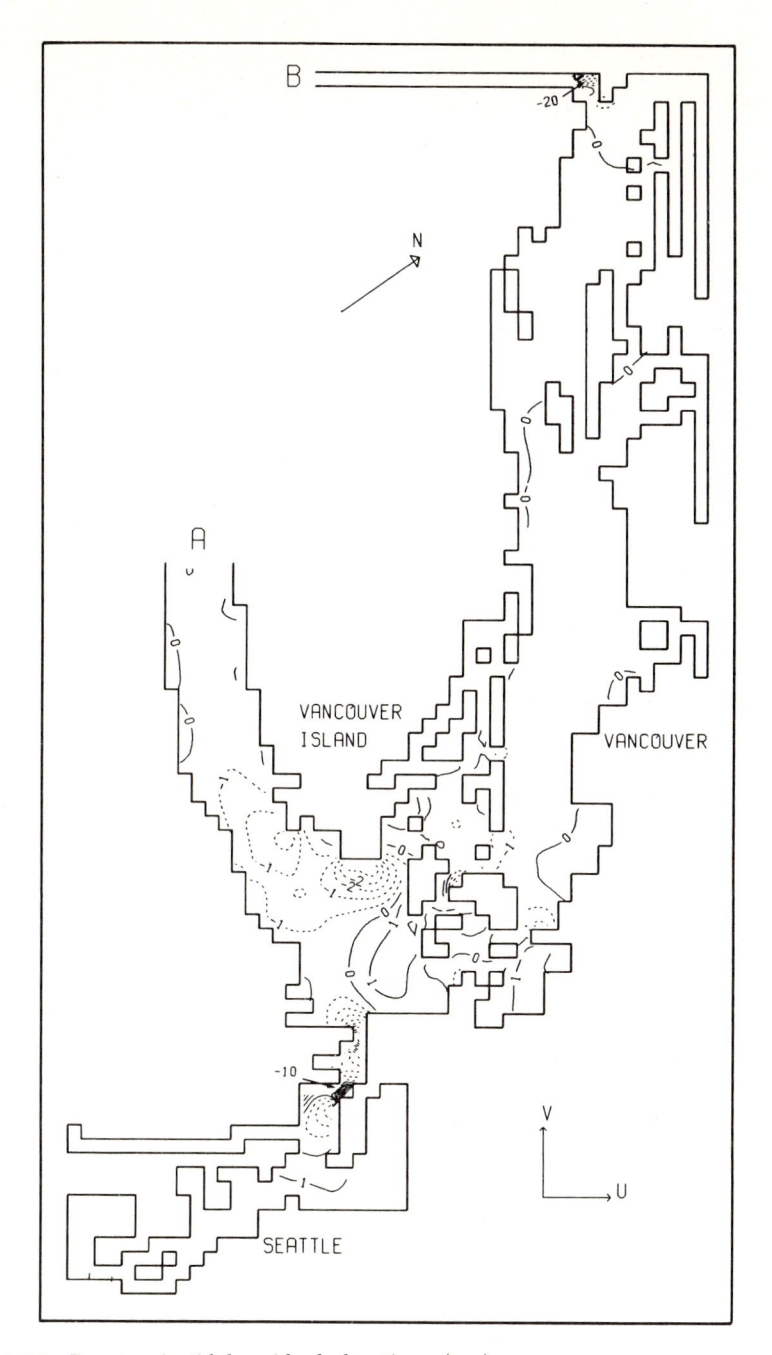

Fig. 13.13 Barotropic tidal residual elevations (cm).

flows generally display the anticipated Coriolis favouring of shores to the right of the direction of motion. In layer 4, distinctive inflow features have developed in the shallow sill region of inner Juan de Fuca Strait and in the southern Strait of Georgia. This

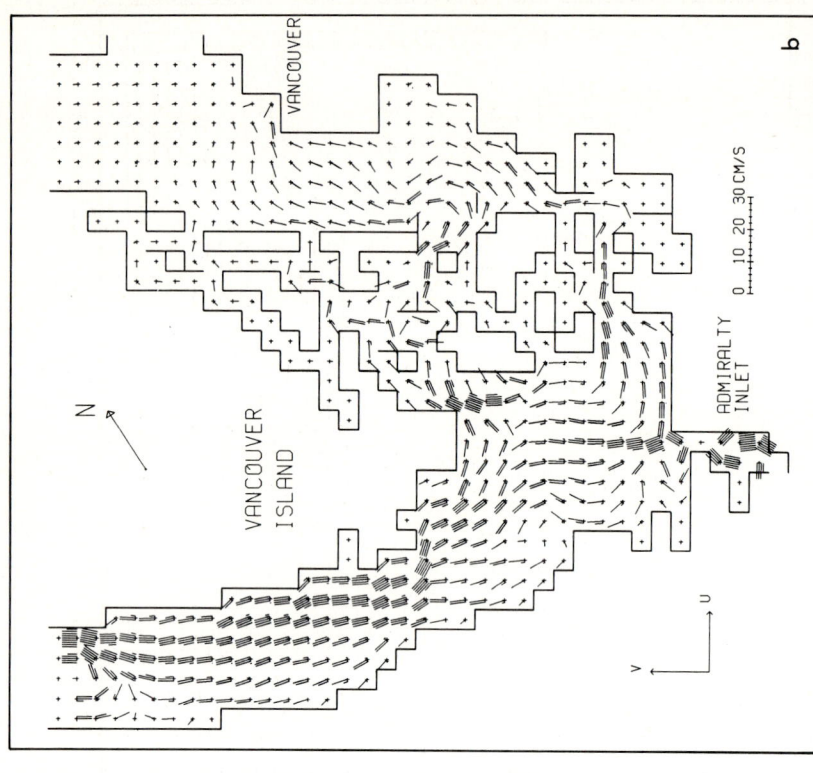

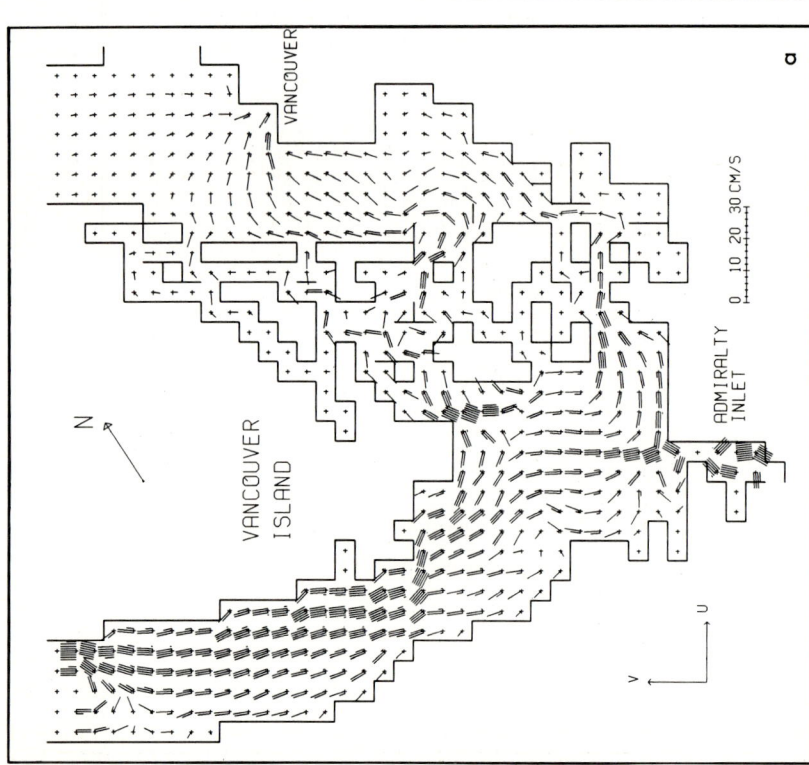

Fig. 13.14 Field of mixed tidal (M_2 and K_1) residual and baroclinic circulations for (a) layer 1 and (b) layer 2 when an initial density field and river discharge, conducive to deep water renewal and estuarine circulation are prescribed for Juan de Fuca Strait and the southern Strait of Georgia.

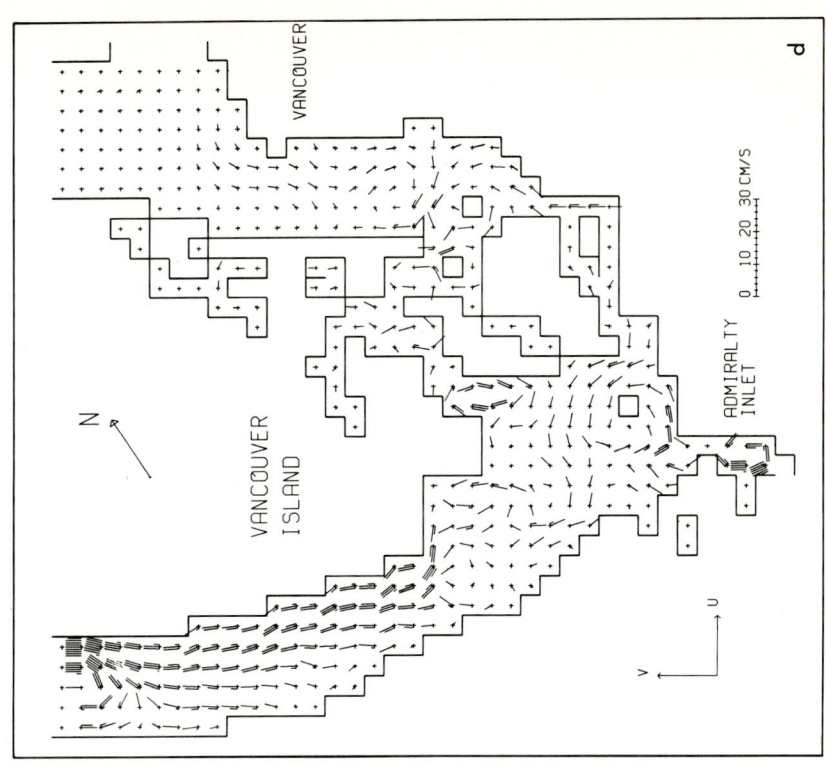

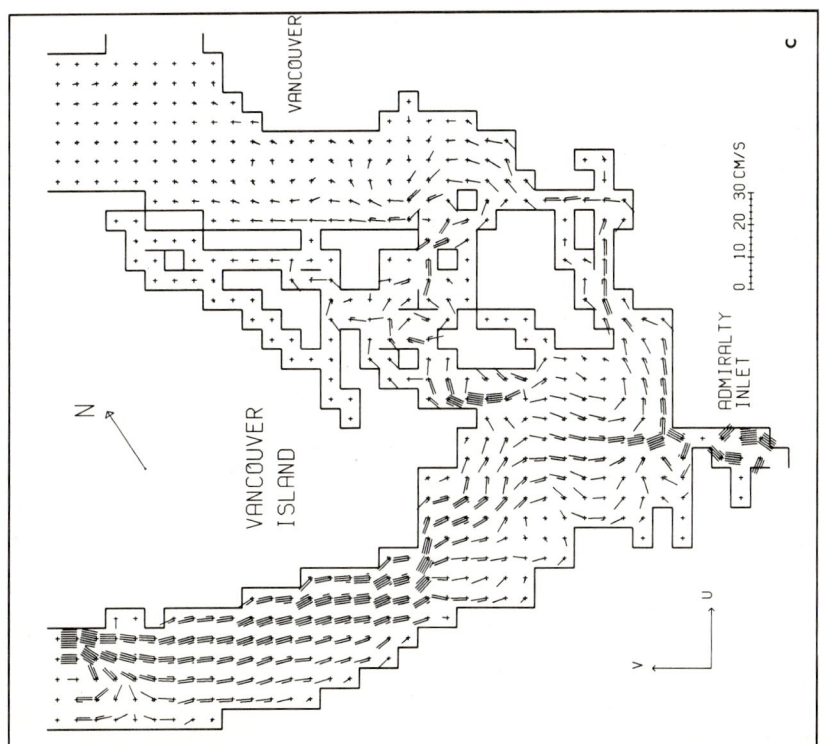

Fig. 13.14 (c) Layer 3 and (d) layer 4.

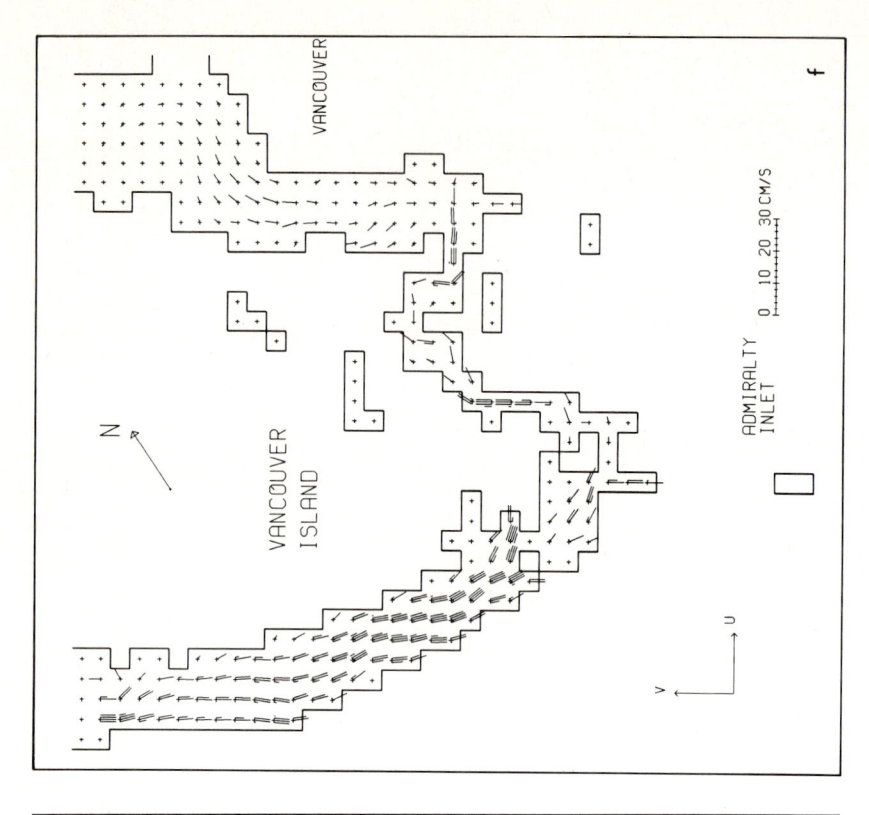

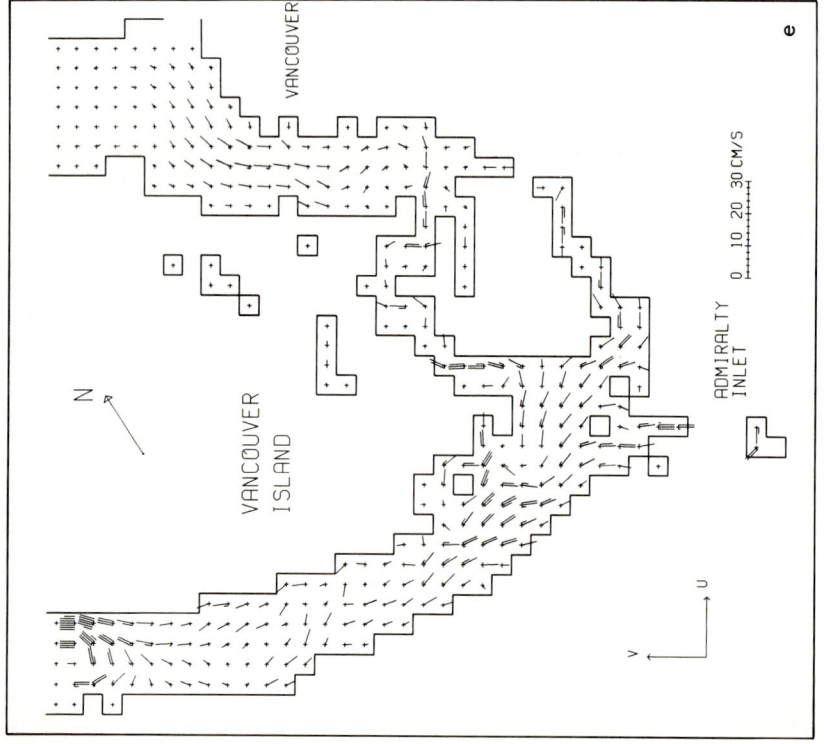

Fig. 13.14 (e) Layer 5 and (f) layer 6.

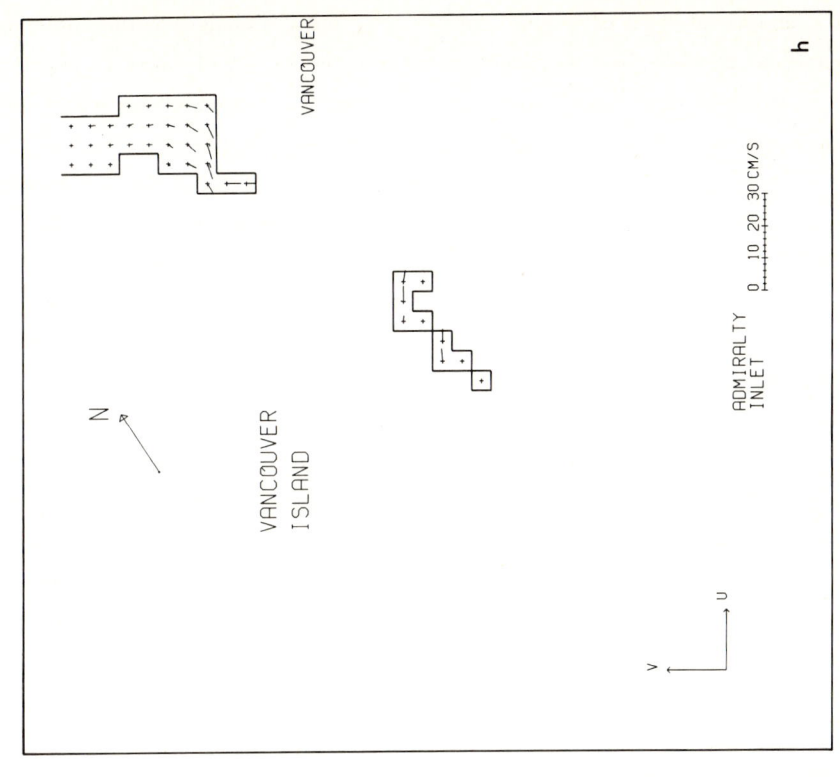

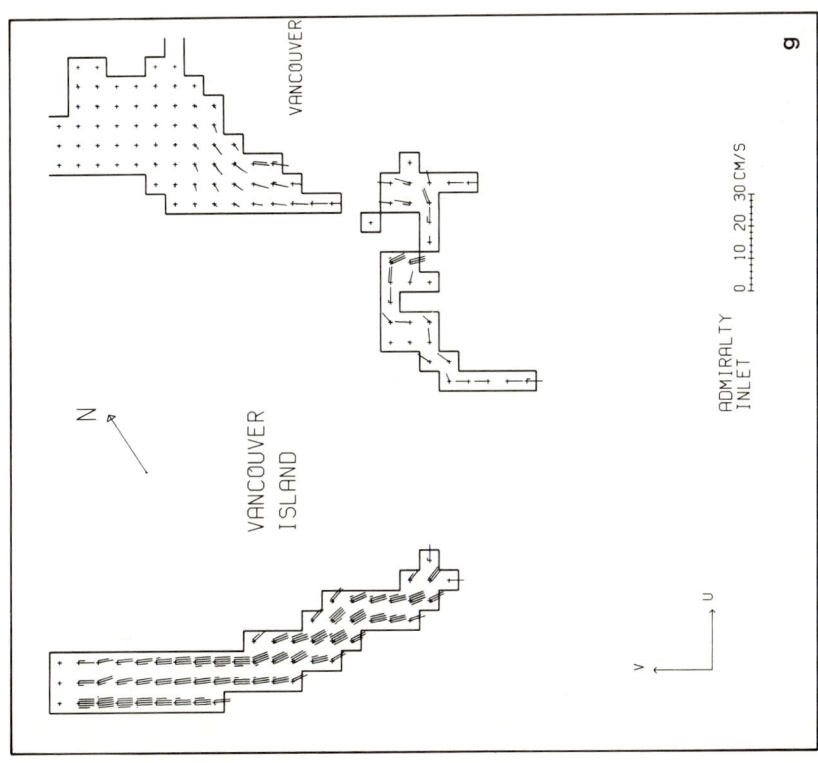

Fig. 13.14 (g) Layer 7 and (h) layer 8.

inflow is increasingly well developed in layers 5, 6, and 7 although, in the latter and in layer 8, counterflows have developed in the southern Strait of Georgia. Near the entrances to the channel constrictions leading seaward these counterflows accord with the barotropic tidal residuals although numerical mixing is also involved.

One feature of interest concerns a modest enhancement of the barotropic tidal residual circulation in layers 1, 2, and 3 directed from the southern end of Rosario Strait towards the Admiralty Inlet ebb jet stream. This, however, would appear to be due to the extensive area of low density water prescribed in Rosario Strait as part of the initial conditions (a consequence of northwest winds of order 35 km/h prior to the aquisition of the cruise data in this part of the system). The dominant surviving features of the purely barotropic tidal residual circulation are eddies 3, 5, and 8 (layers 1–4), eddy 4 (layers 3–5), and the residual jet associated with Admiralty Inlet.

In the remainder of the model (Fig. 13.15a–f) (layers 7 and 8 have been omitted since the larger vector scale used to illustrate the slow deep modelled circulation in the Strait of Georgia is required to resolve relevant features) there is a net outflow in the surface layers and inflow in the deeper layers in the single northern channel. Where this channel opens into the northern Strait of Georgia some evidence of the strong residual tidal eddy is retained while throughout the remainder of the Strait of Georgia net northerly flows occur in layers 1 and 2. In layer 4, roughly about sill depth, an appreciable southerly flow directed away from the northern channel opening has developed. Below this layer there is again a northerly flow, presumably associated with the tidal residual circulation. These flows have interesting qualitative implications with respect to the flow regime at the southern end of Discovery Passage, particularly southerly-directed flows at middepth which are dependent on the relative densities of water entering into the strong turbulent tidal mixing processes at the southern end of the Passage.

In the Puget Sound system, the basic circulation appears to involve the movement of lower density water in the upper layers towards the strong tidal mixing region near the entrance and return flow at depth. Although a significant net slope of the sea surface downwards towards the entrance exists over this period it would still appear inadequate to overcome a small net inflow into the system at this stage of adjustment in the overall density field. The vertical profiles of the velocities on either side of the column of meshes at the entrance, which were averaged to obtain the vector at the centre of the mesh, show a well developed convergence near the surface and divergence at depth. Thus, the residual circulation is dominating the flow field at this location in the model at this time.

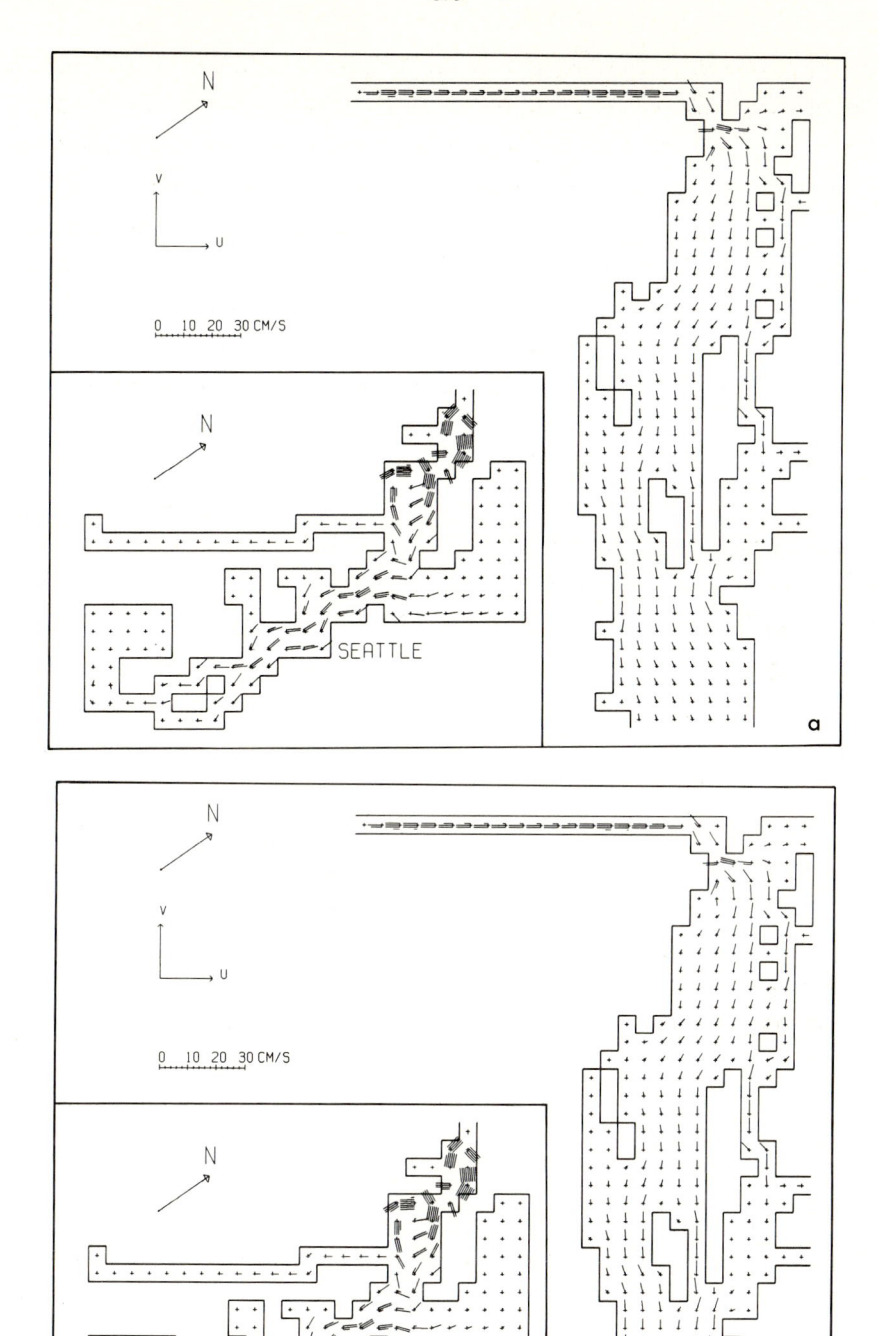

Fig. 13.15 Field of mixed (M_2 and K_1) residual and baroclinic circulations for (a) layer 1 and (b) layer 2 when an initial density field and river discharge, conducive to deep water renewal and estuarine circulation are prescribed for Puget Sound and the northern Strait of Georgia.

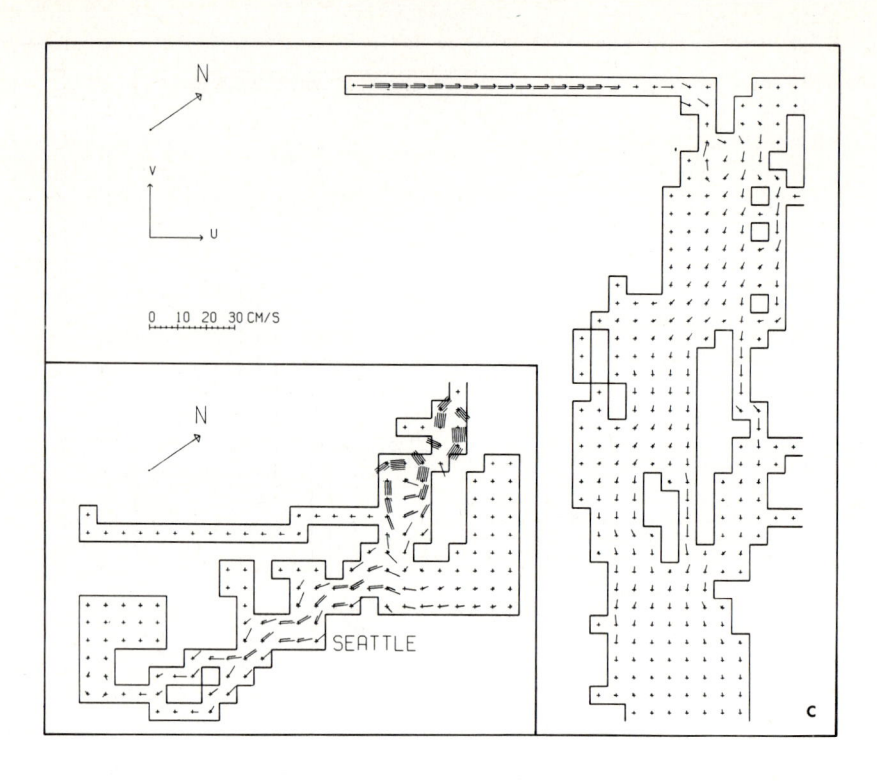

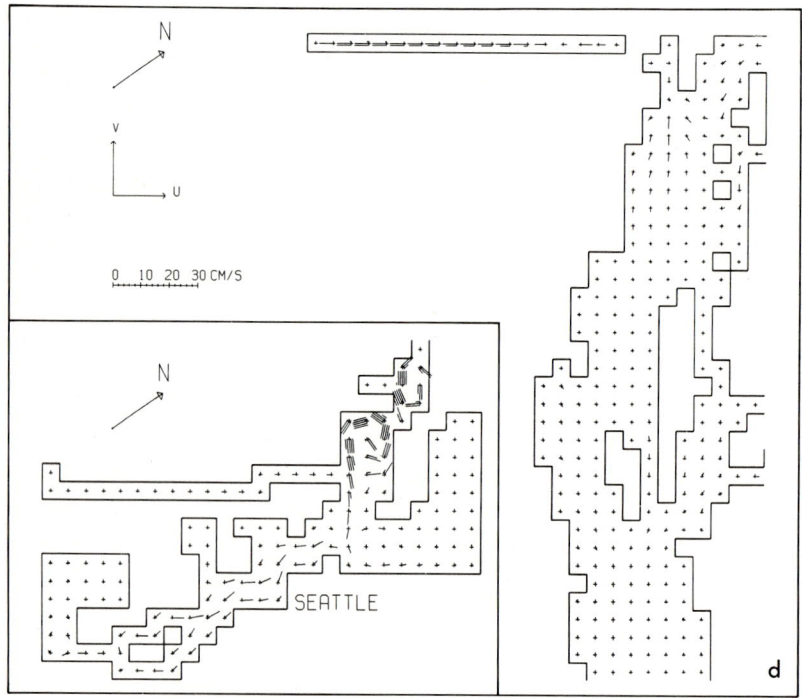

Fig. 13.15 (c) Layer 3 and (d) layer 4.

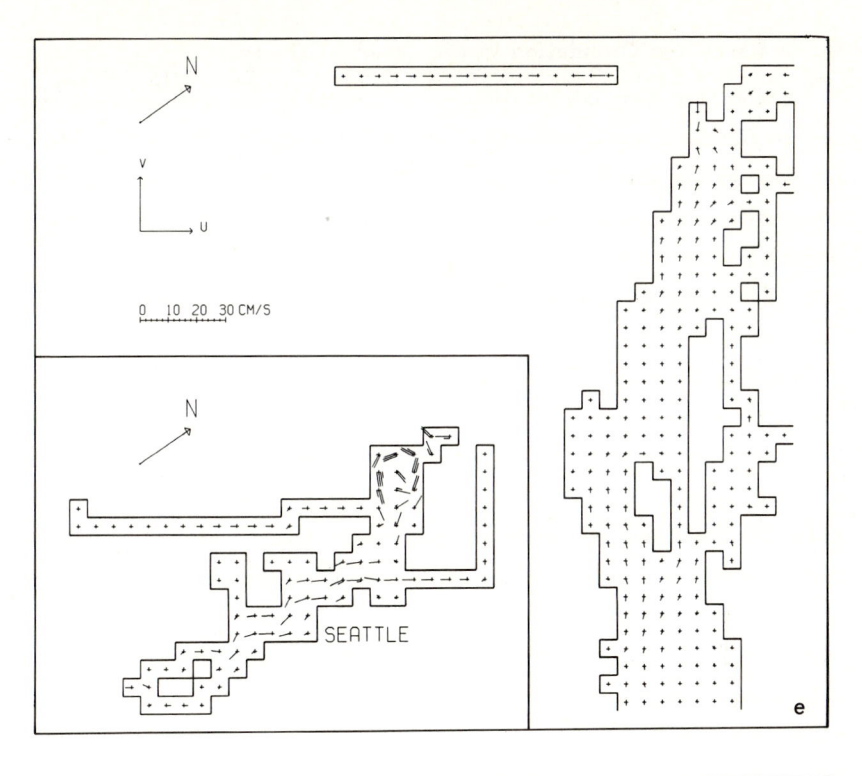

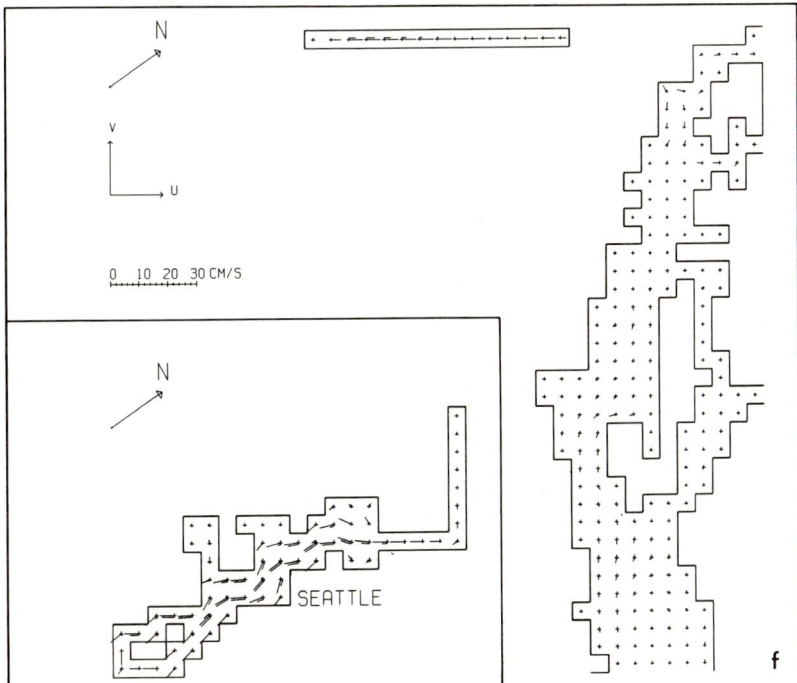

Fig. 13.15 (e) Layer 5 and (f) layer 6.

13.4.3 Baroclinic Circulation in the Strait of Georgia

A topic of primary interest concerns the circulation and replacement of water in the deep semi-enclosed basin of the Strait of Georgia. Diagnostic trials using the simplified and non-rotating model GF5 suggest that deep water replenishment occurs through a slow movement of mixed water along the bottom from the mixing regions at its southern extremity. This results in vertical displacements of water of order 2–3 m/day in the deep basin giving a replacement time of about 3 months, a period in agreement with that inferred from contoured time series of salinities shown in Fig. 1.5a and b.

Using vector scales appropriate to the resolution of the small velocities associated with this circulation, Fig. 13.16 shows the horizontal distributions of residual circulation in the Strait of Georgia. The top three layers are dominated by the movement of estuarine discharge towards the boundary openings leading seaward. In layers 4, 5, and 6 (30–150 m and increasing the vector scale again) there exist clearly developed undercurrents, one moving from the southern openings northward, tending to favour the mainland coast, and a second moving southward from the northern opening, generally tending to favour the Vancouver Island shore. These give rise to a distinct counter-clockwise circulation in the deep central basin of the Strait of Georgia. In layers 7 and 8 the basic pattern changes to one dominated by outflows along the bottom to the north and south of the deep basin.

In summary, the divergent flow in the upper layers (1, 2, and 3) surmounts a convergent flow and general counter-clockwise circulation in the intermediate layers (4, 5, and 6). Below this in the bottom layers there is again a generally divergent flow field. It will be shown that, in point of fact, this latter is an artifact introduced by numerical diffusion.

13.4.4 Residual Surface Elevations

The residual circulation, previously described in Section 13.4, derives both from the effects of tides and from the barotropic and baroclinic pressure gradients. It is these gradients that largely determine the residual circulation in the Strait of Georgia away from the strongly tidal mixing regions at its northern and southern extremities.

The barotropic pressure gradients that drive the estuarine circulation may be inferred from the distribution of modelled residual surface elevations (Fig. 13.17) associated with vector plots previously presented. From a maximum of 30 cm near the mouth of the Fraser River, the slopes fall away to a value of 22 cm along the opposite shore and to a value of 15 cm at the approaches to the northern and southern boundary openings. Throughout the central and northern Strait of Georgia there is a small cross-channel slope of order 1–2 cm/s downward from the mainland coast due to the Coriolis deflection of a net estuarine northward surface flow of river water. A steeper cross-channel slope of order 4–5 cm/s is apparent in Juan de Fuca Strait, consistent

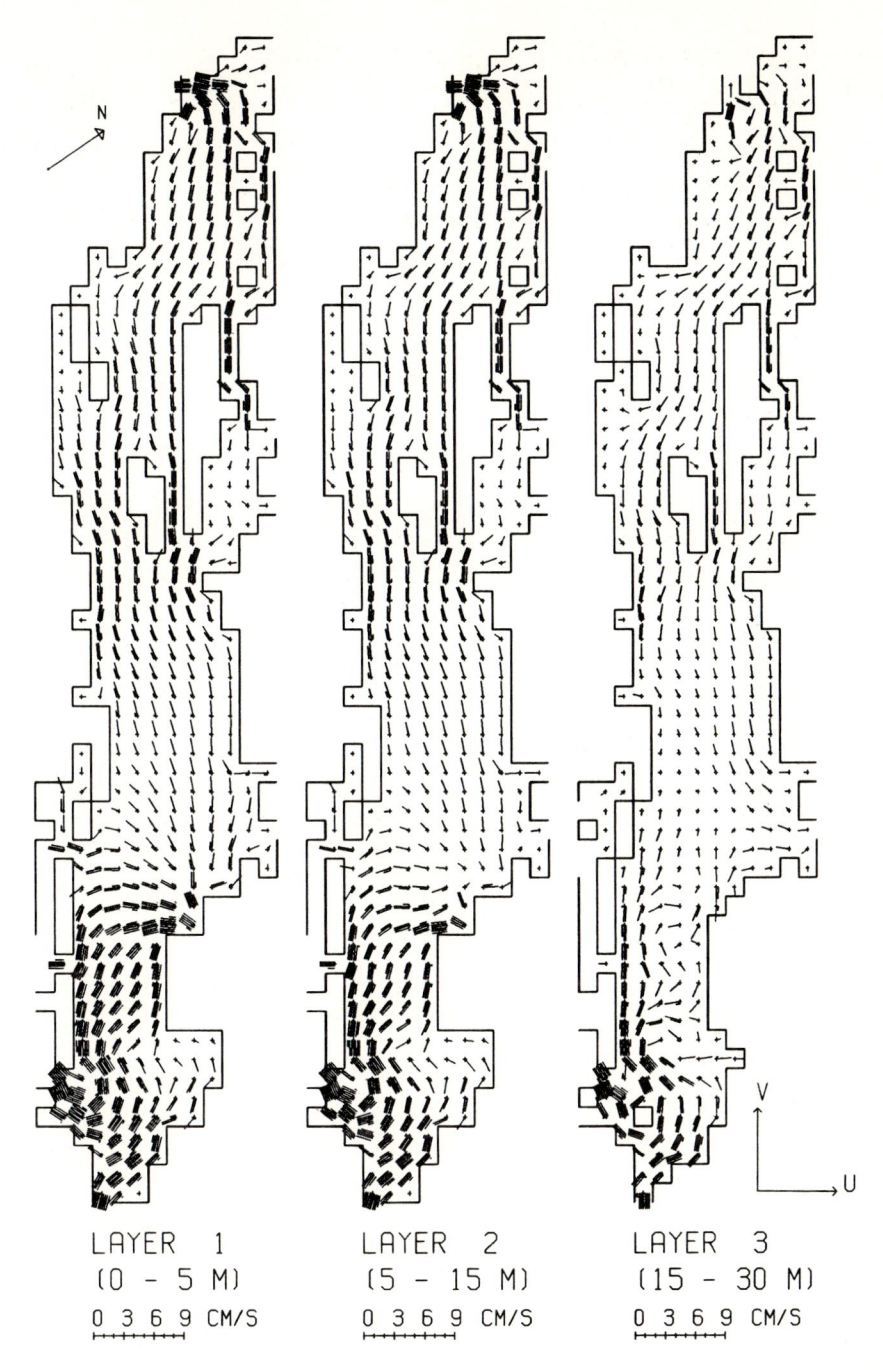

Fig. 13.16 Distribution of horizontal tidal residual and baroclinic circulation in the Strait of Georgia.

with the strong estuarine exchange and deep water intrusion. Marked local depressions of the water surface over the sill in the northern passages and at the entrance to

LAYER 4
(30 - 60 M)
0 1 2 3 CM/S

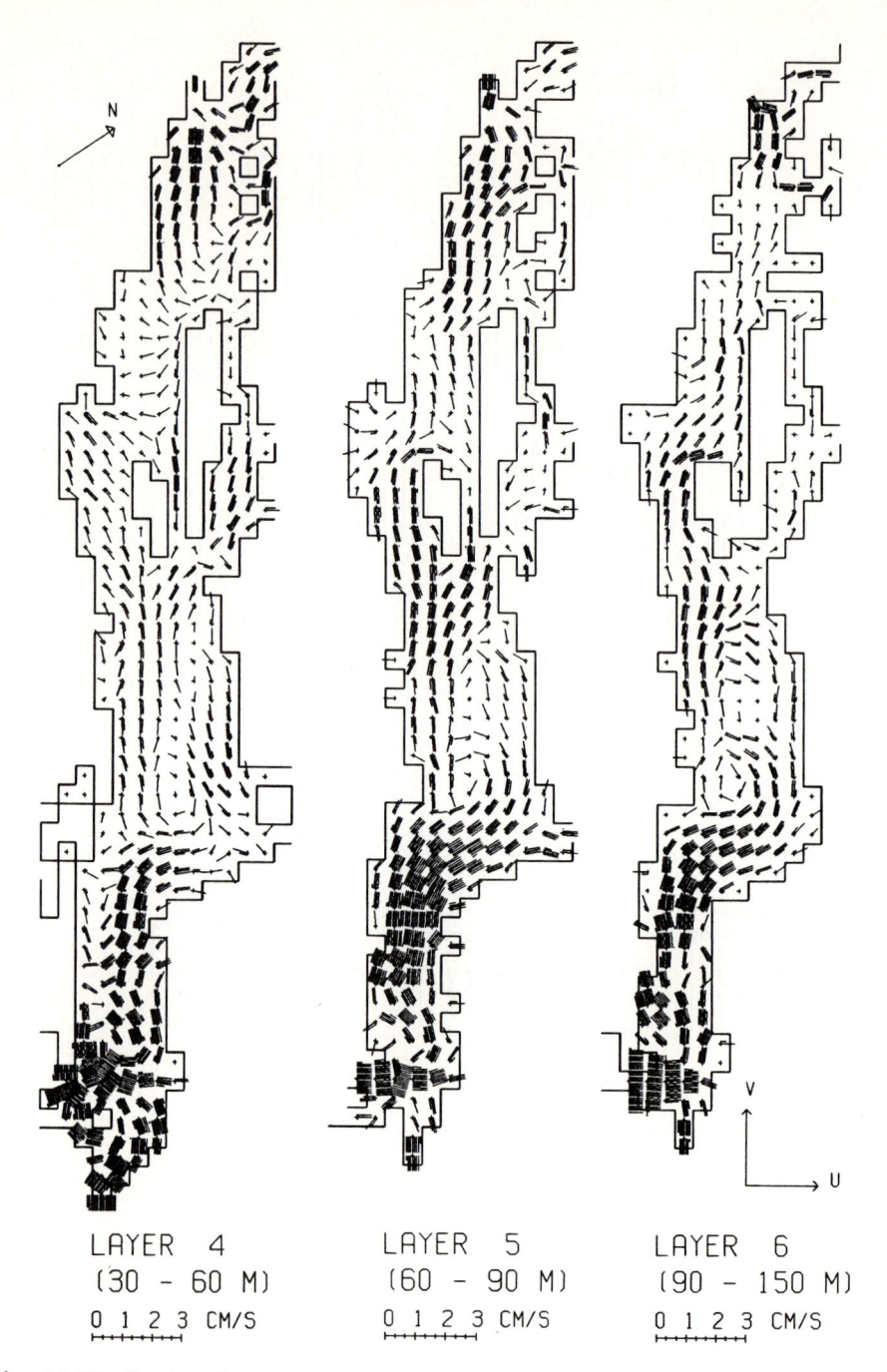

LAYER 5
(60 - 90 M)
0 1 2 3 CM/S

LAYER 6
(90 - 150 M)
0 1 2 3 CM/S

Fig. 13.16 Continued.

Admiralty Inlet are due to the strong tidal streams as illustrated in Fig. 13.13.

The residual surface elevation near Little River is essentially the same as that obtained from the model GF5 (16 cm), a value in reasonable agreement with that

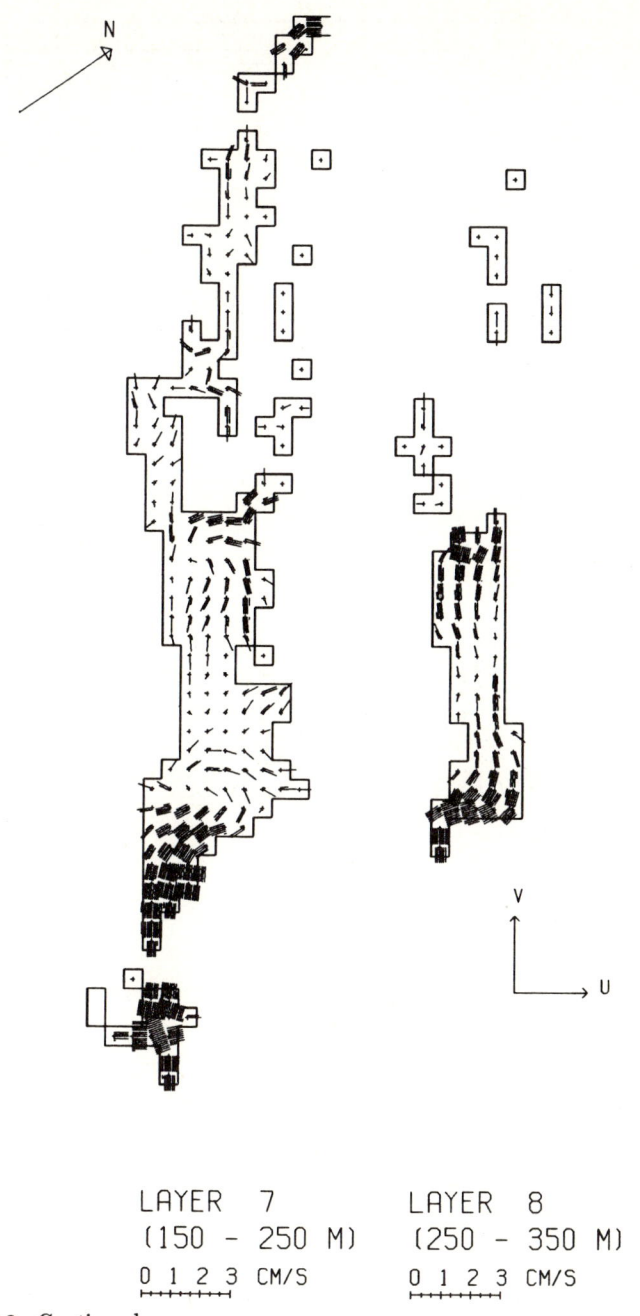

LAYER 7   LAYER 8
(150 - 250 M)   (250 - 350 M)
0 1 2 3 CM/S   0 1 2 3 CM/S

Fig. 13.16 Continued.

obtained from observation (13 ± 4 cm) discussed earlier in Section 12.7.3.5.

13.4.5 Residual Density Distribution

The baroclinic pressure gradients associated with the residual circulation derive

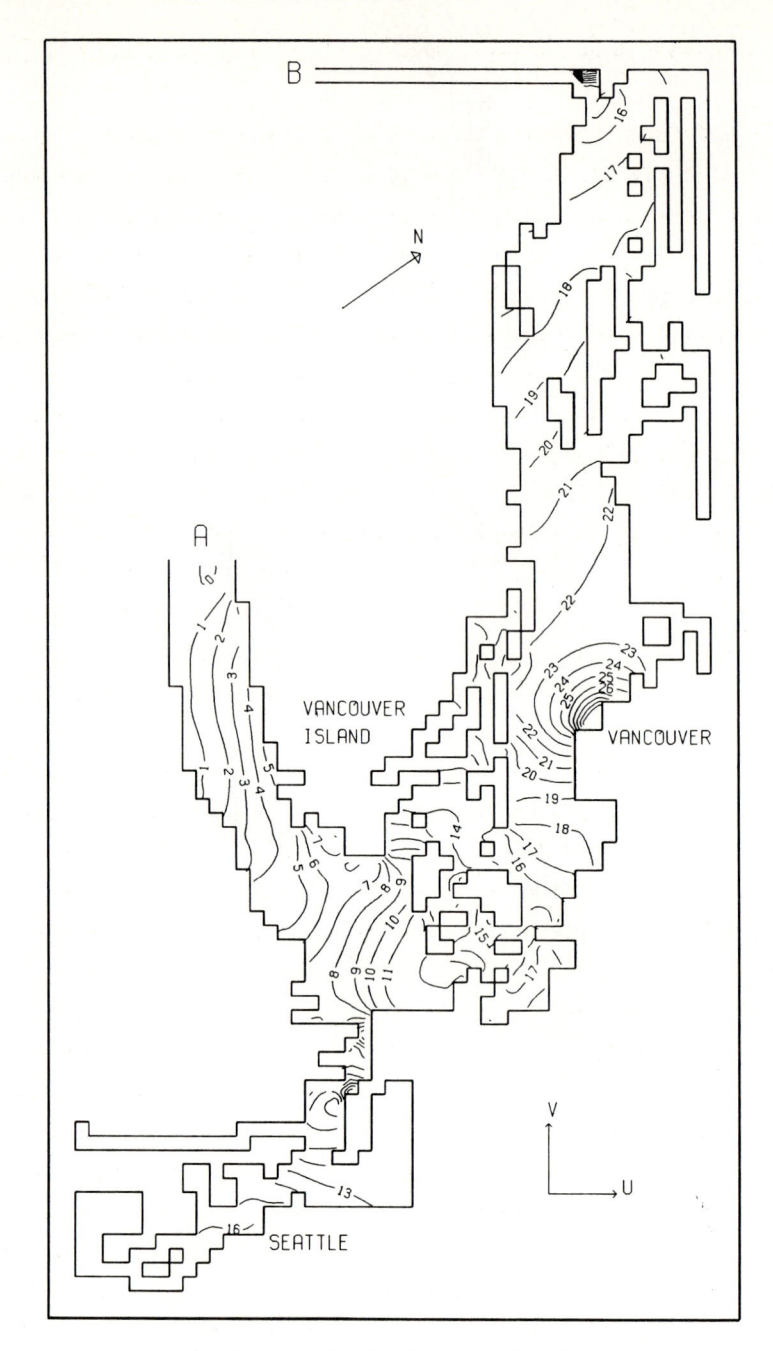

Fig. 13.17 Contoured distribution of residual surface elevations (cm) over the modelled region for the baroclinic case.

from the corresponding time-averaged density field. As for the preceding discussion, it is desirable to present separately the strong horizontal density gradients associated with Juan de Fuca Strait and the upper layers in the Strait of Georgia and the much weaker gradients occurring at depth (layers 5, 6, and 7) in the Strait of Georgia (Fig. 13.18a–g).

In layers 1 and 2 there is a general gradient of increasing density from the river mouth to the open sea. In the region of Rosario Strait there is a substantial area of dilution (minimum $\sigma_t = 17.4$) presumably due to the effect of northwest winds on the 2 days preceding the acquisition of the hydrocasts in this area that were used in prescribing the initial density field. A consequence of this was the presence of slightly lower density surface water at the entrance to the Puget Sound system compared with that occurring just inside the opening, thus favouring a movement of surface water into the Sound. In Juan de Fuca Strait the marked inclination of the contours to the median line of the Strait, indicative of cross-channel slopes of isopycnal surfaces, is consistent with the seaward movement of freshwater. This inclination is also evident in layer 3 and, in addition, appears in the southern part of the Strait of Georgia. Considering the residual circulation, the flow in this part of the model is more restricted to the southwestern shore (Fig. 13.16). As the seaward flows in Juan de Fuca Strait become weaker and tend to be more confined along the Vancouver Island coast the cross-channel inclination of the isopycnals (in layers 5 and 6) is associated with the intrusion of denser ocean water along the Washington shore (Fig. 13.14).

In layer 6, the water in the deep trench of Haro Strait is essentially of the same density as the deep water in the Strait of Georgia. Thus, the movement of mixed water at depth into the Strait of Georgia, as illustrated earlier, is presumably determined by the horizontal density gradients in layers 1–5. The density distribution for layer 8 has been omitted since the values over this limited region are essentially contained in the small σ_t range (23.6–23.7).

Horizontal density distributions in the remainder of the system (Fig. 13.19a–f) clearly show the dilution brought about by Fraser River discharge along the mainland coast in layers 1, 2, and 3. Of particular interest are the increased densities in the northern Strait of Georgia near the opening when compared with the densities at corresponding depths elsewhere in the northern Strait. These indicate the presence of a baroclinic pressure gradient tending to move mixed water southward immediately below its counteraction by the barotropic pressure gradient resulting from the slope of the free surface. In the Puget Sound system, the distinctive feature in the four top layers are gradients of decreasing density inward from the mouth of the Sound.

13.4.6 Residual Vertical Velocities

It will be recalled that the vertical velocity components in this three-dimensional model are computed at the centre of the upper surface of each computational 'box' using the equation of continuity. Positive flows are directed upward. Contours of the time-averaged vertical velocity (m/day) over the upper surface of each computational

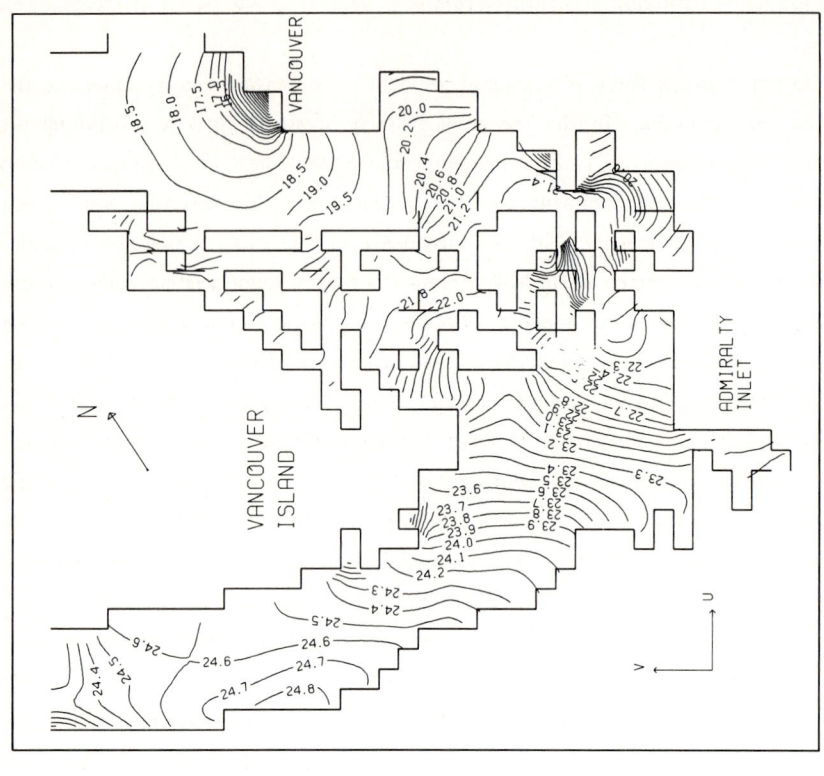

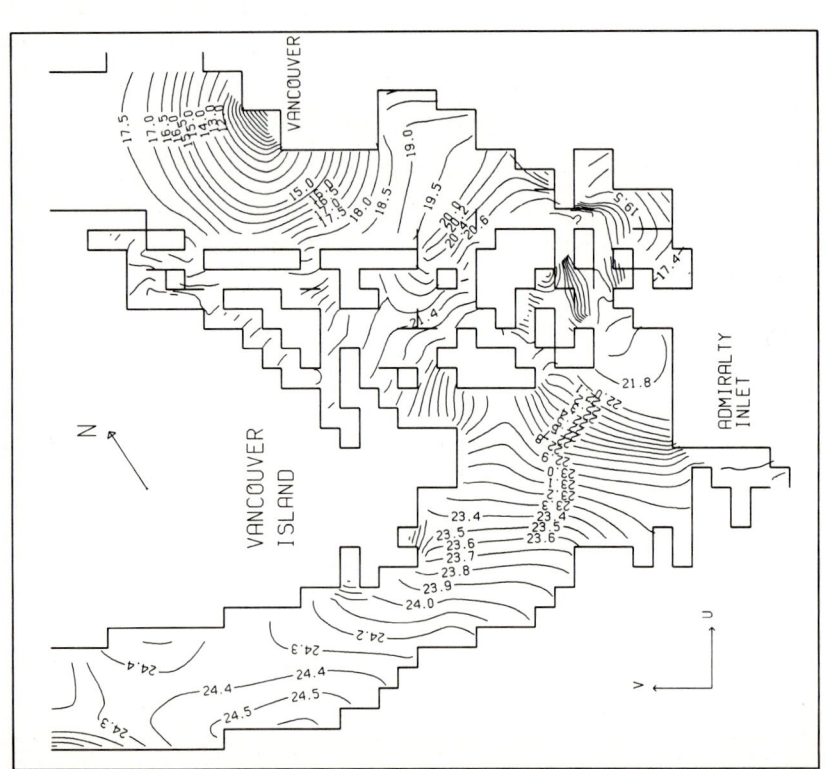

Fig. 13.18 Contoured horizontal distributions of the mean density distributions (σ_t) for the last 25 h in Juan de Fuca Strait and the southern Strait of Georgia for (a) layer 1 and (b) layer 2.

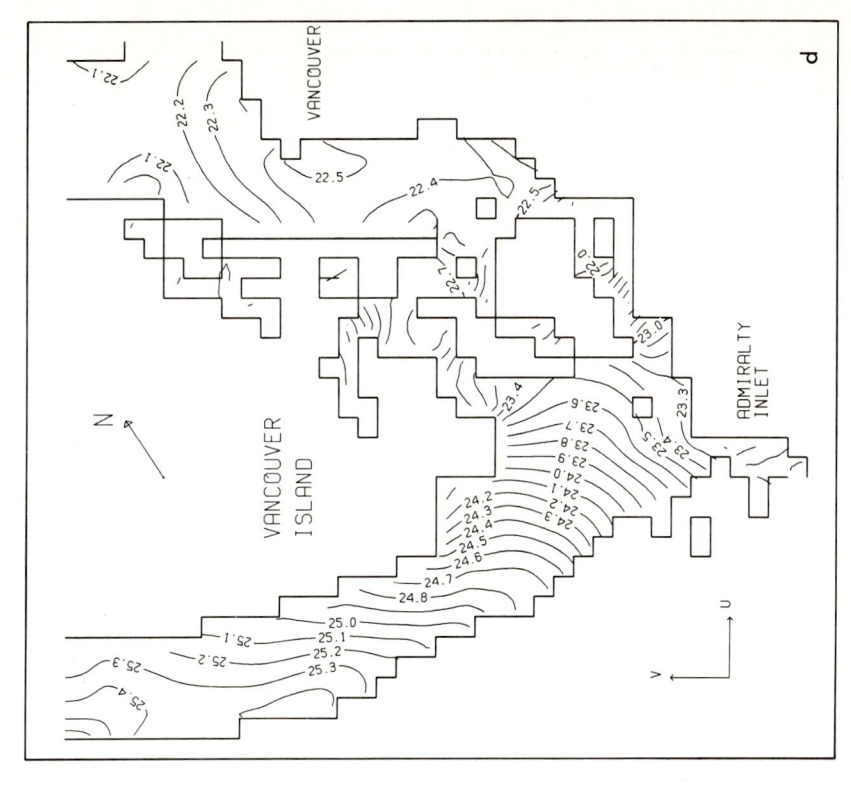

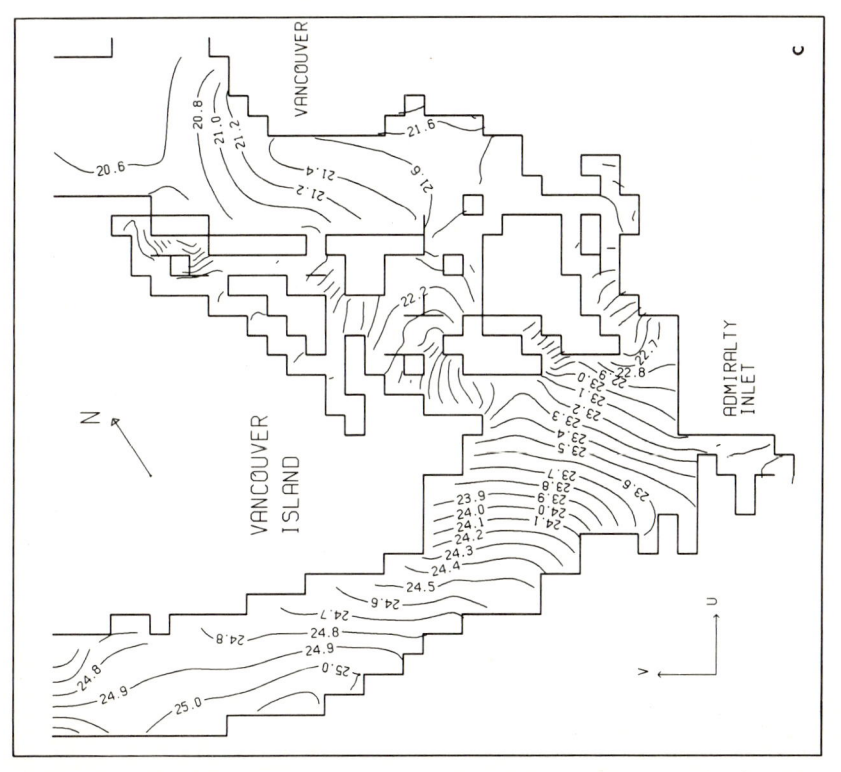

Fig. 13.18 (c) Layer 3 and (d) layer 4. Contoured values are σ_t units.

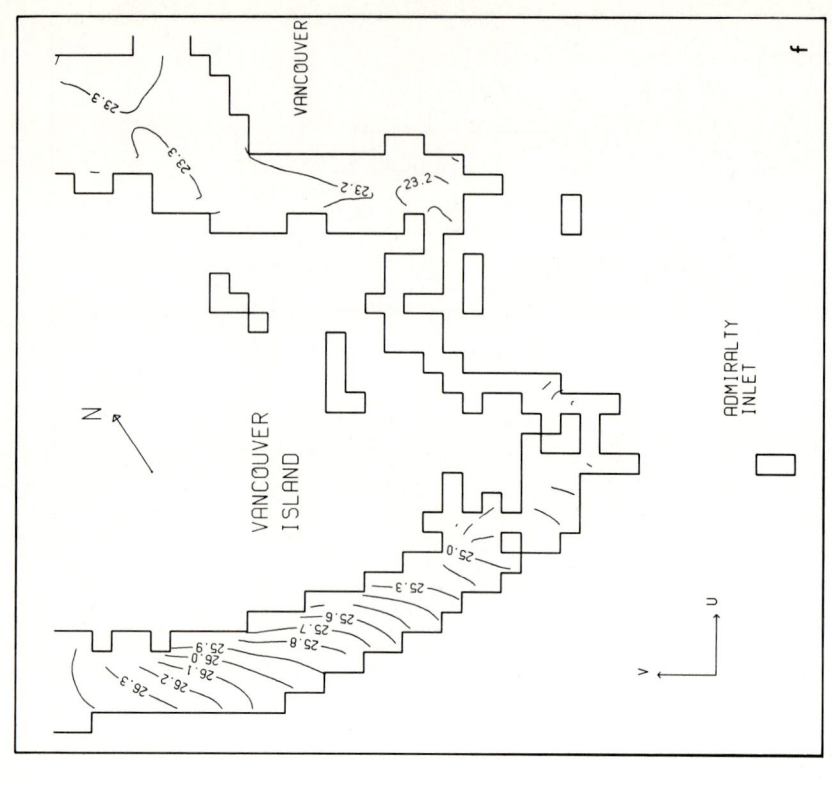

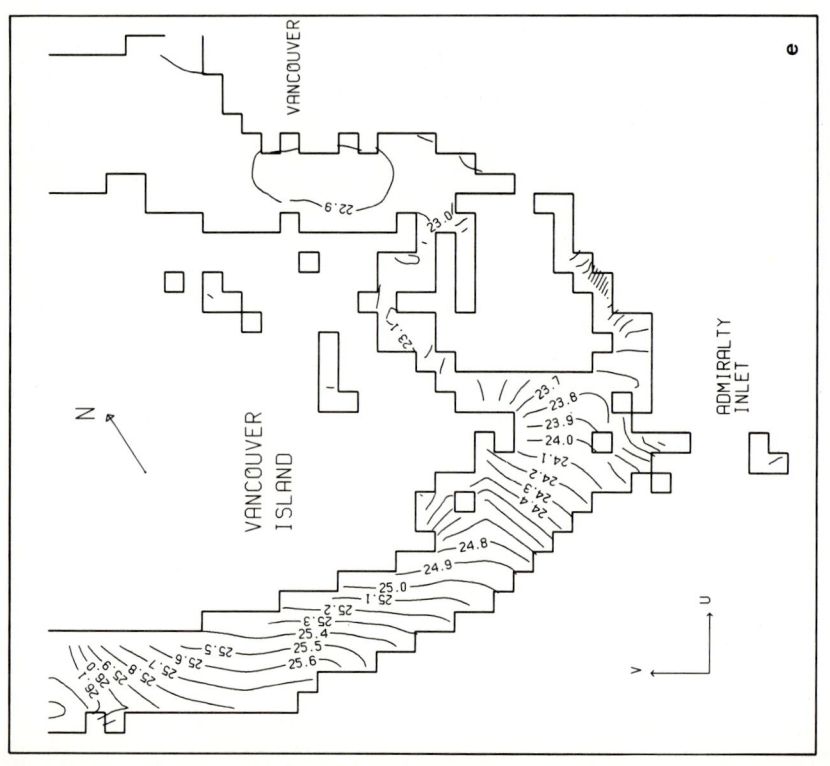

Fig. 13.18 (e) Layer 5 and (f) layer 6. Contoured values are σ_t units.

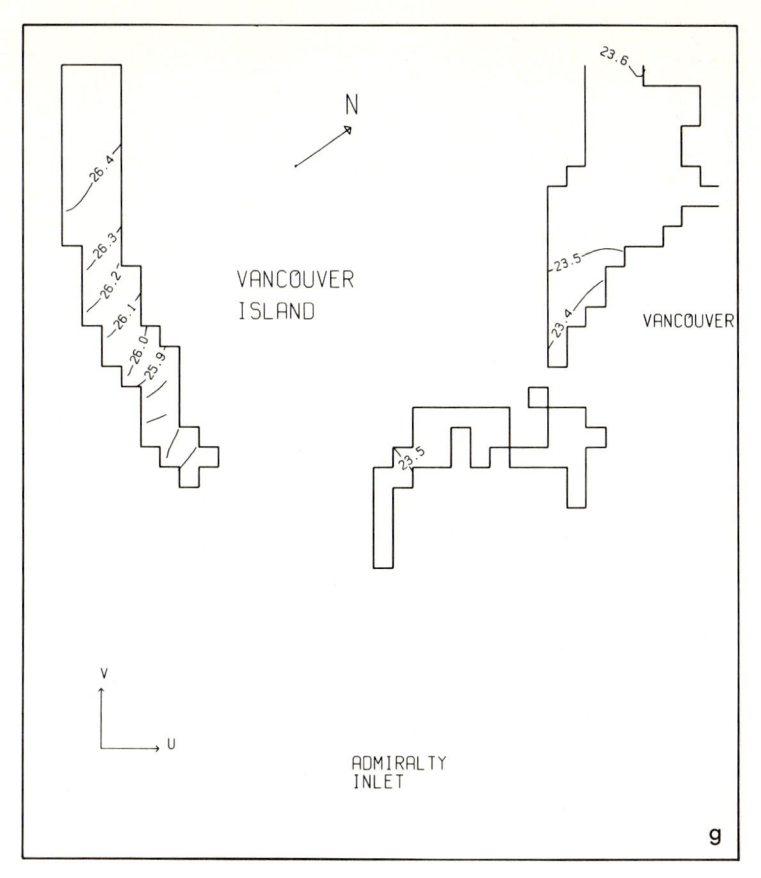

Fig. 13.18 (g) Layer 7. Contoured values are σ_t units.

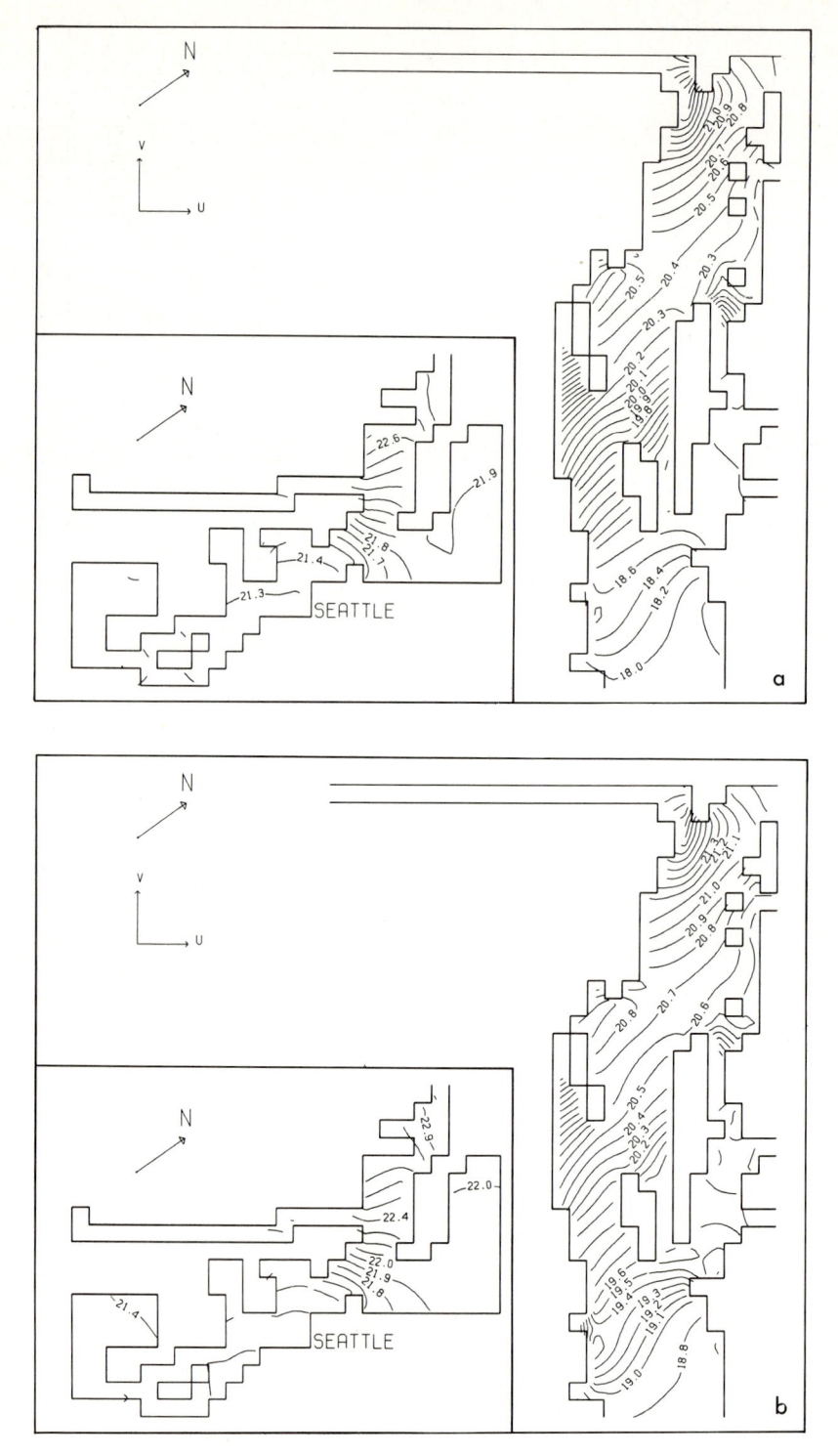

Fig. 13.19 Contoured horizontal distributions of the mean density distributions (σ_t) for the last 25 h for (a) layer 1 and (b) layer 2 in the northern Strait of Georgia and the Puget Sound system.

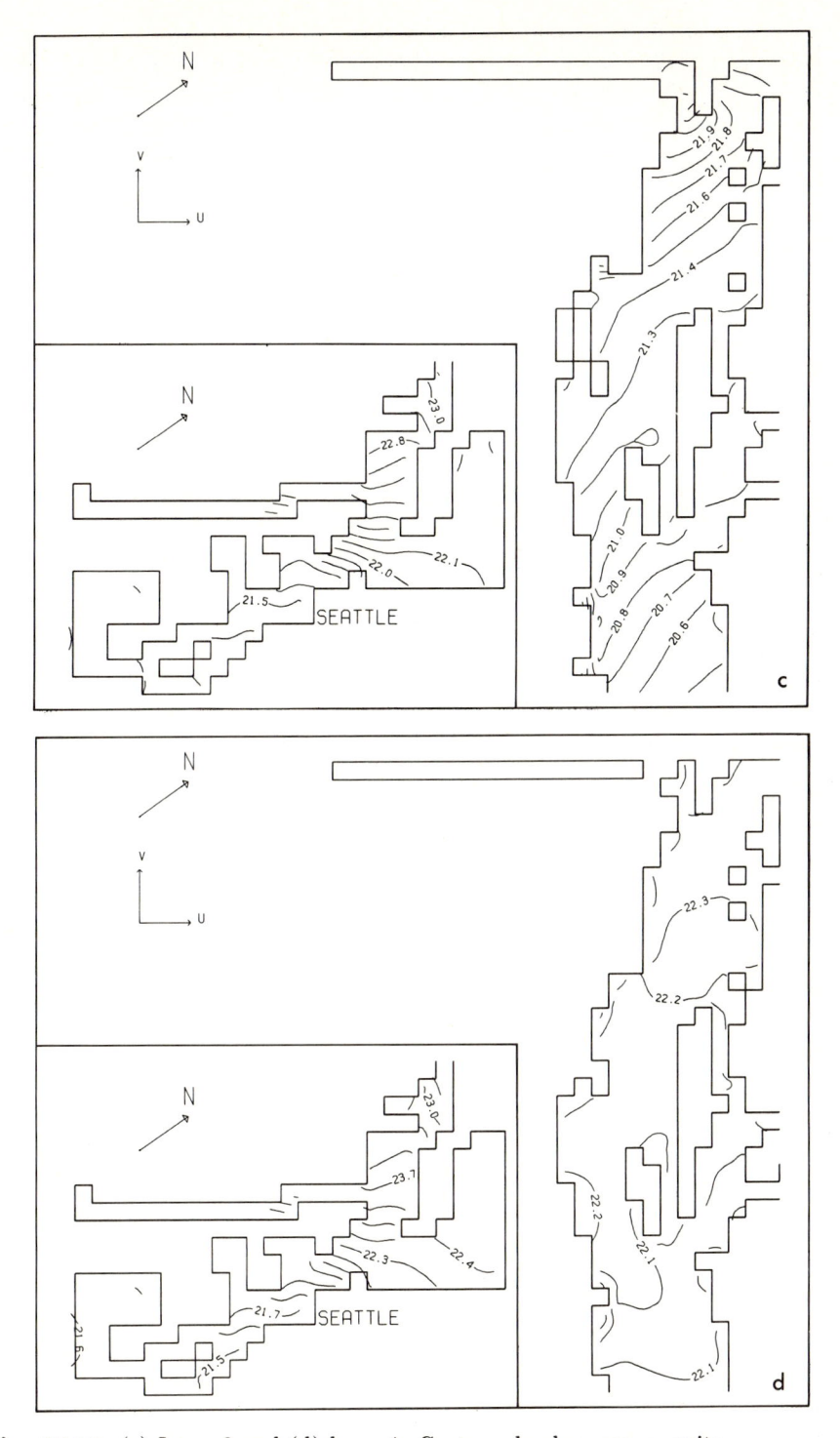

Fig. 13.19 (c) Layer 3 and (d) layer 4. Contoured values are σ_t units.

layer are shown in Fig. 13.20a–h. Positive values are contained by the full line and

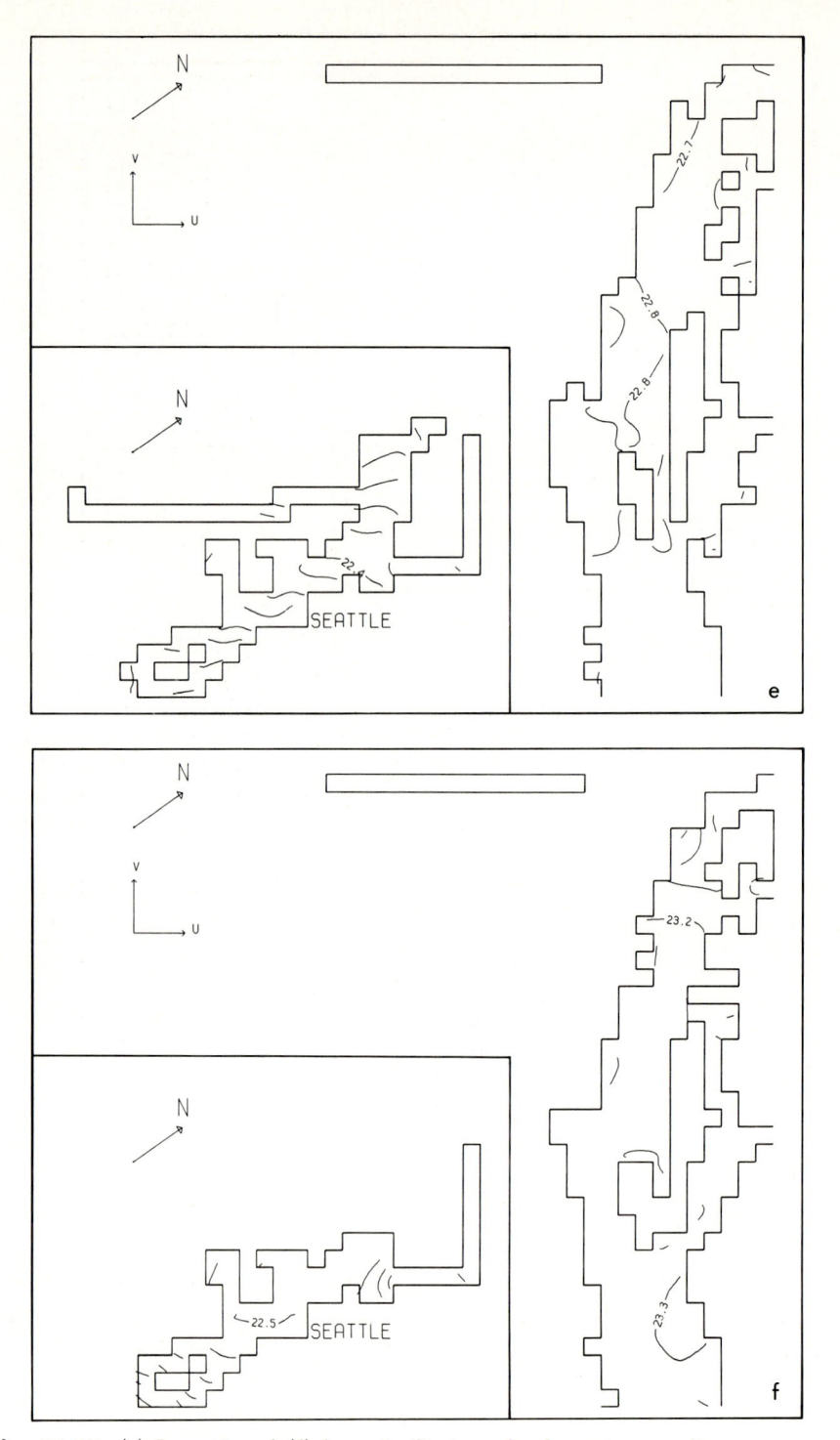

Fig. 13.19 (e) Layer 5 and (f) layer 6. Contoured values are σ_t units.

negative by the dashed line contours. It may be anticipated that the distribution of

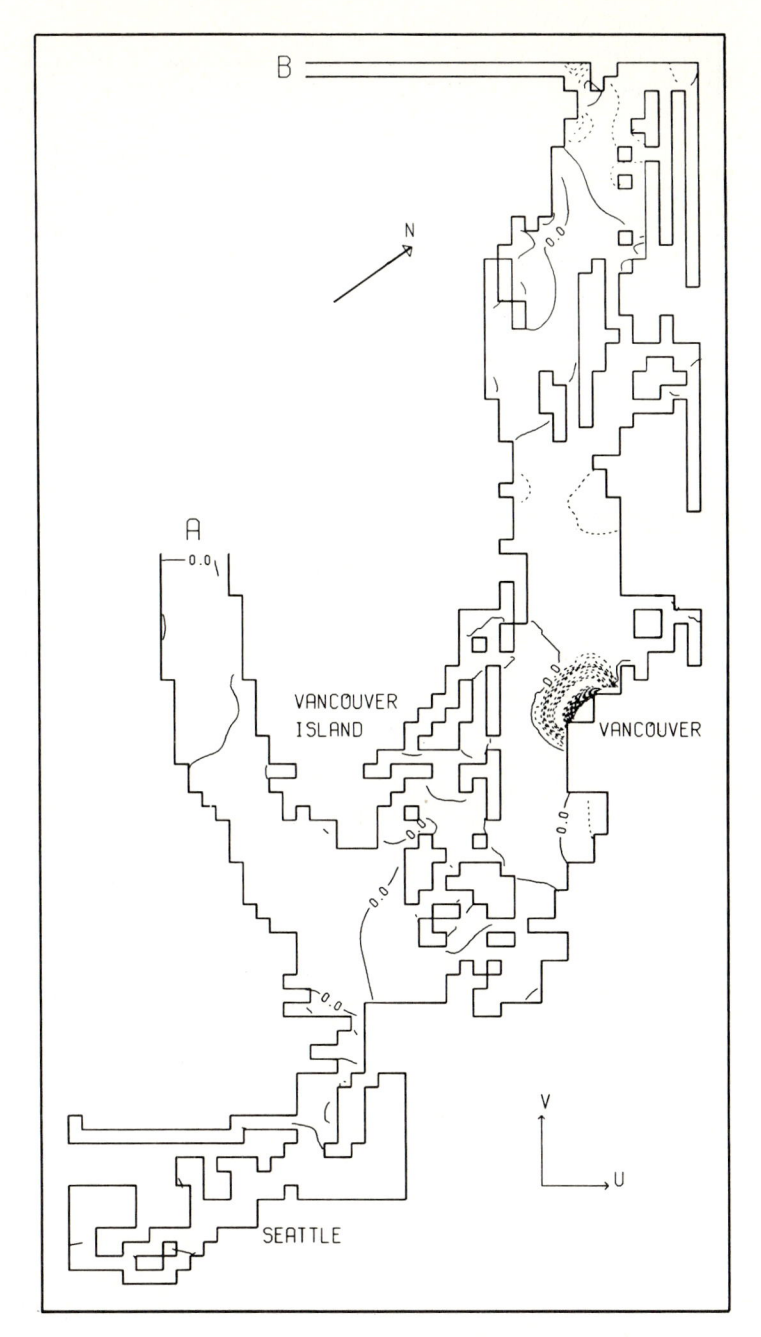

Fig. 13.20a Contours of the time-averaged vertical velocity (m/day) in layer 1. Positive values are shown by solid lines and negative values by dashed lines.

values will be strongly affected by topography.

Vertical velocities at the first level which correspond to zero mean sea level are essentially negligible except for a large negative value (−37 m/day) at the single mesh

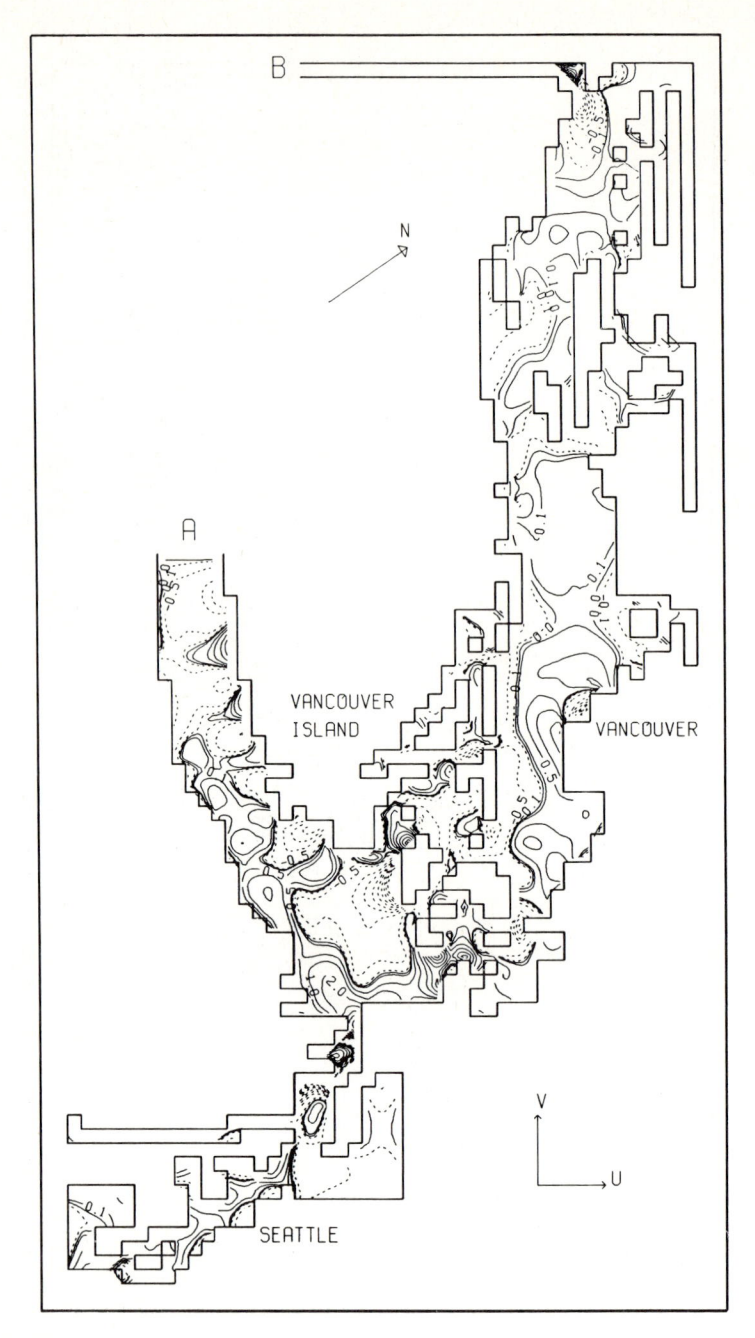

Fig. 13.20b Layer 2. Positive values are shown by solid lines and negative values by dashed lines. Vertical velocity contours are in m/day.

where the Fraser River discharge is introduced in layer 1. This effect is also apparent though much reduced, at this location in layer 2. There has, however, appeared a great deal of additional structure, the major features of which generally cohere in the

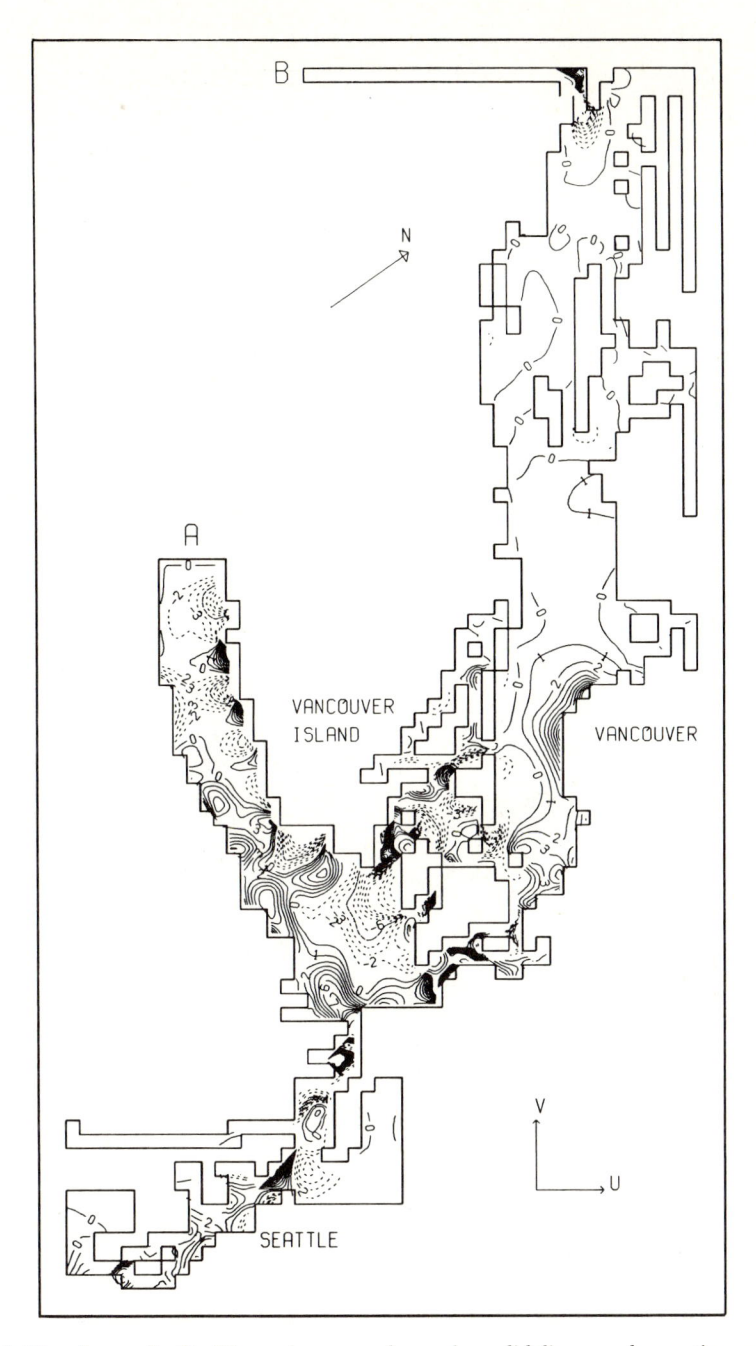

Fig. 13.20c Layer 3. Positive values are shown by solid lines and negative values by dashed lines. Vertical velocity contours are in m/day.

underlying layers.

In Juan de Fuca Strait the contours in the various layers display a cellular structure reminiscent of Poincaré standing wave formation. Such waves, in a simple channel of

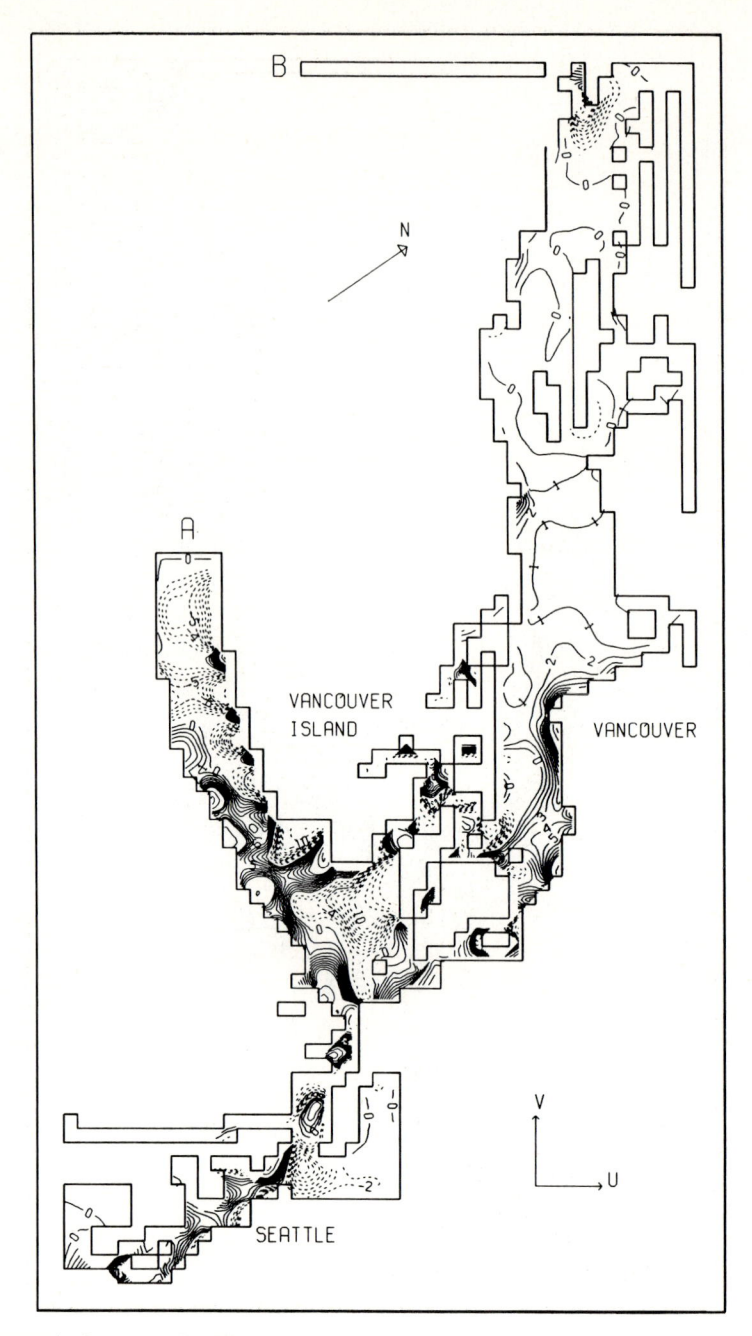

Fig. 13.20d Layer 4. Positive values are shown by solid lines and negative values by dashed lines. Vertical velocity contours are in m/day.

uniform depth and rectangular cross-section, are subject to a boundary limitation such that the wavelength parallel to the boundaries is equal to the channel width (Platzman, 1971). Although the topography in Juan de Fuca Strait is much more complicated, it

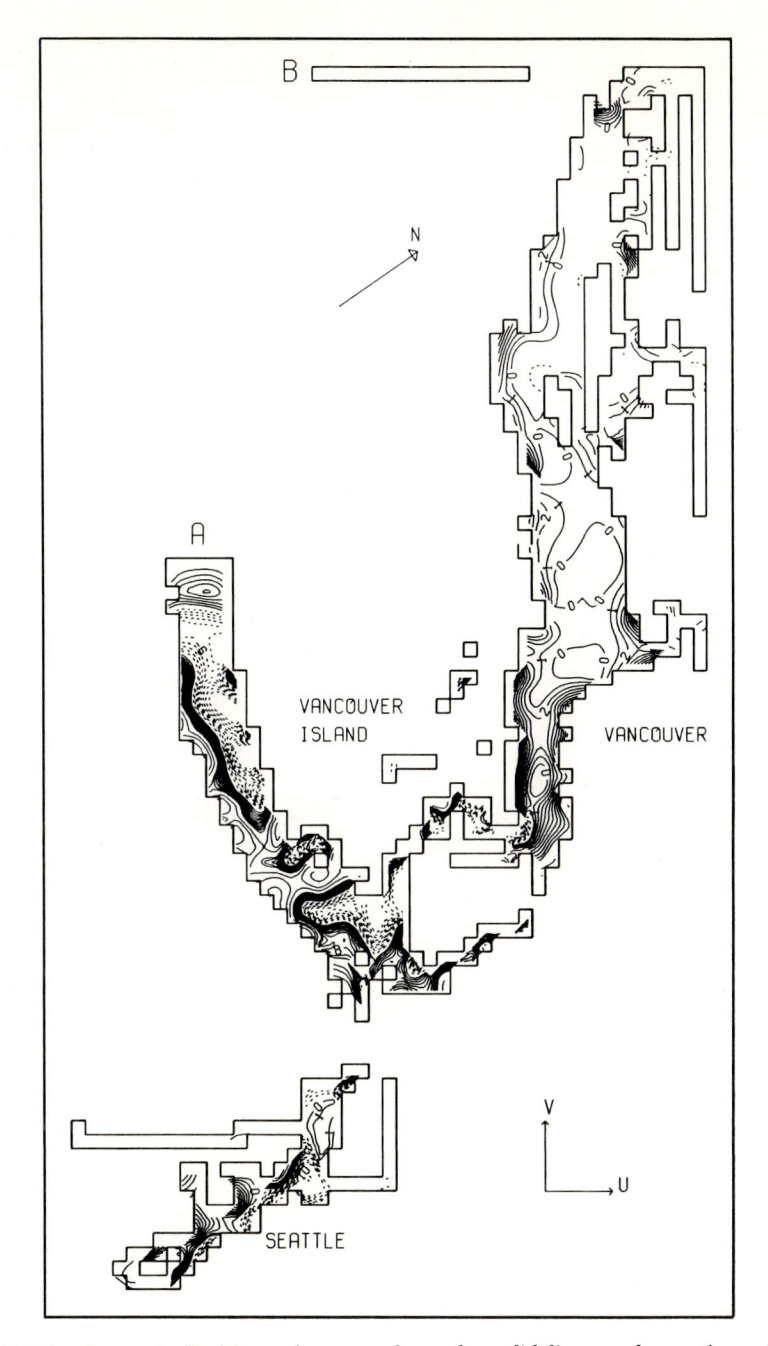

Fig. 13.20e Layer 5. Positive values are shown by solid lines and negative values by dashed lines. Vertical velocity contours are in m/day.

would appear that this relationship is approximately satisfied. It should be noted, however, that the grid meshes per wavelength are such as to provide minimal resolution of such a feature, and that, only with significant amplitude and phase distortion. The

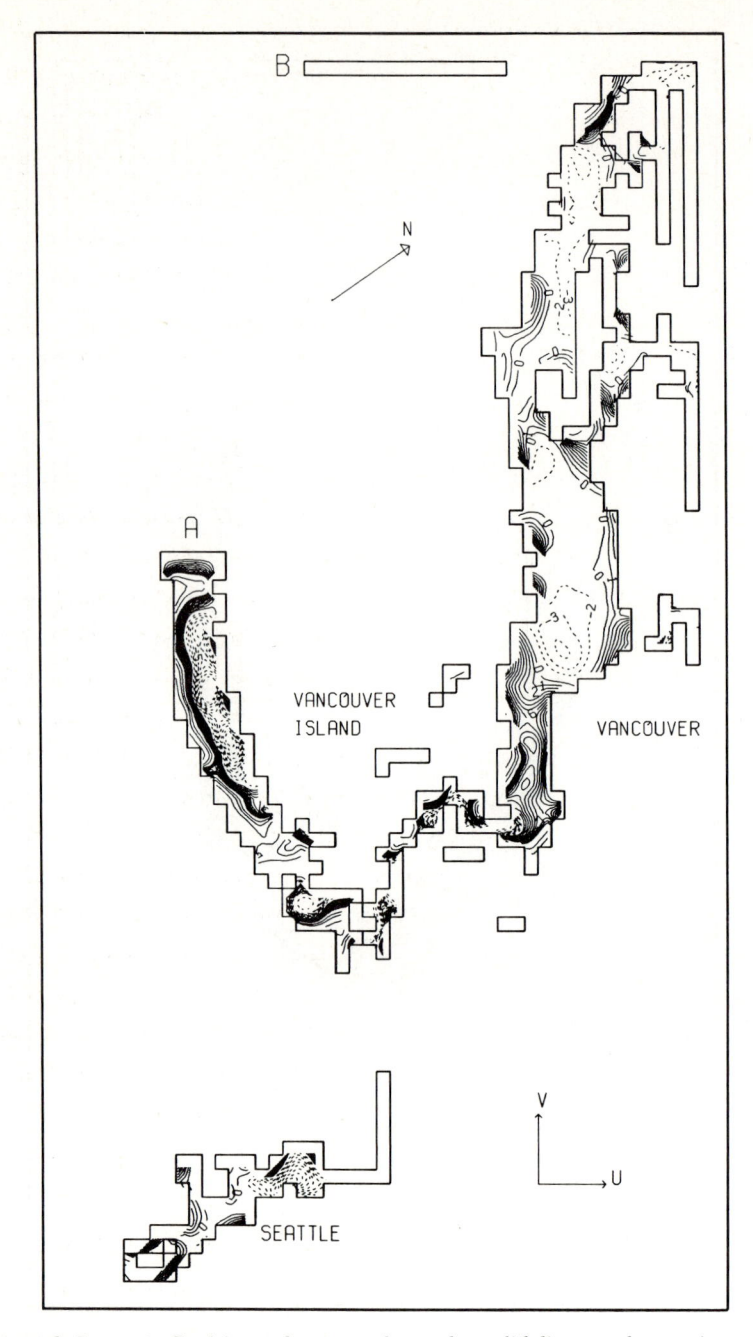

Fig. 13.20f Layer 6. Positive values are shown by solid lines and negative values by dashed lines. Vertical velocity contours are in m/day.

wave pattern is clearly superposed on further topographical features of flow field, notably associated with the downward slope of the sea bottom from the sill in the inner part of Juan de Fuca Strait to the ocean entrance. Thus, general downwelling is associ-

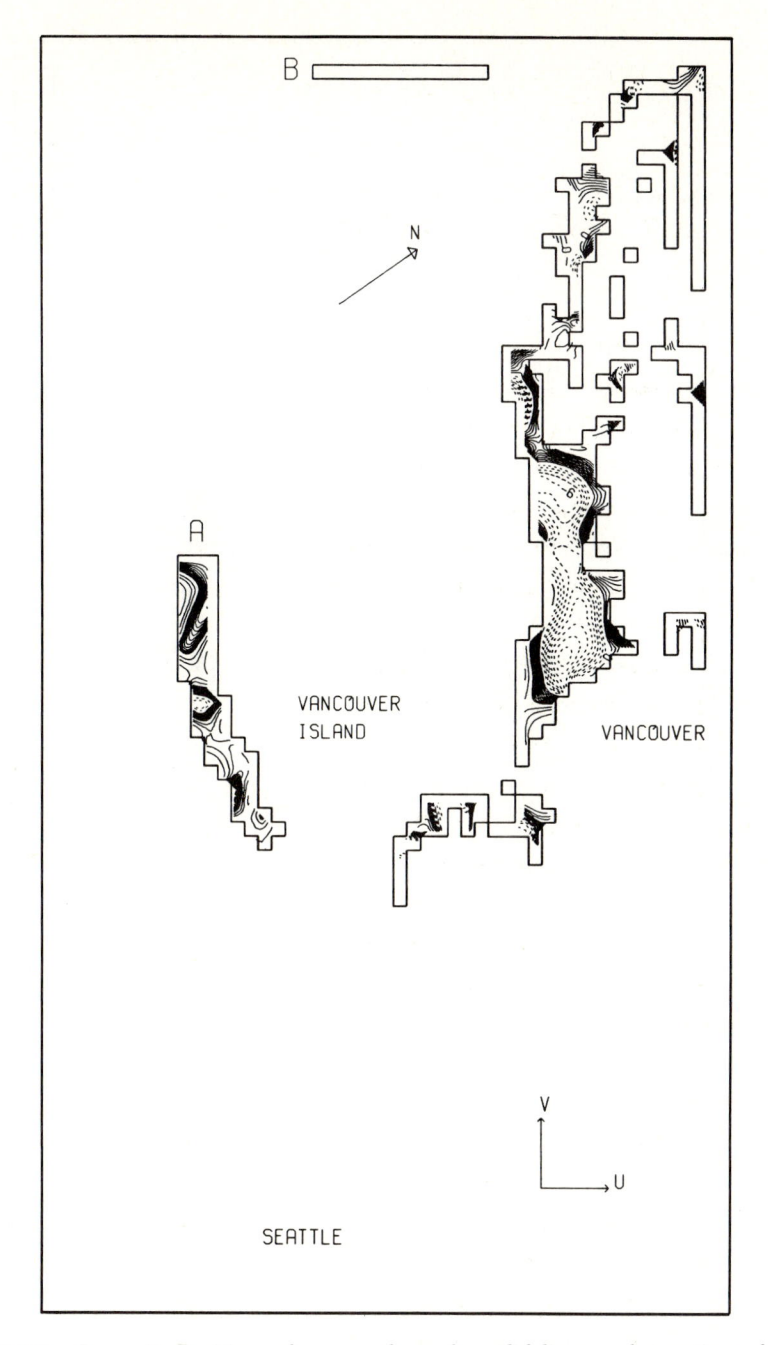

Fig. 13.20g Layer 7. Positive values are shown by solid lines and negative values by dashed lines. Vertical velocity contours are in m/day.

ated with outflow along the Vancouver Island shore; upwelling with the inflow at depth favouring the Washington shore. Inside the sill there is a strong downwelling associated with the depression of the sea bottom leading into the deep trench of Haro Strait.

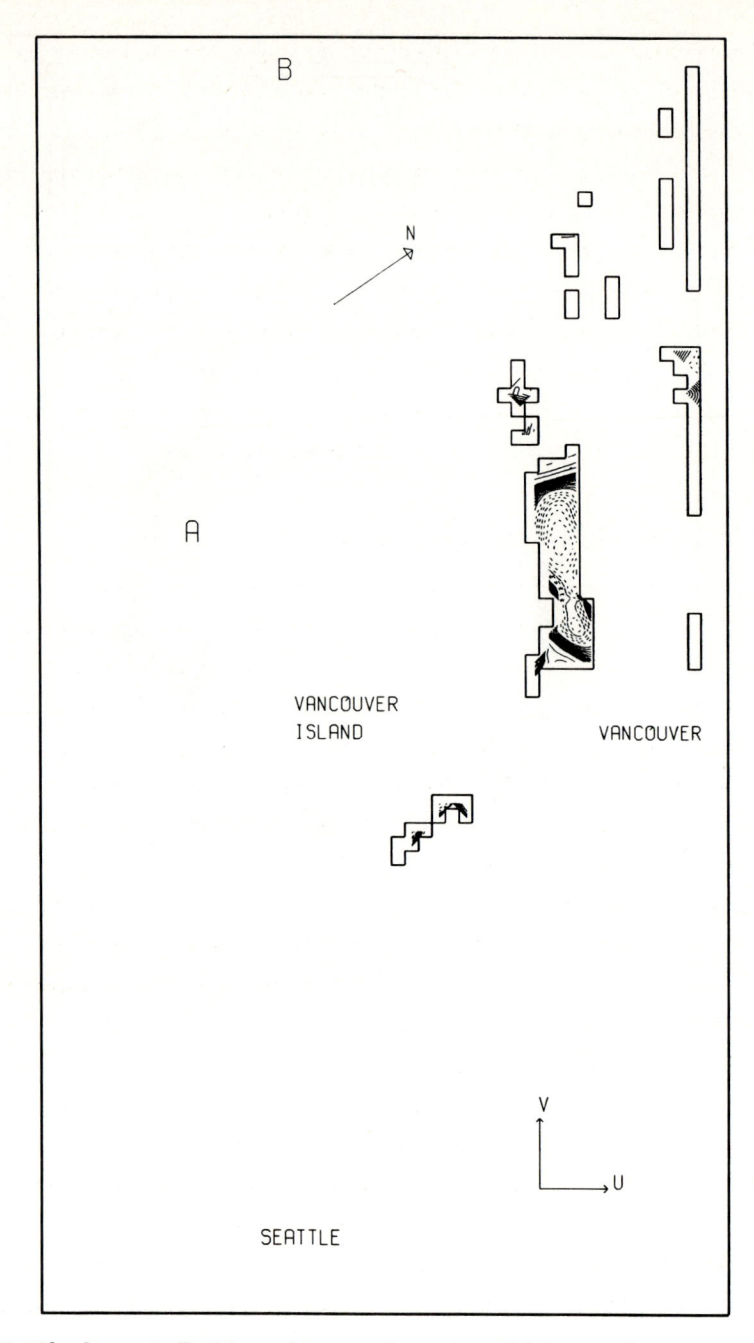

Fig. 13.20h Layer 8. Positive values are shown by solid lines and negative values by dashed lines. Vertical velocity contours are in m/day.

In the central basin and southern Strait of Georgia there is a general slow advection upward in layers 2–5 other than for a limited area of downward advection in layers 2 and 3 along the islands that constitute its southwestern shore. This general vertical advection is consistent with the divergence in the upper layers and convergence in the middepth layers. In layers 5 and 6 there is strong upward vertical advection in the narrow southernmost part of the Strait. In layers 7 and 8 the dominant feature is downward advection in the general region of the central basin, again consistent with the divergence occurring in the model at these depths.

In the northern part of the Strait of Georgia there is a local downward advection in layers 1–4 near the entrance to the northern channel. In layers 5 and 6 there is a strong vertical flow upward at this location. This is consistent with the net horizontal flow away from that location in layers 4 and 5 (Fig. 13.16).

At numerous other locations near the model boundaries there are strong local vertical advections, undoubtedly due to local topography.

13.4.7 Dynamics of Modelled Circulation

The baroclinic residual circulation derives from the balance between the inertial, Coriolis, friction, and pressure gradient terms. The latter term may be readily inferred from the surface slopes and the baroclinic pressure distributions over level surfaces passing at middepth through each of the layers in the model. The baroclinic pressure distributions Juan de Fuca Strait and the Strait of Georgia are shown in Fig. 13.21a–g. These have been computed as the difference between the modelled time-averaged (25 h) pressure at a point on a particular level and the corresponding pressure computed for the depth of that level below mean sea level in a field of unit density.

In layers 1, 2, and 3 there is an overall net pressure gradient seaward. In Juan de Fuca Strait, the isobars are roughly orthogonal to the median line of the Strait in the inner shallower part and roughly parallel in the outer part. This presumably reflects the role of friction in the former and the Coriolis effect in the latter.

The seaward decrease of pressure in the southern Strait of Georgia in layers 1 and 2 is considerably weaker than that occurring in the inner part of Juan de Fuca Strait. This presumably reflects the slow estuarine-type recirculation in the Strait of Georgia compared with the more vigorous circulation associated with the direct deep intrusion of dense water from the continental shelf into Juan de Fuca Strait.

In layers 5, 6, and 7 the general direction of the net pressure gradient along the median line of Juan de Fuca Strait is reversed consonant with the deep inflow occurring at this time. The general decrease in pressure with increasing distance from the mainland coast in the southern Strait of Georgia in layers 4 and 5 is consistent with a partial continuation of this general net inflow. The complex pattern of isobars in layer 4 in the inner part of Juan de Fuca Strait is presumably associated with tidal residual circulation effective in a region of relatively weak horizontal pressure and uncertain flow regime.

The pressure distributions in layers 7 and 8 will be discussed further below. Pres-

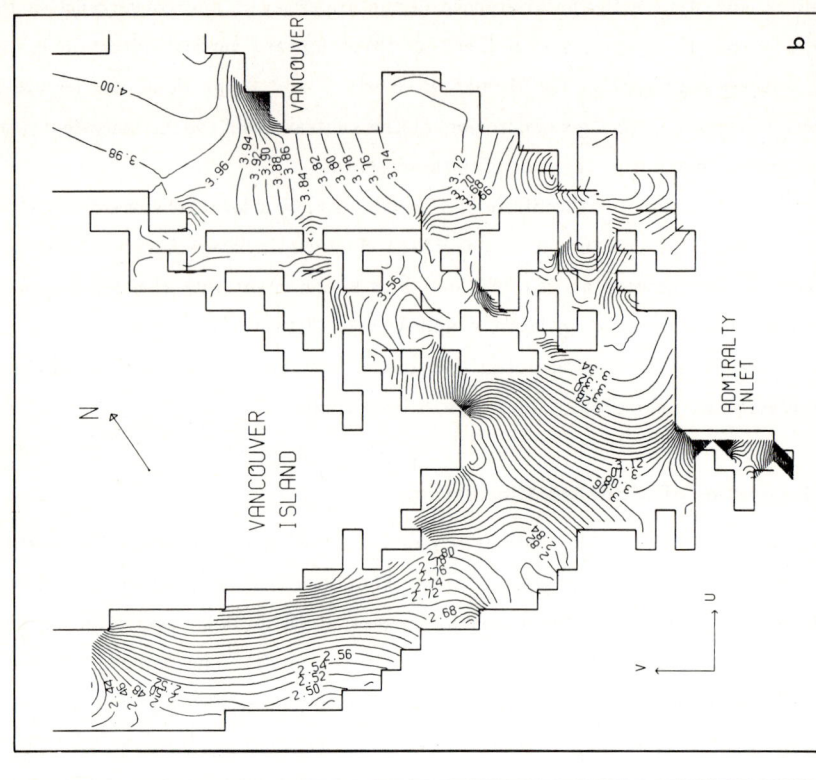

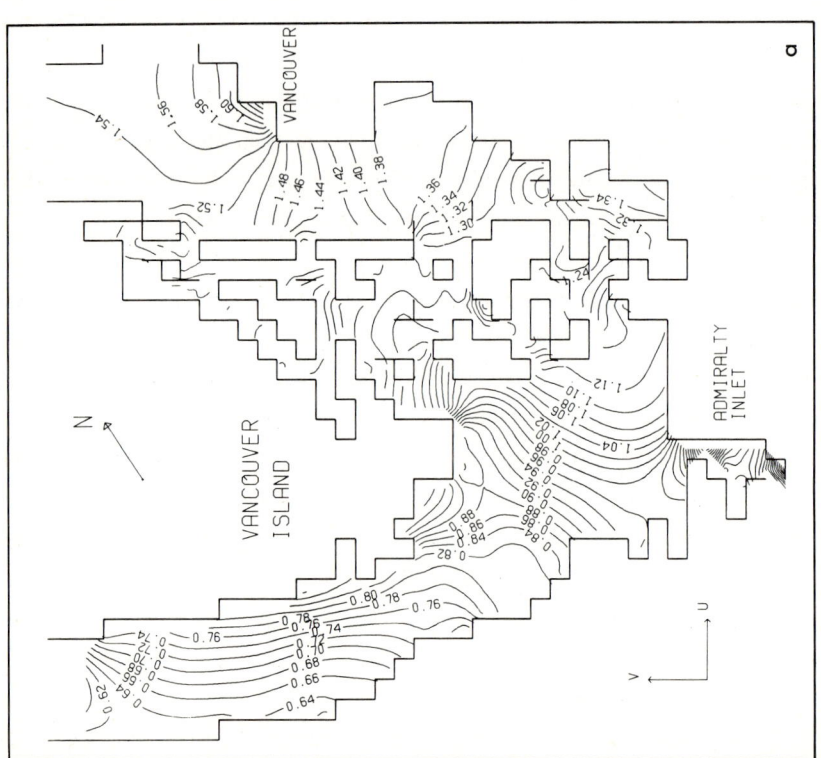

Fig. 13.21 Contoured distributions of pressure below mean sea level on surfaces through the middepth of (a) layer 1 and (b) layer 2 due to the density field in Juan de Fuca Strait and the southern Strait of Georgia. Units are in pascals $\times 10^{-3}$.

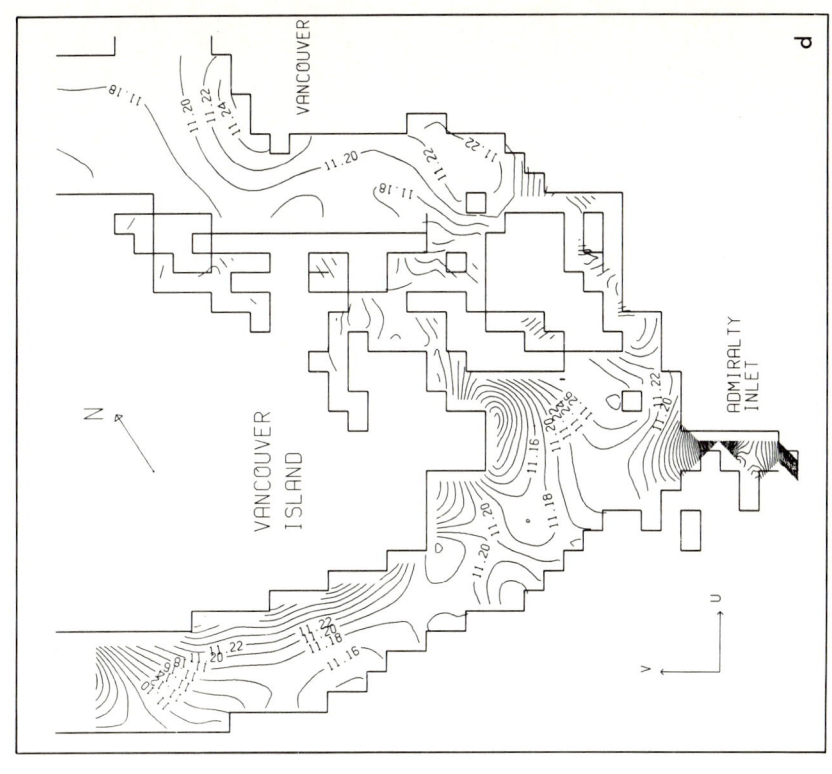

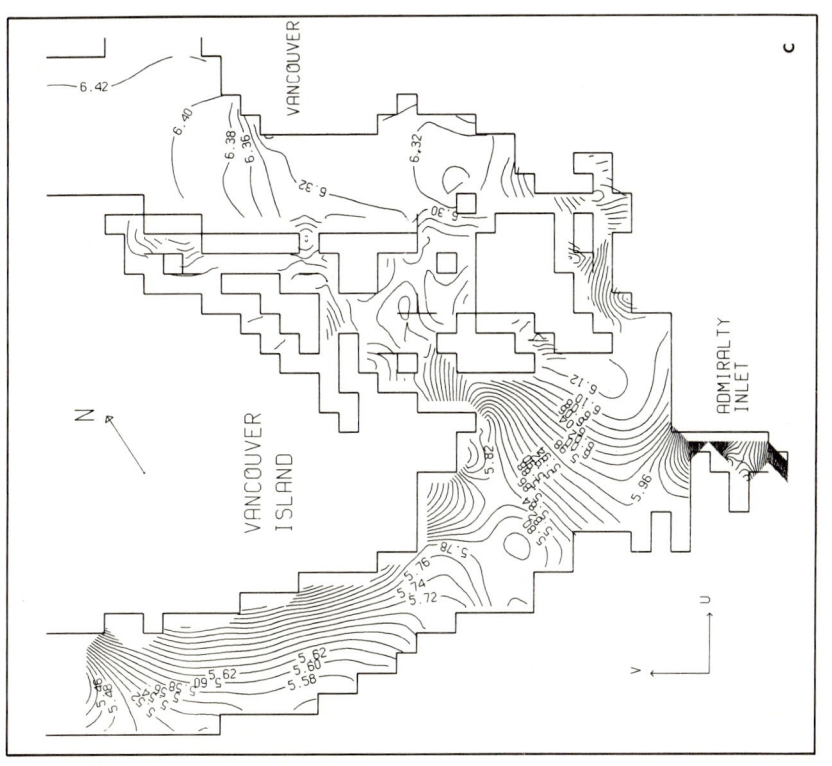

Fig. 13.21 (c) Layer 3 and (d) layer 4. Units are in pascals $\times 10^{-3}$.

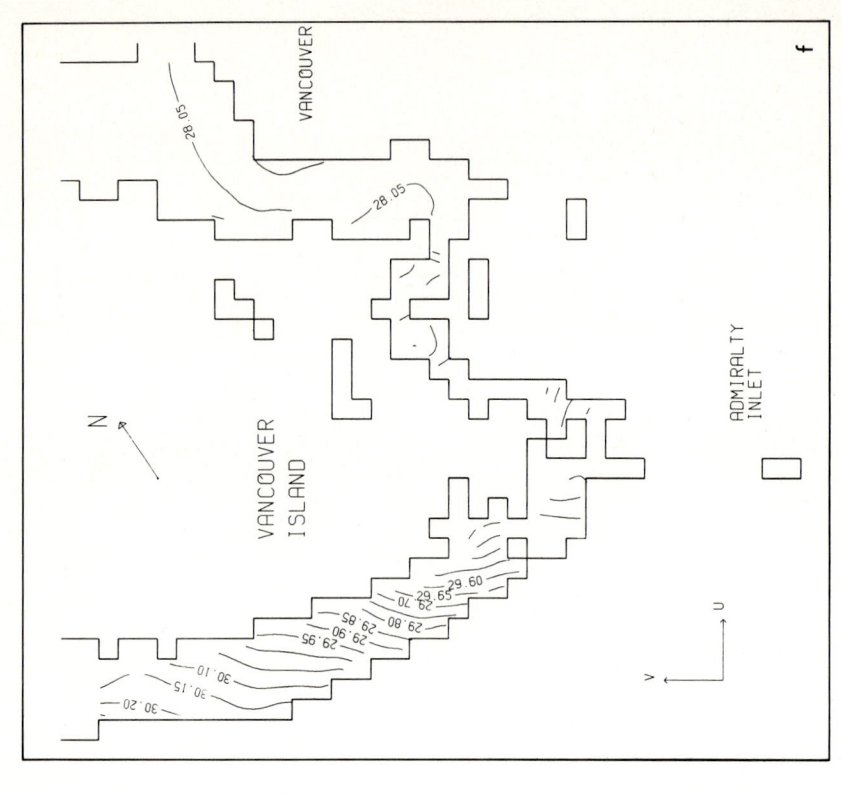

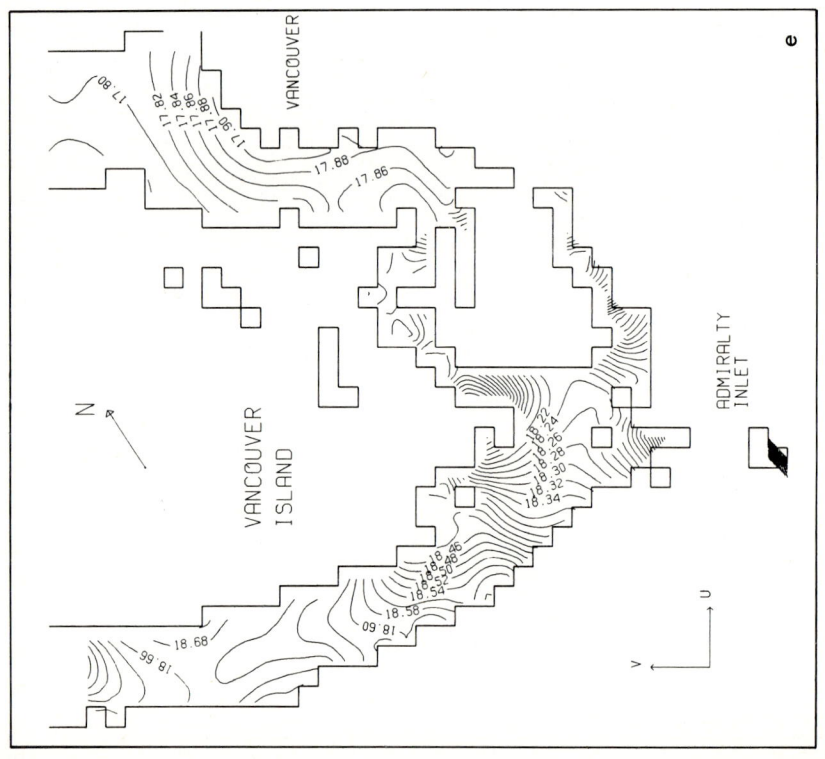

Fig. 13.21 (e) Layer 5 and (f) layer 6. Units are in pascals $\times 10^{-3}$.

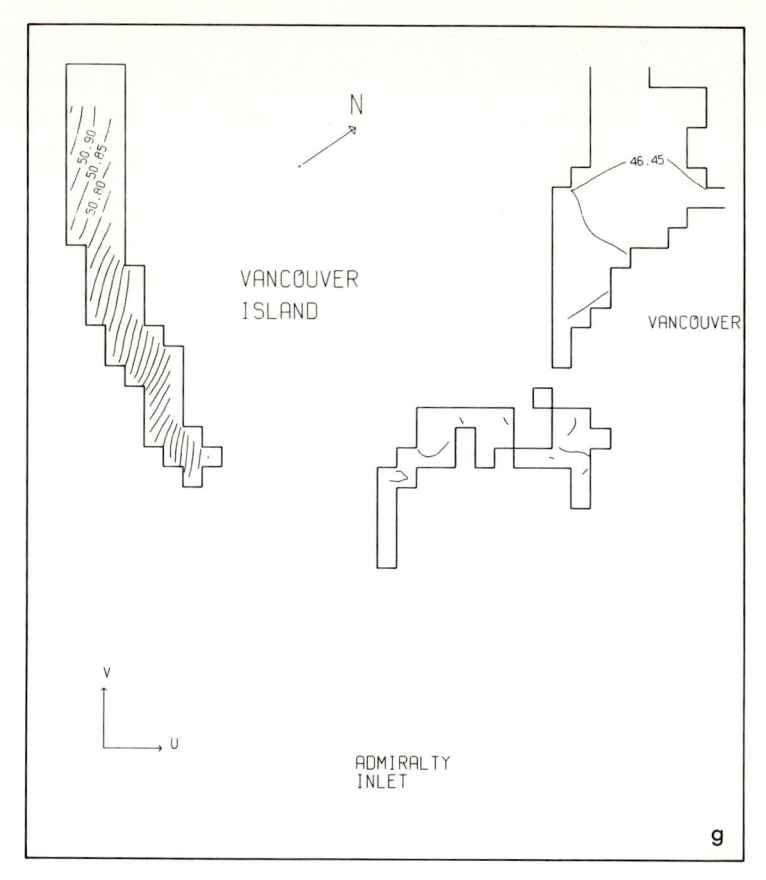

Fig. 13.21 (g) Layer 7. Units are in pascals $\times 10^{-3}$.

sure gradients in the northern Strait of Georgia (Fig. 13.22a–f) are consistent with a general surface flow northward in layers 1 and 2 and with a cross-channel gradient consistent with the Coriolis deflection towards the mainland coast. In layers 3–6, the horizontal pressure differences in the Strait of Georgia are small and will be discussed below in connection with the slow residual circulation in the deep basin.

In the Puget Sound system there is a marked minimum in the elevation of the water surface due to the Bernoulli effect of the strong tidal streams. In the remainder of the Sound in layers 1, 2, and 3 the surface gradients are consistent with flow towards the mouth of the system. However, there is some evidence of cross-channel gradients associated with Coriolis deflection. The gradients are weak in layer 4 while in layers 5 and 6 there is evidence of a reversal in the gradients which is consistent with deep inflow.

The generally weaker horizontal pressure gradients associated with the deep circulation in the Strait of Georgia are now considered. The surface layers have been included for comparison in this set of diagrams (Fig. 13.23). In layer 3 there remains a weak horizontal pressure gradient consistent with Coriolis-deflected estuarine surface discharge towards the northern and southern boundary openings, respectively. In layers 4, 5, and 6 the isobars indicate weak pressure gradients directed, respectively, from the northern and southern boundary openings towards the deep central basin of the Strait. This is consistent with the deep inflows shown in Fig. 13.16. The cross-channel gradients associated with the stronger inflow from the south shows the effect of Coriolis acceleration.

In layer 7, in the southernmost part of the deep basin, the inclination of the isobars accords with the effect of the earth's rotation on the southerly flow in this layer (Fig. 13.16). The other isobars in the deep basin indicate a pressure distribution consistent with the cross-channel flows and presence of rotation.

In the southern half of the Strait there is a pressure gradient acting in the general vicinity of the Fraser River mouth towards the north and south. This gradient continues up to the vicinity of the northern boundary opening where a local reversal occurs. In the northern part of the basin the residual flow direction runs counter to the pressure gradient, presumably a reflection of strong topographic influences. In layer 8 the pressure gradients are directed towards the northern and southern limits of the deep basin, consistent with the flow divergence shown in Fig. 13.16.

13.4.8 Critical Assessment of the Model

A basic pattern of overall residual circulation emerges in which the river discharge, on entering the Strait of Georgia, splits into north- and south-going components. When these reach the strong tidal mixing regions, undercurrents of some density intermediate between these surface waters and intruding ocean water, are formed. These move

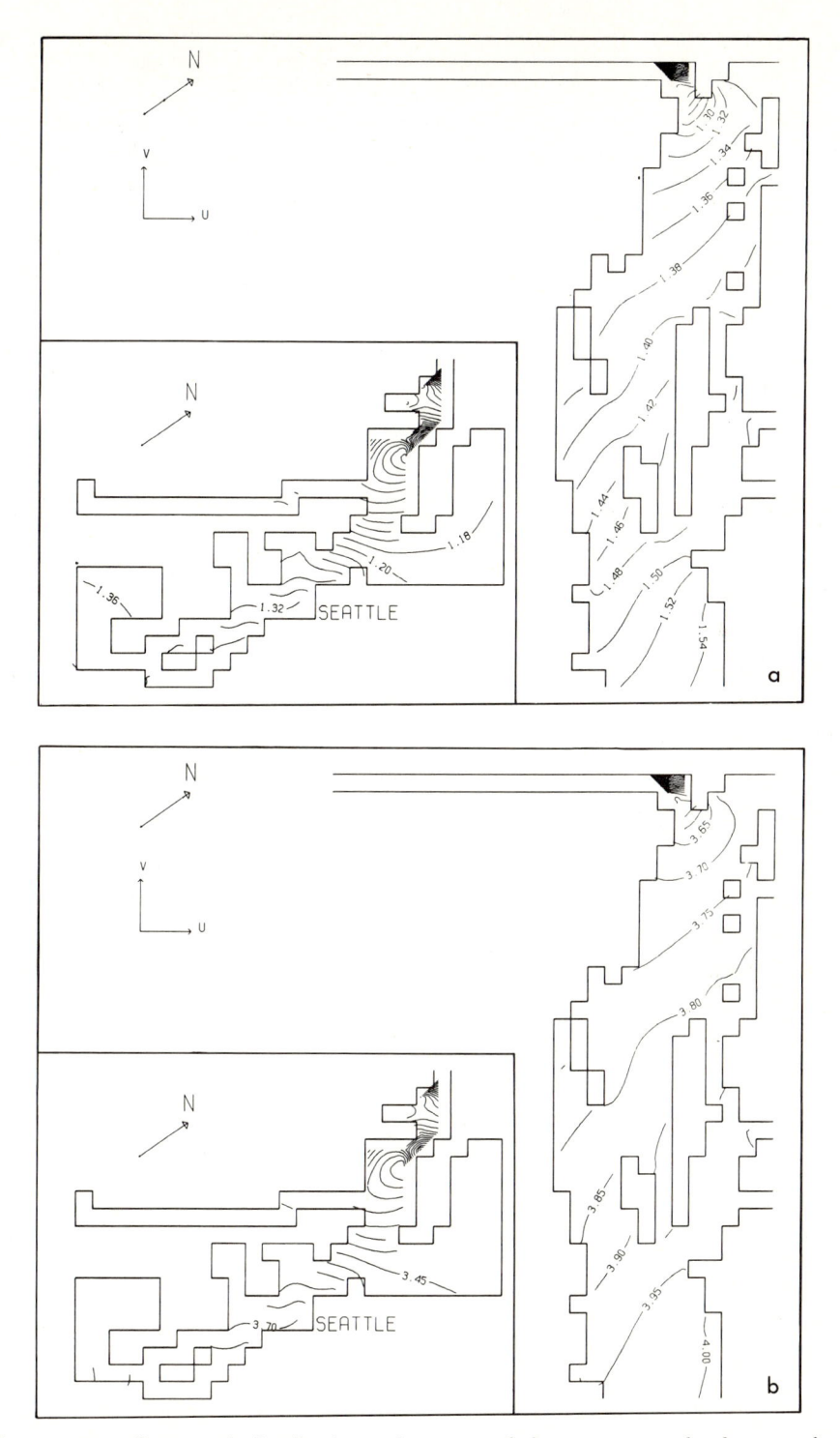

Fig. 13.22 Contoured distributions of pressure below mean sea level on surfaces through the middepth of (a) layer 1 and (b) layer 2 due to the density field in Puget Sound and the northern Strait of Georgia. Units are in pascals $\times 10^{-3}$.

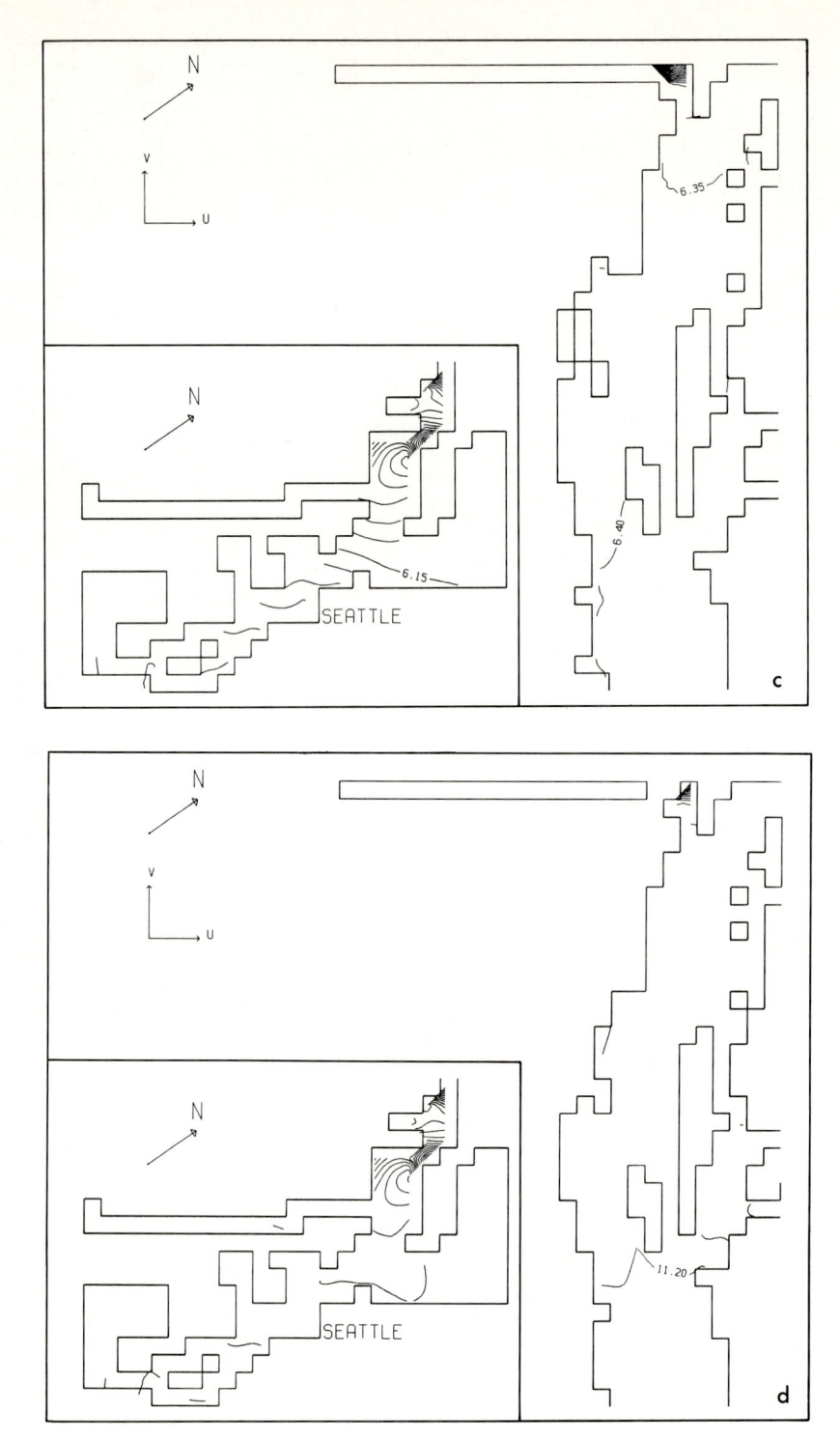

Fig. 13.22 (c) Layer 3 and (d) layer 4. Units are in pascals $\times 10^{-3}$.

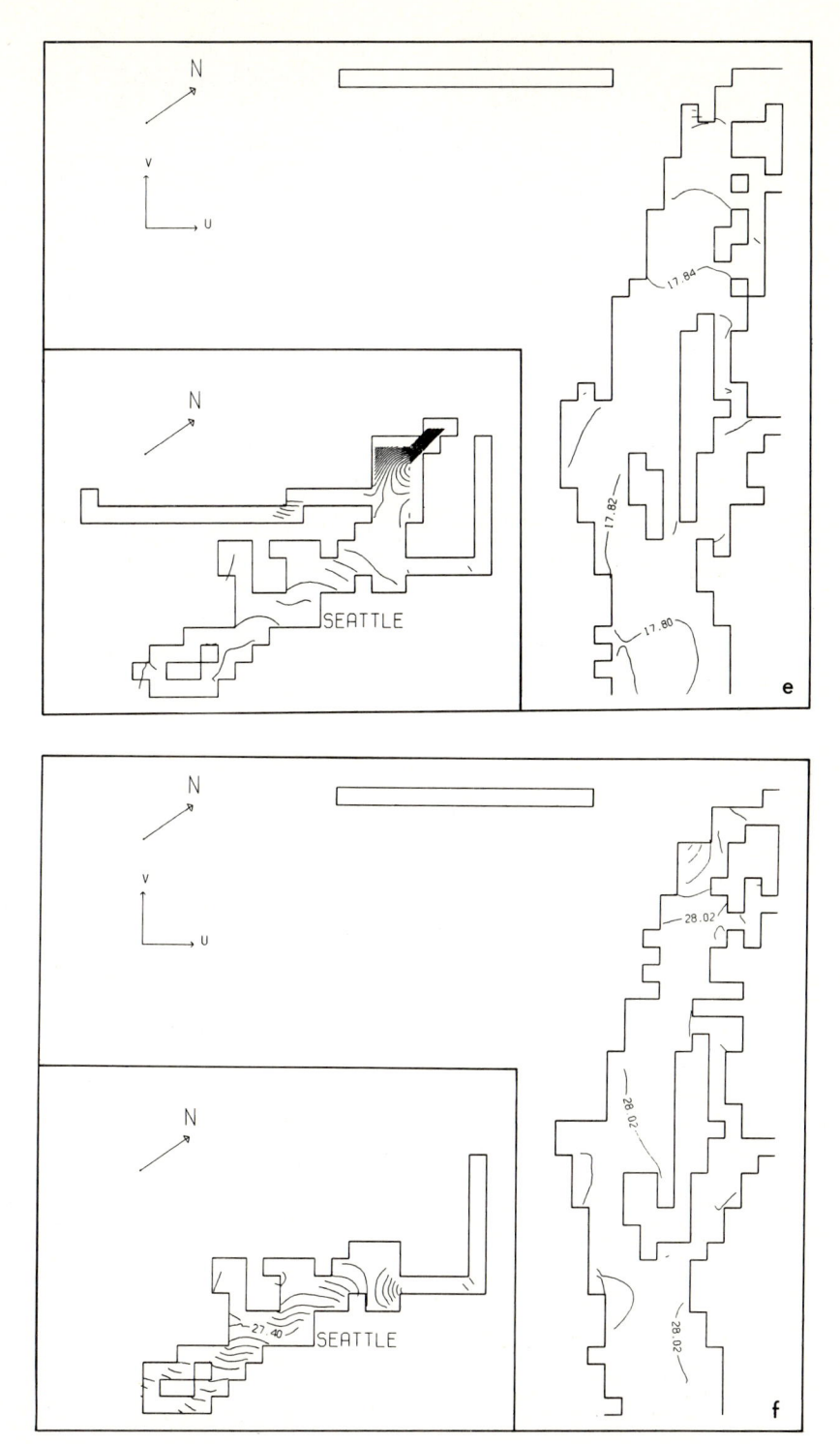

Fig. 13.22 (e) Layer 5 and (f) layer 6. Units are in pascals $\times 10^{-3}$.

Layer 1 (1–5 m)   Layer 2 (5–15 m)

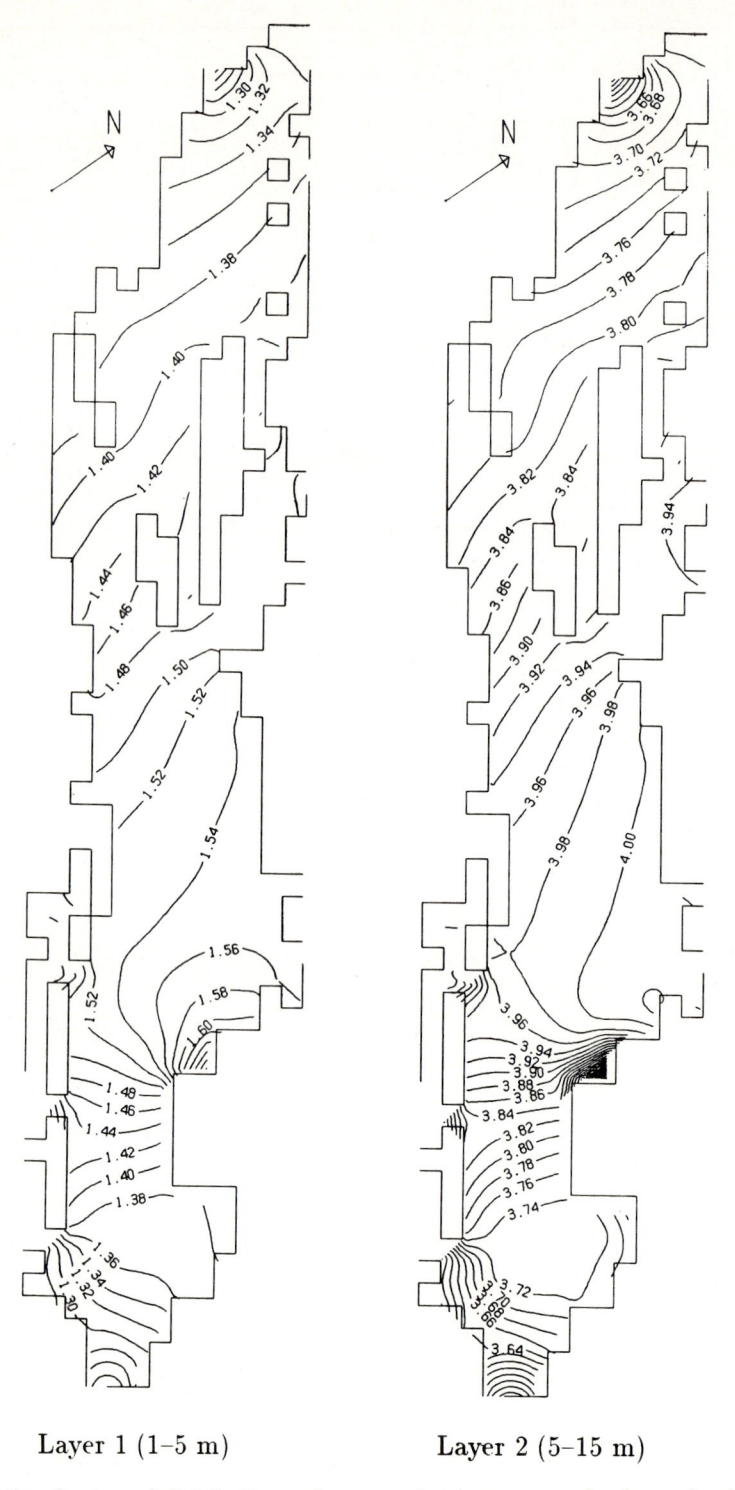

Fig. 13.23 Contoured distributions of pressure below mean sea level over level surfaces through the middepth of each layer due to the density field in the Strait of Georgia. Units are in pascals $\times 10^{-3}$.

Layer 3 (15–30 m)   Layer 4 (30–60 m)

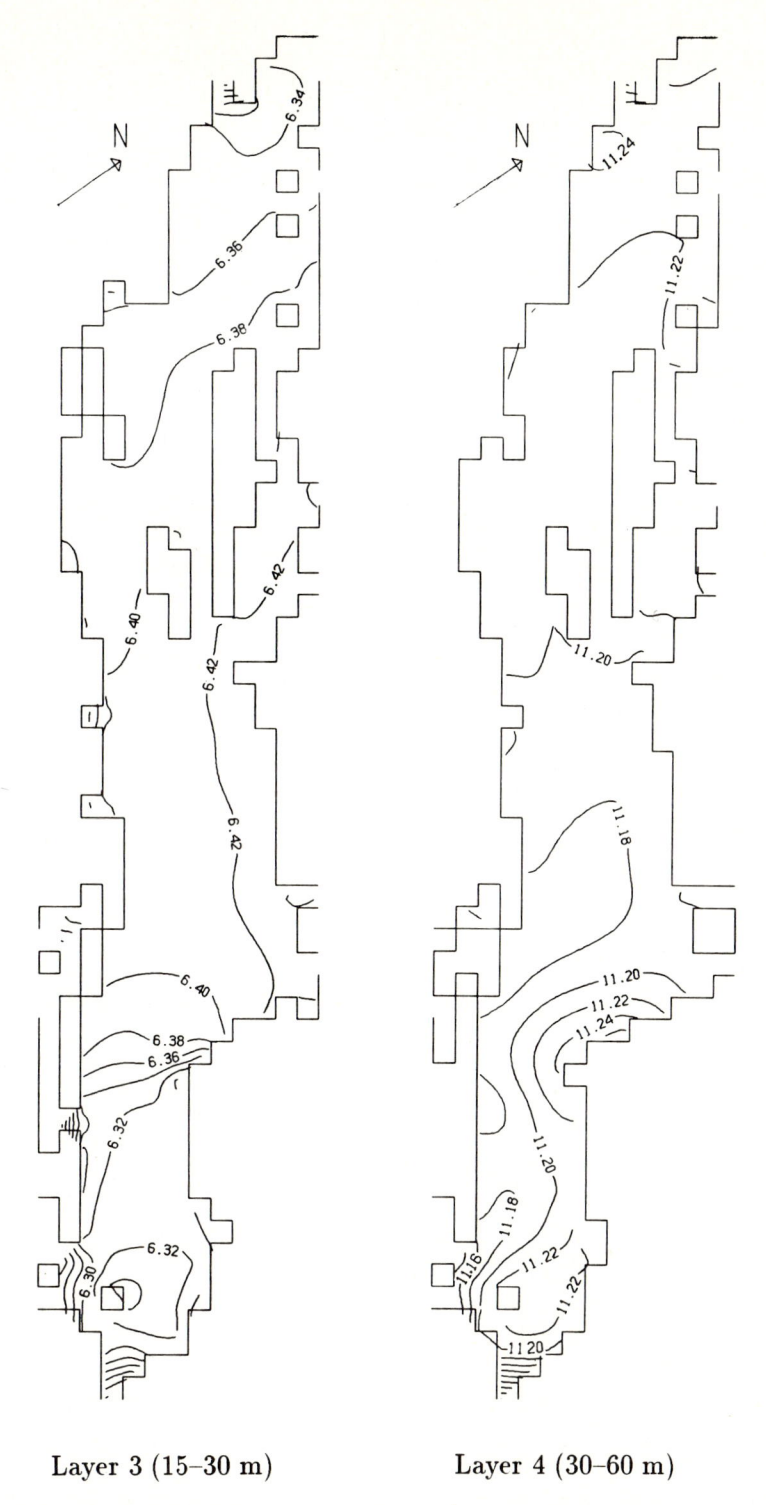

Fig. 13.23 Continued.

Layer 5 (60–90 m)   Layer 6 (90–150 m)

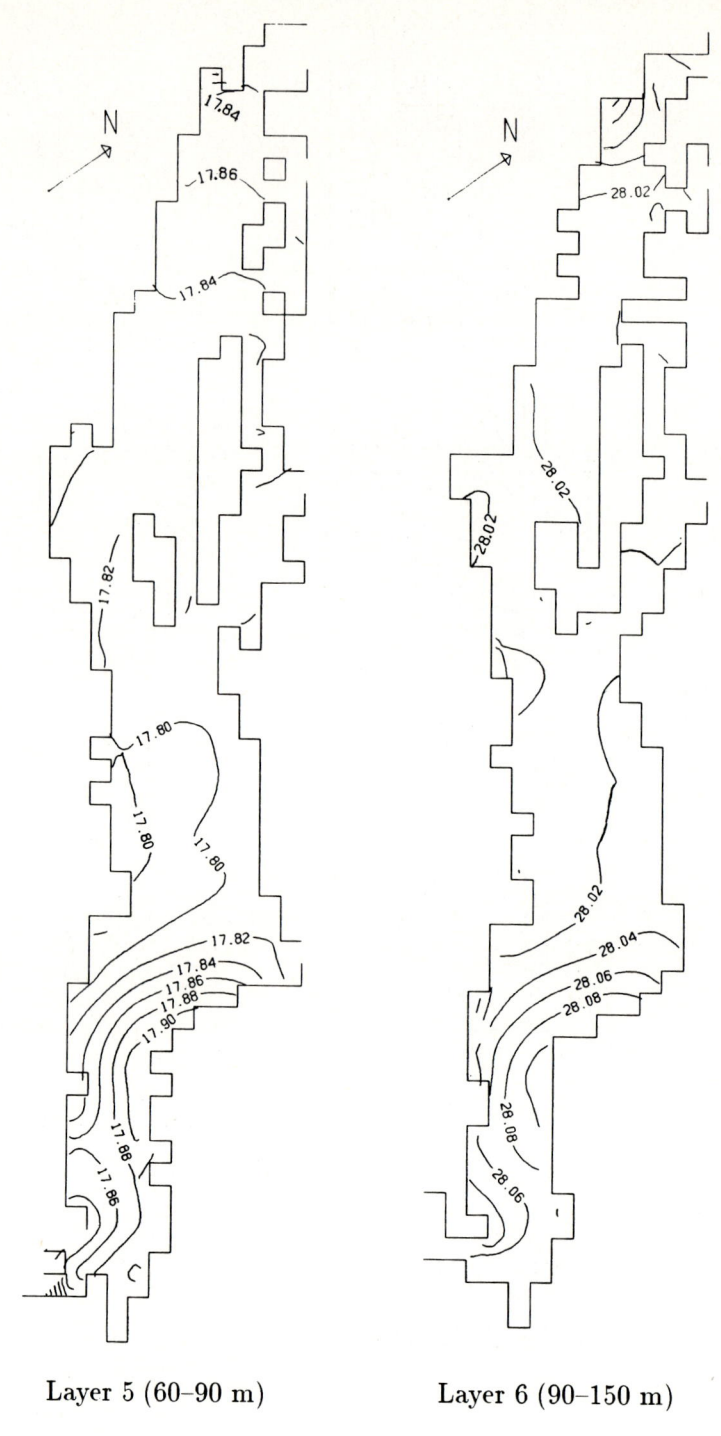

Fig. 13.23 Continued.

Layer 7 (150–250 m)

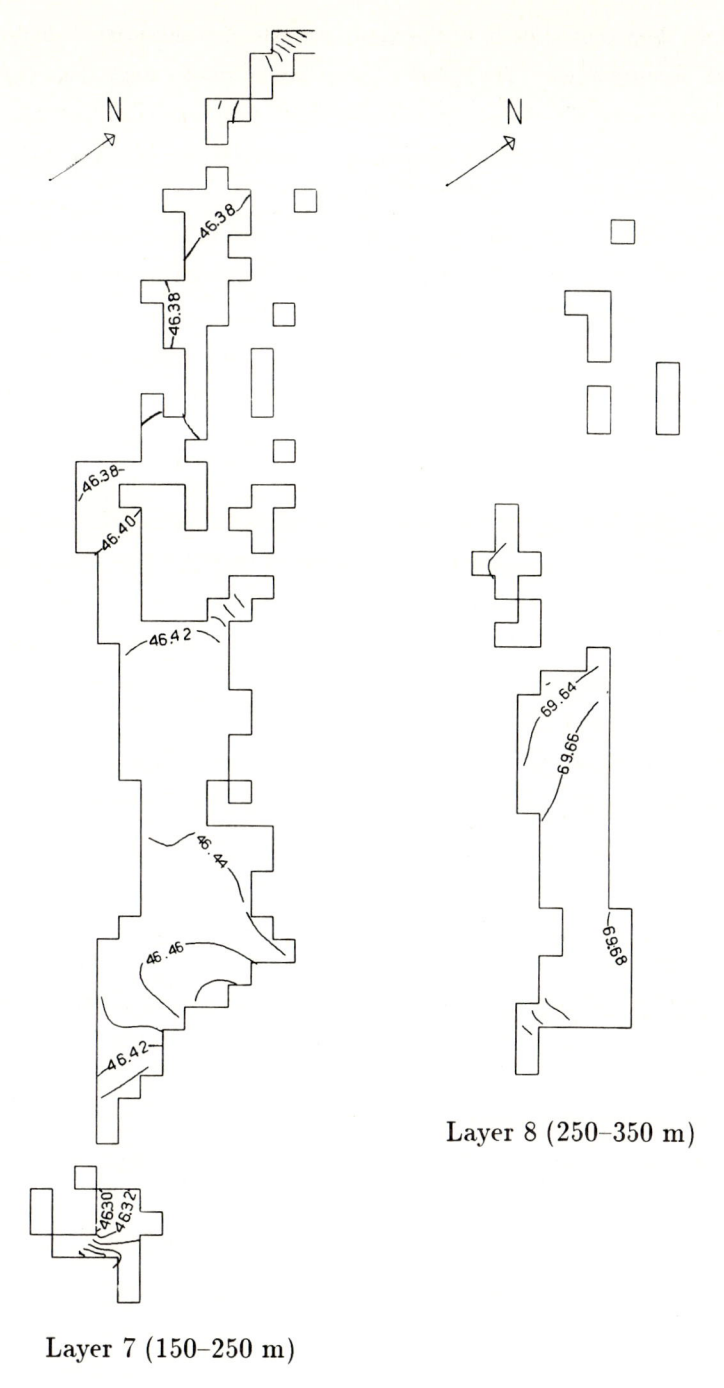

Layer 8 (250–350 m)

Fig. 13.23 Continued.

northward from the southern mixing region and southward from the northern mixing region, each tending to favour the shore to the right of the direction of motion. On entering the deep central basin of the Strait, an extensive subsurface counter-clockwise circulation is established. The question arises as to what extent this residual circulation correctly describes the actual residual circulation associated with the combined processes of estuarine flows, tidal mixing, and deep water renewal. The basic pattern of residual circulation in Juan de Fuca Strait appears to be reasonably consistent with a largely counterbalancing inflow at depth and outflow at the surface as presented earlier in terms of field observations in Chapter 2. Although possible responses of the Strait of Georgia to forcing by tide, wind, and river discharge are undoubtedly many and varied, these would appear to be changes rung upon the basic theme of residual circulation outlined above, a theme clearly detectable in available observations.

The proportion of residual surface flow moving southward from the vicinity of the Fraser River mouth appears larger than might be inferred from the drogue tracks described earlier (see Fig. 1.26). However, these tracks are confined to a shallow (2 m) layer compared to the upper layer of the model (0–5 m). Such a general southerly flow is consistent with the current meter records shown in Fig. 1.27 (p. 38). Applying isentropic analysis, Waldichuk (1957) determined the depth of a density surface ($\sigma_t = 23.5$) which descends from the free surface in the tidal mixing areas to the north and south to depths of order 100 m. From the horizontal distribution of these depths and associated distributions of salinity, temperature, and oxygen on the surface for March and September, a basic pattern of circulation was inferred. Illustrative of this analysis Fig. 13.24 shows the distribution of depth of the $\sigma_t = 23.5$ surface and circulation inferred from this and distributions of scalar properties (not shown here) for March 1953. There is a northward undercurrent along the mainland shore from the San Juan Island passages and a southerly undercurrent along the Vancouver Island shore from the southern end of Discovery Passage. On entering the deep basin of the central Strait of Georgia, these flows appear to form a clockwise eddy. Though the direction of rotation of the eddy is at variance both with the model and extended moored current meter records, the basic overall features of this inferred circulation are clearly consonant with the modelled baroclinic residual circulation at comparable depths. A similar plot for September 1952 shows essentially the same features.

Evidence in moored current meter records of a northward-moving undercurrent from the southern tidal mixing region is now considered. The large counter-clockwise residual eddy (Fig. 7.9, p. 151) driven by the flood tidal jet from Haro Strait is consistent with the inception of such a flow. Records obtained in summer, 1969 from five current meters moored at a depth of 100 m across the Strait of Georgia (line 6, see Fig. 2.1) show persisting net flows (4–7 cm/s) across the Strait at the central and two eastern locations. This is consistent with the sense of rotation of the large residual tidal eddy. The two remaining moorings on the Gulf Islands side of the Strait show flows of similar magnitude moving up into the Strait. This (line 6) agrees well with model predictions

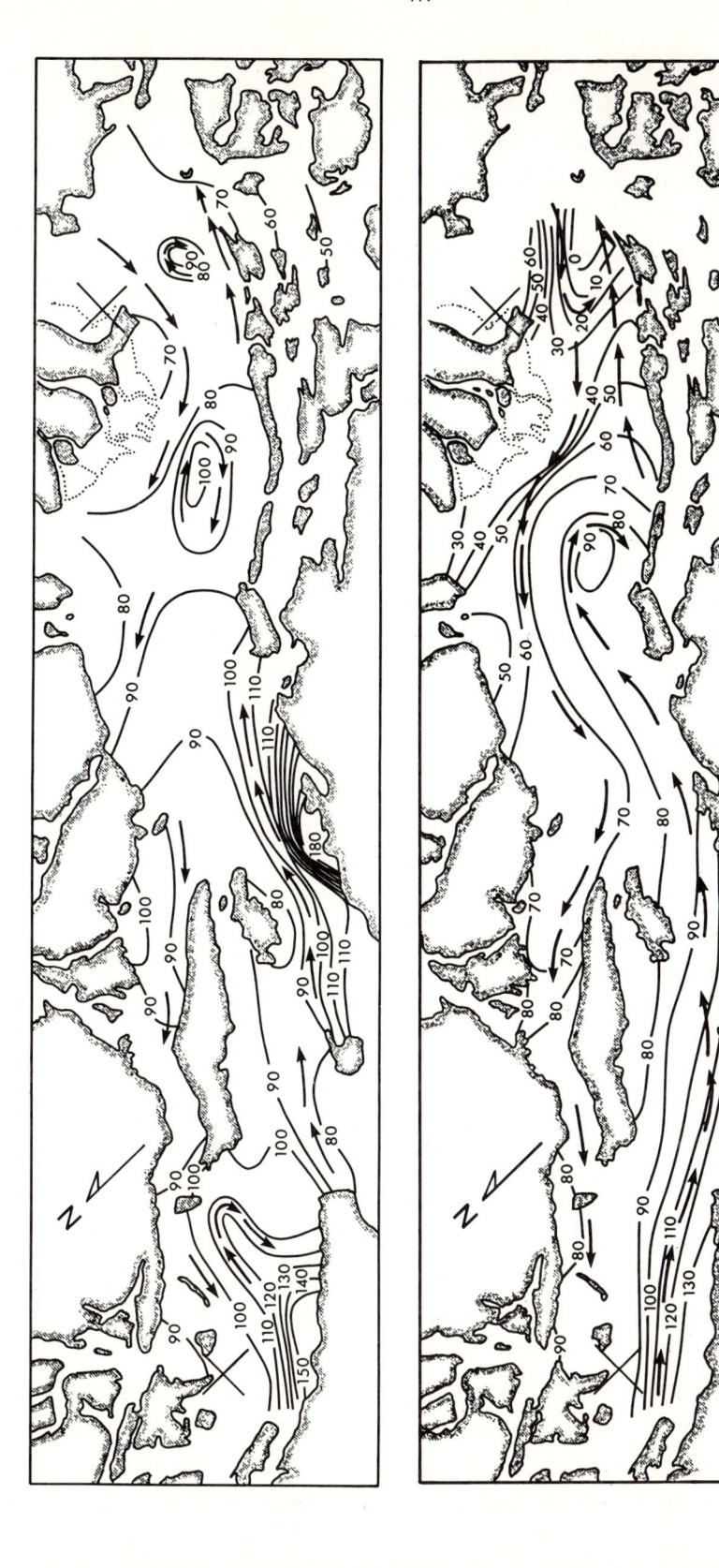

Fig. 13.24 Contoured depths of the isopycnal surface $\sigma_t = 23.5$. Arrows denote residual circulation inferred from these depths and distribution of scalar properties (not shown) on these surfaces for (top) September 1952 and (bottom) March 1953. (from Waldichuk, 1957)

at this location in layer 5 (60–90 m) (Fig. 13.16).

Records obtained at 100 and 180 m off the mouth of the Fraser River (line 5) in the summer of 1969 show a well developed residual flow (10–12 cm/s) directed up the Strait of Georgia past the delta front. Further records obtained at 50 and 100 m near the mainland coast (line 4) in spring, 1969 show residual flows of 3–4 cm/s which are consistent with model predictions. Of particular interest are the residual velocities determined over a 1-yr period from an array of current meters in the southern part of the deep central basin (Chang et al., 1976). Flows at 50 m at the western, central, and eastern moorings (Fig. 13.25) accord well with the flows in this region of layer 4 (Fig. 13.16). The flows at the western and central moorings at 200 m would appear at least qualtitatively consistent with a clockwise eddy in layer 7 (Fig. 13.16) and the strong northward flow at the eastern mooring is again consistent with the model flow in layer 6.

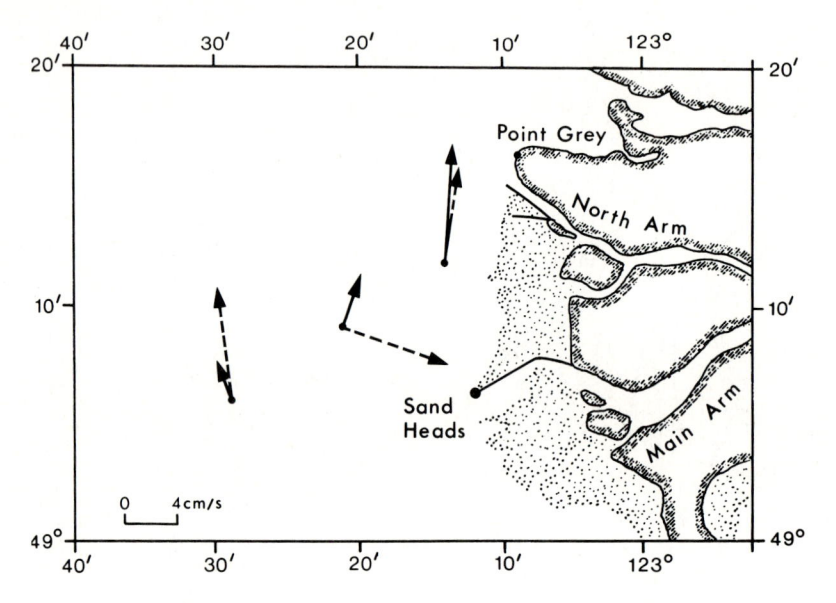

Fig. 13.25 Residual currents obtained from moored current meters at 50 m (dashed) and 200 m (solid) in the central part of the Strait of Georgia for approximately 1 yr.

Considering now the undercurrent moving in a southerly direction along the Vancouver Island shore from the northern opening, a 5-day record obtained at a depth of 80 m 13 km southeast of the opening shows a strong residual flow (10 cm/s) consistent with the model prediction for layer 4 (30–60 m) though this may be anticipated to contain a substantial barotropic tidal residual component. Current meter records obtained at 10 and 24 m off Cape Lazo (see Fig. 1.30) for 1 month (January 1979) show persistent southerly flows of 4–6 cm/s (Buckingham, 1979). For the model, such flows appear in layer 4. Records obtained at a depth of 100 m near the Vancouver Island coast (line 2, Fig. 2.1) in the winter of 1970 show a persistent southerly net flow from 15–100 m at the two moorings nearest to the Vancouver Island shore. The strongest and most consistent velocity occurs at 100 m near the coast and has a value of 8 cm/s. This is reasonably consistent with the model prediction.

The formation of a large counter-clockwise eddy in the southern end of the deep central basin in the Strait of Georgia is strikingly confirmed by some 7 months of moored current meter and cyclesonde observations (Stacey et al., 1987) (Fig. 13.26). This agrees well with model predictions in layers 4, 5, and 6 but not in layers 7 and 8.

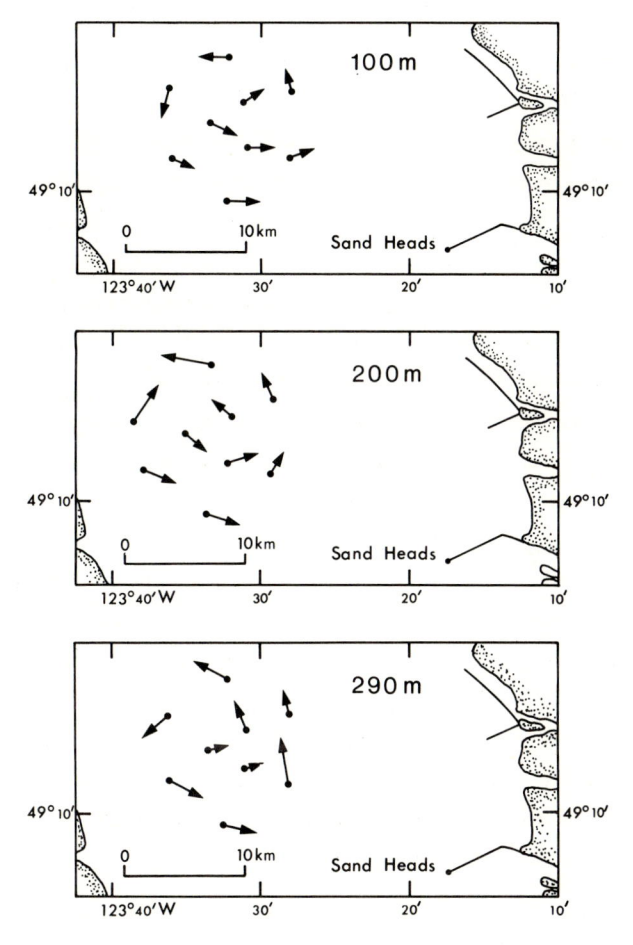

Fig. 13.26 Residual current vectors obtained from approximately 7 months of moored current meter and cyclesonde observations in the southern region of the deep central basin in the Strait of Georgia. (from Stacey et al., 1987)

The basic pattern of modelled residual circulation in the deep central basin of the Strait of Georgia may be characterized in terms of a divergent velocity field in layers 1, 2, and 3, a convergence in layers 4, 5, and 6, and a further divergence in layers 7 and 8. This is confirmed by the distribution of vertical velocities in which there is a net flow from below up to layers 1, 2, and 3 and downward from 6 into layers 7 and 8. This is consistent with the results of the diagnostic salinity trial using the much more simplified non-rotational model GF5. In that case, there was a slow net upward advection of some 2–3 m/day in the deepest part of the basin which yielded a deep water replacement time consistent with that inferred from the time series of salinity observations. It would appear that the divergent flow in layers 7 and 8 in the basin is due to the effects of numerical diffusion. This problem has been referred to earlier in connection with model performance and may be conveniently discussed using modelled and observed vertical profiles of σ_t at locations representative of the horizontal density gradients giving rise to the northern and southern undercurrents, respectively. The locations are shown in Fig. 13.2.

Considering initially the modelled residual circulation in the southern Strait of Georgia, the higher densities in the surface layer at location S3 (Fig. 13.27a) compared to those at S4 are consistent with the inception of the baroclinic pressure exceeding the barotropic pressure gradient due to the slope of the water surface that gives rise to the southern undercurrent moving up into the deep central basin. At greater depths there is a weak reversal in the density differences consistent with the pressure gradients bringing about a southerly net flow in layers 7 and 8 (Fig. 13.16).

The actual observed profiles at the beginning (Fig. 13.27b) and end (Fig. 13.27c) of the month of July at these locations suggest that the southern undercurrent extends down to the sea bottom and furthermore that the model underestimates its magnitude. In the case of the northern undercurrent, the modelled vertical residual σ_t profiles (Fig. 13.27d) at locations S5 and S6 are consistent with the associated driving horizontal pressure gradients. Again, however, there is evidence of a reversal of the weak horizontal density gradient at depth. The observed density gradients at these locations at the beginning (Fig. 13.27e) and end (Fig. 13.27f) of July 1968 suggest that the model underestimates the data at the beginning but is consistent with the latter data. It may be concluded that the southerly-directed flow in layers 7 and 8 in the deep basin is a consequence of numerical diffusion and that the magnitude of the southern undercurrent is underestimated in the three-dimensional model.

13.4.9 Effects of Winds

The model may be employed to provide estimates of the likely response of the summer tidal residual sea levels and baroclinic circulation to winds from the two most common prevailing directions, the southeast and northwest. Following a 5-day period of tidal and baroclinic computation, as described above, a uniform wind field (10 m/s) parallel to the y axis, was imposed over the modelled sea surface for a period of 48 h. Residual elevations and velocities were determined over the last 25 h of computation.

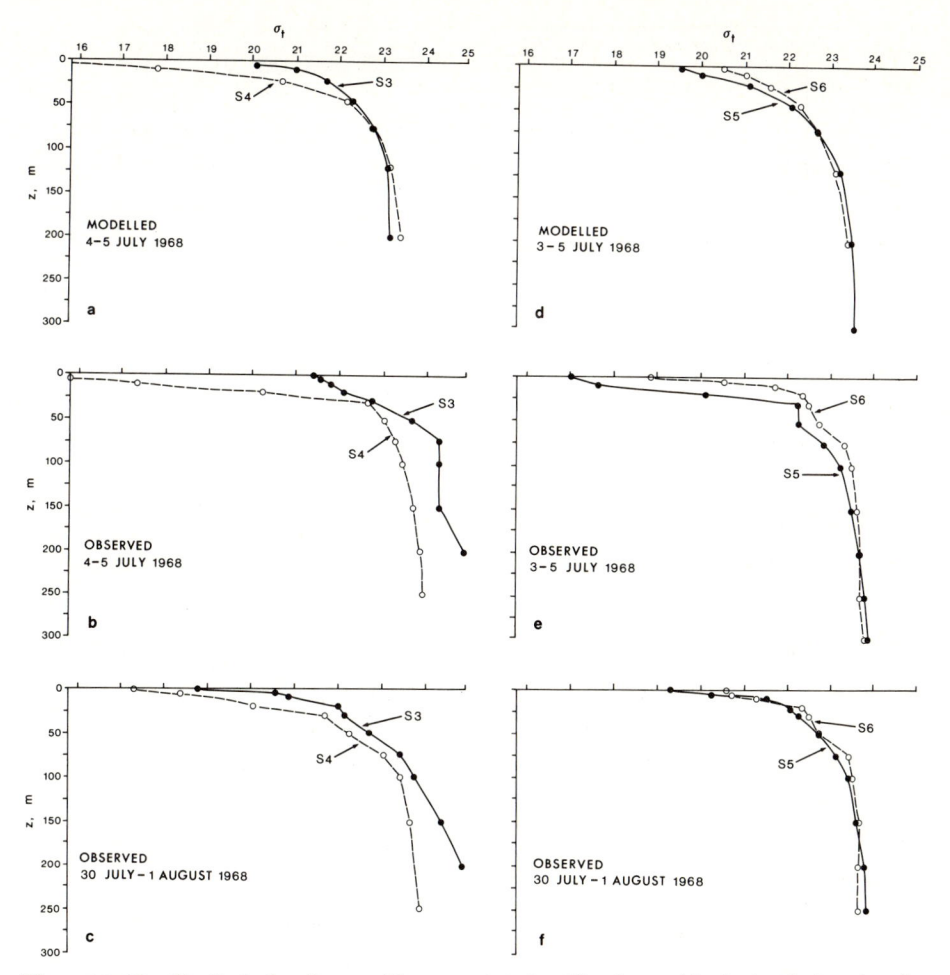

Fig. 13.27 Vertical density profiles associated with the residual circulation in the southern (Stations S3 and S4 in Fig. 13.2) and northern (Stations S5 and S6) circulations.

The positive y direction corresponds approximately to winds from the southeast; the negative direction to winds from the northwest.

Considering initially the effects of the southeasterly wind, the distribution of mean sea surface elevation (Fig. 13.28a) is appreciably altered from that shown in Fig. 13.17. The main effect in the Strait of Georgia is to superimpose a net tilt downward of about 10 cm from the northern boundary opening to those at the south. The net slope from the river mouth northward is thus decreased and that to the south increased. Furthermore, the cross-channel slopes from the mainland coast downward towards Vancouver Island are increased. This increase in the cross-channel slope is also evident in Juan de Fuca Strait from the Vancouver Island coast to the Washington shore.

In the Puget Sound system the modelled mean sea level at its southern end is

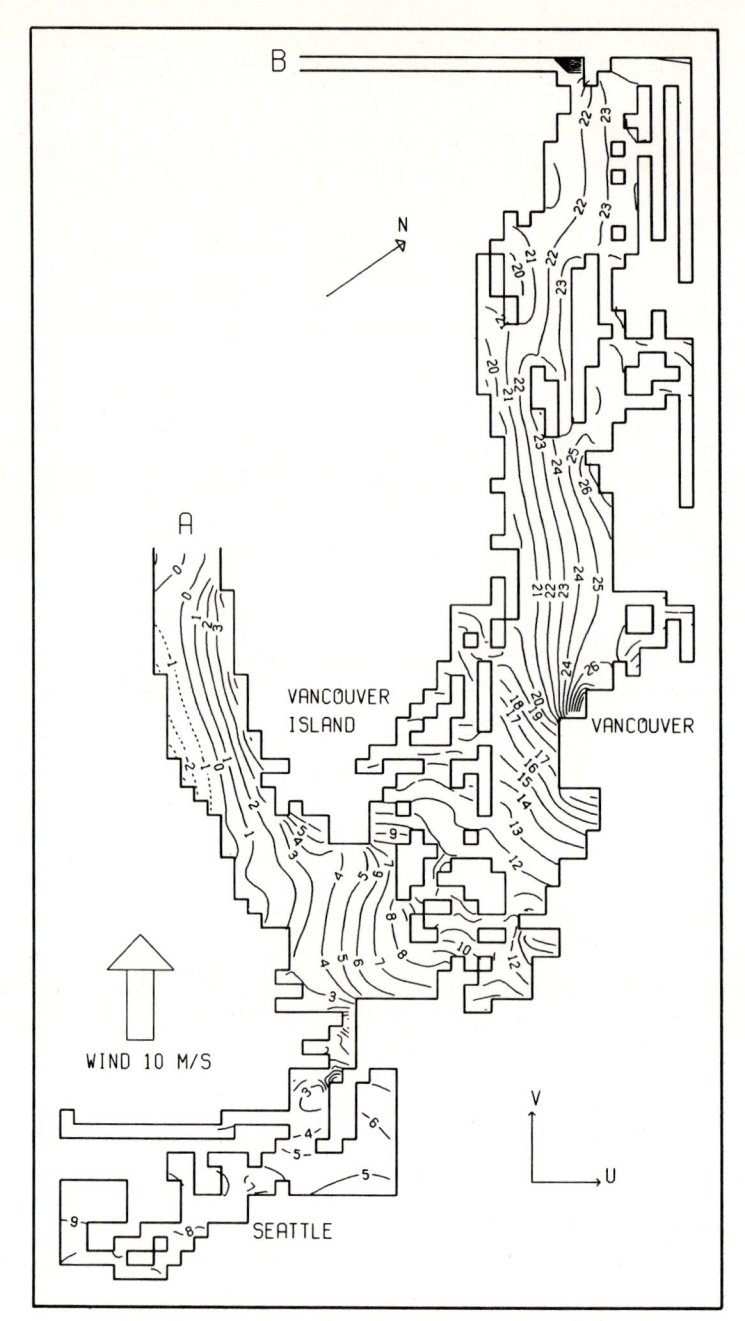

Fig. 13.28a Contoured distributions of net surface elevations (cm) following the application of a uniform wind field at 10 m/s from the southeast.

9 cm in the case of these southeast winds, contrasting with the value of 16 cm for the purely baroclinic case (Fig. 13.17). In the case of the northwest wind, there is a n slope upward of the sea surface from the northern boundary to the southern end of th

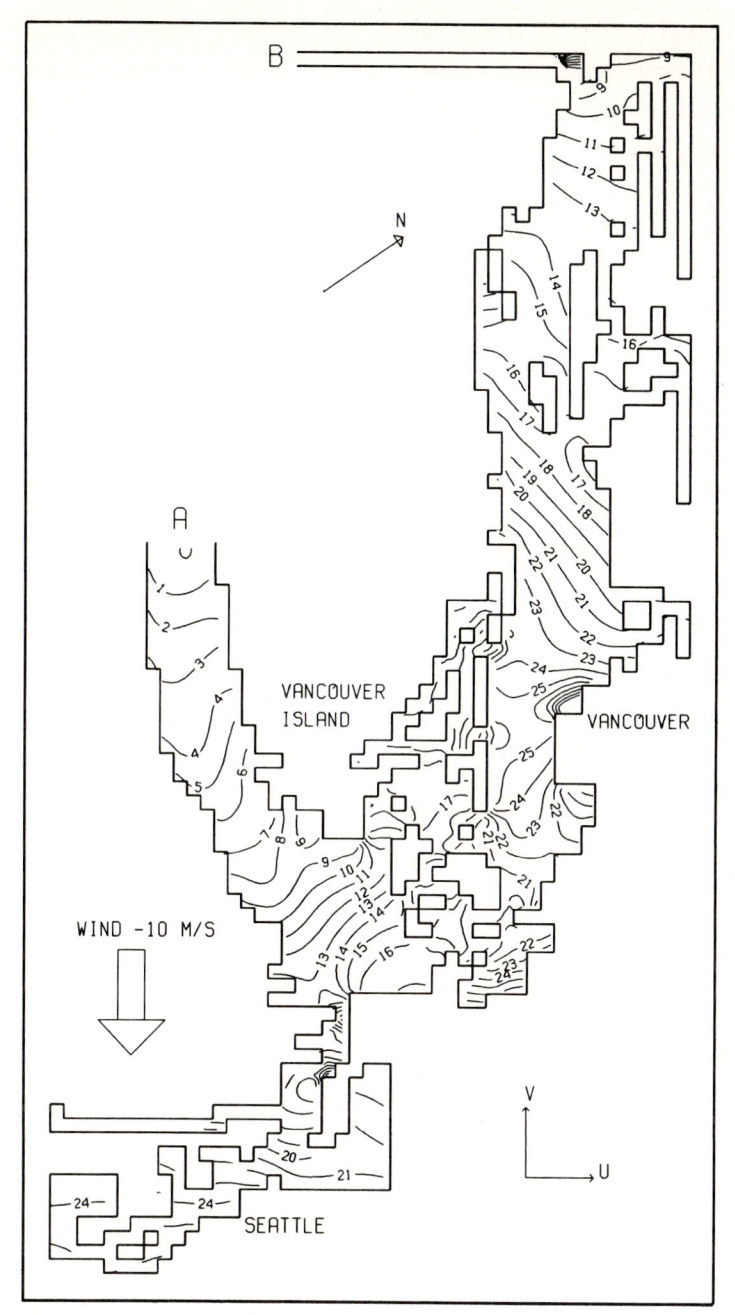

Fig. 13.28b Contoured distributions of net surface elevations (cm) following the application of a uniform wind field at 10 m/s from the northwest.

Strait of about 13 cm and the cross-channel surface slope in the central basin (3–4 cm) is reversed from that either in the absence of wind or for the southeast wind. At the southern end of Puget Sound elevations are of the order of 24 cm.

The results obtained from the model with respect to residual circulation may be conveniently summarized using vertical profiles of horizontal velocity components representative of the basic elements in the overall flow regime, *i.e.*, the turnover occurring in Juan de Fuca Strait, the surface flows away from the Fraser River mouth, the returning subsurface undercurrents, and deep water in the Strait of Georgia. The locations are denoted by P1–10 (Fig. 13.2).

Considering initially the vertical distribution of velocity components essentially parallel to the median axis (Fig. 13.29) at a central representative location (P1) in Juan de Fuca Strait, the tidal residual circulation is limited to a very small net outflow near the surface. For the baroclinic case there is a strong surface outflow and deep inflow. This profile is in reasonable accord with the observed velocities over a cross-section at this location shown earlier (Fig. 1.21, p. 33). The southeast winds greatly increase both the surface outflow and the deep inflow. It should be noted, however, that no account is taken in the southeasterly wind experiment of the effects of these winds over the exposed waters of the continental shelf, which can bring about a surface inflow in Juan de Fuca Strait. The northwest winds, on the other hand, reduce the surface outflow velocity to near zero with the maximum outflow occurring at about 50 m. The inflow at depth is greatly retarded.

At the southern end of Haro Strait (P2) there is a tidal residual current directed inward which is associated with eddy 3 (see Fig. 7.9). This inflow is considerably increased at middepth in the baroclinic case while at the surface a shallow net outflow directed towards Juan de Fuca Strait occurs. The effect of the southeast winds is to produce a net inflow at the surface and to decrease the inflow at depth. The effect of the northwest winds is to greatly enhance both the outflow at the surface and the inflow at depth.

In Rosario Strait (P3) the residual tidal circulation shows a net outflow in the surface layers and inflow in the bottom layer. In the baroclinic case, the surface outflow is slightly increased and the inflow at the bottom is decreased. The effect of southeast winds is to produce a small inflow in the surface and bottom layers while outflows are confined to the layers in between. Northwest winds lead to a major increase in surface outflow in the top three layers.

Considering now the flow in the southern Strait of Georgia, the vertical profiles of velocity parallel to the median line of the Strait are shown at two locations, P4 near the Gulf Islands shore and P5 near the mainland shore. The barotropic residual velocities show a weak flow northward on the mainland side and southward on the Islands side which is consistent with the general sense of rotation in the large barotropic tidal residual eddy 4 (Fig. 7.9). The baroclinic circulation is characterized at both locations by a surface outflow, shallow and weak on the mainland shore and much stronger and deeper on the Vancouver Island side. The slow undercurrent of mixed water moving

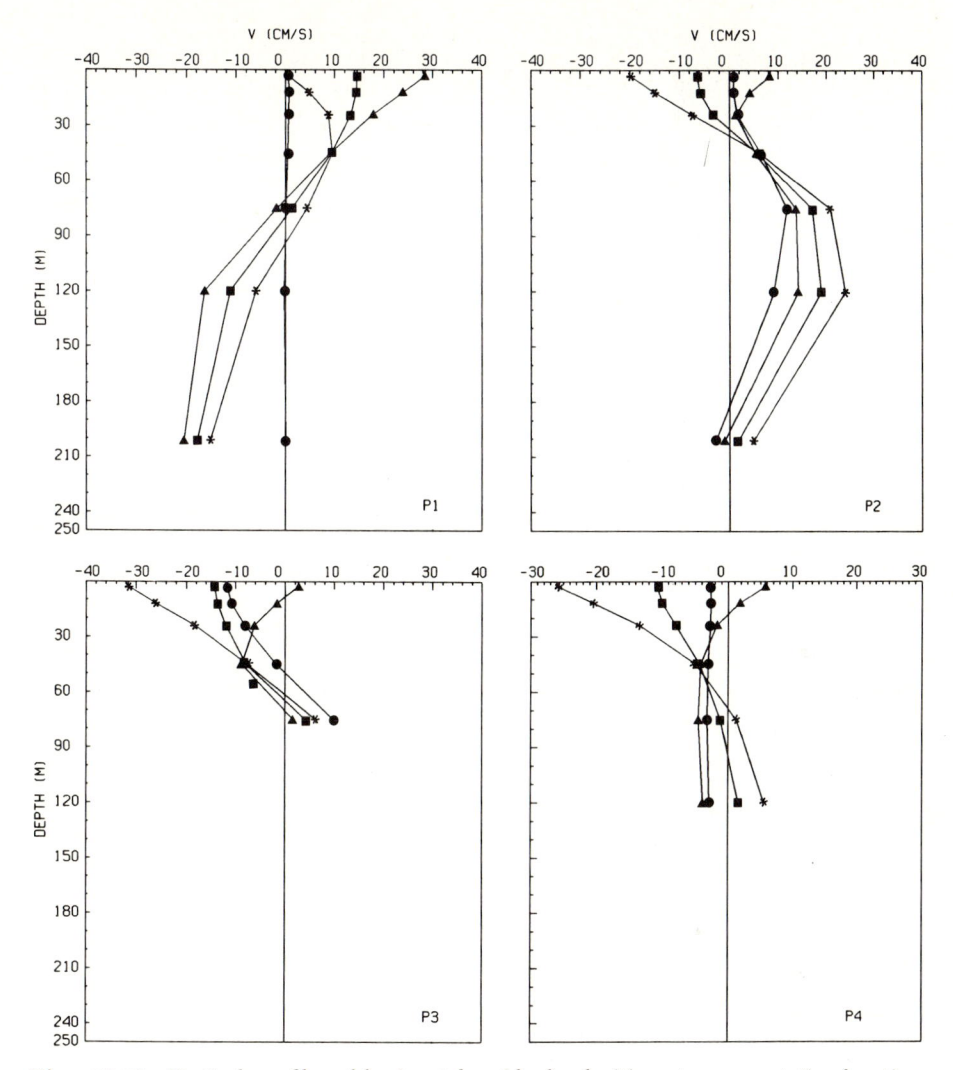

Fig. 13.29 Vertical profiles of horizontal residual velocities at representative locations summarizing the basic net barotropic tidal (circle), baroclinic net circulation (square) and its modification by southeasterly (triangle) and northwesterly (asterisk) winds.

up into the Strait from the southern mixing region extends over the greater part of the water column along the mainland shore but is confined close to the sea bottom near Vancouver Island.

At both locations the southeast winds reverse the basic baroclinic circulation, inflow occurring at the surface and outflow at depth, the inflow being stronger on the mainland side and the outflow stronger along the Vancouver Island shore. Northwest winds enhance the general baroclinic circulation, greatly increasing the surface outflow and also the deep inflow of mixed water, particularly along the mainland shore.

In the deep central basin of the Strait of Georgia the barotropic tidal residual circu-

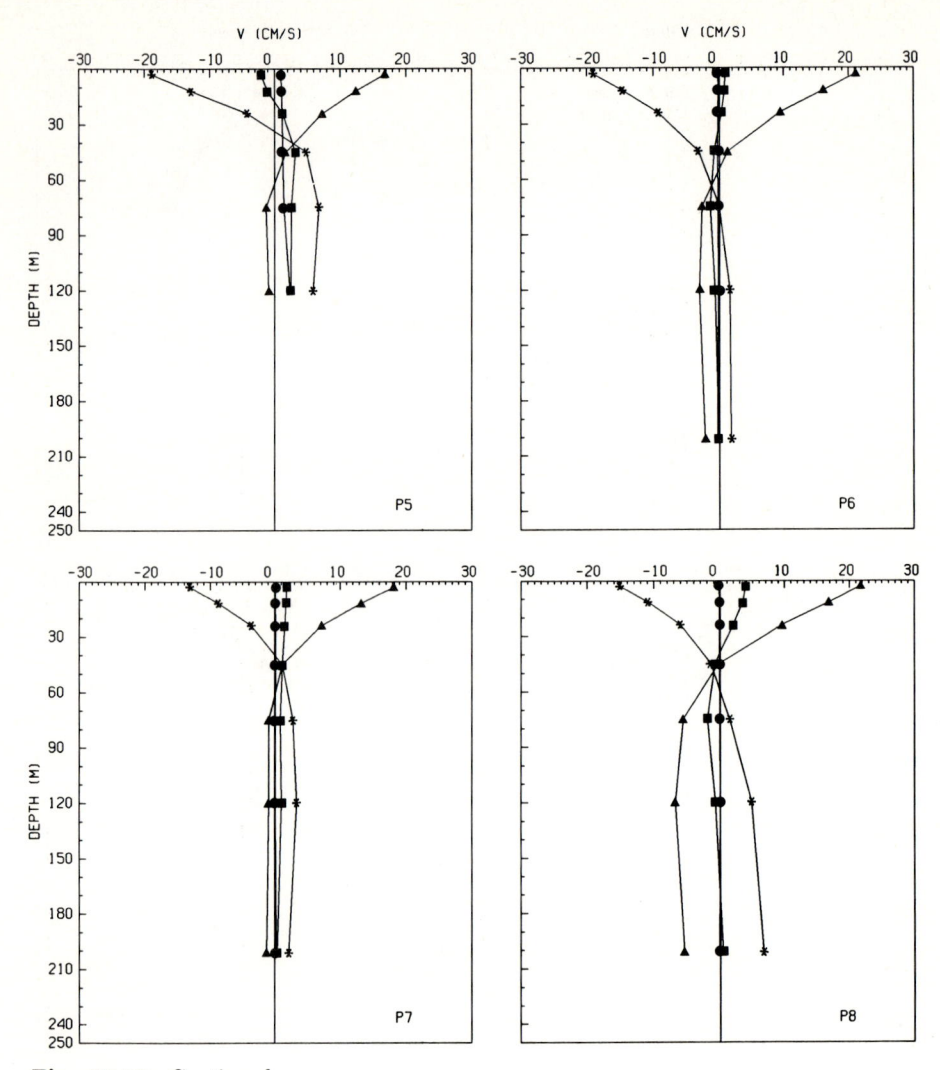

Fig. 13.29 Continued.

lation is negligible. The baroclinic residual circulation involves a general northwesterly flow near the sea surface while at middepth there is a northwesterly flow along the mainland side (P7) and a southeasterly flow along the Vancouver Island side (P6), consistent with the counter-clockwise circulation. This latter is disrupted by the southeast winds effecting a net northwesterly surface and counterflow at depth. In the case of northwest winds there is a net southeasterly flow at the surface and northwesterly one at depth.

Representative of the flows in the northern Strait of Georgia, the profiles at P8 and P9 show negligible barotropic tidal residual circulation but with a northwesterly flow at the surface and southeasterly flow of the undercurrent at depth in the baroclinic

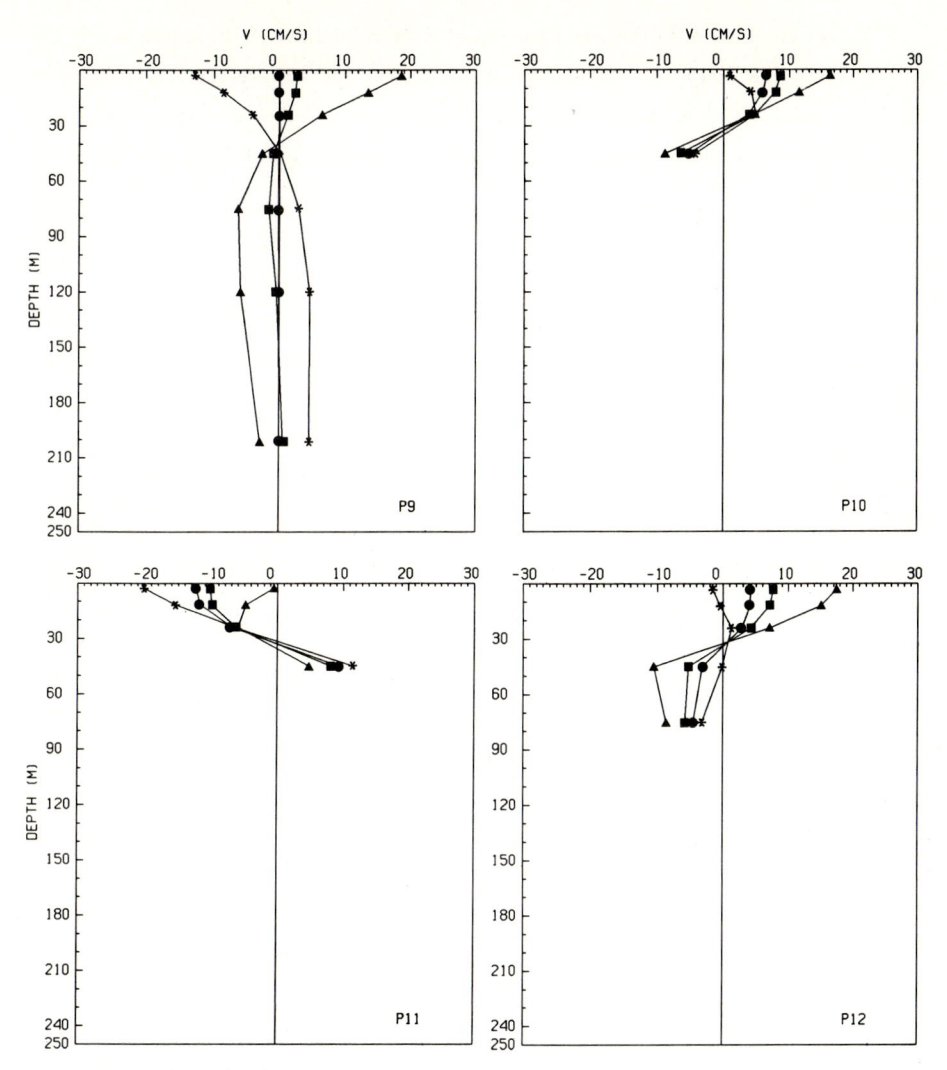

Fig. 13.29 Continued.

case. The effect of southeast winds is to greatly increase the surface flow northward and the undercurrent southward. This process is reversed by northwest winds.

Reference was made earlier to the misleading nature of the current vectors computed from the averaged u and v components at the centre of a mesh in certain locations where the tidal streams are very strong. This may be illustrated by the vertical profiles (P10 and P11) on opposite sides of the single mesh leading into the Puget Sound system. The surface flows are strongly divergent. Although substantially altered by winds in a manner that accords with anticipation, the dominant role of the tidal vertical residual circulation is clearly evident. It remains to determine from observations to what extent this process is realized in nature.

The net circulation at the southern end of Admiralty Inlet (P12) shows a net flow at depth into the Sound for all four cases while near the surface there is a net flow seaward towards Juan de Fuca Strait. The exception is for the case of the northwest winds in which a small net inflow at the surface surmounts a small net outflow above the deeper inflow.

13.4.10 Net Volume Transports

Table 13.5 shows the residual volume transports through the northern and southern boundary openings of the major conveying channels computed in the course of four numerical experiments: (1) barotropic tide alone, (2) barotropic tide plus a prescribed initial density field, (3) experiment 2 is repeated but with the addition of a uniform southeast wind field, (4) experiment 2 is again repeated but with a uniform northwest wind field.

Table 13.5 Residual inflows, outflows, and net flows (km^3/day) through boundary openings A,B,C (Fig. 13.2) obtained over a 25-h period from trials of the model GF6. Inflow towards the model region is positive and outflow away from the model is negative. The Fraser River inflow, other than in the barotropic trial, is included as a constant flow of 0.69 km^3/day.

Trial	Volume Flux	Juan de Fuca Strait	Fraser River	Johnstone Strait
1	Tidal barotropic	+ 2.25	0	+0.30
		− 2.48	0	0
		− 0.23	0	+0.30
2	Tidal baroclinic	+11.95	+0.69	+1.45
		−11.62	+0.69	−2.45
		+ 0.33	+0.69	−1.00
3	Tidal baroclinic, NW wind	+ 7.51	+0.69	+1.57
		− 7.7	+0.69	−2.00
		− 0.19	+0.69	−0.43
4	Tidal baroclinic, SE wind	+16.92	+0.69	+1.34
		−16.69	+0.69	−2.04
		+ 0.23	+0.69	−0.70

In experiment 1 the net outflow is about evenly divided between the northern and southern boundary openings. Of course, this does not imply that the freshwater from the Fraser River is thus proportioned evenly. In the baroclinic case there is a dramatic increase in the net inflows and outflows in Juan de Fuca Strait as a major intrusion of water from the continental shelf brings about a major efflux of brackish water above, the flow being largely compensatory. This was the situation observed over a 5-wk period when the line 11 current meters were in position. At the northern end there is a net outflow closely approximating the net river inflow. When northwest winds are applied, the near-compensatory exchange of the Juan de Fuca Strait boundary opening (experiment 3) is greatly reduced. On the other hand, when southeast winds are applied (experiment 4) this near-compensatory exchange is greatly increased. It should be stressed, however, that no account was taken in these experiments of the effects of the winds over the offshore waters. In all these cases, there is a close balance between the net inflow (including river discharge) and net outflow, indicating an approximate equilibration.

13.4.11 Freshwater Volumes and Flushing Rates

Reference was made in Chapter 1 to the importance of determining freshwater concentrations and flushing rates with respect to practical problems of water management. Some comparisons of results obtained from the model with those obtained from Fraser River discharge measurements and observed salinity fields when a fixed salinity of $33.8^0/_{00}$ for sea-water entering the system is assumed (Waldichuk, 1957) are now presented.

The computed volumes of freshwater in the Strait of Georgia ranged from 110 km^3 in January 1949 to 132 km^3 in August 1950. Using an overall volume of 1,025 km^3 for the Strait, the proportion of freshwater varied from 11–23%. In the case of the model GF6 using the value of modelled volume of the Strait of Georgia (1,050 km^3), the freshwater volume based on the highest value of σ_t (26.67) in the incoming deep ocean water was 163 km^3. The resulting freshwater concentration is 15%, somewhat higher than the maximum value obtained for the summer of 1950. Records of monthly mean river discharges indicate that during the months prior to the determination of these summer values, the flow rates in 1950 were about half those obtained in 1968. Combined with the fact that the subsequent peak discharges were much greater in 1950 than in 1968, these may be attributed to a much increased mountain snowpack in the case of the former.

In the case of Juan de Fuca Strait, though an alternative set of values derived from observations are not immediately available, using a modelled volume of 402 km^3, the freshwater proportion was determined to be 29 km^3. The freshwater concentration is thus about 7%. Comparisons of freshwater concentrations obtained from cruise data in September 1952 (Waldichuk, 1957) indicated those in Juan de Fuca Strait to be about half those occurring over comparable depth intervals in the Strait of Georgia. For the overall model GF6, the total volume is 2,673 km^3 and freshwater volume 360 km^3, yielding a freshwater concentration of 13%.

The data used in these computations consist of the density value averaged over the last 25 h of the baroclinic computation. Employing the corresponding residual flow velocities normal to cross-sections close to the northern and southern boundary openings of the model, estimates of an overall flushing rate and proportioning of freshwater outflow between the northern and southern boundary openings are readily obtained. The net freshwater outflow rate is 0.72 km^3/day. If such a rate were sustained it would thus take 1.4 yr to replace the volume of freshwater in the overall system. This accords well with the value of 1.3 yr estimated for the Strait of Georgia alone (Waldichuk, 1957). Furthermore, the ratio of the net outflow rate to the Fraser River inflow rate is 1.2. This agrees well with the range of values of 0.7 in winter to 1.3 in summer computed directly from river discharge and salinity distributions (Waldichuk, 1957).

A further question of interest concerns the proportioning of this net outflow rate between the northern and southern openings leading to the ocean. In the case of this model computation, 83% of the freshwater leaves through Juan de Fuca Strait with the remaining 17% exiting through Johnstone Strait. This is rather lower than the value given earlier in Chapter 1 (92%) based on observed residual velocities and salinities of cross-sections. However, in view of the approximations with respect to both modelled and observed values, the result would appear to be satisfactory.

The net volume fluxes through the model boundary openings may readily be computed for the four types of trial referred to in the previous section. With respect to the trials including wind effects, it will be recalled that concomitant effects of such winds over the exposed coastal waters are not represented in the conditions prescribed at the model boundary and thus must be considered with reservation.

For the purely barotropic tidal case the net fluxes are small. The net fluxes at the entrance to Juan de Fuca Strait are probably associated with the boundary itself where a net inflow balanced by a net outflow can result from small imbalances between the Coriolis surface slope along the boundaries and prescribed densities required by the modelled velocities normal to that boundary. In the case of Johnstone Strait, there is a small residual circulation in the vertical presumably due to bottom slope. For the three trials referred to below, these barotropic tidal components have been removed from computed fluxes.

In the baroclinic tidal case, the net exchange through Juan de Fuca Strait involves a small increase in outflow over inflow (0.33 km^3/day). The actual volume transports are largely counterbalanced and are of order 10 km^3/day agreeing well with the range (8–14 km^3/day) inferred by Godin *et al.* (1981). The net outflow in Johnstone Strait closely approximates the summed net river inflow and inflow through the Juan de Fuca opening.

13.5 CONCLUSIONS

When the results obtained from this model are compared with field observations, there emerges a coherent picture of the basic residual flow distribution when the estuarine circulation and deep water renewal are fully operative. The results may thus be considered descriptive of the key process in the overall flushing and deep water renewal of the system. It is evident from the distribution of residual sea surface elevations that the Strait of Georgia may be regarded as a semi-enclosed basin from which its freshwater discharge finds primary egress through its southern boundary openings into Juan de Fuca Strait. In terms of subsurface processes, there is a large and essentially compensatory exchange of water in Juan de Fuca Strait with the adjacent ocean, brought about by the movement of high density water onto the continental shelf and its intrusion into the Strait of Georgia. Some of this high salinity water enters into the strongly tidal mixing system of Haro Strait. This gives rise to an undercurrent consisting of a mixture of brackish surface water and entering ocean water into the Strait of Georgia. This undercurrent tends to favour the mainland coast up into the deep central basin of the Strait of Georgia. A somewhat weaker undercurrent results from a similar mixing process in the northern passages and moves southward along the coast of Vancouver Island into the deep basin. The result of these two Coriolis-deflected undercurrents is to produce a marked counter-clockwise residual circulation in the basin. This circulation in the Strait of Georgia can, however, be offset by winds parallel to its major axis. Thus, southeast winds, while enhancing northward surface flow, produce a net southerly flow at depth. In the case of northwest winds, there is a net southerly flow at the surface and a much increased northerly flow at depth. It would appear that the flushing process occurs essentially through a slow vertical advection in agreement with the earlier diagnostic experiments using the simple model GF5. There is some evidence of the existence of Poincaré waves in Juan de Fuca Strait.

14 CONCLUDING DISCUSSION

The waters between Vancouver Island and the mainland are characterized by complex topography, large tides, and a distribution of stratification that varies both regionally and seasonally. There exist a wide range of possible responses, both external and internal, to forcing by winds and to effects from the fresh and salt water boundary openings. Although clearly susceptible of extensive refinement, the results from this program of field and numerical model studies provide a coherent explanation of the basic features dominating the circulation within the system.

In the early stages of model development, effects due to stratification, were omitted and associated energy losses deriving from mixing were subsumed in the tidal models under the effects of regionally varying coefficients of bottom friction, an approach amply jusified by its outcome. The distribution of such coefficients could only be determined through correctly simulating the observed tides and streams within the system. Although the external Rossby radius is of order ten times the width of the broadest conveying channels in the system, the multi-channel numerical model GF1 emphasized the need for inclusion of the effects of the earth's rotation before the barotropic tidal co-oscillations could be satisfactorily resolved.

A two-dimensional model with multiple boundary openings where tidal elevations were prescribed provided excellent results with respect to the observed and modelled surface tides using a conventional overall coefficient of friction (0.003) normally applied to coastal seas but yielded quite unrealistic velocities at the boundary openings. To develop a model of adequate sensitivity for calibration by empirical adjustment of friction coefficients required an appropriate selection of boundary openings at which water levels would be prescribed. It thus proved necessary to include both the mainland inlets and the northern passages as one-dimensional additions to the two-dimensional scheme. Extensive trials employing this overall model (GF2) to simulate the M_2 tidal co-oscillations revealed the need for high frictional dissipation in the region of the Gulf and San Juan Islands and also showed that the average depths used in each 4-km mesh underestimated the speed of wave propagation in Haro Strait. With some deepening of a contiguous line of meshes (thalweg) through the Strait and the use of high frictional dissipation in the overall region of the Gulf and San Juan Islands, it became possible to provide a modelled distribution of M_2 tidal harmonic constants fully in accord with observations. It was discovered that such a configuration of bottom friction did not yield satisfactory results when a K_1 tidal simulation was attempted. Trials including the tidal generating forces in the momentum equations showed these to have a negligible effect on the co-oscillations determined by the ocean tides on the open boundaries. Though the harmonic analyses of tide gauge observations in the region revealed negligibly small compound tides and overtides, it was found that the high frictional dissipation leads to significant interactions between the major tidal constituents in the model. Thus, empirical determination of the appropriate distribution of coefficients of bottom friction can only be effected in at least the presence of both the major semi-diurnal (M_2) and diurnal (K_1) tides. It was then possible to model the mixed tidal co-oscillations throughout this system in a fully satisfactory manner.

Tides in the system are of the mixed type, the two largest constituents, M_2 and K_1 having amplitudes of similar magnitude in the Strait of Georgia. In the case of the M_2 tide there is a degenerate amphidromic system in the inner part of Juan de Fuca Strait whereas the K_1 tidal amplitude and phase increase monotonically from the entrance of Juan de Fuca Strait up to the northern Strait of Georgia. Since the ocean tide traverses the length of Vancouver Island much more rapidly to the sea entrance of the northern passages than do the tides within the shallower confines of the waters between Vancouver Island and the mainland, the large phase and elevation differences across the northern passages that occur for the M_2 tide lead to a larger proportion (17%) of the overall tidal complement entering from the northern end compared with the K_1 tide (7%). Of the incoming M_2 tidal energy flux at the entrance to Juan de Fuca Strait (3.21×10^{16} erg/s), 14.7% is dissipated in the turbulent tidal passages and rapids leading northward from the Strait of Georgia. For the K_1 tide, 16.3% of the incoming tidal energy (1.19×10^{16} erg/s) is dissipated in these passages. Of the energy flux entering the modelled northern passages from the ocean entrance, the dominant proportion is that of the M_2 tide (1.55×10^{16} erg/s); that of the K_1 tide being negligible. At the entrance to the Puget Sound system, the dominant contribution to the energy flux comes from the M_2 tide (0.472×10^{16} erg/s); that of the K_1 tide (0.072×10^{16} erg/s) being of secondary importance.

Although the energy imparted directly to the overall modelled system by the tide-generating forces is small in comparison to that of the tidal co-oscillations, the actual energy imparted is still appreciable (0.38×10^{16} erg/s for the M_2 tide; 0.176×10^{16} erg/s for the K_1 tide). This energy is almost exclusively propagated out into the North Pacific. Though the waters between Vancouver Island and the mainland constitute an energy sink with respect to ocean tides, they constitute an energy source with respect to the work done directly therein by the tide-generating potential.

Tidal streams are strong in Juan de Fuca Strait, the Gulf and San Juan Island passages, and Admiralty Inlet. The streams entering the system through the constrictions and tidal rapids of the northern passages are even greater, reputedly the fastest in the world. In general, the tidal currents are rectilinear throughout most of the system though there is a strong rotational character to the weak tidal streams over a limited area in the northeastern part of the Strait of Georgia. This is clearly related to wave reflection though such reflection is not simple due to the nature of the topography, damping, and partial transmission. Significant rotation is also evident in the tidal current ellipses in the approaches to the southern end of Haro Strait where the incoming tidal streams undergo a sharp deflection by San Juan Island, presumably bringing about a partial reflection. A further point of interest concerns the relative phasing of the tidal elevations and associated streams. Thus, for the diurnal tides the stream maxima generally occur roughly about mean sea level on a rising tide throughout the system. In the case of the semi-diurnal tide and its greater wave resonance, the stream maxima generally occur roughly about mean sea level on a rising tide in the Strait of Georgia but on a falling tide in Juan de Fuca Strait.

Non-linear friction and inertial terms in the momentum equations give rise to significant effects within the system. Thus, the friction terms appreciably change the Greenwich phase distribution of the K_1 tide in the presence of M_2 tidal co-oscillations, due to the stronger combined flow, as distinct from when it is modelled on its own. The effect is even more pronounced in the case of the O_1 tide. Of particular interest is the amphidromic system, centrally located in Juan de Fuca Strait, of the small L_2 tidal constituent which has a frequency coincident with an interaction frequency of the M_2 and N_2 tides. This amphidrome differs markedly from the degenerate amphidromic systems of all the other semi-diurnal tidal constituents. Although the presence or absence of advective accelerations in the momentum equations has little influence on the overall tidal co-oscillations, there is an important effect on the system in terms of tidal residual circulation. In the regions where the tidal streams are strong these residual velocities are comparable with the estimated baroclinic velocities associated with the estuarine circulation and its seasonal modification. It was clearly evident, however, that the 4-km mesh size used in this two-dimensional scheme was too coarse to provide adequate resolution of the major tidal eddies and associated residual circulation. Furthermore, there existed another shortcoming with respect to correctly simulating the phase speed of the tide in Haro Strait.

To resolve these problems, that part of the 4-km mesh scheme including Juan de Fuca Strait, the passages of the San Juan and Gulf Islands, and the southern half of the Strait of Georgia were replaced by a 2-km mesh scheme (GF3). This model has three boundary openings located at the entrances to Juan de Fuca Strait and the Puget Sound system, and across the central part of the Strait of Georgia. Earlier problems with multiple boundary openings and elevations prescribed on the basis of tide gauge observations could largely be obviated by using interpolated elevations obtained from the overall model GF2 along the open boundaries. The computed velocities from GF2 and GF3 in the vicinity of the Strait of Georgia boundary could be compared with those obtained when the elevations prescribed were values interpolated between the predicted tides based on 1 yr of tide gauge observations at each end of the boundary. Large errors resulted in the computed velocity field near the boundary, in some instances by as much as an order of magnitude, when the data based on observation were employed even though the difference in elevation over the length of the line at any time was only 2–3 cm.

Comparisons of observed and computed elevations and velocities showed excellent agreement. The results obtained from these two models were employed as the basis of a tidal current atlas which is now in common highly successful use. Comparison of the fields of residual circulation obtained from these two models, though of similar character, showed the marked smoothing effect on the horizontal velocity gradients brought about by the coarser grid. Observational confirmation of some of these tidal eddies was obtained from drogue tracks and from the associated residual circulation by moored current meter records. In general, it is required that an eddy extend over four or five meshes and be of sufficient strength to be clearly discernible against the background of estuarine circulation and its modification by tides, winds, river discharge, and disturbances on the open boundaries.

With access to improved computing capacity, it became possible to extend the 2-km mesh model GF3 to include the whole system. Though this model (GF7) provided an improved resolution of the tidal streams in the northern Strait of Georgia, the basic reasons for its development involved the provision of barotropic tidal elevations and velocities for entry into the 2-km mesh upper layer model (GF4) of the Fraser River discharge in the Strait of Georgia and furthermore, to establish a consistent overall two-dimensional grid that could readily be converted by machine to a fully three-dimensional operation (the proposed model GF8). The tidal residual circulation in the inner part of Juan de Fuca Strait obtained from this model was essentially the same as that obtained from GF3 except that an unrealistic appreciable net flow from the Puget Sound boundary opening out through Juan de Fuca Strait to the open sea had been eliminated. In general, the tidal calibration of the GF7 model is not as good as that obtained in the earlier models GF2 and GF3, due to the approximate representation of some of the narrower channels when tidal velocities are large. Nevertheless, the values obtained are considered adequate for the baroclinic studies envisaged.

The models of the system described above can be employed to study other phenomena of an essentially barotropic character, notably storm surge, tsunami, and normal mode responses of the system.

The overall barotropic tidal model GF2 was employed to simulate a storm surge that occurred in December 1982 in response to the movement of an intense low pressure system onto the British Columbia coast. The results showed that the direct effect of winds within the modelled region was small and that the surge derived primarily from the propagation into the system of a disturbance generated in the offshore waters. In the process little amplification of the surge occurred. The 'inverted barometer' effect contributed significantly to the surge elevations.

Tsunamis generated elsewhere in the Pacific Ocean appear to be of little consequence in the waters between Vancouver Island and the mainland coast. The primary hazard in this context devolves on tsunamis generated imminently by seismic events within the system itself. On the 23 of June 1946, a tsunami was generated by an earthquake close to the Vancouver Island shore in the northern Strait of Georgia. The GF7 model was used to simulate the propagation of this disturbance through the system. The tsunami propagated with negligible damping in the Strait of Georgia but was attenuated to insignificance in the turbulent passages leading seaward both to the north and south. In the case of an ocean-generated incoming tsunami, observations showed strong attenuation in the region of the San Juan Island passages.

It is evident from the fact that semi-diurnal tides within the system possess a nodal region in the inner part of Juan de Fuca Strait but that the diurnal tides do not. There must be a barotropic normal mode having some intermediate periodicity.

The barotropic model GF2 was employed to determine the response of the system to a summed range of harmonics having periodicities ranging from 1–25 h. The resulting distributions of gain indicated resonant responses at 16-, 5-, and 2-h periods which constitute the fundamental, and what would appear to be the second and third modes, as indicated by approximations obtained from the Merian formula for a semi-enclosed basin. Time series analyses of observed water surface elevations show little energy at these frequencies, presumably because of the relatively sheltered nature of the system.

The barotropic tidal currents considered above, though contributing to a complex pattern of residual circulation, do not directly contribute significantly to any major net exchange of water with the adjacent ocean other than through relatively slow mixing processes over the length of the system. To determine the nature of such a net exchange, it is necessary to take into account the estuarine character of these waters. This presents a profusion of problems, notably in connection with the regionally-varying stratification brought about largely by the changes from place to place in strength of the barotropic tidal streams. Thus, the vertical salinity distribution in Juan de Fuca Strait resembles that of a deep partially-mixed estuary. That of Haro Strait can resemble a deep vertically-homogeneous estuary while the stratification of the Strait of Georgia resembles that of a typical fjord. This introduces considerations of a hybrid layer/fixed-level three-dimensional model. Over and above this, the complex topography and non-linear aspects of the tidal circulation make severe demands on any computational scheme and the determination of the appropriate parameters. The approach adopted, as in the barotropic studies, was to develop the requisite models in parallel with the acquisition of an observational data base adequate to their adjustment and verification in a gradual progress from the simple to the more recondite.

To provide some general appreciation of the magnitudes involved and the numerical problems likely to be encountered, the one-dimensional schematization of Juan de Fuca Strait and Strait of Georgia employed in GF1 was adapted to yield a single channel two-dimensional model (GF5) of variable width and depth and closed at its inner end. When observed salinity fields were prescribed and held fixed throughout a computation, a reasonably convincing picture of the slow movement of large volumes of water that constitute the estuarine circulation emerged. Salient features include realistic net sea levels and velocities and the high degree of refluxing of the surface seaward flow in the Strait of Georgia that results from vigorous tidal mixing in the region of the Gulf and San Juan Islands. There is a reversal of the normal pattern of estuarine circulation in Juan de Fuca Strait by an observed winter density field following strong southerly winds over the exposed coastal waters seaward of the the model boundary and consequent presence of Columbia River water.

Prognostic trials afforded a salutary lesson in the numerical simulation of the density conservation equation. Thus, for example, in one instance the persistent occurrence of catastrophic instability was eventually traced to the sequence of operations in an arithmetic expression. The basic problem originated, however, in the excessive section lengths employed in the grid scheme. This resulted in unrealistically high numerical diffusion when the barotropic tides were included in the baroclinic calculations.

The success of the barotropic tidal models and the results obtained from these exploratory simple baroclinic calculations suggested the possibility of introducing the barotropic tidal elevations and velocities, predetermined by numerical simulation of the tidal co-oscillations in an overall three-dimensional model, into the subsequent baroclinic calculations using that model. The basic assumption is that the strong tides in the system are not significantly affected by the baroclinic circulation whereas this type of circulation is indeed strongly affected by the tides. The stringent limitation on the computational time step brought about by the deep waters of this coastal system, the resulting high speed of free surface long wave propagation, and the Courant-Friedrichs-Lewy criterion might thus be circumvented. This could be done through a coastal sea equivalent of the 'rigid lid' technique employed in deep ocean models for this purpose. The barotropic pressure field exerted by the lid at any time is determined by, and may be computed from, the fields of density and motion then extant within the model. In the tidal coastal application, continuity requirements further mandate inclusion of the changing tidal elevations of the water surface. These are already known and available for such inclusion. There emerges the notion of a 'prescribed lid' fulfilling a similar function to that of the ocean 'rigid lid'. Numerical trials indicate that residual sea levels are essentially determined by the overall field of density in the system and thus commend the applicability of this technique.

Using a flat-bottomed rectangular basin having a length and mean depth equivalent to that of the Strait of Georgia and Juan de Fuca Strait, the laterally-integrated numerical scheme employed in GF5 was used to carry out preliminary trials of this notion. The barotropic tide was generated using an artificially large M_2 astronomical tide-generating potential in the presence of an oscillating internal seiche derived from the prescribed initial density field. Though this seiche was strongly damped due to the attenuation of the driving internal pressure gradients by the effect of numerical diffusion on the density field, it was demonstrated that the Courant-Friedrichs-Lewy criterion could be circumvented in this manner and matching 'free surface' and 'prescribed lid' solutions obtained. It was clear, however, that the methodical extension of this work to include open boundaries, varying topography, and reduction of numerical diffusion would involve a digression of activity incommensurate with the basic aims of this project. Accordingly, attention was turned to another major problem in this region, *i.e.*, that of the tidally and seasonally pulsed jet of Fraser River water and its interaction with the strong tidal cross-streams in the southern Stait of Georgia.

The most striking fact concerning the drogue tracking observations on the Fraser River plume in the 1960's was the simplicity and coherence over periods of up to a day or more, of the paths followed by a group of drogues launched near the river mouth at about the same time. This suggested that a vertically-integrated model of the thin upper layer might be employed to simulate the motions resulting from various combinations of tide, river flow, and wind. It was then assumed that this thin upper layer did not significantly affect the tidal motions in the much deeper water mass below, though itself was much affected.

It is thus possible to take water surface slopes and velocities from an independent solution to the problem of barotropic tidal co-oscillations and apply them to the introduction of tidal effects into the model of the layer. Following extensive numerical trials and field studies, a vertically-integrated model of the upper layer was established having its grid coincident with the Strait of Georgia part of the overall barotropic fine-mesh (2 km) tidal model GF7. Surface slopes and velocities from the latter could thus be introduced with facility into the layer model. In this manner, a considerably longer time step could be achieved (150 s), the wave speed entering into the Courant-Friedrichs-Lewy criterion being that of the internal wave speed rather than that of the much more rapid free surface wave. Conditions at the lower interface of the model require that account be taken of vertical exchanges varying from entrainment, a net movement of mass and volume, to mixing, a net movement of mass but with compensatory exchanges of volume. This was effected through independent calculations of positive and negative components of velocity in a manner determined by vertical velocity and density differences across the bottom of a particular mesh. The most serious violation of the assumed very deep underlying lower layer occurs in the vicinity of the shallow banks near the river mouth. In this case the fixed depth schematization of the shallow banks was retained. Since the tidal motions are known, these can be prescribed, coincident with the free surface of the layer model, in the meshes immediately seaward of the shallow banks model. Thus, in effect, the layer model rises and falls with the tide. If the thickness of the layer is greater than the depth of water on the banks, the flooding stream onto the bank consists of layer water. If the thickness is less, then layer water in the upper part of the column and salt water in the lower part make up the flood stream onto the banks. These components are assumed to be mixed to vertical uniformity, the resulting intermediate salinity water entering the layer on the following ebb tide.

At the boundary openings it is necessary to postulate a condition relating layer thickness to velocity. In the course of numerical experiments it was found that setting the second derivative of the internal Froude number equal to zero normal to the boundary gave satisfactory results. At the boundary openings representing the Main and North Arms of the Fraser River, the river discharge is modulated by the tidal elevations in the Strait to pulse the flow in a manner according to moored current meter measurements near the mouth of Main Arm (Sand Heads).

This model (GF4) was found to be capable of indefinite stable operation, the salinity, layer thickness, and velocity fields responding in a stable manner to the forcing applied. The effect of the initial conditions became minimal after 2 wk of simulation in the case of strong river discharge and 4 wk when the discharge was low. South of Sand Heads where the Strait is narrower and the tidal streams are stronger, the motion in the layer displays the reversing character of the flood and ebb streams. Off Sand Heads, the emergent jet builds to a maximal value at about low water, its major axis directed a little to the south on the ebbing tide but swinging vigorously northward and back towards the mainland coast under deflection by the opposite shore and the earth's rotation, these being augmented further by the flooding tide. About the time of high water, the ebb streams start to pick up immediately south of the region where the pulse of river discharge is still moving northward towards the mainland coast giving rise to a large clockwise eddy at about the time of high water. These basic features of the jet have been confirmed by extensive observations.

When winds parallel to the major axis of the Strait of Georgia are applied, the formation of coastal jets and layer thickening along the shore to the right of the wind direction accord with theoretical anticipation. These trials also show the effects of wind and river flow in the production of local layer deepening. In summary, the model reproduces the basic features of the plume motions in accord with existing observations and provides an extensive basis of hypotheses for the planning of further field work.

The primary shortcoming of the model involves horizontal velocities which are persistently lower than those obtained from drogue tracks. This is, in part, probably due to the fixed depth of the drogue elements (2 m) as distinct from the variable thickness and the vertically-integrated velocities in the model layer. The most probable major contributory factor, however, is the effect of deep intruding higher salinity water into the Strait of Georgia and consequent increase in surface outflow, a feature entirely omitted from the layer model at this stage of development.

To assess the combined effects of river discharge and deep water renewal in the Strait of Georgia, it is necessary to undertake a fully three-dimensional numerical model. Accordingly, the two-dimensional mesh employed in the vertically-integrated tidal model GF2 was machine-converted to a full three-dimensional model (GF6) having eight fixed computational levels. To minimize problems at open boundaries the entire system was represented on this 4-km grid with two boundary openings corresponding to those employed in GF2. An extensive program of numerical trials was undertaken to assess the performance of this scheme, already successfully employed in North Sea applications, but in the markedly different context for the waters between Vancouver Island and the mainland.

Initial tests addressed aspects of the model's performance. The first major problem concerned the suppression of non-linear instability. This arose from the advective accelerations in the momentum equations where strong tidal streams encounter marked changes in the topography. Such instability could readily be suppressed by the use of high horizontal eddy viscosity. The values of the latter, however, become excessive in that they control energy dissipation within the system rather than the bottom friction shown earlier to be so significant in simulating the tides and associated non-linear interactions. Use of upwind differencing in the formulation of the advective accelerations or use of eddy viscosities functionally dependent on strain rates in the modelled fluid merely served to introduce equivalent amounts of unwanted energy dissipation.

A reasonable avenue of solution is suggested by the fact that, as in the case of the earlier barotropic models, the advective accelerations, though of key importance to the tidal residual circulation, have negligible influence on the overall barotropic tidal co-oscillations. Trials of the three-dimensional model GF6 indicated that advective accelerations can be computed, stored but not actively included in the momentum equations, over the course of a barotropic tidal simulation then successfully prescribed as a tidal stress in subsequent baroclinic calculations. The horizontal tidal residual circulation that results accords with that predicted by the barotropic models, substantially confirmed by observation, but still subject to the limitation of the 4-km mesh.

In addition to this horizontal tidal residual circulation, the three-dimensional scheme displays a vigorous vertical residual circulation in the presence of strong tidal streams and sharp changes in bottom topography. Although consistent with theoretical anticipation, the performance of the model in this regard awaits observational confirmation.

This problem is closely related to the role of vertical eddy viscosity in the model. It is clearly evident from model trials that very large numerical vertical eddy viscosities must be employed if, for example, the effect of high bottom friction employed in Haro Strait is to affect the velocities higher up in the water column. Only then can a vertical velocity profile be achieved that is in approximate accord with observation.

Further shortcomings devolve on the formulation of the density conservation equation which entails an unacceptable degree of numerical diffusion. This basically derives from the small values of the Courant number $u\Delta t/\Delta x$ resulting primarily from the small time step and large mesh size. Even so, the actual turbulent diffusive processes play a major role in this system and quite remarkable results have been obtained despite the present crudity of formulation employed in simulating changes in the density field. A variety of options is available to bring about a closer approximation to a Courant number of unity. Thus, it will be necessary in subsequent work to employ a reduced mesh size (2 km) which will improve matters. Furthermore, larger time steps can be employed in the density computation or perhaps a time step can to some degree be selected as a function of some overall velocity characteristic of the system at a particular time.

The preceding discussion of the three-dimensional numerical model GF6 has dwelt heavily upon problems and shortcomings. It remains to summarize the remarkable results already furnished by this model despite the present approximate formulation and adjustment.

Use of a prescribed initial summer density field when Fraser River flow was high and dense water was intruding at depth from the continental shelf into Juan de Fuca Strait and the Strait of Georgia provided a realistic quantitative description of these major processes crucial to the estuarine circulation and flushing of the system. In Juan de Fuca Strait there is an outflow of surface water favouring the Vancouver Island shore and an inflow of water at depth favouring the Washington side. The latter upwells over the sill in the inner part of Juan de Fuca Strait. The division of the Fraser River discharge into north- and south-going components, the formation of mixed water in the strongly tidal passages leading seawards from this semi-enclosed basin, the resulting Coriolis-deflected subsurface currents moving northward along the mainland shore and southward along the Vancouver Island shore from their respective originating mixing sites, and the resulting extensive counter-clockwise circulation in the deep basin of the central Strait of Georgia all accord with an extensive body of field data accumulated over many years. The success of the demonstration is largely dependent on the use of the density field in a semi-diagnostic context. It is thus of interest to consider the results obtained from the laterally-integrated model GF5 and the three-dimensional GF6, both of which evince behaviour substantially contingent on the same density field. The longitudinal distribution of residual surface elevations is strikingly similar in both models, indicating the dominant role of the overall density field in their determination and consequent driving of the surface circulation. The resulting north- and south-going components of residual surface flow respond to these surface gradients. The large slow return flows at depth towards the central basin in the diagnostic trials of GF5 accord with the undercurrents from both southern and northern mixing regions, respectively, which feature prominently in the residual velocity fields of the Strait of Georgia in GF6.

In comparing these results with those obtained from the upper layer model GF4, it is remarkable that in prescribing a similar river flow, an initial uniform layer salinity, a fixed salinity in the hypothetical lower layer, and a fixed intermediate salinity for inflows at the boundary openings, there should result a stable residual salinity distribution in reasonable accord with that occurring at the sea surface in the observations used to prescribe the initial density (or salinity) field in the model GF6. The residual slopes of the sea surface from the vicinity of the river mouth to the mixing regions at the seaward openings of the Strait of Georgia are also in reasonable accord. Absolute values of the sea surface elevations, however, differ markedly, those in the GF6 model being some 15 cm higher. This reflects the presence of a process, not included in the GF4 model but present in GF6. This process is the effect of the deep density field and its slow replenishment of deep water in the Strait of Georgia.

The results obtained from these studies afford a striking demonstration of the role that can be played by numerical models of this type, with their capacity to approximate real topography and non-linear processes, in the ongoing dialectic between theory and observation, in the movement from description to explanation.

Future work requires resolution of the numerical problems outlined above, notably in connection with the role of vertical eddy viscosity, layer spacing, and residual circulation. This must be supported by appropriate observations. A promising location for such field studies would be the region where Discovery Passage opens out into the Strait of Georgia. Time series vector and scalar records from moored instruments should permit resolution of the barotropic tidal residual and baroclinic components of the motion. Further areas of numerical development concern the reduction of the effects of numerical diffusion in the density computations and economizing model performance. In this connection, the implementation of a wholly implicit method of solving the three-dimensional finite-difference equations has been developed at the University of Hamburg. This can permit much greater latitude in selecting the time step used in the computations resulting in much reduced computing costs. This development should be evaluated in the context of the complex coastal and bottom topography and non-linear aspects of the Georgia/Fuca system. An alternative approach could involve the extension of the 'prescribed lid' method to a fully three-dimensional application.

A feature of particular interest emerging from these studies concerns the relation between sea levels and baroclinic processes within the system. Improved accuracy in establishing the relative elevations of gauge datums that derive from present advances in surveying methods indicate that comparisons of observed and modelled sea levels could play a very important role in verifying major baroclinic processes within the system whether associated with river discharge, winds, or disturbances on the open boundaries. A significant application of this would involve 'hindcasting' in connection with major events in biological records.

The basic aim of this work is the development of a hybrid layer/fixed-level three-dimensional numerical model capable of simulating the fjordic type layer phenomena in the Strait of Georgia as well as those associated with the weaker vertical and horizontal scalar gradients elsewhere in the system. Thus, the vertically-integrated overall fine-grid (2 km) model of the system (GF7) can readily be converted by machine to a fully three-dimensional model (GF8). The layer model GF4 will be incorporated into the upper level of the Strait of Georgia in this model. In this context it will be necessary to determine appropriate conditions to be applied at a moving horizontal boundary (front) between the layer and the fixed-level numerical schemes. This will constitute the final hybrid model (GF9) of this series. The potential applications of vector and scalar data from such a model are highly significant. The Strait of Georgia is a semi-enclosed basin sustaining high commercial and recreational traffic densities while constituting a region of high biological resources, the latter being rapidly augmented by the methods of aquaculture. With expanding population and associated domestic and industrial waste discharge, it is necessary to implement water management decisions which can at once protect and fully utilize this immense and valuable national resource. In this context it will be necessary to provide real-time information on the winds at the sea surface and concomitant moored instrument monitoring at strategic sea locations to provide information on such meteorological events that are likely, for example, to lead to oceanographic conditions which could wipe out stocks at an aquaculture installation. Similarly, the operation of truly predictive forecast models of currents could greatly reduce the search area in rescue operations. An on-line version of the model could provide regular forecasts of surface current and wave conditions. Studies of the type that have been described abundantly demonstrate the attainability of these goals.

14.1 CONCLUDING SUMMARY

On the basis of the foregoing, the following thematic summary of basic elements in the system and its overall circulation is presented. The waters between Vancouver Island and the mainland consist essentially of an outer strait (Juan de Fuca) joined by a complex of strongly tidal channels to an inner strait (Georgia). The latter consists of a deep central basin with a major freshwater source located on the mainland coast to the south while connected to the open sea to the north by an extended system of passages and highly constricted by tidal rapids. Adjacent to, and within, the complex of passages between the inner and outer straits high frictional dissipation brings about significant non-linear interactions between tidal harmonic constituents. In the inner part of Juan de Fuca Strait eddies resulting from tidal topographic interactions of the strong tidal streams give rise to an extensive and complex pattern of tidal residual circulation. At the southern end of the Strait of Georgia the tidal eddy associated with an entering flood jet from Haro Strait generates large internal tides and an extensive residual eddy circulation extending as far as the mainland coast. In the northern Strait of Georgia the entering flood jet from Discovery Passage generates a smaller tidal residual eddy.

The major freshwater source, the Fraser River, is located on the mainland coast in the southern part of the inner strait. Stratification in the inner strait is characterized by a shallow brackish layer and resembles that of a fjord estuary. The combination of buoyancy and Coriolis deflection of the river discharge brings about a maximum in mean sea levels along the mainland coast immediately north of the river mouth. The resulting barotropic pressure gradients that result are stronger for those directed towards the southern route seaward compared to those directed to the north. Below and to the south of the immediate region of Coriolis-deflected river flow, the southerly-directed surface pressure gradient moves brackish water towards the southern mixing region. Under the influence of the earth's rotation, this seaward-moving flow favours the southwestern shores of the inner strait. In the southern mixing region this brackish water is mixed with deeper saline water advancing into the system through the outer strait. The resulting internal pressure gradient gives rise to an undercurrent which, under the influence of Coriolis deflection, favours the mainland coast. In the constricted passages to the north, strong mixing brings about an internal pressure field in the northern part of the inner strait giving rise to a second Coriolis-deflected weaker undercurrent moving in a southeasterly direction along the Vancouver Island shore. When these two oppositely-directed currents move into the deep central basin of the inner strait, a counter-clockwise circulation results at depth. Continuity requirements are satisfied by a slow vertical advection of deeper water into the shallower divergent flow which responds to the barotropic pressure gradients established by the downward slopes of the sea surface from a region along the mainland coast, just north of the river mouth, towards the strong tidal channels leading seaward.

Two major seasonal changes occur in the sytem. One is associated with the spring thaw in the Canadian Rockies and associated freshet of the the main freshwater source. The other is concerned with the shift in offshore winds and transition from downwelling along the open coast under prevailing southerly winds in winter to summer upwelling under the influence of predominantly northwest winds. The movement of deep dense water onto the continental shelf continues into Juan de Fuca Strait as a deep intrusion favouring the Washington shore accompanied by a largely compensatory shallower outflow favouring the Vancouver Island shore. The entry of this higher-salinity water in the mixing passages separating the inner and outer straits brings about a slow intrusion of more saline water into the deep basin of the Strait of Georgia with a consequent replenishment occurring throughout the summer and fall.

The basic pattern of circulation is modified by wind events. In the case of the outer strait, fall and winter southerly winds can move Columbia River water along the exposed Washington coast into the mouth of Juan de Fuca Strait which is located halfway between the two major freshwater sources, the Fraser and the Columbia Rivers. This brings about episodic reversals of the estuarine circulation in the outer strait primarily in the winter months. The direct local action of the northwest winds in the outer strait can have a similar effect. The estuarine surface outflow is halted or reversed, outflow then occurring below the depth of wind influence due to a buildup of residual sea level in the inner part of the outer strait.

In the case of the inner strait, southeast winds in winter bring about northwesterly surface flow into the semi-enclosed northern Strait of Georgia where the net tilt in sea level from north to south along the median axis of the Strait brings about a barotropic pressure gradient that tends to enhance the undercurrent moving in a southeasterly direction along the Vancouver Island shore. Conversely, northwest winds move surface waters into the southern mixing region where the net tilt in sea level from south to north along the median axis of the Strait tends to enhance the northwesterly undercurrent moving up along the mainland coast.

It has been shown that all these features can readily be approximated with relatively crude, and in the baroclinic case, 'untuned' numerical models. It would thus appear entirely feasible to develop an overall hybrid layer/fixed-level three-dimensional numerical model in which GF7 would be connected to a fully three-dimensional (fixed level) operation (GF8). This would be followed by the insertion of the surface layer model GF4. Positioning of the lateral boundary of the layer in the top level of the three-dimensional fixed level scheme would be determined from scalar and vector elements of the computed solution. Such a model would be of a high utility in providing information of manifold application to the management of these waters and also, boundary data requisite to detailed local numerical studies pertinent to the development of aquaculture or coastal engineering installations.

REFERENCES

Arakawa, A. and V.R. Lamb. 1981. A potential enstrophy and energy conserving scheme for the shallow water equations. *Mon. Wea. Rev.* **109**: 18–36.

Backhaus, J.O. 1980. Simulation von Bewegungsvorgängen in der Deutschen bucht. Deutsche Hydrographische Zeitschrift. Ergänzungsheft Reihe B, Nr. 15: 22–28.

Backhaus, J.O., P.B. Crean, and Do Kyu Lee. 1987. On the application of the three-dimensional numerical model to the waters between Vancouver Island and the mainland coast of British Columbia and Washington state. 149–176. **In:** Three-Dimensional Coastal Ocean Models. N.S. Heaps (ed.). A.G.U., Washington, D.C.

Baines, P.G. 1982. On internal tide generation models. *Deep Sea Res.* **29**: 307–338.

Barber, F.G. 1958. Currents and water structure in Queen Charlote Sound, British Columbia. 196–199. **In:** The Proceedings of the Ninth Pacific Congress, 1957. Fisheries Research Board No. 519, reprint.

Barker, M.L. 1974. Water resources and related land uses Strait of Georgia–Puget Sound basin. Department of Environment Geographical Paper 56. 55 pp.

Barnes, C.A., A.C. Duxbury, and B.A. Morse. 1972. Circulation and selected properties of the Columbia River effluent at sea. 41–80. **In:** The Columbia River Estuary and Adjacent Ocean Waters: Bioenvironmental Studies. A.T. Pruter and D.L. Alverson (eds.). University of Washington Press, Washington.

Bjerknes, V., J. Bjerknes, H. Solberg, and T. Bergeron. 1934. *Hydrodynamic Physique.* Vol. **II**, Chap. XI, 457–491. Les Presses Universitaires de France, Paris.

Blumberg, A.F. 1975. A Numerical Investigation into the Dynamics of Estuarine Circulation. Technical Report 9, Chesapeake Bay Institute, Johns Hopkins University, Maryland. 110 pp.

British Columbia Pilot. 1965. British Columbia Coast (South Portion) Vol. 1 and (North Portion) Vol. 2, seventh edition. Canadian Hydrographic Service, Victoria.

Buckingham, W.R., 1979. Oceanographic studies for proposed outfall near Cape Lazo, B.C. Dobrocky Seatech Ltd., Victoria, B.C. 119 pp.

Canadian Hydrographic Service. 1972a. Data Record of Current Observations, Strait of Georgia, Section 2, Cape Lazo to Grief Point, 1970. Manuscript Report Series 9.

Marine Sciences Directorate, Pacific Region, Department of the Environment, Ottawa. 88 pp.

Canadian Hydrographic Service. 1972b. Data Record of Current Observations, Strait of Georgia, Section 4, Gabriola Island to Gower Point, 1969–1972. Manuscript Report Series 10. Marine Sciences Directorate, Pacific Region, Department of the Environment, Ottawa. 153 pp.

Canadian Hydrographic Service. 1973a. Data Record of Current Observations, Strait of Georgia, Section 6, Samuel Island to Point Roberts, 1969–1970. Manuscript Report Series 12. Marine Sciences Directorate, Pacific Region, Department of the Environment, Ottawa. 96 pp.

Canadian Hydrographic Service. 1973b. Data Record of Current Observations, Strait of Georgia, Section 3, Northwest Bay to McNaughton Point, 1968–1969. Manuscript Report Series 13. Marine Sciences Directorate, Pacific Region, Department of the Environment, Ottawa. 106 pp.

Canadian Hydrographic Service. 1983. Current Atlas. Juan de Fuca Strait to Strait of Georgia. Department of Fisheries and Oceans, Ottawa. 211 pp.

Cannon, G.A. and N.P. Laird. 1978. Variability of currents and water properties from year-long observations in a fjord estuary. 515–535. **In:** Hydrodynamics of Estuaries and Fjords. J.C.J. Nihoul (ed.). Elsevier Scientific Publishing Company, Amsterdam. 546 pp.

Chang, P., S. Pond, and S. Tabata. 1976. Subsurface currents in the Strait of Georgia, west of Sturgeon Bank. *J. Fish. Res. Bd. Can.* **33**: 2218–2241.

Cordes, R.E., S. Pond, B.R. de Lange Boom, P.H. LeBlond, and P.B. Crean. 1980. Estimates of entrainment in the Fraser River plume, British Columbia. *Atmosphere-Ocean* **18**: 15–26.

Crean, P.B. 1976. Numerical model studies of tides between Vancouver Island and the mainland coast. *J. Fish. Res. Bd. Can.* **33**: 2340–2344.

Crean, P.B. 1978. A numerical model of barotropic mixed tides between Vancouver Island and the mainland and its relation to studies of the estuarine circulation. 238–313. **In:** Hydrodynamics of Estuaries and Fjords. J.C.J. Nihoul (ed.). Elsevier Scientific Publishing Company, Amsterdam.

Crean, P.B. 1983. The numerical approach to simulating the estuarine circulation in a strongly tidal deep coastal sea between Vancouver Island and the mainland. 89–116. **In:** Proceedings of the Third International Conference on Applied Mathematical Modelling. Erschienen im Eigenverlag des Institut für Meereskunde der Universität, Hamburg.

Crean, P.B. and A. Ages. 1971. Oceanographic Records from Twelve Cruises in the Strait of Georgia and Juan de Fuca Strait, 1968. Volumes 1–5. Department of

Energy, Mines and Resources, Marine Sciences Branch, Victoria, British Columbia. 389 pp.

Crean, P.B., W.S. Huggett, and M. Miyake. 1979. Data Report of STD Observations, Volume 1: Strait of Juan de Fuca 1973. Pacific Marine Science Report 79-13. Institute of Ocean Sciences, Patricia Bay, Sidney, B.C. 184 pp.

Crean, P.B. and A.G. Lewis. 1976. Monthly Sections – Strait of Juan de Fuca, June–December 1973. Data Report 39 (Provisional). Institute of Oceanography, University of British Columbia, Vancouver, B.C. 184 pp.

Crean, P.B. and M. Miyake. 1976. STD Sections, Strait of Juan de Fuca, March-April 1973. Data Report 38 (Provisional). Institute of Oceanography, University of British Columbia, Vancouver, B.C. 229 pp.

Doodson, A.T. and H.D. Warburg. 1941. Admiralty Manual of Tides. Her Majesty's Stationery Office, London. 270 pp.

Dronkers, J.J. 1964. Tidal Computations. North-Holland Publishing Company, Amsterdam. 518 pp.

Dronkers, J.J. 1969. Tidal computations for rivers, coastal estuaries and seas. *J. Hydraulics Div., ASCE* **95**: 29–77.

Dunbar, D. 1985. A Numerical Model of Stratified Circulation in a Shallow-Silled Inlet. Ph.D. Thesis. University of British Columbia, Vancouver B.C. 273 pp.

Ebbesmeyer, C.C. and C.A. Barnes. 1980. Control of a fjord basin's dynamics by tidal mixing in embracing sill zones. *Est. Coast. Mar. Sci.* **11**: 311–330.

Elliot, J.A. 1976. A Numerical Model of the Internal Circulation in a Branching Tidal Estuary. Special Report 54. Chesapeake Bay Institute, Johns Hopkins University, Maryland. 110 pp.

Filloux, J.H. 1973. Tidal patterns and energy balance in the Gulf of California. *Nature* **243**: 217-221.

Flather, R.A. 1987. A tidal model of the northeast Pacific. *Atmosphere-Ocean* **25**: 22–45.

Flather, R.A. and N.S. Heaps. 1975. Tidal computations for Morcambe Bay. *Geophys. J. R. Astr. Soc.* **42**: 489–517.

Foreman, M.G.G. 1977. Manual for Tidal Heights Analysis and Prediction. Pacific Marine Science Report 77-10. Institute of Ocean Sciences, Victoria, British Columbia. 101 pp.

Foreman, M.G.G. 1978. Manual for Tidal Current Analysis and Prediction. Pacific Marine Science Report 78-6. Institute of Ocean Sciences, Victoria, British Columbia. 70 pp.

Freeland, H.J. and K.L. Denman. 1982. A topographically controlled upwelling center off southern Vancouver Island. *J. Mar. Res.* **40**: 1069–1093.

Fromm, J.E. 1969. High speed computing in fluid dynamics. The Physics of Fluids Suppl. 11: II.3–II.12.

Gade, H.G. and A. Edwards. 1980. Deep water renewal in fjords. 453–489. **In:** Fjord Oceanography. H.J. Freeland, D.M. Farmer, and C.D. Levings (eds.). Plenum Publishing Corp., N.Y.

Garcia, A.W. and J.R. Houston. 1975. Type 16 Flood Insurance Study: Tsunami Predictions for Monterey and San Francisco Bays and Puget Sound. Technical Report H-75-17, Hydraulics Laboratory, U.S. Army Engineers Waterways Experiment Station, Vicksburg, Mississippi. 263 pp.

Garvine, R.W. and J.D. Monk. 1974. Frontal structure of a river plume. *J. Geophys. Res.* **79**: 2251–2259.

Geyer, W.R. and G.A. Cannon. 1982. Sill processes related to deep water renewal in a fjord. *J. Geophys. Res.* **87**: 7985–7996.

Giovando, L.F. and S. Tabata. 1970. Measurements of Surface Flow in the Strait of Georgia by Means of Free-Floating Current Followers. Fisheries Research Board of Canada Technical Report No. 163. 60 pp.

Godin, G. 1972. The Analysis of Tides. University of Toronto Press, Toronto. 264 pp.

Godin, G., J. Candela, and R. de la Paz-Vela. 1981. An analysis and interpretation of the current data collected in the Strait of Juan de Fuca in 1973. *Marine Geodesy* **5**: 273–302

Grace, S.F. 1936. Friction in the tidal currents of the Bristol Channel. *Geophys. Supp. M.N.R. Astr. Soc.* **3**: 388.

Grace, S.F. 1937. Friction in the tidal currents of the English Channel. *Geophys. Suppl. M.N.R. Astr. Soc.* **4**: 133.

Hamilton, P. 1975. A numerical model of the vertical circulation of tidal estuaries and its application to the Rotterdam waterway. *Geophys. J. Roy. Astr. Soc.* **40**: 1–21.

Hansen, W. 1956. Theorie zur Errechnung des Wasserstandes und der Strömungen in Randmeeren nebst Anwendungen. *Tellus*, **8**: 287–300.

Hansen, W. 1961. Hydrodynamical methods applied to oceanographic problems. 25–34. **In:** Proceedings of the Symposium on Mathematical-Hydrodynamical Methods of Physical Oceanography. Institut für Meereskunde der Universität, Hamburg, 1962.

Harris, R.A. 1904. Measuring tides and currents at sea. *Science*, **19**: 704–707.

Heisanken, W. 1921. Uber den einfluss der gezeiten auf die säkulare accelereation des mondes. *Ann. Acad. Sci. Fennicae*, **A18**: 1–84.

Hendershott, M.C. and A. Speranza. 1971. Co-oscillating tides in long, narrow bays; the

Taylor problem revisited. *Deep-Sea Res.* **18**: 959–980.

Henry, R.F. 1982. Automated Programming of Explicit Shallow-Water Models: Part I. Linearized Models with Linear or Quadratic Friction. Canadian Technical Report of Hydrography and Ocean Sciences No. 3. Institute of Ocean Sciences, Department of Fisheries and Oceans, Sidney, B.C. 70 pp.

Herlinveaux, R.H. and L.F. Giovando. 1969. Some Oceanographic Features of the Inside Passage between Vancouver Island and the Mainland of British Columbia. Journal of the Fisheries Research Board of Canada Technical Report No. 142. 48 pp.

Holbrook, J.R., G.A. Cannon, and D. Kachel. 1983. Two-year observations of coastal-estuarine interaction in the Strait of Juan de Fuca. 411–426. **In:** Coastal Oceanography. H. Gade, H. Svendsen, and A. Edwards (eds.). Plenum Press, N.Y.

Holbrook, J.R., R.D. Muench, and G.A. Cannon. 1980. Seasonal observations of low-frequency atmospheric forcing in the Strait of Juan de Fuca. 305–317. **In:** Fjord Oceanography. H.J. Freeland, D.M. Farmer, and C.D. Levings (eds.). Plenum Publishing Corp., N. Y.

Hough, S.S. 1898. On the application of harmonic analysis to the dynamical theory of tides. Part II. On the general integration of Laplace's dynamical equations. *Phil. Trans. Roy. Soc. Lond.* **A 191**: 138–185.

Houston, J.R. and A.W Garcia. 1974. Type 16 Flood Insurance Study: Tsunami Predictions for Pacific Coastal Communities. Technical Report H-75-3. U.S. Army Engineer Waterways Experiment Station, ACE, Vicksburg, Mississippi. 139 pp.

Huggett, W.S., J.F. Bath, and A. Douglas. 1976a. Data Record of Current Observations. Johnstone Strait, 1973. Vol. XIV. Institute of Ocean Sciences, Sidney, B.C. 155 pp.

Huggett, W.S., J.F. Bath, and A. Douglas. 1976b. Data Record of Current Observations. Juan de Fuca Strait, 1973. Vol. XV. Institute of Ocean Sciences, Patricia Bay, Victoria, B.C. 169 pp.

Ianiello, J.P. 1979. Tidally induced residual currents in estuaries of variable breadth and depth. *J. Phys. Ocean.* **9**: 962–974.

Jeffreys, H. 1921. Tidal friction in shallow seas. *Phil. Trans. Roy. Soc. Lond.* **A 221**: 239–264.

Kelvin, Lord W., 1879. On gravitational oscillations of rotating water. *Proc. Roy. Soc. Edin.* **10**: 92–109, Papers, **4**: 141–148.

Kendrew, W.G. and D. Kerr. 1955. The Climate of British Columbia and Yukon Territory. Meteorological Division, Department of Transport. Toronto. 222 pp.

Kuipers, H. and C.B. Vreugdenhil. 1973. Calculations of Two-Dimensional Horizontal Flow. Report on Basic Research S 163. Part 1, Delft Hydraulics Laboratory, The Netherlands. 64 pp.

Le Prévost, T.C. 1976. Theoretical analysis on the structure of the tidal wave spectrum in shallow water areas. *Mem. Soc. Roy. Sci. de Liege.*, **X**: 97–111.

Lighthill, M.J. 1957. River Waves. Naval Hydrodynamics Publ. 515. National Academy of Sciences – National Research Council.

Miller, G.R. 1966. The flux of tidal energy out of the deep oceans. *J. Geophys. Res.* **71**: 2485–2489.

Möfjeld, H.O. and L.H. Larsen. 1984. Tides and Tidal Currents of the Inland Waters of Western Washington. NOAA Technical Memo. ERL PMEL-56. Pacific Marine Environmental Laboratory, Seattle. 52 pp.

Mortimer, C.H. 1963. Frontiers in Physical Limnology with Particular Reference to Long Waves in Rotating Basins. 9–42. **In:** Proceedings of the 6th Conference on Great Lakes Research, Great Lakes Research Division, University of Michigan, Publ. 9.

Mortimer, C.H. 1974. Mitt. Int. Ver. Theor. Angew. Limol. **20**: 124–197.

Mungall, J.C.J. 1973. Numerical Tidal Models with Unequal Grid Spacing. Ph.D. Thesis, University of Alaska, College, Alaska. 213 pp.

Murty, T.S. 1977 Seismic Sea Waves – Tsunamis. Fish. Res. Bd. Can., Bull. No. 198, Ottawa. 337 pp.

Murty, T.S. 1984. Storm Surges – Meteorological Ocean Tides. Can. Bull. Fish. Aq. Sci., Bull. No. 212, Ottawa. 906 pp.

Murty, T.S. and P.B. Crean. 1985. A reconstruction of the tsunami of June 23, 1946 in the Strait of Georgia. 121–125. **In:** Proceedings of the International Tsunami Symposium, Institute of Ocean Sciences, Sidney, B.C.

O'Brien, G.G, M.A. Hyman, and S. Kaplan. 1950. A study of the numerical solution of partial differential equations. *J. Math. Phys.* **29**: 223–251.

Parker, B.B. 1977. Tidal Hydrodynamics in the Strait of Juan de Fuca–Strait of Georgia. U.S. Department of Commerce, NOAA. Technical Report NOS 69. Washington, D.C. 56 pp.

Paul, J.F. and W. Lick. 1974. A numerical model of thermal plumes and river discharges. 445–455. **In:** Proceedings of the 17th Conference on Great Lakes Research, Int. Assoc. Great Lakes Res., McMaster University, Hamilton, Ontario.

Petrauskas, C. and L.E. Borgman. 1971. Frequencies of Crest Heights for Random Combinations of Astronomical Tides and Tsunamis Recorded at Crescent City, California. Technical Report HEL 16-8. Hydraulic Engineering Laboratory, University of California, Berkeley, Calif.

Platzman, G. W. 1971. Ocean tides and related waves. 239–291. **In:** Mathematical Problems in the Geophysical Sciences. W. H. Reid (ed.). Am. Math. Soc., Providence, Rhode Island.

Proehl, J.A. and M. Rattray. 1984. Low-frequency response of wide deep estuaries to non-local atmospheric forcing. *J. Phys. Ocean.* **14**: 904–921.

Proudman, J. 1953. Dynamical Oceanography. Methuen and Co. Ltd., London. 409 pp.

Redfield, A.C. 1950a. Note on the circulation of a deep estuary – the Juan de Fuca-Georgia Straits. Proceedings of the Colloquium on the Flushing of Estuaries. Woods Hole Oceanographic Institution Rept. No. 50-37: 175–177.

Redfield, A.C. 1950b. The analysis of tidal phenomena in narrow embayments. Papers in Physical Oceanography and Meteorology. Vol. XI, No. 4. Massachussetts Institute of Technology and Woods Hole Oceanographic Institution. 206 pp.

Richardson, D.J. 1964. The solution of two-dimensional hydrodynamic equations by the method of characteristics. 295–318. **In:** Methods on Computational Physics, Vol. 3. B. Alder, S. Fernbach, and M. Rotenberg (eds.). Academic Press, N.Y.

Roache, P.J. 1972. Computational Fluid Dynamics. Hermosa Publishers, Albuquerque, N.M. 446 pp.

Rogers, G.C. and H.S. Hasegawa. 1978. A second look at the British Columbia earthquake of June 23, 1946. *Bull. Seis. Soc. Amer.* **678**: 653–676.

Rossiter, J.R. and G.W. Lennon. 1965. Computations of tidal conditions in the Thames Estuary by the initial value method. *Proc. Inst. Civil Engrs.* **31**: 25–56.

Schumacher, J.D., C.A. Pearson, R.L. Charnell, and N.P. Laird. 1978. Regional response to forcing in southern Strait of Georgia. *Est. Coast. Mar. Sci.* **7**: 79–91.

Simons, T.J. 1980. Circulation Models of Lakes and Inland Seas. Canadian Bulletin of Fisheries and Aquatic Sciences, Bull. 200, Ottawa. 146 pp.

Smagorinsky, J. 1963. General circulation experiments with the primitive equations. I. The basic experiment. *Mon. Wea. Rev.* **91**: 99–164.

Stacey, M.W., S. Pond, P.H. LeBlond, H.J. Freeland, and D.M. Farmer. 1987. An analysis of the low-frequency current fluctuations in the Strait of Georgia from June 1984 until January 1985. *J. Phys. Ocean.*, **17**: 326–342.

Stronach, J.A. 1977. Observational and Modelling Studies of the Fraser River Plume. Ph.D. Thesis. University of British Columbia, Vancouver, British Columbia. 221 pp.

Stronach, J.A. and P.B. Crean. 1978. Data Report, Upper Layer of the Strait of Georgia. No. 44. Institute of Oceanography, University of British Columbia, Vancouver. 284 pp.

Stronach, J.A., J.A. Helbig, and N.E.J. Boston. 1984. Tidal and Residual Currents in the Strait of Georgia – Juan de Fuca Strait System. Beak Consultants Ltd, Vancouver, B.C. 213 pp.

Sündermann, J. 1971. Die hydrodynamisch-numerische Berechnung der Vertikalstruktur von Bewegungsvorgängen in Kanälen und Becken. Mitteilungen des Institut für

Meereskunde der Universität Hamburg, XIX. 104 pp.

Tabata, S., L.F. Giovando, J.A. Strickland, and J. Wong. 1970. Current Velocity Measurements in the Strait of Georgia, 1967. Technical Report No. 169, Fisheries Research Board of Canada, Pacific Biological Station, Nanaimo, British Columbia. 245 pp.

Tabata, S. and J.A. Strickland. 1972a. Summary of Oceanographic Records Obtained from Moored Instruments in the Strait of Georgia, 1969–1970. Current Velocity and Seawater Temperature from Station H-06. Pacific Marine Science Report 72-7. Environment Canada. 132 pp.

Tabata, S. and J.A. Strickland. 1972b. Summary of Oceanographic Records Obtained from Moored Instruments in the Strait of Georgia, 1969–1970. Current Velocity and Seawater Temperature from Station H-16. Pacific Marine Science Report 72-8. Environment Canada. 144 pp.

Tabata, S. and J.A. Strickland. 1972c. Summary of Oceanographic Records Obtained from Moored Instruments in the Strait of Georgia, 1969–1970. Current Velocity and Seawater Temperature from Station H-26. Pacific Marine Science Report 72-9. Environment Canada. 141 pp.

Tabata, S., B. Thomas, and D. Ramsden. 1986. Annual and interannual variability of steric sea level along line P in the northeast Pacific Ocean. *J. Phys. Ocean.* **16:** 1378–1398.

Taylor, G.I. 1919. Tidal friction in the Irish Sea. *Phil. Trans. Roy. Soc. A.* **220:** 1–33.

Taylor, G.I. 1921. Tidal oscillations in gulfs and rectangular basins. *Proc. London Math. Soc.* **20:** 148–181.

Thomson, K.A. and G.C. Louttit. 1986. Spatial Correlations and Internal Modes at the Nanoose Underwater Test Range. Dobrocky Seatech Ltd., Sidney, B.C. 65 pp.

Thomson, R.E. 1976. Tidal currents and estuarine circulation in Johnstone Strait, British Columbia. *J. Fish. Res. Bd. Can.* **33:** 2242–2264.

Thomson, R.E. 1977. Currents in Johnstone Strait, British Columbia: Supplemental data on the Vancouver Island side. *J. Fish. Res. Bd. Can.* **34:** 697–703.

Thomson, R.E. 1981. Oceanography of the British Columbia Coast. Canadian Special Publication of Fisheries and Aquatic Sciences 56. 291 pp.

Thomson, R.E. and W.S. Huggett. 1980 M_2 baroclinic tides in Johnstone Strait, British Columbia. *J. Phys. Ocean.* **10:** 1509–1539.

Turner, J.S. 1973. Buoyancy Effects in Fluids. Cambridge University Press, Cambridge. 367 pp.

Waldichuk, M. 1957. Physical oceanography of the Strait of Georgia, British Columbia. *J. Fish. Res. Bd. Can.* **15:** 1065–1102.

Waldichuk, M. 1964. Daily and Seasonal Sea Level Oscillations on the Pacific Coast of Canada. 181–201. **In:** Studies in Oceanography, Hidaka Jubilee Committee (eds.). University of Tokyo.

Webster, I. 1977. A Physical Oceanographic Study of Haro Strait – Summer 1976. Dobrocky Seatech Ltd., Victoria, B.C. Prepared for the Coastal Zone Oceanographic Group, Institute of Ocean Sciences, B.C. 103 pp.

Whitham, G.B. 1974. Linear and Non-linear Waves. J. Wiley and Sons, N.Y. 636 pp.

Zimmerman, J.F.T. 1981. Dynamics, diffusion and geomorphological significance of tidal residual eddies. *Nature* **290**: 549–554.

APPENDIX I

BAROCLINIC MOTION WITH A PRESCRIBED LID IN A CHANNEL OF UNIFORM SECTION

Basic Equations (Elliot, 1976)

Continuity Equation

$$\frac{\partial}{\partial x}(Bu) + \frac{\partial}{\partial z}(Bw) = 0. \tag{1}$$

Salt Balance Equation

$$\frac{\partial}{\partial t}(Bs) + \frac{\partial}{\partial x}(Bus) + \frac{\partial}{\partial z}(Bws) - \frac{\partial}{\partial x}\left(BK_x \frac{\partial s}{\partial x}\right) - \frac{\partial}{\partial z}\left(BK_z \frac{\partial s}{\partial z}\right) = 0. \tag{2}$$

Momentum Equation

$$\frac{\partial}{\partial t}(Bu) + \frac{\partial}{\partial x}(Buu) + \frac{\partial}{\partial z}(Buw) - \frac{\partial}{\partial x}\left(BN_x \frac{\partial u}{\partial x}\right) - \frac{\partial}{\partial z}\left(BN_z \frac{\partial u}{\partial z}\right) = -\frac{B}{\rho}\frac{\partial p}{\partial x} + BX. \tag{3}$$

Hydrostatic Pressure

$$\frac{\partial p}{\partial z} = \rho g \tag{4}$$

where

- x, z = co-ordinates in the undisturbed surface along the channel axis and vertically down, respectively,
- u, w = corresponding breadth-averaged components of velocity, $u = 1/B \int_a^b \tilde{u}\, dy$, $w = 1/B \int_a^b \tilde{w}\, dy$, $a = a(x, z)$, $b = b(x, z)$, and $B = b - a = B(x, z)$,
- $\tilde{u}, \tilde{w}(x, y, z, t)$ = components of actual velocity,
- s = salinity,
- t = time,
- p = pressure,
- ρ = water density, in general, $\rho = \rho(x, z, t)$,
- K_x, K_z = coefficients of horizontal and vertical diffusion, respectively,
- N_x, N_z = coefficients of horizontal and vertical eddy viscosity, respectively,
- X = body force, i.e., tide-generating force = $X(x, t)$.

The continuity equation can be vertically integrated to give

$$\int_{-\zeta}^{h} \frac{\partial}{\partial x}(Bu)\, dz + \int_{-\zeta}^{h} \frac{\partial}{\partial z}(Bw)\, dz = 0.$$

Thus,

$$\frac{\partial}{\partial x}\int_{-\zeta}^{h}(Bu)\,dz - \left[Bu\frac{\partial h}{\partial x}\right]_{z=h} - \left[Bu\frac{\partial \zeta_s}{\partial x}\right]_{z=-\zeta} + [Bw]_h - [Bw]_{-\zeta} = 0,$$

and using the surface and bottom conditions and neglecting cross-channel variations,

$$\frac{D}{Dt}(z+\zeta(x,t))=0 \quad \Rightarrow \quad w(-\zeta)+\frac{\partial \zeta}{\partial t}+u(-\zeta)\frac{\partial \zeta}{\partial x}=0,$$

$$\frac{D}{Dt}(z-h(x))=0 \quad \Rightarrow \quad w(h)-u(h)\frac{\partial h}{\partial x}=0$$

where h is the undisturbed water depth. This gives

$$\frac{\partial}{\partial x}\int_{-\zeta}^{h} Bu\,dz + B(-\zeta)\frac{\partial \zeta}{\partial t} = 0. \tag{5}$$

Alternative Assumptions and Approximations

1. Barotropic Motion

If $\rho = $ constant, integration of Eq. (4) gives

$$p = \int_{-\zeta}^{z} \rho g\,dz' = \rho g(z+\zeta) + \text{contant}.$$

Also, since at $z = -\zeta$, $p = p_a$ the surface atmospheric pressure

$$p = p_a + \rho g(z+\zeta). \tag{6}$$

Thus, eliminating p from Eq. (3),

$$\frac{\partial}{\partial t}(Bu) + T(u) = -\frac{B}{\rho}\frac{\partial p_a}{\partial x} - Bg\frac{\partial \zeta}{\partial x} + BX \tag{7}$$

where

$$T(u) = \frac{\partial}{\partial x}(Buu) + \frac{\partial}{\partial z}(Buw) - \frac{\partial}{\partial x}\left(BN_x\frac{\partial u}{\partial x}\right) - \frac{\partial}{\partial z}\left(BN_z\frac{\partial u}{\partial z}\right). \tag{8}$$

The set of working equations would then consist of (1), (5), (7), and (8).

2. General Free Motion

In this case $\rho = \rho(x, z, t)$, so integrating Eq. (4)

$$p = p_a + g\int_{-\zeta}^{z}\rho\,dz'$$

$$= p_a + g(z+\zeta)\bar{\rho} \tag{9}$$

where

$$\bar{\rho} = \frac{1}{z+\zeta}\int_{-\zeta}^{z}\rho\,dz'.$$

Eliminating p from Eq. (3) gives

$$\frac{\partial}{\partial t}(Bu) + T(u) = -\frac{B}{\rho}\frac{\partial p}{\partial x}$$

$$= -\frac{B}{\rho}\frac{\partial p_a}{\partial x} - Bg\frac{\bar{\rho}}{\rho}\frac{\partial \zeta}{\partial x} - Bg\frac{(z+\zeta)}{\rho}\frac{\partial \bar{\rho}}{\partial x} + BX. \qquad (10)$$

The working set of equations would then be (1), (2), (3), (8), and (10) with the addition of an equation of state relating density and salinity, such as (Hamilton, 1975)

$$\rho = \rho_0(1 + \alpha s) \qquad (11)$$

where ρ_0 is the density of freshwater

$$\alpha = 7.8 \times 10^{-4}.$$

Retaining the free surface in this case admits surface gravity waves with the consequent limitation on time step in numerical calculations. Alternative assumptions allow this limitation to be relaxed.

3. Baroclinic Motion with a Rigid Lid

Suppose $\rho = \rho(x, z, t)$ but the motion is confined by a rigid lid at $z = 0$. Then integrating Eq. (4), as before,

$$p = g\int_0^z \rho\, dz' + \text{constant}.$$

The constant of integration is determined from the pressure at $z = 0$,

$$= p_a + p_0$$

where p_0 is the pressure on the lid due to the fluid motions, so

$$p = p_a + p_0 + g\int_0^z \rho\, dz'$$

$$= p_a + p_0 + gz\bar{\rho}_0 \qquad (12)$$

where

$$\bar{\rho}_0 = \frac{1}{z}\int_0^z \rho\, dz'.$$

Eliminating pressure from Eq. (3) gives

$$\frac{\partial}{\partial t}(Bu) + T(u) = -\frac{B}{\rho}\frac{\partial p}{\partial x} + BX$$

$$= -\frac{B}{\rho}\frac{\partial p_a}{\partial x} - \frac{B}{\rho}\frac{\partial p_0}{\partial x} - \frac{Bgz}{\rho}\frac{\partial \bar{\rho}_0}{\partial x} + BX. \qquad (13)$$

An additional equation is now required to determine p_0, the pressure on the lid. This is derived from Eq. (13) by integrating vertically from $z = 0$ to $z = h$, then eliminating with respect to x,

$$\frac{\partial}{\partial x}\int_0^h \frac{B}{\rho}\frac{\partial p_0}{\partial x}\, dz' = -\frac{\partial}{\partial x}\int_0^h \frac{\partial}{\partial t}(Bu)\, dz' - \frac{\partial}{\partial x}\int_0^h T(u)\, dz'$$

$$- \frac{\partial}{\partial x}\int_0^h \frac{B}{\rho}\frac{\partial p_a}{\partial x} - g\frac{\partial}{\partial x}\int_0^h \frac{Bz}{\rho}\frac{\partial \bar{\rho}_0}{\partial x}\, dz' + \frac{\partial}{\partial x}\int_0^h XB\, dz'. \qquad (14)$$

Now, since $p_0 = p_0(x, t)$, the left hand side becomes

$$\frac{\partial}{\partial x}\left(\frac{\partial p_0}{\partial x} \cdot \int_0^h \frac{B}{\rho}\, dz'\right).$$

Changing the order of differentiation and integration and using Eq. (1) in the first term on the right hand side,

$$-\frac{\partial}{\partial x}\int_0^h \frac{\partial}{\partial t}(Bu)\, dz' = -\int_0^h \frac{\partial}{\partial x}\frac{\partial}{\partial t}(Bu)\, dz' - \left[\frac{\partial}{\partial t}(Bu)\right]_h \frac{\partial h}{\partial x}$$

$$= -\int_0^h \frac{\partial}{\partial t}\frac{\partial}{\partial x}(Bu)\, dz' - \left[\frac{\partial}{\partial t}(Bu)\right]_h \frac{\partial h}{\partial x}$$

$$= \int_0^h \frac{\partial}{\partial t}\frac{\partial}{\partial z}(Bw)\, dz' - \left[\frac{\partial}{\partial t}(Bu)\right]_h \frac{\partial h}{\partial x}$$

$$= \left[\frac{\partial}{\partial t}(Bw)\right]_h - \left[\frac{\partial}{\partial t}(Bw)\right]_h - \left[\frac{\partial}{\partial t}(Bu)\right]_h \frac{\partial h}{\partial x}.$$

Making use of the bottom condition,

$$\frac{D}{Dt}(z - h(x)) = w(h) - u(h)\frac{\partial h}{\partial x} = 0 = \frac{\partial}{\partial t}\left(w(h) - u(h)\frac{\partial h}{\partial x}\right),$$

$$-\frac{\partial}{\partial x}\int_0^h \frac{\partial}{\partial t}(Bu)\, dz' = -\left[\frac{\partial}{\partial t}(Bw)\right]_0 = -B\frac{\partial}{\partial t}w(0).$$

Also,

$$\frac{\partial}{\partial x}\int_0^h BX\, dz' = \frac{\partial}{\partial x}\left(X\int_0^h B\, dz'\right).$$

Equation (14) can thus be written

$$\frac{\partial}{\partial x}\left(\frac{\partial p_a}{\partial x}\cdot\int_0^h \frac{B}{\rho}\, dz'\right) = -B\frac{\partial}{\partial t}w(0) - \frac{\partial}{\partial x}\int_0^h T(u)\, dz'$$

$$-\frac{\partial}{\partial x}\left(\frac{\partial p_a}{\partial x}\cdot\int_0^h \frac{B}{\rho}\, dz'\right) - g\frac{\partial}{\partial x}\int_0^h \frac{B}{\rho}\cdot z'\frac{\partial \overline{\rho_0}}{\partial x}\, dz'$$

$$+ \frac{\partial}{\partial x}\left(X\int_0^h B\, dz'\right). \tag{15}$$

The working set of equations would then be (1), (2), (11), (13), and (15).

4. Baroclinic Motion with a Prescribed Lid

Suppose $\rho = \rho(x, z, t)$ but the motion is confined by a prescribed lid at $z = -\zeta_s(x, t)$ where the surface disturbance ζ_s is associated with tides or other divergent motion. Then integrating Eq. (4), as before,

$$p = g\int_{-\zeta_s}^z \rho\, dz' + \text{constant}.$$

The constant of integration is determined from the pressure at $z = -\zeta_s$

$$= p_a + p_s,$$

where p_s is the pressure on the lid due to the fluid motions, so

$$p = p_a + p_s + g \int_{-\zeta_s}^{z} \rho\, dz'$$
$$= p_a + p_s + g(z + \zeta_s)\bar{\rho}_s \qquad (16)$$

where

$$\bar{\rho}_s = \frac{1}{z + \zeta_s} \int_{-\zeta_s}^{z} \rho\, dz'.$$

Then eliminating p from Eq. (3) gives

$$\frac{\partial}{\partial t}(Bu) + T(u) = -\frac{B}{\rho}\frac{\partial p}{\partial x} + BX$$
$$= -\frac{B}{\rho}\frac{\partial p_a}{\partial x} - \frac{B}{\rho}\frac{\partial p_s}{\partial x} - \frac{Bg}{\rho}\bar{\rho}_s\frac{\partial \zeta_s}{\partial x} - \frac{Bg}{\rho}(z+\zeta_s)\frac{\partial \bar{\rho}_s}{\partial x} + BX. \qquad (17)$$

An additional equation is now required to determine p_s, the pressure on the lid. This is derived from Eq. (17) by integrating vertically from $z = -\zeta_s$ to $z = h$, then differentiating with respect to x,

$$\frac{\partial}{\partial x}\int_{-\zeta_s}^{h}\frac{B}{\rho}\frac{\partial p_s}{\partial x}\,dz' = -\frac{\partial}{\partial x}\int_{-\zeta_s}^{h}\frac{\partial}{\partial t}(Bu)\,dz' - \frac{\partial}{\partial x}\int_{-\zeta_s}^{h}T(u)\,dz' - \frac{\partial}{\partial x}\int_{-\zeta_s}^{h}\frac{B}{\rho}\frac{\partial p_a}{\partial x}\,dz'$$
$$- g\frac{\partial}{\partial x}\int_{-\zeta_s}^{h}\frac{B}{\rho}\frac{\partial}{\partial x}(z+\zeta_s)\bar{\rho}_s)\,dz' + \frac{\partial}{\partial x}\int_{-\zeta_s}^{h}BX\,dz'. \qquad (18)$$

Now, since $p_s = p_s(x,t)$, the left hand side becomes

$$\frac{\partial}{\partial x}\left(\frac{\partial p_s}{\partial x}\int_{-\zeta_s}^{h}\frac{B}{\rho}\,dz'\right).$$

Changing the order of differentiation and integration and using Eq. (1) in the first term on the right hand side,

$$-\frac{\partial}{\partial x}\int_{-\zeta_s}^{h}\frac{\partial}{\partial t}(Bu)\,dz' = -\int_{-\zeta_s}^{h}\frac{\partial}{\partial x}\frac{\partial}{\partial t}(Bu)\,dz' - \left[\frac{\partial}{\partial t}(Bu)\right]_{h}\frac{\partial h}{\partial x} - \left[\frac{\partial}{\partial t}(Bu)\right]_{-\zeta_s}\frac{\partial \zeta_s}{\partial x}$$
$$= -\int_{-\zeta_s}^{h}\frac{\partial}{\partial t}\frac{\partial}{\partial x}(Bu)\,dz' - \left[\frac{\partial}{\partial t}(Bu)\right]_{h}\frac{\partial h}{\partial x} - \left[\frac{\partial}{\partial t}(Bu)\right]_{-\zeta_s}\frac{\partial \zeta_s}{\partial x}$$
$$= +\int_{-\zeta_s}^{h}\frac{\partial}{\partial t}\frac{\partial}{\partial z}(Bw)\,dz' - \left[\frac{\partial}{\partial t}(Bu)\right]_{h}\frac{\partial h}{\partial x} - \left[\frac{\partial}{\partial t}(Bu)\right]_{-\zeta_s}\frac{\partial \zeta_s}{\partial x}$$
$$= \frac{\partial}{\partial t}([Bw]_h - [Bw]_{-\zeta_s}) - \left[\frac{\partial}{\partial t}(Bu)\right]_{h}\frac{\partial h}{\partial x} - \left[\frac{\partial}{\partial t}(Bu)\right]_{-\zeta_s}\frac{\partial \zeta_s}{\partial x}$$
$$= \frac{\partial}{\partial t}\left(B_h\left(w_h - u_h\frac{\partial h}{\partial x}\right)\right) - \frac{\partial}{\partial t}\left(B_{-\zeta_s}\left(w_{-\zeta_s} + u_{-\zeta_s}\frac{\partial \zeta_s}{\partial x}\right)\right).$$

Making use of the surface and bottom conditions,

$$\frac{D}{Dt}(z + \zeta_s(x,t)) = w(-\zeta_s) + u(-\zeta_s)\frac{\partial \zeta_s}{\partial x} + \frac{\partial \zeta_s}{\partial t} = 0$$

and

$$\frac{D}{Dt}(z - h(x)) = w(+h) - u(h)\frac{\partial h}{\partial x} = 0,$$

$$-\frac{\partial}{\partial x}\int_{-\zeta_s}^{h}\frac{\partial}{\partial t}(bu)\,dz' = +\frac{\partial}{\partial t}\left(B_{-\zeta_s}\frac{\partial \zeta_s}{\partial t}\right).$$

Equation (18) can thus be written

$$\frac{\partial}{\partial x}\left(\frac{\partial p_s}{\partial x}\int_{-\zeta_s}^{h}\frac{B}{\rho}\,dz'\right) = \frac{\partial}{\partial t}\left(B_{-\zeta_s}\frac{\partial \zeta_s}{\partial t}\right) - \frac{\partial}{\partial x}\int_{-\zeta_s}^{h}T(u)\,dz'$$

$$-\frac{\partial}{\partial x}\left(\frac{\partial p_a}{\partial x}\int_{-\zeta_s}^{h}\frac{B}{\rho}\,dz'\right) - g\frac{\partial}{\partial x}\int_{-\zeta_s}^{h}\frac{B}{\rho}\frac{\partial}{\partial x}(z+\zeta_s)\bar{\rho}_s\,dz'$$

$$+\frac{\partial}{\partial x}(X\int_{-\zeta_s}^{h}B\,dz'). \tag{19}$$

The working set of equations would then be (1), (2), (11), (17), and (19).

Now, if the prescribed lid elevation ζ_s is derived from a barotropic solution with constant density ρ_s, it follows from Eq. (7) that

$$Bg\frac{\partial \zeta_s}{\partial x} = -\frac{B}{\rho_s}\frac{\partial p_a}{\partial x} - \frac{\partial}{\partial t}(Bu_s) - T_s(u_s) + BX \tag{20}$$

where the subscript denotes a quantity deriving from the barotropic solution. Then using (20) to eliminate the surface gradient from (17),

$$\frac{\partial}{\partial t}(Bu) + T(u) = -\frac{B}{\rho}\frac{\partial p_a}{\partial x} - \frac{B}{\rho}\frac{\partial p_s}{\partial x} - \frac{\bar{\rho}_s}{\rho}\left[-\frac{B}{\rho_s}\frac{\partial p_a}{\partial x} - \frac{\partial}{\partial t}(Bu_s) - T_s(u_s) + BX\right]$$

$$-\frac{Bg}{\rho}(z+\zeta_s)\frac{\partial \bar{\rho}_s}{\partial x} + BX,$$

i.e., $$\frac{\partial}{\partial t}(Bu) - \frac{\bar{\rho}_s}{\rho}\frac{\partial}{\partial t}(Bu_s) + T(u) - \frac{\bar{\rho}_s}{\rho}T_s(u_s) = -\frac{B}{\rho}\frac{\partial p_a}{\partial x}\left(1 - \frac{\bar{\rho}_s}{\rho_s}\right)$$

$$-\frac{B}{\rho}\frac{\partial p_s}{\partial x} - \frac{Bg}{\rho}(z+\zeta_s)\frac{\partial \bar{\rho}_s}{\partial x} + BX\left(\frac{1-\bar{\rho}_s}{\rho}\right). \tag{21}$$

Since the non-linear terms in Eq. (21) include the total currents, interaction between the baroclinic motion and the divergent motion associated with the prescribed lid is taken into account. In this case (21) would replace (17) in the working set of equations.

Using the vertically-integrated continuity equation for the barotropic motion,

$$B(-\zeta_s)\frac{\partial \zeta_s}{\partial t} = -\frac{\partial}{\partial x}\int_{-\zeta_s}^{h}Bu_s\,dz.$$

The first term on the right hand side of Eq. (19) can be replaced by

$$\frac{\partial}{\partial t}\left(-\frac{\partial}{\partial x}\int_{-\zeta_s}^{h}Bu_s\,dz'\right) = -\frac{\partial}{\partial x}\left(\frac{\partial}{\partial t}\int_{-\zeta_s}^{h}Bu_s\,dz'\right)$$

so that (19) can be integrated with respect to x to give

$$\frac{\partial p_s}{\partial x}\int_{-\zeta_s}^{h}\frac{B}{\rho}\,dz' = -\frac{\partial}{\partial t}\int_{-\zeta_s}^{h} Bu_s\,dz' - \int_{-\zeta_s}^{h} T(u)\,dz' - \frac{\partial p_a}{\partial x}\int_{-\zeta_s}^{h}\frac{B}{\rho}\,dz'$$
$$ - g\int_{-\zeta_s}^{h}\frac{B}{\rho}\frac{\partial}{\partial x}\left((z+\zeta_s)\bar{\rho}_s\right)dz' + \text{constant} + X\int_{-\zeta_s}^{h} B\,dz'. \quad (19a)$$

If the basin has a closed end at $x = 0$, say, where $u = 0$ and $\int_{-\zeta_s}^{h} Bu_s\,dz' = 0$, from (17)

$$\frac{B}{\rho}\frac{\partial p_s}{\partial x} = -T(u) - \frac{B}{\rho}\frac{\partial p_a}{\partial x} - g\frac{B}{\rho}\frac{\partial}{\partial x}\left((z+\zeta_s)\bar{\rho}_s\right) + BX \quad \text{at} \quad x = 0.$$

It follows that the constant of integration in (19a) must then be zero.

5. Baroclinic Motion in a Channel of Uniform Section

A special case in Section 4, studied by Crean (1983), took $B = $ constant and neglected atmospheric forcing. Then (16) reduces to

$$p = p_s + g(z + \zeta_s)\bar{\rho}_s \quad (22)$$

where

$$\bar{\rho}_s = \frac{1}{z + \zeta_s}\int_{-\zeta_s}^{h}\rho\,dz',$$

and Eq. (17) becomes

$$\frac{\partial u}{\partial t} + T(u) = -\frac{1}{\rho}\frac{\partial p}{\partial x} + X$$
$$ = -\frac{1}{\rho}\frac{\partial p_s}{\partial x} - g\frac{\bar{\rho}_s}{\rho}\frac{\partial \zeta_s}{\partial x} - \frac{g}{\rho}(z+\zeta_s)\frac{\partial \bar{\rho}_s}{\partial x} + X \quad (23)$$

where

$$T(u) = \frac{\partial}{\partial x}(uu) + \frac{\partial}{\partial z}(uw) - \frac{\partial}{\partial x}\left(N_x\frac{\partial u}{\partial x}\right) - \frac{\partial}{\partial z}\left(N_z\frac{\partial u}{\partial z}\right). \quad (24)$$

The equation for pressure on the lid (19) simplifies to

$$\frac{\partial}{\partial x}\left(\frac{\partial p_s}{\partial x}\int_{-\zeta_s}^{h}\frac{1}{\rho}\,dz'\right) = \frac{\partial^2}{\partial t^2}\zeta_s - \frac{\partial}{\partial x}\int_{-\zeta_s}^{h} T(u)\,dz'$$
$$ - g\frac{\partial}{\partial x}\int_{-\zeta_s}^{h}\frac{1}{\rho}\frac{\partial}{\partial x}\left((z+\zeta_s)\bar{\rho}_s\right)dz' + \frac{\partial}{\partial x}((h+\zeta_s)X). \quad (25)$$

If ζ_s derives from a barotropic solution, then it satisifies a simplified form of (20),

$$g\frac{\partial p_s}{\partial x} = -\frac{\partial u_s}{\partial x} - T_s(u_s) + X \quad (26)$$

so that Eq. (21) becomes

$$\frac{\partial u}{\partial t} - \frac{\bar{\rho}_s}{\rho}\frac{\partial u_s}{\partial t} + T(u) - \frac{\bar{\rho}_s}{\rho}T_s(u_s) = -\frac{1}{\rho}\frac{\partial p_s}{\partial x} - g(z+\zeta_s)\frac{1}{\rho}\frac{\partial \bar{\rho}_s}{\partial x} + \left(1 - \frac{\bar{\rho}_s}{\rho}\right)X. \quad (27)$$

In addition, the continuity and salt balance equations simplify to give

$$\frac{\partial u}{\partial x} + \frac{\partial w}{\partial z} = 0, \tag{28}$$

$$\frac{\partial s}{\partial t} + \frac{\partial}{\partial x}(us) + \frac{\partial}{\partial z}(ws) - \frac{\partial}{\partial x}\left(K_x \frac{\partial s}{\partial x}\right) - \frac{\partial}{\partial z}\left(K_z \frac{\partial s}{\partial z}\right) = 0. \tag{29}$$

Using the vertically-integrated continuity equation for the barotropic motion,

$$\frac{\partial \zeta_s}{\partial t} = \frac{\partial}{\partial x} \int_{-\zeta_s}^{h} u_s \, dz',$$

so that the first term on the right hand side of Eq. (25) can be replaced by

$$\frac{\partial}{\partial t}\left(-\frac{\partial}{\partial x}\int_{-\zeta_s}^{h} u_s \, dz'\right) = -\frac{\partial}{\partial x}\left(\frac{\partial}{\partial t}\int_{-\zeta_s}^{h} u_s \, dz'\right).$$

Then integrating Eq. (25) with respect to x gives

$$\frac{\partial p_s}{\partial x} \int_{-\zeta_s}^{h} \frac{1}{\rho} \, dz' = -\frac{\partial}{\partial t} \int_{-\zeta_s}^{h} u_s \, dz' - \int_{-\zeta_s}^{h} T(u) \, dz'$$
$$- g \int_{-\zeta_s}^{h} \frac{1}{\rho} \frac{\partial}{\partial x}\left((z+\zeta_s)\bar{\rho}_s\right) dz' + \text{constant} + (h+\zeta_s)X. \tag{25a}$$

If the basin has a closed end at $x = 0$, say, where $u = 0$ and $\int_{-\zeta_s}^{h} u_s \, dz' = 0$, it follows from Eq. (23) that

$$\frac{1}{\rho}\frac{\partial p_s}{\partial x} = -T(u) - \frac{g}{\rho}\frac{\partial}{\partial x}\left((z+\zeta_s)\bar{\rho}_s\right) + X \quad \text{at} \quad x = 0.$$

It follows that the constant of integration in Eq. (25a) must then be zero.

AUTHOR INDEX

Arakawa, A., 357
Backhaus, J.O., 336, 338, 342
Baines, P.G., 344
Barber, F.G., 45
Barker, M.L., 5, 6
Barnes, C.A., 14, 15, 20
Bjerknes, V., 167
Blumberg, A.F., 303
British Columbia Pilot, 63, 64
Buckingham, W.R., 418
Canadian Hydrographic Service, 6, 38, 121, 122, 147
Cannon, G.A., 22
Chang, P., 39, 89, 418
Cordes, R.E., 32, 216
Crean, P.B., 17, 28, 46–48, 89, 216, 291, 311, 313, 356, 460
Doodson, A.T., 89, 91, 160
Dronkers, J.J., 54, 56, 57, 74–76, 79, 91, 93, 98, 108, 109
Dunbar, D., 313
Ebbesmeyer, C.C., 22
Elliot, J.A., 303, 304, 454
Filloux, J.H., 79
Flather, R.A., 6, 72, 210
Foreman, M.G., 140, 310
Freeland, H.J., 14, 16
Fromm, J.E., 326
Gade, H.G., 344
Garcia, A.W., 195, 199
Garvine, R.W., 30
Geyer, W.R., 22, 26
Giovando, L.F., 29, 33, 34
Godin, G., 17, 20, 98, 321, 430
Grace, S.F., 69, 79
Hamilton, P., 304, 456

Hansen, W., 72, 338
Harris, R.A., 5
Heisanken, W., 79
Hendershott, M.C., 347
Henry, R.F., 188, 190
Herlinveaux, R.H., 38
Holbrook, J.R., 13, 20, 25, 302, 329, 369
Hough, S.S., 167
Houston, J.R., 195–197, 199, 202
Huggett, W.S., 22, 43, 99, 101, 351, 369
Ianiello, J.P., 358
Jeffreys, H., 79
Kelvin, Lord W., 168
Kendrew, W.G., 4
Kuipers, H., 117, 149
Le Prévost, T.C., 91
Lighthill, M.J., 220
Miller, G.R., 79
Möfjeld, H.O., 6, 22
Mortimer, C.H., 13
Mungall, J.C.J., 76
Murty, T.S., 167–169, 172, 178, 187, 188, 193
O'Brien, G.G, 55
Parker, B.B., 6, 24
Paul, J.F., 333
Petrauskas, C., 200
Platzman, G. W., 398
Proehl, J.A., 13, 302
Proudman, J., 53, 54, 63, 71, 89
Redfield, A.C., 5, 37, 70, 79, 302
Richardson, D.J., 133
Roache, P.J., 217
Rogers, G.C., 187, 188, 191, 193
Rossiter, J.R., 54
Schumacher, J.D., 22, 37, 38

Simons, T.J., 13, 255, 330
Smagorinsky, J., 357
Stacey, M.W., 37, 39, 419
Stronach, J.A., 35, 48, 210, 216
Sündermann, J., 338
Tabata, S., 14, 29, 39, 310, 329
Taylor, G.I., 53, 79, 351
Thomson, K.A., 39
Thomson, R.E., 1, 33, 42, 43, 101, 160
Turner, J.S.. 215
Waldichuk, M., 7, 13, 37, 40, 42, 44, 45,
   302, 316, 322, 416, 417, 429, 430
Webster, I., 26
Whitham, G.B., 220
Zimmerman, J.F.T., 146

SUBJECT INDEX

Active Pass, 77, 80 140
Admiralty Inlet, 20, 24, 26, 65, 66, 68, 69, 351, 364, 369, 378, 384, 428, 433
amphidromes, 4, 6, 94, 165
amphidromic point, 347,
amphidromic region, 87, 117
amphidromic system, 91, 93
Arran Rapids, 40
atmospheric pressure, 4, 178, 179, 182–185, 324, 328, 329
baroclinic circulation, 155, 311, 369, 370, 374, 379, 382, 383, 420, 424, 425, 437
barotropic tidal circulation, 425
barotropic tidal forcing, 207, 210
Bernoulli effect, 369, 408
Bernoulli equation, 147
Bernoulli function, 291
boundary conditions, 13, 44, 53, 57, 59, 116, 117, 125, 129, 133, 135, 140, 168, 193, 207, 210, 216, 217, 219, 220, 233, 305, 308,
boundary openings, 62, 67, 74, 79, 82–91, 93, 107, 109–113, 115, 117, 125, 133, 135–137, 140, 149–151 165, 178, 181–185, 205, 215, 217, 219, 221, 228, 233, 260, 277, 291, 308, 336, 343, 344, 346, 347, 353, 354, 357, 382, 408, 421, 428–432, 434, 435, 439, 440, 442
bottom deformation, 189
Burrard Inlet, 81, 82, 85, 193
coastal jets, 222, 260, 310, 439
Columbia River, 12–15, 20, 44, 302, 310, 313, 315, 335, 437
Comox, 12, 191, 193
computation of energy dissipation, 108–110, 114
   entrainment, 32, 205, 207, 209, 210, 214–217, 220, 233, 255, 300, 438
   depletion, 207, 209, 210, 214, 215, 255
computation of equilibrium tidal forcing, 113, 114
computation of kinetic energy, 107–109
computation of potential energy, 107–109
constituents, 6, 7, 60, 61, 79. 83-86, 88–94, 97, 102, 103, 105, 110, 112–114, 117, 124, 125, 140, 142, 143, 146, 149, 311
continental shelf, 13, 14, 17, 38, 199, 403, 424, 429, 431, 445
convergence, 27, 29, 255, 316, 317, 319, 364, 378, 403, 420

Coriolis deflection, 14, 29, 34, 35, 38, 205, 219, 222, 223, 228, 233, 255, 260, 277, 300, 310, 316, 382, 408

cross-channel flow, 28, 408

cross-channel slope, 65, 382, 387, 408, 421, 423

cross-channel velocity, 98, 118, 146, 177

cruises in 1968, 12, 45, 46

current ellipses, 39, 53, 123, 160–166, 434

current meter moorings, 17, 22, 28, 37–39, 42, 45, 46, 48–50, 93 98, 100, 129, 320, 337, 351, 361, 416, 419, 435

current profiles, 98

cyclesonde observations (records), 36, 337, 419

damped seiche, 330, 333, 334, 438

damped waves, 5, 88, 191

damping, 347, 434, 436

damping factor, 70

deceleration, 32

deep water, 12, 22, 26, 38, 64, 291, 294, 301, 315, 323, 335

deep water renewal (replenishment), 336, 357, 369, 374, 379, 382, 416, 420, 431, 440, 442

density distribution, 179, 291, 308, 387, 388, 392

density field, 238, 291, 310, 311, 313–315, 322, 333, 337, 357, 358, 363, 364, 369, 374, 378, 379, 386, 387, 403, 409, 412, 428, 437, 438, 441, 442

density gradient, 205, 304, 327, 386, 387, 420

density intrusion, 22, 389, 403

density stratification (layer), 100, 101, 214

density surfaces, 217, 389

derivation of equations, 334

diagnostic trials (cases), 313, 323, 325, 328, 329, 358, 382, 420, 431, 442

Discovery Passage, 7, 40, 42, 57, 60, 61, 63, 66, 83, 154, 156, 166, 228, 233, 238, 260, 277, 347, 354, 378, 416, 442, 444

divergence, 214, 316, 317, 319, 333, 358, 364, 378, 403, 408, 420

Dodd Narrows, 140

drainage area, 2

drogue tracking, 32, 35–37, 49, 293, 438

drogue tracks, 124, 158, 293, 416, 435, 440

earthquake, 187–189, 191–194, 201, 436

eddies, 13, 34–36, 39, 42, 120, 124, 135, 146–149, 152, 154–160, 222, 298, 336, 357, 358, 364, 378, 416, 418, 424, 434, 435
   residual, 26, 27, 149, 152–155, 358, 444
   residual tidal, 369, 444
   tidal, 40, 45, 49, 129, 136, 146, 153, 166, 444

eddy formation, 147, 222, 223, 369, 416, 419, 439
   advective terms, 147, 148
   diffusion, 342, 357

diffusivity, 313, 325
stress, 341
viscosity, 147, 148, 210, 214, 303, 311, 336, 338, 346, 354–358, 361–363, 440–442
vorticity, 147, 148
effects of rotation, 13, 65, 160, 299, 408, 432, 444
energy flux, 79, 105, 109–114, 165, 166, 190, 346, 351, 353–355
equatorial tides, 7
equilibrium forcing, 113
estuarine circulation, 12, 13, 20, 26, 43, 49, 302, 310, 314, 319, 320, 323, 335, 336, 379, 382, 430, 431, 434, 435, 437, 441, 445
estuarine flow reversal, 13, 20, 35, 302, 320
finite-difference equations, 210, 212, 304, 308, 338, 442
finite-difference formulations, 54, 56, 74–76, 79, 89, 107–109, 133, 147, 189, 210, 222, 305–307, 322, 323, 332, 338, 341
flow field, 34, 35, 133, 147, 148, 156, 217, 277, 357, 382, 400
Fraser River, 2, 7, 14, 17, 26, 27, 29, 30, 32, 34–38, 44, 115, 120, 126 217, 290, 319 428, 439, 444
delta, 165, 218
discharge, 205, 216, 219–224, 228, 233, 255, 260, 277, 290–293, 295, 296, 299–301, 310, 311, 322, 374, 387, 396, 429, 435, 441
mouth, 27–30, 32–35, 48, 118, 120, 122, 124, 126, 166, 205, 222, 308, 316–318, 323, 325, 335, 382, 408, 416, 424
plume, 29, 32–36, 45, 126, 291, 316, 438
freshwater proportion (fraction), 320–322, 428, 430
freshwater volume, 43, 44, 321, 429
frictional adjustment, 53, 75, 79, 84, 87-89, 91, 93
frictional dissipation, 5, 57, 64, 76, 79, 82–84, 86-89, 91, 105, 107-110, 117, 118, 121 140, 165, 311, 314, 344, 347, 432, 444
Froude number, 36, 210, 217, 219, 220, 439
Gabriola Pass, 140
gain, 168–170, 172, 436
Georgia/Fuca system, 44, 69, 70, 107, 114, 302, 310, 319, 329, 330, 334, 354, 442
governing equations, 133, 160, 199, 206, 303, 334
Gulf and San Juan Islands 1, 3, 12, 13, 22, 37, 40, 80, 85, 91, 105, 108, 115, 116, 118, 121, 122, 125, 140, 149, 165, 169, 290, 296, 321, 361, 364, 418, 424, 432–434, 436, 437,
Haro Strait, 2, 3, 22, 24, 26, 49, 56, 60, 61, 63, 64, 70, 87–89, 107, 117, 120, 122, 125, 128, 135, 140, 158–160, 166, 304, 305, 321, 322, 361–363, 387, 401, 416, 424, 431, 432, 434, 436, 441, 444
Howe Sound, 64, 65, 68, 82, 85, 166, 172, 193, 228, 233, 238, 255, 277, 297, 298
implicit method, 442
implicit procedures, 342
interaction with tide (of surge), 13
interaction with tide (of tsunami), 13, 197–202

interfacial boundaries, 214
internal modes, 172, 175–177
internal response, 13
internal seiche, 330, 331, 333, 334, 344, 438,
internal tides, 13, 39, 89, 444
Jervis Inlet, 64, 65, 68, 82, 85, 193
jet of freshwater (river water), 12, 35, 205, 222, 290, 291, 438
Johnstone Strait, 7, 36, 40, 43–45, 60, 66, 87, 100, 101, 126, 347, 430
junction between two- and one-dimensional implicit schemes, 75-77
Kelvin wave, 13, 14, 79, 87, 93, 160, 177, 302, 347
Lagrangian interpolation, 77
layer thickness, 205, 208, 209, 214–216, 219–221, 223, 228, 233, 238, 260, 293, 294, 299,
    300, 301, 342, 344, 439, 440, 444
Little River, 86, 95, 102, 182, 185, 186, 328–330, 347, 350, 384
longitudinal circulation, 308, 310,
Main Arm, 27–29, 216, 217, 294, 297, 439
mesh Reynolds number, 325, 326, 328
narrow pass schematization, 77–79
neap tides, 7
non-linear instability, 355, 362, 440
non-linear interactions, 79, 84, 88-91, 93, 140, 361, 440, 444
non-linear region, 116, 125
non-linear terms, 61, 101, 146, 199, 200, 214, 434
normal modes, 167, 168, 172, 185, 435
North Arm, 26, 216, 439
northern passages, 2, 36, 40, 41, 61, 63, 65, 66, 71, 82, 84, 87, 96, 105, 107, 109, 112
    113, 138, 140, 317, 364, 383, 431–433
numerical diffusion, 322, 326, 328, 336, 342, 357, ,358, 364, 382, 420, 437, 438, 441, 442
numerical formulation, 210, 218, 255
numerical grid scheme, 56, 73, 129, 137, 218, 304, 309
numerical stability, 55, 190, 301, 322
observed and calculated (modelled) sea levels, 179, 181, 182, 184, 328, 329, 443
Okisollo Channel 40, 139, 140
one-dimensional explicit scheme, Puget Sound, 81
one-dimensional implicit scheme, Burrard Inlet, 81
one-dimensional implicit scheme, Howe Sound, 82
one-dimensional implicit scheme, Jervis Inlet, 82
one-dimensional implicit scheme, northern passages, 82
overall system, 1, 71, 129, 330, 364, 430
phase lag, 57, 60 61, 63, 64, 66, 68–70, 83, 88, 91, 185, 347–349, 351, 356
phase shift, 32
Poincaré mode, 13, 160
Poincaré wave, 160, 397, 431

Point Atkinson, 6, 28, 31, 90, 91, 95, 102, 110, 168, 182, 185, 140, 217, 222, 223, 347, 349, 350

Porlier Pass, 80, 87, 140

prescribed lid, 329–331, 333, 334, 438, 442

pressure distribution, 308, 403, 408, 409, 430

pressure gradient, 28, 129, 132, 178, 205–208, 218–221, 228, 291, 301, 310, 314–316, 331, 333, 334, 339, 358, 360, 382, 385, 387, 403, 408, 420, 438, 444, 445

prognostic trials (calculations), 322–326, 437

proportioning of freshwater discharge between boundary openings, 44, 430

Puget Sound, 1, 4, 6, 19, 20, 22, 64–66, 68–70, 71, 74, 77, 81, 85, 87-89, 91, 94, 96, 102, 105, 107–109, 112, 113, 121, 140, 146, 149, 152, 187, 193–195, 197, 200, 201, 344, 346, 347, 354–357, 364, 369, 370, 378, 379, 392, 408, 409, 423, 424, 427, 433–435

Race Rocks, 17, 18, 20, 49, 93, 98, 100, 135, 159

residual circulation, 13, 45, 129, 146–149, 156–158, 160, 205, 290, 302, 304, 311–314, 325, 326, 337, 344, 356, 362–364, 369, 378, 379, 382, 385, 387, 408, 416, 419–421 424, 430, 431, 434–436, 442, 444

residual currents, 36, 148, 159, 290, 311, 418, 424

residual current profiles, 322, 317

residual current vectors, 419,

residual density distribution, 385

residual sea level, 323, 327, 328, 420, 437, 445

residual velocities, 290, 312, 318, 320, 324, 327, 335, 356, 357, 364, 424, 430, 434, 442

residual vertical velocity, 387

rigid lid, 329–333, 437

river discharge (flow), 2, 7, 13, 26, 27, 29, 30, 32, 34–38, 43, 44, 48, 49, 115, 126, 137

Rosario Strait, 2, 4, 20, 22, 24, 56, 63, 64, 87, 88, 105, 107, 140, 304, 305, 369, 378, 387, 424

Rossby radius, 432

run-up, 194, 197, 199–202

salinity field, 12, 221, 223, 233, 238, 260, 291, 293, 301, 302, 310–315, 317, 322, 323, 326, 328, 332, 335, 429, 437

salinity profiles, 17, 21, 29, 39, 40, 215

salt intrusion, 17, 18, 126

salt wedge (tongue), 27, 47, 126, 205, 217, 220

Sand Heads, 12, 27, 29, 34, 185, 439

schematization, 56–58, 82, 83, 105, 139, 140, 153, 220, 304, 308, 354, 437, 439

sea level, 14, 178, 179, 182–185, 308, 310, 321, 325, 327–330, 369, 437, 443
    mean, 14, 45, 61, 85, 110, 178–181, 198, 199, 310, 316, 325, 328–330, 369, 395, 403, 404, 409, 412, 421, 434, 444

Seattle, 1, 5, 86, 91, 96, 102, 146, 182, 185, 186, 201, 347, 350

semi-implicit scheme, 210

sensitivity trials, 139

Seymour Narrows, 40, 42, 61, 65, 66, 68, 75, 83, 101, 102, 139, 140

shallow banks, 28, 115, 122, 165, 166, 205, 216, 218, 220, 438
spring tides, 7
stability criteria, 190, 191, 342
stability restrictions, 75
storm surge propagation, 184, 185, 436
subsurface water, 36
surface elevation, 72, 102, 104, 117, 129, 181, 191, 291, 302, 308, 311–315, 318, 320, 323, 324, 332, 333, 336, 352, 369, 382, 384, 421–423, 431, 436, 441, 442
surface water, 7, 12, 26, 37, 40, 42, 98, 219, 302, 309, 310, 313, 315, 322, 329, 335, 387, 417, 431, 441, 445
surge elevation, 181–184, 186, 436
tidal jet, 154
tidal cross-flows, 300
tidal damping, 91, 192
tidal forcing, 107, 220, 290, 334, 357
tidal mixing, 7, 13, 20, 22, 26, 36, 40, 126, 205, 214, 215, 217, 218, 221, 228, 302, 310, 317, 378, 408, 416, 437
tidal modulation, 13, 30, 32, 47
tidal propagation, 71, 87, 140, 316
tidal velocity profiles, 311, 313, 327
time series of isopycnals, 14
time series off the Fraser River mouth, 29, 30
time series of salinity, 17, 20, 23, 30, 38–40, 382, 420
time series of sigma-t, 16
time series of temperature, 30
time series of tsunami amplitude, 193, 198
time series of velocity, 45, 216
time series of wind, 25
time step, 55, 75, 81, 117, 132, 147, 190, 191, 206, 212, 214, 219, 293, 301, 304, 331–334 342, 357, 438, 441, 442
Tofino, 182–184, 328–330
tropic tides, 7
tsunami, 13, 167, 169, 187–189, 191–202, 435, 436
tsunami propagation, 187, 189, 193, 194, 199
two-dimensional scheme, 71, 75–78, 80, 87, 115
undercurrents, 13, 382, 408, 416, 418, 420, 424, 425, 427, 431, 442, 444, 445
upwelling, 38, 401, 441, 444
Vancouver, 1, 4, 172, 182–185, 188, 193
vectors,
    current, 37, 42, 148, 149, 152, 153, 157
    residual tidal velocity vector, 360
    residual tidal velocity vector field, 148
    residual velocity, 314, 315, 318, 322, 324, 326

tidal residual current, 365–368

velocity, 29, 30, 84, 117, 120–122, 135, 149, 156, 166, 238, 255, 260, 290–299, 313, 351, 352

velocity field, 117, 118, 135, 224, 228, 331

velocity field, 118, 121, 123, 124, 135, 147, 210, 233, 238, 260, 293, 295, 301, 322, 323, 330, 331, 333, 344, 420, 435, 439, 442

velocity profiles, 39, 40, 48, 98, 312, 313, 316, 317, 319, 320, 323, 325

vertical mixing, 79, 216, 219, 233, 308, 326, 327, 344, 358

vertical profiles of density, 172, 176, 358, 359, 421

vertical profiles of horizontal velocity, 424, 425

vertical profiles of sigma, 420

vertical profiles of velocity, 361–364, 378, 424, 441

vertical velocity, 209, 215, 217, 219, 232, 233, 237, 238, 242, 246, 250, 254, 259, 260, 264, 268, 272, 276, 277, 281, 289, 311, 315, 316, 333, 337, 339, 361, 387, 395–401, 420, 438

Victoria, 6, 12, 64, 129, 140, 152, 182, 188, 193, 328, 347, 350

volume exchanges, 61, 66, 85, 320

volume flux, 87, 105, 319–321, 428, 430

volume transport, 24, 77, 78, 81, 82, 85, 87, 105–107, 132, 135, 315, 320, 346, 347, 428, 430

wind effects, 4–6, 12–14, 20, 29, 34–37, 49, 126, 178, 179, 185, 222, 233, 238–289, 292–294, 299–302, 315, 328, 335, 357, 363, 420–431, 436, 439

wind forcing, 220, 233, 255, 290

interactions, 223, 293

no wind, 229–238

no wind forcing, 228, 290

wind field, 183, 186, 223, 293, 337, 363, 420, 422, 423, 428

winds, 12–14, 20, 29, 34–37, 49, 59, 183–186, 205, 209, 214, 215, 222, 223, 233, 238–289, 292–294, 299, 300–302, 313, 315, 328, 329, 363, 373, 387, 416, 420–431, 435, 437–439, 443–445

wind stress, 178, 185, 189, 210, 214, 219, 222, 238, 338, 341

Yuculta Rapids 40, 139, 140